Bioinformatics –
From Genomes to Therapies

Edited by
Thomas Lengauer

1807–2007 Knowledge for Generations

Each generation has its unique needs and aspirations. When Charles Wiley first opened his small printing shop in lower Manhattan in 1807, it was a generation of boundless potential searching for an identity. And we were there, helping to define a new American literary tradition. Over half a century later, in the midst of the Second Industrial Revolution, it was a generation focused on building the future. Once again, we were there, supplying the critical scientific, technical, and engineering knowledge that helped frame the world. Throughout the 20th Century, and into the new millennium, nations began to reach out beyond their own borders and a new international community was born. Wiley was there, expanding its operations around the world to enable a global exchange of ideas, opinions, and know-how.

For 200 years, Wiley has been an integral part of each generation's journey, enabling the flow of information and understanding necessary to meet their needs and fulfill their aspirations. Today, bold new technologies are changing the way we live and learn. Wiley will be there, providing you the must-have knowledge you need to imagine new worlds, new possibilities, and new opportunities.

Generations come and go, but you can always count on Wiley to provide you the knowledge you need, when and where you need it!

William J. Pesce
President and Chief Executive Officer

Peter Booth Wiley
Chairman of the Board

Bioinformatics – From Genomes to Therapies Volume 1

The Building Blocks: Molecular Sequences and Structures

Edited by
Thomas Lengauer

BICENTENNIAL 1807 WILEY 2007 BICENTENNIAL

WILEY-VCH Verlag GmbH & Co. KGaA

The Editor

Prof. Dr. Thomas Lengauer
Max-Planck-Institute
for Informatics
Stuhlsatzenhausweg 85
66123 Saarbrücken
Germany

Library of Congress Card No.:
applied for

British Library Cataloguing-in-Publication Data:
A catalogue record for this book is available from the British Library.

Bibliographic information published by the Deutsche Nationalibliothek
The Deutsche Nationalbibliothek lists this publication in the Deutsche Nationalbibliografie; detailed bibliographic data are available in the Internet at http://dnb.d-nb.de

Printed in the Federal Republic of Germany
Printed on acid-free paper

Cover Design: Schulz Grafik-Design, Fussgönheim
Composition: Steingraeber Satztechnik GmbH, Ladenburg
Printing: betz-Druck GmbH, Darmstadt
Bookbinding: Litges & Dopf Buchbinderei GmbH, Heppenheim

ISBN: 978-3-527-31278-8

For Sybille, Sara and Nico

Contents

Volume 1

Bioinformatics - From Genomes to Therapies Vol. 1. Edited by Thomas Lengauer

ISBN: 978-3-527-31278-8

Volume 2

Volume 3

Preface

This book is a substantially expanded sequel to the book *Bioinformatics – From Genomes to Drugs* that appeared in 2002. Since the publication of the predecessor book the field of bioinformatics has experienced continuing and substantially accelerated growth in terms of the volume and diversity of available molecular data, as well as the development of methods for analyzing and interpreting these data. This book is a reflection of the dynamic maturation of the field. Like its predecessor, it discusses bioinformatics in the context of pharmaceutical and medical challenges pertaining to the understanding, diagnosis and therapy of diseases. The previous book covered bioinformatics issues accompanying the stages from the collection of genomic data across the elucidation of the molecular basis of disease and the identification of target proteins for drug design to the search for leads for potential drugs. This book extends this schema in various ways. First, the process line from genome to drug is extended downstream towards the optimization of drug leads and further towards the personalization of drug therapies, which is also beginning to be supported with bioinformatics methods. Second, the book covers the field in substantially more breadth. The different types of available data are discussed more comprehensively and in more detail. On the sequence side, two chapters on RNA have been added. The bioinformatics analysis of evolutionary relationships is addressed in several chapters. The discussion of protein structure has been significantly expanded. There are new sections on molecular networks, mRNA expression data and protein function, covering several chapters each. The disease-specific part of the book has also been expanded, including discussions of bacterial and viral infections. Finally, several chapters on informatics technologies employed for bioinformatics are included.

Bioinformatics is continuing to present one of the grand challenges of our times. It has a large basic research aspect, since we cannot claim to be close to understanding biological systems on an organism or even cellular level. At the same time, the field is faced with a strong demand for immediate solutions, because the genomic and postgenomic data that are being collected harbor

Bioinformatics - From Genomes to Therapies Vol. 1. Edited by Thomas Lengauer

ISBN: 978-3-527-31278-8

many biological insights whose deciphering can be the basis for dramatic scientific and economical success, and is promising to have large impact on society.

The book is directed at readers who are interested in how bioinformatics can spur biological and medical innovation towards understanding, diagnosing and curing diseases. The book is designed to be useful to readers with a variety of backgrounds. Biologists, biochemists, pharmacologists, pharmacists and medical doctors can get an introduction into basic and practical issues of the computer-based part of handling and interpreting genomic, postgenomic and clinical data. In particular, many chapters point to bioinformatics software and data resources which are available on the Internet (often at no cost), and make an attempt at classifying and comparing those resources. For computer scientists and mathematicians, the book contains an introduction to the biological background and the necessary information in order to begin appreciating the difficulties and wonders of modeling complex biochemical and biomolecular issues by computer. Since the book caters to a readership with widely varying backgrounds, it also contains chapters with a diverse makeup. There are chapters that put the biology in the foreground and only sketch methodical issues, and a smaller number of chapters in which the algorithmic and statistical content dominates. By and large, the way in which the chapters are written reflects the viewpoint from which the authors, and that also often means the world-wide research community, approaches the respective topic.

The book contains a name and a subject index. A methodical index is integrated inside the subject index and points to those sections that present the master introductions to the quoted computational methods.

The world's leading experts have contributed their expertise and written largely autonomous chapters on the specific topics of this book. In order to render added coherence to the book, the chapters contain a large number of cross-references to aid in relating the topics of different chapters to each other. In a few cases, overlap between the chapters has been allowed to ensure the independent readability of the chapters.

I am grateful to the many people who helped make this book possible. Above all, I thank the 91 authors of contributed chapters who have shown extraordinary commitment during the draft and revision stages of their text. Ruth Christmann spent many hours helping me to master the logistic feat of collecting the texts, encouraging authors to keep to their commitments, handling versions and completing revisions with a special focus on reference lists. Joachim Büch kept the website for book authors alive and well maintained during the preparation and production process. Ray Loughlin did a superb job on copy-editing the book. Frank Weinreich and later Steffen Pauly were always responsive partners on the side of the publisher. Finally, I would like to express my deep gratitude to my wife Sybille and my children Sara and Nico who had to cope with my physical or mental absence too much while the project was ongoing. They gave the most for receiving the least.

Saarbrücken *Thomas Lengauer*
October 2006

List of Contributors

Bissan Al-Lazikani
Inpharmatica Ltd
1 New Oxford Street
London WC1A 1NU
UK

Russ B. Altman
Stanford University Medical Center
Department of Genetics
300 Pasteur Drive, Lane L301
Stanford, CA 94305-5120
USA

Ron D. Appel
Swiss Institute of Bioinformatics
CMU - 1, rue Michel Servet
1211 Geneva 4
Switzerland

Karl-Heinz Baringhaus
Sanofi-Aventis Deutschland GmbH
Chemical Science / Drug Design
65926 Frankfurt
Germany

Niko Beerenwinkel
Harvard University
Program for Evolutionary Dynamics
1 Brattle Square
Cambridge, MA 02138
USA

Asa Ben-Hur
Colorado State University
Department of Computer Science
222 University Services Center
601 South Howes Street
Fort Collins, CO 80523 USA
USA

Richard Bonneau
New York University
Department of Biology/Computer Science
Center for Comparative Functional Genomics
100 Washington Square East
New York, NY 10003-6688
USA

Philip E. Bourne
University of California-San Diego
Department of Pharmacology
9500 Gilman Drive
La Jolla, CA 92093-0505
USA

Guillaume Bourque
Genome Institute of Singapore
60 Biopolis Street, #02-01, Genome,
Singapore 138672
Singapore

Bioinformatics - From Genomes to Therapies Vol. 1. Edited by Thomas Lengauer

ISBN: 978-3-527-31278-8

Yannick Brunner
University of Geneva
Biomedical Proteomics Research Group
Department of Structural Biology and Bioinformatics
Centre Médical Universitaire
1, rue Michel Servet
1211 Genève 4
Switzerland

Douglas Lee Brutlag
Stanford University
Department of Biochemistry
Beckman Center, B400, Mail Code 5307
Stanford, CA 94305-5307
USA

Richard R. Copley
University of Oxford
Wellcome Trust Centre for Human Genetics
Roosevelt Drive
Oxford OX3 7BN
UK

Martin Däumer
The University Hospital of Cologne
Institute for Virology
Fürst-Pückler-Str. 56
50935 Köln
Germany

Jörg Degen
University of Hamburg
Center for Bioinformatics Hamburg (ZBH)
Bundesstrasse 43
20146 Hamburg
Germany

Francisco S. Domingues
Max-Planck-Institute for Informatics
Computational Biology and Applied Algorithmics
Stuhlsatzenhausweg 85
66123 Saarbrücken
Germany

Roland L. Dunbrack, Jr.
Institute for Cancer Research
Fox Chase Cancer Center
333 Cottman Avenue
Philadelphia, PA 19111
USA

Ingo Ebersberger
Heinrich-Heine-University Düsseldorf
Bioinformatics
Universitätsstrasse 1, Geb. 25.02.02
40225 Düsseldorf
Germany

Kevin W. Eliceiri
University of Wisconsin
Laboratory for Optical and Computational Instrumentation
Department of Molecular Biology
1675 Observatory Drive
Madison, WI 53706
USA

Edward J. Feil
University of Bath
Department of Biology and Biochemistry
Bath BA2 7AY
UK

David Fell
Oxford Brookes University
School of Biological and Molecular Sciences
Headington Campus
Gipsy Lane
Oxford OX3 0BP
UK

Dawn Field
Centre for Ecology and Hydrology, Oxford
Molecular Evolution and Bioinformatics Section
Mansfield Road
Oxford OX1 3SR
UK

Anna Gaulton
Pfizer Global Research and Development
Pfizer Ltd
Ramsgate Road
Sandwich
Kent CT13 9NJ
UK

Adam Godzik
The Burnham Institute
10901 North Torrey Pines Road
La Jolla, CA 92037
USA

Ilya G. Goldberg
National Institute on Aging, Gerontology Research Center
Image Informatics and Computational Biology Unit
Laboratory of Genetics
5600 Nathan Shock Drive
Baltimore, MD 21224-6825
USA

Johannes Goll
Forschungszentrum Karlsruhe
Institute for Toxicology and Genetics
Hermann-von-Helmholtz-Platz 1
76344 Eggenstein-Leopoldshafen
Germany

Brian J. Haas
The Institute for Genomic Research
9712 Medical Center Drive
Rockville, MD 20850
USA

Arndt von Haeseler
Center for Integrative Bioinformatics
Max F. Perutz Laboratories
Dr. Bohr Gasse 9
1030 Vienna
Austria

Anthony Hasseldine
Department of Pharmacology and Biological Chemistry
Mount Sinai School of Medicine
One Gustave L. Levy Place
New York, NY 10029
USA

Anja von Heydebreck
Max-Planck-Institute for Molecular Genetics
Ihnestrasse 63–73
14195 Berlin
Germany

Andreas Hildebrandt
Center for Bioinformatics
Building E11
P.O. Box 151150
66041 Saarbrücken
Germany

Harry Hochheiser
Towson University
Department of Computer and Information Sciences
7800 York Road, Room 425
Towson MD 21252
USA

Ivo L. Hofacker
University of Vienna
Institute for Theoretical Chemistry
Währingerstr. 17
1090 Vienna
Austria

Andrew Hopkins
Pfizer Global Research and Development
Pfizer Ltd
Ramsgate Road
Sandwich
Kent CT13 9NJ
UK

Wolfgang Huber
EMBL Outstation - Hinxton
European Bioinformatics Institute
Wellcome Trust Genome Campus
Hinxton
Cambridge, CB10 1SD
UK

Duncan Hull
University of Manchester
School of Computer Science
Oxford Road
Manchester M13 9PL
UK

Daniel Huson
University of Tübingen
Faculty of Computer Science
Chair of Algorithms in Bioinformatics
Sand 14
72076 Tübingen
Germany

Ravi Iyengar
Department of Pharmacology and Biological Chemistry
Mount Sinai School of Medicine
One Gustave L. Levy Place
New York, NY 10029
USA

Li Jin
Fudan University
Laboratory of Theoretical Systems Biology
School of Life Science
Handan Road 220
Shanghai 200433
China

Andreas Kämper
Max-Planck-Institute for Informatics
Computational Biology and Applied Algorithmics
Stuhlsatzenhausweg 85
66123 Saarbrücken
Germany

Teri E. Klein
Stanford University Medical Center
Department of Genetics
300 Pasteur Drive, Lane L301
Stanford, CA 94305-5120
USA

Oliver Kohlbacher
University Tübingen
Wilhelm Schickard Institute for Computer Science
Division for Simulation of Biological Systems
Sand 14
72076 Tübingen
Germany

Dennis Kostka
Max-Planck-Institute for Molecular Genetics
Ihnestrasse 63-73
14195 Berlin
Germany

Martin Krallinger
Protein Design Group (PDG)
National Biotechnology Center (CNB)
Campus Universidad Autónoma (UAM)
Cantoblanco 28049 (Madrid)
Spain

David C. Kulp
University of Massachusetts
Bioinformatics Research Laboratory
Computer Science Department
140 Governors Drive
Amherst, MA 01003
USA

Zoé Lacroix
Arizona State University
Scientific Data Management Laboratory
P.O. Box 875706
Tempe, AZ 85287-5706
USA

Jerry Lanfear
Pfizer Global Research and Development
Pfizer Ltd
Ramsgate Road
Sandwich
Kent CT13 9NJ
UK

Thomas Lengauer
Max-Planck-Institute for Informatics
Computational Biology and Applied Algorithmics
Stuhlsatzenhausweg 85
66123 Saarbrücken
Germany

Hans-Peter Lenhof
Saarland University
Center for Bioinformatics
Building E11
P.O. Box 151150
66123 Saarbrücken
Germany

Azi Lipshtat
Department of Pharmacology and Biological Chemistry
Mount Sinai School of Medicine
One Gustave L. Levy Place
New York, NY 10029
USA

Phillip Lord
University of Manchester
School of Computer Science
Oxford Road
Manchester M13 9PL
UK

Claudio Lottaz
Max-Planck-Institute for Molecular Genetics
Ihnestrasse 63-73
14195 Berlin
Germany

Xing Jian Lou
Stanford University Medical Center
Department of Genetics
300 Pasteur Drive, Lane L301
Stanford, CA 94305-5120
USA

Bertram Ludäscher
University of California, Davis
Department of Computer Science
One Shields Avenue
Davis, CA 95616
USA

Avi Ma'ayan
Department of Pharmacology and Biological Chemistry
Mount Sinai School of Medicine
One Gustave L. Levy Place
New York, NY 10029
USA

François Major
Université de Montréal
Institute for Research in Immunology and Cancer
Computational and Theoretical Biology
2900, boulevard Édouard-Montpetit
Pavillon Marcelle-Coutu, Quai 20
Montreal QC H3T 1J4
Canada

Hans Matter
Sanofi-Aventis Deutschland GmbH
Chemical Science / Drug Design
65926 Frankfurt
Germany

Christian von Mering
University of Zurich
Institute of Molecular Biology
Bioinformatics Group
Winterthurerstrasse 190
8057 Zurich
Switzerland

William Stafford Noble
University of Washington
Department of Computer Science and Engineering
1705 NE Pacific Street
Seattle, WA 98195-7730
USA

John Overington
Inpharmatica Ltd
1 New Oxford Street
London WC1A 1NU
UK

Patricia M. Palagi
Swiss Institute of Bioinformatics
Proteome Informatics Group
CMU - 1, rue Michel Servet
1211 Geneva 4
Switzerland

Gaia Paolini
Pfizer Global Research and Development
Pfizer Ltd
Ramsgate Road
Sandwich
Kent CT13 9NJ
UK

Dariusz Przybylski
Columbia University
CUBIC, Department of Biochemistry and Molecular Biophysics
1130 St. Nicholas Ave
New York, NY 10032
USA

John Quackenbush
Dana-Farber Cancer Institute
Department of Biostatistics and Computationsl Biology
44 Binney Street, Sm822
Boston, MA 02115
USA

Jörg Rahnenführer
Max-Planck-Institute for Informatics
Computational Biology and Applied Algorithmics
Stuhlsatzenhausweg 85
66123 Saarbrücken
Germany

Matthias Rarey
University of Hamburg
Center for Bioinformatics Hamburg (ZBH)
Bundesstrasse 43
20146 Hamburg
Germany

Udo Reichl
Otto-von-Guericke-University
Chair of Bioprocess Engineering
Universitätsplatz 2
39106 Magdeburg
Germany

Knut Reinert
Freie Universität Berlin
Institute for Computer Science
Computational Molecular Biology
Takustrasse 9
14195 Berlin
Germany

Ingo Reulecke
University of Hamburg
Center for Bioinformatics Hamburg (ZBH)
Bundesstrasse 43
20146 Hamburg
Germany

Didier Rognan
Bioinformatics Group
Laboratoire de Pharmacochimie de la Communication Cellulaire
CNRS UMR 7081
74, route du Rhin, B.P.24
67401 Illkirch
France

Kirsten Roomp
Max-Planck-Institute for Informatics
Computational Biology and Applied Algorithmics
Stuhlsatzenhausweg 85
66123 Saarbrücken
Germany

Burkhard Rost
Columbia University
CUBIC, Department of Biochemistry and Molecular Biophysics
1130 St. Nicholas Ave
New York, NY 10032
USA

Steven L. Salzberg
University of Maryland
Center for Bioinformatics and
Computational Biology
3125 Biomolecular Sciences Bldg
College Park, MD 20742
USA

Jean-Charles Sanchez
University of Geneva
Biomedical Proteomics Research
Group
Department of Structural Biology
and Bioinformatics
Centre Médical Universitaire
1, rue Michel Servet
1211 Genève 4
Switzerland

Heiko A. Schmidt
Center for Integrative Bioinformatics
Max F. Perutz Laboratories
Dr. Bohr Gasse 9
1030 Vienna
Austria

Stefan Schuster
Friedrich Schiller University
Department of Bioinformatics
Ernst-Abbe-Platz 2
07743 Jena
Germany

Yuri Sidorenko
Max-Planck-Institute for Dynamics
of Complex Technical Systems
Bioprocess Engineering
Sandtorstrasse 1
39106 Magdeburg
Germany

Ingolf Sommer
Max-Planck-Institute for Informatics
Computational Biology and Applied
Algorithmics
Stuhlsatzenhausweg 85
66123 Saarbrücken
Germany

Rainer Spang
Max-Planck-Institute for Molecular
Genetics
Ihnestrasse 63-73
14195 Berlin
Germany

Peter F. Stadler
University of Leipzig
Bioinformatics Group
Department of Computer Science
and Interdisciplinary Center for
Bioinformatics
Härtelstr. 16-18
04107 Leipzig
Germany

Robert Stevens
University of Manchester
School of Computer Science
Oxford Road
Manchester M13 9PL
UK

Paul Swift
Centre for Ecology and Hydrology,
Oxford
Molecular Evolution and
Bioinformatics Section
Mansfield Road
Oxford OX1 3SR
UK

Martin S. Taylor
University of Oxford
Wellcome Trust Centre for Human Genetics
Roosevelt Drive
Oxford OX3 7BN
UK

Philippe Thibault
Université de Montréal
Institute for Research in Immunology and Cancer
Computational and Theoretical Biology
2900, boulevard Édouard-Montpetit
Pavillon Marcelle-Coutu, Quai 20
Montreal QC H3T 1J4
Canada

Peter Uetz
The Institute for Genomic Research
9712 Medical Center Drive
Rockville, MD 20850
USA

Alfonso Valencia
Spanish National
Cancer Research Centre (CNIO)
Structural and Computational
Biology Program (S-CompBio)
Melchor Fernandez Almagro, 3
28029 Madrid
Spain

Martin Vingron
Max-Planck-Institute for Molecular Genetics
Ihnestrasse 63-73
14195 Berlin
Germany

Thomas Werner
Genomatix Software GmbH
Bayerstr. 85a
80335 München
Germany

Gareth Wilson
Centre for Ecology and Hydrology, Oxford
Molecular Evolution and Bioinformatics Section
Mansfield Road
Oxford OX1 3SR
UK

Chris Wroe
University of Manchester
School of Computer Science
Oxford Road
Manchester M13 9PL
UK

Momiao Xiong
The University of Texas HSC at Houston
Human Genetics Center
School of Public Health
7000 Fannin, Suite 1200
Houston, TX 77030
USA

Michael Q. Zhang
Watson School of Biological Sciences
Cold Spring Harbor Laboratory
1 Bungtown Road
Cold Spring Harbor, NY 11724
USA

Part 1 Introduction

1
Bioinformatics - From Genomes to Therapies

Thomas Lengauer

1 Introduction

In order to set the stage for this book, this chapter provides an introduction to the molecular basis of disease. We then continue to discuss modern biological techniques with which we have recently been empowered to screen for molecular drugs targets as well as for the drugs themselves. The chapter finishes with an overview of the organization of the book.

2 The Molecular Basis of Disease

Diagnosing and curing diseases has always been and will continue to be an art. The reason is that man is a complex being with numerous facets, many of which we do not and probably will never understand. Diagnosing and curing diseases has many aspects, include biochemical, physiological, psychological, sociological and spiritual aspects.

Molecular medicine reduces this variety to the molecular aspect. Living organisms, in general, and humans, in particular, are regarded as complex networks of molecular interactions that fuel the processes of life. This "molecular circuitry" has intended modes of operation that correspond to healthy states of the organism and aberrant modes of operation that correspond to diseased states. The main goal of molecular medicine is the identification of the molecular basis of a disease, i.e. to answer the question: "What goes wrong in the molecular circuitry?". The goal of therapy is to guide the biochemical circuitry back to a healthy state. The molecular approach has already proven

Bioinformatics - From Genomes to Therapies Vol. 1. Edited by Thomas Lengauer

ISBN: 978-3-527-31278-8

its effectiveness for understanding diseases, and is dramatically enhanced by genomics and proteomics technology [5]. It is the prime purpose of this book to explore the contributions that this technology, particularly its computational aspect, can have to advancing molecular medicine.

As already noted, the molecular basis of life is composed of complex biochemical processes that constantly produce and recycle molecules, and do so in a highly coordinated and balanced fashion. The underlying basic principles are quite alike throughout all kingdoms of life, even though the processes are much more complex in highly developed animals and the human than in bacteria, for instance. Figure 1 gives an abstract view of such an underlying biochemical network, the *metabolic network* of a bacterial cell (the intestinal bacterium *Escherichia coli*) – it affords an incomplete and highly simplified account of the cell's metabolism, but it nicely visualizes the view of a living cell as a biochemical circuit. The figure has the mathematical structure of a *graph*. Each dot (*node*) stands of a small organic molecule that is metabolized within the cell. Alcohol, glucose and ATP are examples for such molecules. Each line (*edge*) indicates a chemical reaction. The two nodes connected by the edge represent the substrate and the product of the reaction. The colors represent the role that the respective reaction plays in metabolism. These roles include the construction of molecular components that are essential for life – nucleotides (red), amino acids (orange), carbohydrates (blue), lipids (light blue), etc. – or the breakdown of molecules that are not helpful or even harmful to the cell. Other tasks of chemical reactions in a metabolic network pertain to the storage and conversion of energy. (The blue cycle in the center of Figure 1 is the citric acid cycle.) A third class of reactions facilitates the exchange of information in the cell or between cells. This includes the control of when and in what way genes are expressed (*gene regulation*), as well as such tasks as the opening and closing of molecular channels on the cell surface, and the activation or deactivation of cell processes such as replication or apoptosis (programmed cell death). The reactions that regulate cellular processes are often collectively called the *regulatory network*. Recently, molecular networks that facilitate the propagation of signals within the cell are being selectively called *signal transduction networks*. Figure 1 only includes metabolic reactions, without any regulatory reactions or signal transduction cascades. Of course, all molecular networks of a cell are closely intertwined and many reactions can have metabolic as well as regulatory aspects. In general, much more is known about metabolic than regulatory networks, even though many relevant diseases involve regulatory rather than metabolic dysfunction.

The metabolic and regulatory networks can be considered as composed of partial networks that we call *pathways*. Pathways can fold in on themselves, in which case we call them *cycles*. A metabolic pathway is a group of reactions that turns a substrate into a product over several steps (pathway) or recycles

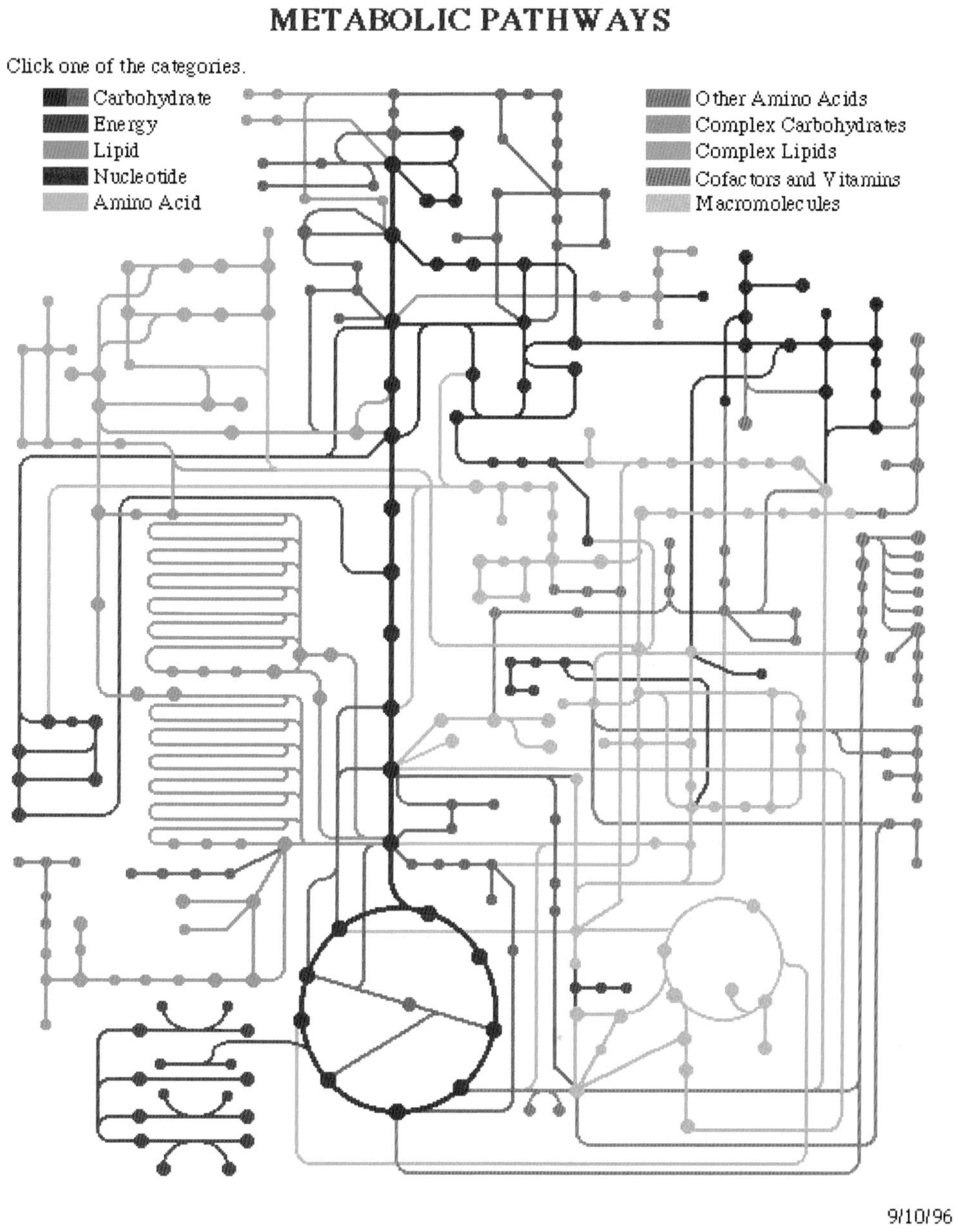

Figure 1 Abstract view of part of the metabolic network of the bacterium *E. coli* (from http://www.genome.ad.jp/kegg/kegg.html).

a molecule by reproducing it in several steps (cycle). *The glycolysis pathway* (the sequence of blue vertical lines in the center of Figure 1) is an example of a pathway that decomposes glucose into pyruvate. The *citric acid cycle* (the blue cycle directly below the glycolysis pathway in Figure 1) is an example of a cycle that produces ATP – the universal molecule for energy transport. Metabolic cycles are essential in order that the processes of life do not accumulate waste or deplete resources. (Nature is much better at recycling than man.)

There are several ways in which Figure 1 hides important details of the actual metabolic pathway. In order to discuss this issue, we have extracted a metabolic cycle from Figure 1 (see Figure 2). This cycle contributes to cell replication; more precisely, ;t is one of the motors that drive the synthesis of thymine – a molecular component of DNA. In Figure 2, the nodes of the metabolic cycle are labeled with the respective organic molecules and the edges point in the direction from the substrate of the reaction to the product. Metabolic reactions can take place spontaneously under physiological conditions (in aqueous solution, under room temperature and neutral pH). However, nature has equipped each reaction (each line in Figure 1) with a specific molecule that catalyzes that reaction. This molecule is called an *enzyme* and, most often, it is a protein. An enzyme has a tailor-made binding site for the transition state of the catalyzed chemical reaction. Thus, the enzyme speeds up the rate of that reaction tremendously, by rates of as much as 10^7. Furthermore, the rate of a reaction that is catalyzed by an enzyme can be regulated by controlling the effectiveness of the enzyme or the number of enzyme molecules that are available.

How does the enzyme do its formidable task? As an example, consider the reaction in Figure 2 that turns dihydrofolate (DHF) into tetrahydrofolate (THF). This reaction is catalyzed by an enzyme called *dihydrofolate reductase (DHFR)*. The surface of this protein is depicted in Figure 3. One immediately recognizes a large and deep pocket that is colored blue (representing its negative charge). This pocket is a *binding pocket* or *binding site* of the enzyme, and it is ideally adapted in terms of geometry and chemistry so as to bind to the substrate molecule DHF and present it in a conformation that is conducive for the desired chemical reaction to take place. In this case, this pocket is also

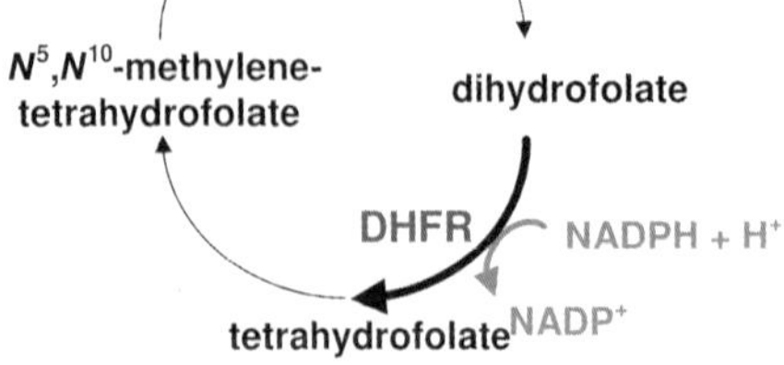

Figure 2 A specific metabolic cycle.

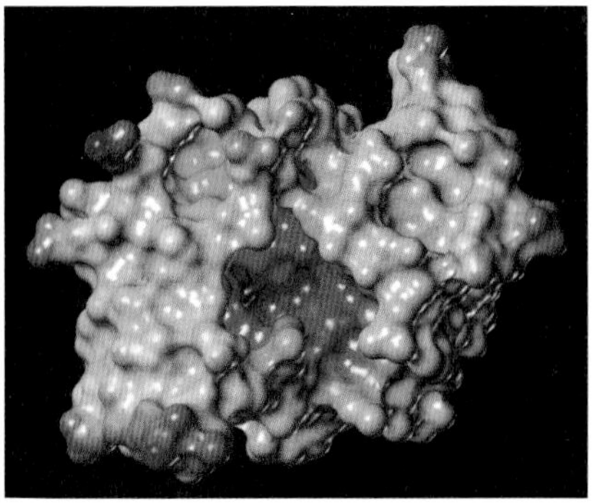

Figure 3 The 3-D structure of DHFR colored by its surface potential. Positive values are depicted in red, negative values in blue.

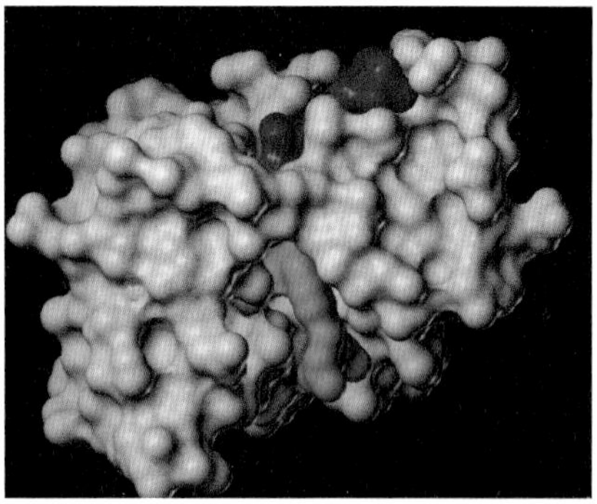

Figure 4 DHFR (gray) complexed with DHF (green) and NADPH (red).

where the reaction is catalyzed. We call this place the *active site*. (There can be other binding pockets in a protein that are far removed from the active site.)

There is another aspect of metabolic reactions that is not depicted in Figure 1 – many reactions involve *cofactors*. A cofactor is an organic molecule or a metal ion that has to be present in order for the reaction to take place. If the cofactor is itself modified during the reaction, we call it a *cosubstrate*. In the case of our example reaction, we need the cosubstrate NADPH for the reaction to happen. The reaction modifies DHF to THF and NADPH to $NADP^+$. Figure 4 shows the molecular complex of DHFR, DHF and NADPH before the reaction happens. After the reaction has been completed, both organic molecules dissociate from DHFR and the original state of the enzyme is recovered.

Now that we have discussed some of the details of metabolic reactions, let us move back to the global view of Figure 1. We have seen that each of the edges in Figure 1 represents a reaction that is catalyzed by a specific protein. (However, the same protein can catalyze several reactions.) In *E. coli* there are an estimated 1500 enzymes [6]; in human there are thought to be about least twice as many. The molecular basis of a disease lies in modifications of the action of these biochemical pathways. Some reactions do not happen at their intended rate (e.g. in gout), resources that are needed are not present in sufficient amounts (vitamin deficiencies) or waste products accumulate in the body (Alzheimer's disease). In general, imbalances induced in one part of the network spread to other parts. The aim of therapy is to replace the aberrant processes with those that restore a healthy state. The most desirable fashion in which this could be done would be to control the effectiveness of a whole set of enzymes in order to regain the metabolic balance. This set probably involves many, many proteins, as we can expect many proteins to be involved in manifesting the disease. Also, each of these proteins would have to be regulated in quite a specific manner. The effectiveness of some proteins would possibly have to be increased dramatically, whereas other proteins would have to be blocked entirely, etc. It is obvious that this kind of therapy involves a kind of global knowledge of the workings of the cell and a refined pharmaceutical technology that is far beyond what man can do today and for some time to come.

3 The Molecular Approach to Curing Diseases

For this reason, the approach of today's pharmaceutical research is far more simplistic. The aim is to regulate a single protein. In some cases we aim at completely blocking an enzyme. To this end, we can provide a drug molecule that effectively competes with the natural substrate of the enzyme. The drug molecule, the so-called *inhibitor*, has to be made up such that it binds more strongly to the protein than the substrate. Then, the binding pockets of most enzyme molecules will contain drug molecules and cannot catalyze the desired reaction in the substrate. In some cases, the drug molecule even binds very tightly (covalently) to the enzyme (suicide inhibitor). This bond persists for the remaining lifetime of the protein molecule. Eventually, the deactivated protein molecule is broken down by the cell and a new identical enzyme molecule takes its place. Aspirin is an example of a suicide inhibitor. The effect of the drug persists until the drug molecules themselves are removed from the cell by its metabolic processes and no new drug molecules are administered to replace them. Thus, one can control the effect of the drug by the time and dose it is administered.

There are several potent inhibitors of DHFR. One of them is *methotrexate (MTX)*. Figure 5 shows MTX (color) both unbound (left) and bound (right) to DHFR (gray). MTX has been administered as an effective cytostatic cancer drug for over two decades.

There are many other ways of influencing the activity of a protein by providing a drug that binds to it. Drugs interact with all kinds of proteins:

- With receptor molecules that are located in the cell membrane and fulfill regulatory or signal transduction tasks.
- With ion channels and transporter systems (again protein residing in the cell membrane) that monitor the flux of molecules into and out of the cell.

The mode of interaction between drug and protein does not always have the effect of blocking the protein. In some cases, the drug mimics a missing small molecule that is supposed to activate a protein. We call such drugs agonists.

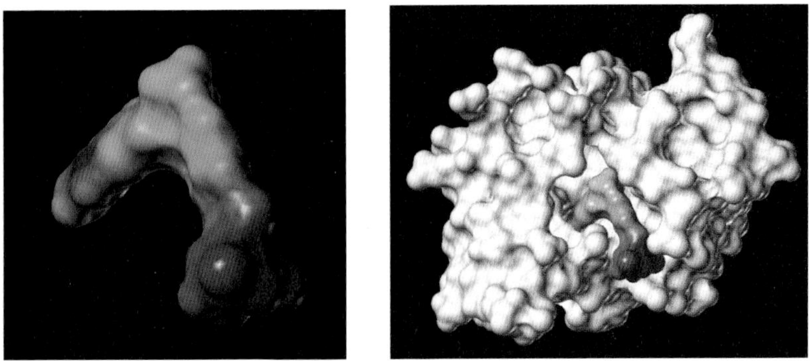

(a) (b)

Figure 5 MTX (colored by its surface potential, see Figure 3): (a) unbound, (b) bound to DHFR (gray).

In general, we are looking for drugs that bind tightly to their protein target (effectiveness) and to no other proteins (selectivity).

Most drugs that are on the market today modify the enzymatic or regulatory action of a protein by strongly binding to it as described above. Among these drugs are long-standing, widespread and highly popular medications, and more modern drugs against diseases such as AIDS, depression or cancer. Even the lifestyle drugs that have come into use in recent years, e.g. Viagra and Xenical, belong to the class of protein inhibitors.

In this view, the quest for a molecular therapy of a disease decomposes into three parts:

- *Question 1: Which protein should we target?* As we have seen, there are many thousands of candidate proteins in the human. We are looking for one that, by binding the drug molecule, provides the most effective remedy of the disease. This protein is called the *target protein*.

- *Question 2: Which drug molecules should be used to bind to the target protein?* Here, the molecular variety is even larger. Large pharmaceutical companies have compound archives with millions of compounds at their disposal. Every new target protein raises the question of which of all of these compounds would be the best drug candidate. Nature uses billions of molecules. With the new technology of combinatorial chemistry, where compounds can be synthesized systematically from a limited set of b uilding blocks, this number of *potential* drug candidates is also becoming accessible to the laboratory.

- *Question 3: Given a choice of different drugs to administer to a patient, in order to alleviate or cure a specific disease, what is the best selection of drugs to give to that individual patient?* Questions 1 and 2 have been posed without the specifics of an individual patient in mind. Target protein and drug were selected for all putative patients collectively. With Question 3 we are entering the more advanced stage of *personalized medicine* – we want to understand the different ways in which different patients react to the same drug.

Question 3 has only come into the focus of research recently. The inclusion of the discussion of this question presents a major new feature of this book over its predecessor.

We will now give a short summary of the history of research on all three questions.

4 Finding Protein Targets

Let us start our discussion of the search for target proteins by continuing our molecular example of DHFR/MTX. As mentioned, DHFR catalyzes a reaction that is required for the production of thymine – a component of DNA. Thus, blocking DHFR impairs DNA synthesis and, therefore, cell division. This is the reason that MTX, an inhibitor of DHFR, is administered as a cytostatic drug against cancer. Is DHFR the "right" target protein in this context? The frank answer to this question must be "no". DHFR is active in every dividing cell, tumor cells as well as healthy cells. Therefore, MTX impairs the division of all dividing cells that it can get to. This is the cause of the serious side-effects of the drug such as loss of hair and intestinal lining. We see that in this case the limits of the therapy are mostly dictated by the choice of the wrong target protein. Why then is this protein chosen as a target? The answer to this question is also very simple: we cannot find a better one. This example shows how central the search for suitable target proteins is for developing effective drug therapies.

Target proteins could not really realistically be searched for until a few years ago. Historically, few target proteins were known at the time that the respective drug had been discovered. The reason is that new drugs were developed by modifying natural metabolites or known drugs, based on some intuitive notion of molecular similarity. Each modification was immediately tested in the laboratory either *in vitro* or *in vivo*. Thus, the effectiveness of the drug could be assessed without even considering the target protein. To this day, all drugs that are on the marketplace worldwide target an estimated set of not much more than 500 proteins [3]. Thus, the search for target proteins is definitely the dominant bottleneck of current pharmaceutical research.

Today, new experimental methods of molecular biology, the first versions of which were developed just a few years ago, provide us with a fundamentally new way – the first systematic way – of looking for protein targets. The basis for all of these methods was the technological progress made in the context of the quest for sequencing the human genome [1]. Based on this

Figure 6 A DNA chip (from http://cmgm.stanford.edu/pbrown/explore/).

technology, additional developments have been undertaken to be able to measure the amount of expressed genes and proteins in cells. We exemplify this progress using a specific DNA chip technology [2]; however, the general picture extends to many other experimental methods under development.

Figure 6 shows a DNA chip that provides us with a differential census of the gene expressed by a yeast cell in two different cell states – one governed by the presence of glucose (green) and one by the absence of glucose (red). In effect the red picture is that of a starving yeast cell, whereas the green picture shows the "healthy" state. Each bright green dot indicates a protein that is manufactured (expressed) in high numbers in the "healthy" state. Each bright red dot indicates a protein that is expressed in high numbers in the starving cell. If the protein occurs frequently in both the healthy and the starving state, the corresponding dot is bright yellow (resulting from an additive mixture of the colors green and red). Dark dots indicate proteins that are not frequent, the tint of the color again signifies whether the protein occurs more often in the healthy cell (green), equally often in both cells (yellow) or more often in the diseased cell (red).

At this point, the exactly nature of the experimental procedures that generate the picture in Figure 6 is of secondary importance. What is important is how much information is attached to the colored dots in the picture. Here, we can make the following general statements.

(i) The identity of the protein is determined by the coordinates of the colored dot. We will assume, for simplicity, that dots at different locations also represent different proteins. (In reality, multiple dots that represent the same protein are introduced, on purpose, for the sake of calibration.) The exact arrangement of the dots is determined before the chip is manufactured. This involves identifying a number of proteins to be represented on the chip and laying them out on the chip surface. This layout is governed by boundary conditions and preferences of the experimental procedures, and is not important for the interpretation of the information

(ii) Only rudimentary information is attached to each dot. At best, the experiment reveals the complete sequence of the gene or protein. Sometimes, only short segments of the relevant sequence are available.

Given this general picture, the new technologies of molecular biology can be classified according to two criteria, as shown in the following two subsections.

4.1 Genomics versus Proteomics

In genomics, it is not the proteins themselves that are monitored, but rather we screen the expressed genes whose translation ultimately yields the respective proteins. In proteomics, the synthesized proteins themselves are monitored. The chip in Figure 6 is a DNA chip, i.e. it contains information on the expressed genes and, thus, only indirectly on the final protein products. The advantage of the genomics approach is that genes are more accessible experimentally and easier to handle than proteins. For this reason, genomics is ahead of proteomics, today. However, there also are disadvantages to genomics. First, the expression level of a gene need not be closely correlated with the concentration of the respective protein in the cell, although the latter figure may be more important to us if we want to elicit a causal connection between protein expression and disease processes. Even more important, proteins are modified post-translationally (i.e. after they are manufactured). These modifications involve glycosylation (attaching complex sugar molecules to the protein surface) and phosphorylation (attaching phosphates to the protein surface), for instance, and they lead to many versions of protein molecules with the same amino acid sequence. Genomics cannot monitor these modifications, which are essential for many diseases. Therefore, it can be expected that, as the experimental technology matures, proteomics will gain importance over genomics (see also Chapter 45).

4.2 Extent of Information Available on the Genes/Proteins

Technologies vary widely in this respect. The chip in Figure 6 is generated by a technology that identifies (parts of) the gene sequence. We are missing information on the structure and the function of the protein, its molecular interaction partners, and its location inside the metabolic or regulatory network of the organism. All of this information is missing for the majority of the genes on the chip.

There are many variations on the DNA chip theme. There are technologies based on so-called *expressed sequence tags (ESTs)* that tend to provide more inaccurate information on expression levels and various sorts of microarray techniques (see Chapter 24). All technologies have in common that the data they produce require careful quality control (Chapter 25). In general, it is simpler to distinguish different disease states from gene expression data (Chapter 26) than to learn about the function of the involved proteins from these data (Chapter 27). Proteomics uses different kinds of separation techniques, e.g. chromatography or electrophoresis combined with mass spectrometry, to analyze the separated molecular fractions (see Chapter 28). As is the case with genomics, proteomics technologies tend to generate information on the sequences of the involved proteins and on their molecular weight, and possibly information on post-translational modifications such as glycosylation and phosphorylation. Again, all higher-order information on protein function is missing. It is not feasible to generate this information exclusively in the wet laboratory – we need bioinformatics to make educated guesses here. Furthermore, basically all facets of bioinformatics that start with an assembled sequence can be of help. This includes the comparative analysis of genes and proteins (Chapter 37), protein structure prediction (Chapters 9–13), protein function prediction (Chapters 30–34), analysis and prediction of molecular interactions involving proteins (Chapters 16 and 17) as well as bioinformatics for analyzing metabolic and regulatory networks (Chapters 20–22). This is why all of bioinformatics is relevant for the purpose of this book.

If, with the help of bioinformatics, we can retrieve enough information on the molecular networks that are relevant for a disease, then we have a chance of composing a detailed picture of the disease process that can guide us to the identification of possible target proteins for the development of an effective drug. Note that the experimental technology described above is universally applicable. The chip in Figure 6 contains all genes of the (fully sequenced) organism *Saccharomyces cerevisiae* (yeast). The cell transition analyzed here is the diauxic shift – the change of metabolism upon removal of glucose. However, we could exchange this with almost any other cell condition of any tissue of any conceivable organism. The number of spots that can be put on a single chip goes into the hundreds of thousands. This is enough to put all of the human genes on a single chip. Also, we do not have to restrict ourselves to disease conditions; all kinds of environmental conditions (temperature, pH, chemical stress, drug treatment, diverse stimuli, etc.) or intrinsic conditions (presence or absence of certain genes, mutations, etc.) can be the subject of study.

The paradigm of searching for target proteins in genomics data has met with intense excitement from the pharmaceutical industry, which has invested heavily in this field over recent years. However, the first experiences have been sobering. It seems that we are further away from harvesting novel target proteins from genomics and proteomics data than we initially thought. However, in principle, a suitable novel target protein can afford a completely new approach to disease therapy and a potentially highly lucrative worldwide market share. For a critical review of the target-based drug development process, see Sams-Dodd [7].

Providing adequate bioinformatics support for finding new target proteins is a formidable challenge that is the focus of much of this book. However, once we have a target protein, our job is not done.

5 Developing Drugs

If the target protein has been selected, we are looking for a molecule that binds tightly to the relevant binding site of the protein. Nature often uses macromolecules, such as proteins or peptides, to inhibit other proteins. However, proteins do not make good drugs – they are easily broken down by the digestive system, they can elicit immune reactions and they cannot be stored for a long period of time. Thus, after an initial excursion into drug design based on proteins, pharmaceutical research has basically gone back to looking for small drug molecules. Here, one idea is to use a peptide as the template for an appropriate drug (peptidomimetics).

Due to the lack of fundamental knowledge of the biological processes involved, the search for drugs was, until recently, governed by chance. However, as long as chemists have thought in terms of chemical formulae, pharmaceutical research has attempted to optimize drug molecules based on chemical intuition and on the concept of molecular similarity. The basis for this approach is the lock-and-key principle formulated by Emil Fischer [4] over 100 years ago. Figures 3 to 5 illustrate that principle: in order to bind tightly, the two binding molecules have to be complementary to each other both sterically and chemically (colors in Figures 3 and 5). The drug molecule fits into the binding pocket of the protein like a key inside a lock. The lock-and-key principle has been the dominating paradigm in drug research ever since its proposal. It has been refined to include the phenomenon of induced fit, by which the binding pocket of the protein undergoes subtle steric changes in order to adapt to the geometry of the drug molecule.

For most of the past century the structure of protein-binding pockets has not been available to the medicinal chemist. Even to this day the structure of the target protein will not be known for many pharmaceutical projects for some time to come. For instance, many diseases involve target proteins that reside in the cell membrane and we cannot expect the three-dimensional (3-D) structure of such proteins to become known soon. If we have no information on the structure of the protein-binding site, drug design is based on the idea that molecules that are similar in terms of composition, shape and chemical features should bind to the target protein with comparable strength. The respective drug-screening procedures are based on comparing drug molecules, either intuitively or, more recently, systematically with the computer. The resulting search algorithms are very efficient and allow searching through compound libraries with millions of entries (Chapter 18).

As 3-D protein structures became available, the so-called *rational* or *structure-based* approach to drug development was invented, which exploited this information to develop effective drugs. Rational drug design is a highly interactive process with the computer originally mostly visualizing the protein structure and allowing queries on its chemical features. The medicinal chemist interactively modified drug molecules inside the binding pocket of the protein at the computer screen. As rational drug design began to involve more systematic optimization procedures interest arose in *molecular docking*, i.e. the prediction of the structure and binding affinity of the molecular complex involving a structurally resolved protein and its binding partner (Chapter 16). Synthesizing and testing a drug in the laboratory used to be comparatively expensive. Thus, it was of interest to have the computer suggest a small set of highly promising drug candidates. After an initial lead molecule has been found that binds tightly to the target protein, secondary drug properties have to be optimized that maximize the effectiveness of the drug and minimize side-effects (Chapter 19).

With the advent of *high-throughput screening* the binding affinity of as many as several hundred thousands drug candidates to the target protein can now be assayed within a day. Furthermore, *combinatorial chemistry* allows for the systematic synthesis of molecules that are composed of preselected molecular groups that are linked with preselected chemical reactions. The number of molecules that is accessible in such a combinatorial library can, in principle, exceed many billions. Thus, we need the computer to suggest promising sublibraries that promise to contain a large number of compounds that bind tightly to the protein (Chapters 16 and 18).

As in the case of target finding, the new experimental technologies in drug design require new computer methods for screening and interpreting the voluminous data assembled by the experiment. These methods are seldom considered part of bioinformatics, since the biological object, i.e. the target protein, is not the focus of the investigation. Rather, people speak of *chemin-formatics* – the computer aspect of medicinal chemistry. Whatever the name, it is our conviction that both aspects of the process that guides us from the genome to the drug have to be considered together and we will do so in this book.

6 Optimizing Therapies

How is it that different patients react differently to the same drug? Reasons for this phenomenon can be manifold. Some are easier to investigate with methods of modern biology and bioinformatics than others. Here, we distinguish between infectious diseases and other diseases.

The molecular basis of any infectious disease is the interplay of a usually large population of a pathogen with the human host. The pathogen takes advantage of the human host or, in the case of virus, even hijacks the infected cells of the patient. Chapter 23 relates a story about the interplay of a viral pathogen with the infected host cell.

With infectious diseases, the drug often targets proteins of the infecting pathogen rather than the human host. The reason is the hope that drugs for such targets harbor less serious side-effects for the patient. However, in all infectious diseases, there is a constant battle going on between the host, whose immune system tries to eradicate the pathogen, and the pathogen that tries to evade the immune system. If the disease is treated with drugs, the administered drugs impose an additional selective pressure on the pathogen. On the road to resistance the pathogen constantly changes its genome and, thus, also the shape of the target proteins for drug therapy. Changes that are beneficial for the pathogen are those that render the drugs less effective, i.e. the pathogen becomes resistant. The results of this process are widely known. With bacteria, we observe increasingly resistant strains against antibiotic therapies (Chapter 41). With viral diseases such as AIDS the drug therapy has to be adapted continually to newly developing resistant strains within the patient (Chapter 40). Therapy selection must be individualized, in both cases, at least by taking the present strain of the pathogen into account and, at best, by also considering the individual characteristics of the host. Since the pathogen is a much simpler organism than the human host, the former is significantly easier than the latter.

Although the drug acts on its intended protein target, the drug has to find its way to the site of action and, eventually, has to be metabolized or excreted again. Along that path there are multiple ways in which the drug can interact with the patient. The resulting side-effects depend on the molecular and genetic status of the individual patient. Furthermore, the protein target often has different functions, such that its inhibition or agonistic activation can incur side-effects on molecular processes that were not intended to be changed. Again, the form and magnitude of such side-effects depends on the individual patient. This process of bringing about different reactions to drugs in different patient is much harder to analyze. The reason is that larger, often widespread, networks of interactions in the patient have to be taken in account. Analyzing them necessitates complex and accurately assembled patient histories and diverse molecular data that are seldom collected in today's clinical practice. Therefore, this approach to personalized medicine is still in its infancy (Chapter 39).

Another issue with diseases is the genetic predisposition of the human individual to the disease. Monogenetic diseases have been known for a long time and are relatively easy to analyze. Here, a defect in a single gene gives rise to the disease. However, these diseases are rare, in general. The major diseases like cancer, diabetes, and inflammatory and neurodegenerative diseases are based on a complex interplay between environmental and genetic factors with probably many genes involved. With data on the genomic differences between individuals just coming into being, the analysis of the genetic basis for complex diseases is embarking on a route that hopefully will lead to more effective means of prognosis, diagnosis and therapy.

7 Organization of the Book

This book is composed of three volumes. It is organized along the line from the genotype to the phenotype.

Volume 1: *The building blocks: sequences and structures.* This volume discusses the analysis of the basic building blocks of life, such as genes and proteins.

Volume 2: *Getting at the inner workings: molecular interactions.* This volume concentrates on the "switches" of the biochemical circuitry, the molecular interactions, as well as the circuits composed by these switches, the biochemical networks. In the former context, it partly also ventures into applied issues of drug design and optimization.

Volume 3: *The Holy Grail: molecular function.* This volume ties the elements provided by the first two volumes together and attempts to draw an integrated picture of molecular function – as far as we can do it today. The volume also discusses ramifications of this picture for the development and administration of drug therapies.

Each volume is subdivided in parts that are summarized below. The total book has 11 parts. *Volume 1* covers Parts 1–4.

Part 1 consists only of this chapter, and gives an introduction to the field and an overview of the book.

Part 2, consisting of Chapter 2, discusses bioinformatics support for assembling genome sequences. This is basic technology which is required to arrive at the genome sequence data that are the basis for much of what follows in the book. Major advances have been made in this area, especially during the finishing stages of completing the human genome sequence. The field has not lost its importance as we are embarking on sequencing many complex genomes, including over a dozen mammalian genomes. Furthermore, the technology is employed in projects that sequence closely related species, such as over a dozen species of *Drosophila*, in order to obtain a more effective database for functional genomics[1]. The authors of the chapter were part of the team that developed the assembler for the draft of the human genome sequence generated by Celera Genomics.

Part 3 is on molecular sequence analysis and comprises Chapters 3–8. Chapter 3 introduces the basic statistical and algorithmic technology for aligning molecular sequences. This technology forms the basis of much that is to follow. The author of the chapter has made seminal contributions to the field. Chapter 4 discusses methods for inferring ancestral histories from sequence data. This is one of the mainstays of comparative genomics. Similar to people, one can learn a lot about genes and proteins from looking at their ancestors and relatives, arguably more so with today's methods than from inspecting the gene or protein by itself. This attributes particular significance to this chapter in the context of this book. The authors of the chapter have made important contributions to the development of methods for inferring phylogenies and applied them to analyzing the evolution of *Homo sapiens*. Chapter 5 discusses the first major step from the genotype to the phenotype, i.e. the identification of protein-coding genes. The author of the chapter has developed one of the leading gene-finding programs. The ongoing debate on exactly what is the number of genes in the human chromosome years after the first draft of the human genome sequence was available shows that the issue of this chapter is still quite up-to-date. Furthermore, genes are a primary unit of linkage between the human genome and disease, as Chapter 38 discusses. Going into the gene's structure, most of the linkage with disease happens not in the coding regions of the genes that affect the structure of the coded protein. In general, proteins are far too well refined to be tampered with. Mostly, changing a protein means death to the individual

1) see http://preview.flybase.net/docs/news/announcements/drosboard/GenomesWP2003.html for the respective community white paper and only a few severe diseases (such as sickle cell anemia, Huntington's chorea or cystic fibrosis) are linked to changes in the coding regions of genes. More subtle influences of the genotype on disease involve polymorphisms in the noncoding regulatory regions of the disease gene that do not affect the structure of the protein, but the mechanism and level of its expression. This lends special importance to Chapter 6, which addresses bioinformatics methods for analyzing these regions. The author of the chapter has led the development of a widely used set of software tools for analyzing regulatory regions in genomes. The analysis of regulatory regions ventures into the more difficult to analyze noncoding regions of genes. However, the really dark turf of the human genome is presented by the long and mysterious repetitive sections. Up to 40% of the human genome is covered with these regions whose relevance (or irrelevance?) is under hot debate, especially since some of these regions seem to harbor potential silenced retroviral genes that may become active again at some suitable or unsuitable time. The identification of these regions (although not the elucidation of their function) is discussed in Chapter 7. The authors of this chapter have made seminal contributions and provided widely used software for computational gene finding, genome alignment and repeat finding. Chapter 8, finally, discussed the algorithmic and statistical basis of analyzing major genome reorganizations that happened as the kingdoms of life evolved, and that include splitting, fusing, mixing and reshuffling at a chromosomal level. Again, we are just beginning to understand the evolutionary role of these transactions. The author of this chapter has provided important contributions to the methodical and biological side of the field, many of them together with David Sankoff and Pavel Pevzner.

Part 4 of the book is on molecular structure prediction and comprises Chapters 9–15. The part starts with a chapter on a half-way approach to protein structure prediction which only aims at identifying the regions of secondary structure of the protein (α-helices and β-strands) and related variants. The resulting information on protein structure is very important in its own right and, in addition, helps guide or verify tertiary structure prediction. The authors of the chapter have made seminal contributions to protein structure prediction starting in the early 1990s that increased the prediction accuracy significantly (from around 65 to well over 70%).

The most promising approach to identifying the fold of a protein, today, selects a template protein from a database of structurally resolved proteins and models the structure of the protein under investigation (the target protein) after that of the template protein with sequence alignment methods. If the sequence similarity between the template and the target is high enough (roughly 40% or larger), then this alignment can even serve as a scaffold for providing a full-atom model of the protein structure. The respective structure prediction method is called homology-based modeling and is described in Chapter 10.

The author of this chapter has developed one of the most advanced homology-based structure prediction tools to date. If the sequence similarity between the template and target is below 40% then generating full-atom models for the target using the template structure becomes increasingly difficult and risky. In such low-sequence-similarity ranges aligning the backbone of the target protein to that of the template protein becomes the critical issue. If this is done correctly, one obtains a 3-D model of target backbone that can serve as an aid for structural classification of the target protein. Chapter 11 describes this process. The author of Chapter 11 has codeveloped a well-performing Internet server for this structural alignment task.

Homology-based modeling can only rediscover protein structures since it models the target on the basis of a known template structure. In *de novo* structure prediction, we try to come up with the structure of the protein, even if it is novel and has never been seen before. This subject is still a major challenge for the field of computational biology, but significant advances have been made in the past 10 years by David Baker's group (University of Washington, Seattle, WA) and the author of the chapter was one of the major contributors in this context. Today, there are several projects that aim at resolving protein structures globally, e.g. over whole proteomes. The approach is a combination of experimental structure resolution of a select set of proteins that promise to crystallize easily and fold into new structures, and homology-modeling other proteins using the thus increased template set. Chapter 13 describes these structural genomics projects. One author of the chapter codirects the Protein Data Bank (the main repository for publicly available proteins structures) and the other directs a major structural genomics initiative.

The last two chapters discuss structure prediction of another important macromolecule in biology – RNA. In contrast to DNA, which basically folds into a double-helical structure, RNA is structurally diverse. There is a well-understood notion of secondary structure in RNA, i.e. the scaffold that is formed by base pairs within the same RNA chain. This algorithmically and biologically well-developed field is presented in Chapter 14. The authors of the chapter have contributed a major software package for analyzing RNA secondary structures. The last chapter in this part looks at tertiary structure prediction for RNA, a comparatively much less mature field, and its author is one of the major experts in that field, worldwide.

Volume 2 covers Parts 5–7. Based on the knowledge about molecular building blocks afforded by Volume 1, Volume 2 ventures into questions of molecular function.

Part 5 starts by considering atomic events in molecular networks, i.e. the interactions between pair of molecules. Molecular interactions are important in two ways. First, understanding which molecules bind in an organism, when and how, is fundamental for understanding of the dynamic basis of life.

Second, as we have seen in the first parts of this chapter, modifying molecular interactions in the body with drugs is the main tool for pharmaceutical therapy of diseases. Drugs bind to target proteins. Understanding the interactions between a drug and its target protein is a prerequisite for rational and effective drug therapy. Part 5 addresses both these questions. The part comprises four chapters. Chapter 16 discusses protein–ligand docking, with the implicit understanding that the ligands of interest are mostly drugs or drug candidates. The chapter discusses how to computationally dock known ligands into structurally resolved protein-binding sites and also how to computationally assemble new ligands inside the binding site of a protein. The senior author is the developer of one of the most widely used protein–ligand docking tools, worldwide. Chapter 17 discusses molecular docking if both docking partners are proteins. This problem is of lower pharmaceutical relevance, as most drugs are small molecules and not proteins, but of high medical relevance, as the basis of a disease can often be an aberration of protein–protein binding events. Furthermore, the chapter also discusses protein–DNA docking, which is at the heart of gene regulation. (Here, the protein is a transcription factor binding to its site along the regulatory region of a gene, for instance.) The authors of this chapter have developed advanced software for protein–protein docking. The last two chapters in this part discuss problems in finding drugs. As described above, the drug design process is decomposed into a first step, in which a lead structure is sought, and a second step, in which the lead is optimized with respect to secondary drug properties. If the binding site of the target protein is resolved structurally, lead finding can be done by docking (Chapter 16). Otherwise, one takes a molecule that is known to bind to the binding site of the target protein as a reference and searches for similar molecules as drug candidates. Here, the notion of similarity must be defined suitably such that similar molecules have similar characteristics in binding to the target protein. Chapter 18 discusses this type of drug screening. Finally, Chapter 19 addresses the optimization of drug leads. The authors of Chapter 19 are from the pharmaceutical industry. They are experts in applying and advancing methods for drug optimization in the pharmaceutical context.

Part 5 has advanced considerably beyond fundamental research questions and into pharmaceutical practice.

In *Part 6* we take a step back towards fundamental research. This part discusses the biochemical circuitry that is composed of the kind of molecular interactions that were the subject of Part 5. Understanding these molecular networks is clearly the hallmark of understanding life's processes, in general, and diseases and their therapies, in particular. However, the understanding of molecular networks is in its infancy, and is not advanced enough, in general, to be directly applicable to pharmaceutical and medical practice. Still, the vision is to advance along this line and the four chapters in this part present various aspects of this process. Chapter 20 is on metabolic networks, the kind discussed in a little more detail in the beginning of this chapter. Metabolic networks are quite homogeneous with respect to the roles of the participating molecules. In general, we have a substrate that is converted to a product by a chemical reaction that is catalyzed by an enzyme, possibly with the aid of a cofactor. This homogeneity makes metabolic networks especially amenable to theoretical analysis. In addition, much is known about the topology (connection structure) of metabolic networks. However, we are still lacking much of the kinetic data needed to accurately simulate the dynamics of metabolic networks. The chapter presents methods for analyzing networks both statically and dynamically. The authors are among the main methodical contributors to the analysis of metabolic networks, worldwide. Chapter 21 analyzes gene regulation networks. These networks are more heterogeneous, since they incorporate different kinds of interactions – direct interactions, as when transcription factors bind to the regulatory regions of genes, and indirect interactions, as when transcription factors regulate the expression of genes that code for other transcription factors. Furthermore, proteins, as well as DNA and RNA, are involved in gene regulation. Inferring gene regulation networks necessitates much genomic information which is just on the verge of becoming available and, thus, the field is less mature than the area of analyzing metabolic networks. The author of Chapter 21 is one of the prime experts in the field of analyzing gene regulation networks. A very special type of molecular networks is concerned with transmission of information inside the cell. Usually, these signaling networks can be analyzed in terms of smaller modules than regulatory or metabolic networks. The special methods for analyzing these networks are presented in Chapter 22 by a group of outstanding experts in the field. Chapter 23 finally moves beyond the single cell and discusses interactions between a viral pathogen and its infected host cell – a major step from basic research to its application in a medical setting. This is a very young field and the author is one of its main proponents.

Part 7 is focused on a special types of experimental data that form the basis of much research (and debate) today – expression data. We have discussed the microarray (mRNA) expression data in the chapter above, when we addressed the quest of finding new target proteins for drug therapy. Expression data were the first chance to venture beyond the genome, which is the same in all cells of an organism, and analyze the differences between different cells, tissues and cell states. Therefore, these data have a special relevance for advancing molecular medicine and this justifies dedicating a separate part of the book with five chapters to them. Chapter 24 gives a summary of the whole field, from the experimental side of the technology of measuring mRNA expression and its implications on computational analysis methods to the bioinformatics methods themselves. Since expression data are typically quite noisy, with many sources of variance residing both in the technology (which can be improved, in principle) and the underlying biology (which can and should not be changed), issues of quality control of the data play a prominent role in this chapter. The author is a global expert in the field of analysis of expression data. The following four chapters go into more detail on computational issues. Chapter 25 presents statistical methods for pretreating the data so as to arrive at an optimally interpretable dataset and it is written by a leading group of researchers in the area. The following two chapters discuss two fundamentally different kinds of analysis of mRNA expression data. Chapter 26 discusses methods that analyze and group different datasets (microarrays), generated under different circumstances (e.g. from different patients or from the same patient at different time points). Such methods afford the distinction of healthy from sick individuals as well as the analysis of disease type and disease progression, thus providing effective help in disease diagnosis. Chapter 27 groups data differently. Here, we are not interested in distinguishing different experiments, but in understanding the roles of (groups of) genes in, say, the progression of a disease that has been monitored with a sequence of microarray experiments. The results of the analysis are supposed to afford insight into the disease process and clues for drug therapy. This is a much harder task than just grouping microarray datasets and it has turned out that it cannot be solved, in general, just on the basis of expression data. Therefore, this chapter also prepares for later chapters that discuss the analysis of gene and protein function in a more general context (Part 8). The authors of Chapters 26 and 27 participate in a joint German national project that aims both at advancing the methods, and at applying them to biological and medical datasets. mRNA expression data (so-called transcriptomics data, because the data assess the expression level of mRNA transcripts of genes) have the advantage of being generated comparatively easily, due to the homogeneous structure of DNA (to which the mRNA is backtranscribed before measuring expression levels). However, these data correlate only weakly with the expression level of the actual functional unit, i.e. the synthesized and post-translationally modified protein. Measuring expression directly at the protein level is a more direct approach, but experimentally significantly more challenging. Therefore, the state of the field of proteomics, which analyzes protein expression directly, is behind that of transcriptomics, as far the experimental side is concerned. Nevertheless, proteomics is rapidly emerging, with several promising experimental technologies and the respective computational methods for data assembly/analysis. Chapter 28 presents the state of this field. It is written by a leading academic group engaged in software development for the field of proteomics.

Volume 3 builds on Volumes 1 and 2, and aims at embarking along an integrated picture of molecular function, and its consequences for the development and administration of drug therapies. The volume covers Parts 8–11.

Part 8 comprises eight chapters and is devoted to molecular (mostly protein) function. We have has already addressed aspects of molecular function (e.g. the chapters on molecular interactions and molecular networks, as well as the chapters on expression data), and, along the way, it has become increasingly evident that molecular function is a colorful term that has many aspects and whose elucidation relies on many different kinds of experimental data. In fact, molecular function is such an elusive notion that we dedicate a special chapter to discussing exactly this term, and the way it is and should be coded in the computer, respectively. This Chapter 29 is written by two authors that are main proponents of advancing the state of ontologies for molecular biology. Then we dedicate four chapters to inferring information on protein function from different kinds of data: sequence data (Chapter 30), protein interaction data that are based on special experimental technologies that can measure whether proteins bind to each other or not, and do so proteome-wide, in the most advanced instances (Chapter 31), genomic context data, affording an analysis based on the comparison of genomes of many species (Chapter 32), and molecular structure data (Chapter 33). Since all of these data still do not cover protein function adequately, we add another chapter that addresses methods for inferring aspects of protein function directly from free text in the scientific literature (Chapter 34). Chapter 35 presents methods for fusing all the various kinds of information gathered by the methods presented in the preceding chapters to arrive at a balanced account of the available knowledge on the function of a given protein. Finally, Chapter 36 discusses the druggability of targets, i.e. the adequacy of proteins to serve as a target for drug design. This quality encompasses properties such as a suitable shape of the binding pocket to suit typical drug molecules and a certain uniqueness of the shape of the binding pocket, such that drugs that bind to this pocket avoid binding to other proteins that are not targets for the drug. Again, all of these chapters are written by outstanding proponents of the respective fields.

With Parts 1–8 we have covered the space from the genotype (the genome sequence) to the phenotype (the molecular function). However, we can still take additional steps to making all of this knowledge work in applied medical settings. This is the topic of *Part 9*. To this end, Part 9 focuses on the analysis of relationships and differences between genomes. In the first chapter, Chapter 37, the topic is rolled up in a general fashion by asking the question: "What can we learn from analyzing the differences between genomes?". Then, we focus on the medically most relevant differences between genomes of individuals of the same species. Specifically, we are interested in the human and in pathogens infecting the human. Chapter 38 discusses what we can learn from genetic differences between people about disease susceptibility. Chapter 39 then addresses the topic of personalized medicine: how can we learn from suitable molecular and clinical data how a patient reacts to a given drug treatment? The final two chapters address the evolution of pathogens in the human host (mostly to become resistant against the host's immune system and drug treatment). Chapter 40 discusses viral pathogens, specifically HIV, the virus that leads to AIDS. Chapter 41 covers the bacterial world. The authors of all chapters have made seminal contributions to the topic they are describing.

Part 10 is an accompanying section of the book that addresses important informatics technologies that drive the field of computational biology and bioinformatics. There are three chapters. Chapter 42 is on data handling. Chapter 43 discusses visualization of bioinformatics data; here, molecular structures are not the center of the discussion, since their visualization is in a quite mature state, but we focus on microscopic images data, molecular networks and statistical bioinformatics data. Chapter 44 focuses on acquiring the necessary computational power for performing the analysis from computer networks (intranets and the Internet). There is a special research community that provides the progress in the underlying informatics technologies and the authors of these chapters are outstanding proponents of this community.

In *Part 11*, finally, Chapter 45 addresses in a cursory manner emerging trends in the field that were too new at the time of the conceptualization of the book to receive full chapters, but have turned out to become relevant issue at the time that the book was written. Thus, this chapter gives a cautious and anecdotal look into the future of the field of bioinformatics.

The goal of this book is to provide an integrated and coherent account of the available and foreseeable computational support for the molecular analysis of diseases and their therapies. The authors that have contributed to the book represent the leading edge of research in the field. We hope that the book serves to further the understanding and application of bioinformatics methods in the fields of pharmaceutics and molecular medicine.

References

1 COLLINS, F. S., E. D. GREEN, A. E. GUTTMACHER AND M. S. GUYER. 2003. A vision for the future of genomics research. Nature **422**: 835–47.

2 DERISI, J. L., V. R. IYER AND P. O. BROWN. 1997. Exploring the metabolic and genetic control of gene expression on a genomic scale. Science **278**: 680–6.

3 DREWS, J. 2000. Drug discovery: a historical perspective. Science **287**: 1960–4.

4 FISCHER, E. 1894. Einfluss der Configuration auf die Wirkung der Enzyme. Ber. Dt. Chem. Ges. **27**: 2985–93.

5 PAPAVASSILIOU, A. G. Clinical practice in the new era. A fusion of molecular biology and classical medicine is transforming the way we look at and treat diseases. EMBO Rep. 2001. **2**: 80–2.

6 RILEY, M., T. ABE, M. B. ARNAUD, et al. 2006. *Escherichia coli* K-12: a cooperatively developed annotation snapshot – 2005. Nucleic Acids Res. **34**: 1–9.

7 SAMS-DODD, F. 2005. Target-based drug discovery: is something wrong? Drug Discov. Today **10**: 139–47.

Part 2 Sequencing Genomes

2
Bioinformatics Support for Genome-Sequencing Projects

Knut Reinert, Daniel Huson

1 Introduction

Even though the landscape of molecular data in biology has diversified significantly in the past decade, DNA sequence data still remain the principle basis of data collection and contribute to most bioinformatics analyses. Also, the field of bioinformatics was propelled to its current magnitude mostly by the rapid development in DNA-sequencing technology. Since experimental technology only allows the reading of short stretches of DNA, encompassing just a few hundred basepairs, the assembly of these pieces into contiguous chromosomes is still a major computational challenge.

In this chapter we first describe current assembly strategies for large genomes in Section 2. We then present some of the main algorithm problems and their treatment in Section 3 and give an overview of existing assemblers in Section 4.

2 Assembly Strategies for Large Genomes

2.1 Introduction

Humans have always been fascinated by the "secret of life", i.e. the question of how new organisms come into existence, how they develop from "nothing"? What is and where is the "blueprint", the set of instructions that determines the genesis of an animal or plant? In the course of the last century, science has begun to unravel parts of the puzzle. We know now that the instructions to build a complex organism are contained in each of its cells, encoded by a

Bioinformatics - From Genomes to Therapies Vol. 1. Edited by Thomas Lengauer

ISBN: 978-3-527-31278-8

simple, yet fascinating mechanism. In 1928, Frederick Griffith, and, later, in 1944, Oswald Avery and coworkers pointed out that DNA (consisting of four very simple biochemical building blocks named adenine, cytosine, guanine and thymine) plays a vital role in heredity. In 1953, Francis Crick and James Watson discovered the double-helix structure of DNA which suggested that a simple linear sequence of nucleic acids gives rise to an intricate code for describing the blueprint of life. It was not until 1961 that researchers revealed the *genetic code* that employs codons, nonoverlapping triplets of nucleotides, to form a redundant code for the 20 amino acids that are the basic building blocks of proteins.

For a long time it was unthinkable to determine the actual sequence of nucleotides of a DNA molecule, i.e. to *sequence* a fragment of DNA. In the 1970s, a number of different approaches to sequencing DNA were pursued, and the method developed by Fred Sanger and his group prevailed. This method and other advances in biotechnology led to the sequencing of the 49kb-bacteriophage λ genome in 1982. For this work Fred Sanger was awarded the Nobel prize in chemistry in 1980, together with Walter Gilbert and Paul Berg. In the late 1980s, the question arose whether to attempt to determine the sequence of the human genome [42,48], a formidable technological challenge, given the huge size of the genome of approximately 3 billion base pairs. As a consequence, the Human Genome Project (HGP) [30] was established in 1990 to tackle the problem, armed with with a 15-year plan and a budget of approximately US$ 3 billion.

A major milestone in genome sequencing was achieved in 1995, when the 1.8-Mb genome of *Haemophilus influenza* was completed [14]. This was followed by the sequencing of other genomes, among them most notably that of yeast [35]. A main scientific issue in the 1990s was whether large eukaryotic genomes could be sequenced using a global "whole-genome shotgun assembly" (WGS) approach or whether such genomes needed to be broken down into smaller pieces and each piece sequenced separately. The assembly of the genome of the fruit fly (*Drosophila melanogaster*) in 2000 [38], of the human (*Homo sapiens*) in 2001 [54] and of the mouse (*Mus musculus*) in 2002 [37] demonstrated that the WGS approach is indeed feasible, and WGS has now become the leading paradigm.

The sequence of the human genome is of immense medical and biological importance. Significant advances in sequencing technology (in particular, the invention of the capillary gel sequencer), the availability of sufficient computational power and storage technology, and the existence of an appropriate algorithmic approach [56] inspired the founding of a private company, Celera Genomics, in 1998 with the stated goal of sequencing the human genome at a low cost and within a very short time.

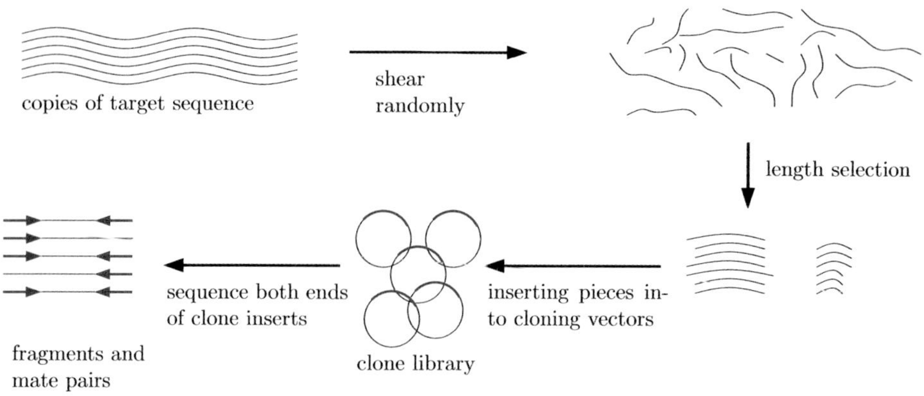

Figure 1 Experimental protocol of paired-end shotgun sequencing.

This sparked off intense competition to produce a first assembly of the sequence of the human genome as quickly as possible, which led to the publication of two papers in February 2001 that both describe a draft sequence of the human genome [25, 54].

All major sequencing projects are based on the same experimental technique, called *shotgun sequencing*. This technique is based on automated gel sequencers that use electrophoresis and fluorescent markers to determine the sequence of the nucleotides. The ability of these machines to read consecutive pieces of DNA degrades quickly with the length of the sequence and today a sequencing machine can read up to around 1000 consecutive base pairs of a fragment of DNA, depending on the degree of accuracy required. The sequence of a fragment determined in this way is called a *read*. The fragments are sampled from a stretch of DNA that is often referred to as the *source* sequence (the sequence that we take the fragment from) or as the *target* sequence (the sequence we want to reconstruct from the reads).

To determine the content of a long source sequence, one produces many copies of the source sequence (e.g. through cloning or growing colonies from a single progenitor) and then randomly breaks them into smaller pieces. Pieces of a given length are selected and one or both ends of such pieces are read by the automated sequencers. If both ends are read one does not only obtain the sequence at both ends of the piece of DNA, but also information about the relative orientation and distance of the two reads. This variant of shotgun sequencing was named *paired-end* shotgun sequencing by Myers and Weber [56], who also were the first to recognize the importance of collecting paired-end reads for sequence assembly. A pair of reads with associated distance information is called a *mate-pair*. Note that not all reads are in mate-pairs since the sequencing of one of the two mates can fail (see Figure 1 for an illustration

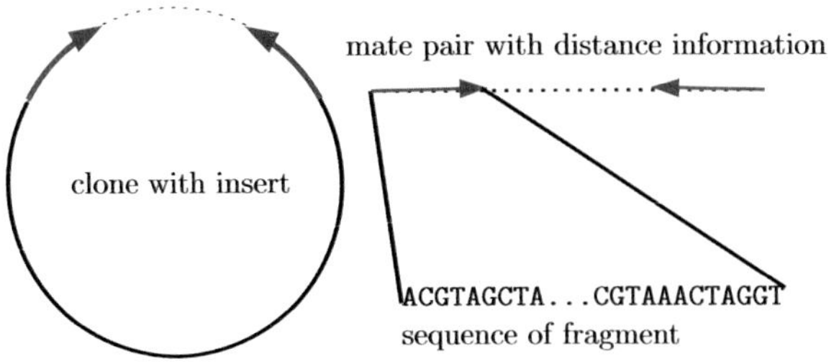

Figure 2 Fragments and mate-pairs.

of the shotgun sequencing process and Figure 2 for a more detailed illustration of a mate-pair).

The read and mate-pair information together with quality estimates of the data is fed into a computer program called an *assembler* that will attempt to reconstruct the original DNA source sequence. Note that there is no information on the location of any given read in the source sequence. However, by construction, many of the reads will come from overlapping regions of the source sequence and the first step in sequence assembly is to search for overlap alignments of high similarity between different reads. The pattern of overlaps between reads can be used to string together longer pieces of contiguous sequence, called *contigs*. The mate-pair information can then be used to order and orient sets of contigs with respect to each other, thus producing scaffolds. This process is called *sequence assembly*, and the resulting set of contigs and scaffolds is called an *assembly*. See Figure 3 for an illustration of the process.

Obviously, the large size of genomes makes sequence assembly a very difficult computational problem. Moreover, there are a number of additional difficulties. The read and mate-pair data contain errors and since DNA is a

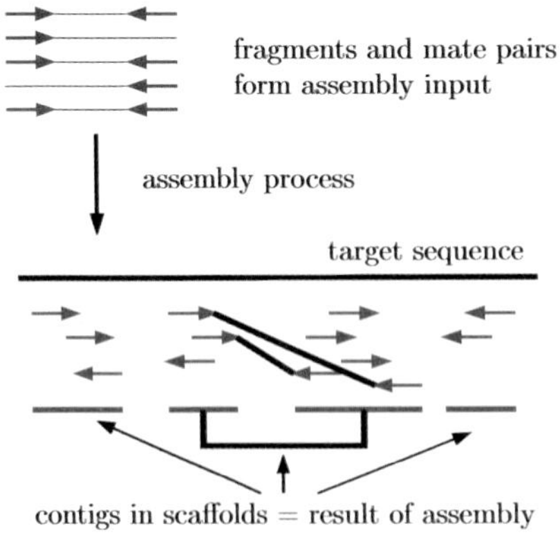

Figure 3 Pair-end reads are assembled into contigs based on how the reads overlap with each other. The contigs are then organized into scaffolds using the mate-pair information.

double-stranded molecule we do not know which strand a read is from. Also, a read may be chimeric, i.e. it may be the result of a fusion of two different pieces from different parts of the source sequence. Another problem is caused by polymorphism. If reads are taken from genomes of different individuals of the same species, they usually differ, in the case of humans at a rate of about 1 in 1000 bp. Even single organisms can be diploid or polyploid (i.e. they contain more than one copy of the same chromosome). Hence, if the data is acquired from different individuals that are not inbred, one must deal with a mixture of reads that come from seemingly different genomes. The largest difficulty is due to the fact that DNA sequences contain many different repeats of different size and fidelity. The detection and analysis of repeats is discussed in Chapter 7.

2.2 Properties of the Data

In this section we discuss some of the properties and error rates of the data generated in large genome-sequencing projects.

2.2.1 Reads, Mate-pairs and Quality Values

Sequencing large genomes is expensive, and over the past 10 years there has been a strong focus on developing faster, cheaper and more accurate ways of determining the sequence of DNA molecules. This includes substantial improvements in methods for DNA shearing, plaque and colony pickers, DNA template preparation systems, and, above of all, huge improvements in the throughput and data quality of automated sequencers (for a review, see Refs. [33,34]).

Most of the modern sequencers employ different fluorescent markers to distinguish between the four types of nucleotides. After a prefix of the fragment has left the sequencer, the marker attached to the last base of the prefix is excited by a laser and the resulting signal is measured. This analog measurement is converted into a digital base call. Each base determined in this way is assigned a quality value, given by $q = -10 \cdot \log_{10}(p)$, where p is the estimated error probability for the base [13]. For example, a quality value of 10 corresponds to an error rate of 1 in 10, whereas a quality value of 30 corresponds to an error rate of 1 in 1000. The value q is usually encoded in a single character that is stored together with the base character. Due to the nature of the sequencing process, it is clear that the distribution of the quality values is not uniform over the length of a read. The middle part usually has the best quality, whereas the quality drops at both ends of a read [12,13].

Older sequencers were slab-based and parallel sequencing lanes on an agarose gel were often mis-tracked, thereby generating incorrect mate-pairs. Modern capillary-based sequencers have eliminated this problem, but even with these machines human error (rotating or mislabeling of sequencing plates) can result in a wrong association of mate-pairs, leading to chimeric mate-pairs of unrelated reads. The error rate for mis-association of mate-pairs used to be about 10% for the older slab-based sequencers, but is now about 1%. Still, most assembly algorithms insist on the presence of more than one mate-pair to infer the relative ordering of two contigs.

In order to generate many copies of a fragment before sequencing, cloning vectors such as plasmids are used. The fragment is incorporated into the cloning vector and a sufficient number of copies is extracted after cloning.

A *spur read* is a read that aligns only partially to other reads from the same region of the source sequence. Spur reads can be the result of chimeric fragments that are obtained when two unrelated fragments fuse together during the creation of a clone library. They may also arise when fragments are contaminated with DNA from the linker or cloning vector.

To address these problems, the reads and mate-pairs obtained in the shotgun sequencing process are subjected to preprocessing steps that try to detect and remove most of the mentioned artifacts. For example, in a process called *vector and quality trimming* all reads are computationally inspected for pieces of the cloning vector genome and any traces of cloning vector sequence are removed. In addition, the quality values can be used to compute the expected number of errors in a window of the read. Any region (usually at the beginning or end of a read) for which this number is too high is then discarded. Such preprocessing steps will remove many of the artifacts, but not all. Hence, an assembly algorithm has to be able to cope with these problems to some degree.

2.2.2 **Physical Maps**

A physical map (see Ref. [47] for a description of the physical map used in the assembly of the human genome) of a genome G is given by the physical location of certain markers along G. The markers are used for navigation and can also be used for anchoring an assembly at its genomic coordinates. If parts of the target sequence are stored in clone libraries, then the correct order of the markers can be used to infer the order of the clones.

One can distinguish between two different families of methods for constructing a physical map [44]:

(i) *Restriction mapping*. Here one uses restriction enzymes to digest the DNA and then uses the lengths of the restriction fragments to reconstruct the positions of the restriction sites along the sequence. However, this technique works only for quite short genomic pieces.

(ii) *Fingerprint mapping*. Here we have a set of overlapping clones that we want to order based on common fingerprints. Therefore, one needs a set of clones that covers the target sequence redundantly. To determine which pairs of clones overlap with each other, we compute a fingerprint for each clone in such a way that overlapping clones have very similar fingerprints. The overlap information is used to order both the markers and the clones.

Fingerprints can be derived in a number of ways. One approach is to digest the DNA with a suitable restriction enzyme (e.g. *Hin*dIII was used in Ref. [25]) and use the restriction fragment sizes as a fingerprint. Alternatively, a whole restriction map of a clone can be used as a fingerprint.

Another way to obtain fingerprints is to use *STS* markers, which are short (200–500 bp) DNA sequences that occur exactly once in the given genome and are detectable by polymerase chain reaction (PCR). A number of other entities can also be used as markers. An EST is an *expressed sequence tag* that is derived from a cDNA [32]. It can be detected via a hybridization experiment or by PCR. The point is that one needs reliable, (essentially) unique markers, the presence of which is easily tested for. The assumption is that two clones overlap if they share a common set of markers.

Since the process of obtaining the fingerprints is error prone, it is very difficult to obtain a complete and accurate physical map of an entire genome. Physical maps are believed to have a high error rate of 10–20% [11] which makes the construction of a (correct) minimum tiling path a daunting task.

2.3 Assembly strategies

Given their higher complexity and larger size, it is not surprising that eukaryotic genomes are much more difficult to assemble than prokaryotic genomes. The assembly of a prokaryotic genome has become a routine task, whereas the assembly of a eukaryotic genome remains difficult. In the large sequencing projects of the last decade two different strategies were employed to determine the sequence of large eukaryotic genomes, i.e. the clone-by-clone (CBC) approach, which was used by the HGP to produce their assembly of the human genome [25], and the WGS approach, which was originally applied only to small genomes and was extended to large eukaryotic genomes by researchers at Celera Genomics [38,54].

Both approaches are based on shotgun sequencing technology, but differ in an essential preparatory step. In the CBC approach, the target sequence is broken up into a redundant collection of overlapping pieces of an easily manageable size of approximately 100–150 kb. DNA molecules of this size can be incorporated into a vector such as a BAC and they are often referred to as *bacterial artificial chromosome (BAC)* clones. The problem of determining the sequence of a BAC clone is easily solved by using shotgun sequencing and subsequent assembly. For the human genome this step reduces the problem of assembling 3 Gb to approximately 40 000 small assembly problems, each of size around 100–150 kb. This is done with the help of available physical maps and computer programs [8,29].

The CBC approach has a number of advantages. Each individual assembly problem is easily solved, since the data sets are small and contain only local repeats. In a joint effort, work can be distributed by assigning different clones to different institutions for sequencing and assembly. The assembly itself is easier and can often be done without mate-pair information. However, during the course of the HGP it became evident that these advantages come at a high price. First, the physical maps used to place the location of assembled BAC clones are incomplete and have very high error rates. Second, since overlaps of the BACs are required to determine their order, there is a certain amount of redundant sequencing necessary, which results in higher costs. Third, one needs to construct many individual libraries of sequences for both the individual BAC clones and all their fragments. This allows the introduction of many artifacts; in particular, the creation of chimeric BAC clones. Fourth, it turned out that for the final assembly mate-pairs are necessary to improve the local ordering of contigs (see also description of current assemblers in Section 4). Finally, the assignment of sequencing a subset of the clones to different institutions using different protocols and standards leads to data of uneven quality.

The WGS approach is very bold. Rather than reducing the genome assembly problem to a large set of small BAC clone assembly problems, in this approach the shotgun strategy is applied to the whole genome. This method has essentially the opposite advantages and disadvantages of the CBC approach. The computational problem of assembling the reads is by no means trivial, requiring sophisticated algorithms, sufficient mate-pair information in the input and substantial computational resources. In particular, the assembler software has to cope with the full set of repetitive elements. However, the problem of mapping the resulting contigs and scaffolds to the genomic axis is not significantly more difficult than in the CBC approach. WGS data is much less effected by uneven sampling. The main advantage of this approach is that it is far easier to automate. Only very few libraries need to be created and sequenced, and all sequenced data is processed in a single computation, usually in an incremental fashion.

There was much debate over whether the WGS approach could possibly work for large eukaryotic genomes [17, 56]. However, the feasibility of the WGS approach in conjunction with paired-end reads as input was demonstrated by the assemblies of *D. melanogaster* [38], *H. sapiens* [54] and *M. musculus* [4, 26, 37]. WGS is now the predominant approach, and most current assembly programs are based on it (see Table 1 for an overview).

3 Algorithmic Problems and their Treatment

The sequence assembly problem is to reconstruct the sequence of a target DNA molecule from read and mate-pair information, in the presence of errors and repeats. The simplest mathematical formulation of this problem is the "shortest common super-string" (SCS) problem. Given a set of strings as input, the task is to find string s that contains all input strings as substrings and is shortest among all such super-strings.

Although this formulation is an extreme simplification of the sequencing problem, it is known to be NP-hard [15] and thus is believed to be impossible to solve optimally for large instances. A more sophisticated approach is to cast the problem as a maximum-likelihood problem [41], but this has not led to a deterministic approximation algorithm.

Current assemblers were developed using an engineering approach and are not designed to optimize some explicitly stated mathematical objective function. In this section we will discuss the fundamental tasks that any assembler program must address and we will outline some of the algorithmic approaches that are employed. The process of sequence assembly must address the following fundamental tasks:

(i) *Computation of overlaps in the presence of repeats.* To determine the layout of the reads on the genomic axis each assembly algorithm is based on the fact that the sequencing is redundant, in the sense that any given position in the sequence is covered by an average of x reads, where x is usually between 3 and 12. The value of x is called the *x-coverage*. Reads that were sampled from overlapping locations in the source sequence will exhibit a high scoring overlap alignment. The goal here is to determine which pairs of reads overlap. Unfortunately, reads may also exhibit a high-scoring overlap alignment if they stem from different instances of a repeat in the source sequence.

(ii) *Layout of reads.* Based on the overlap information, a second fundamental task is to determine a layout of the reads that overlap in a consistent way. This amounts to reconstructing nonrepetitive parts of the target sequence.

(iii) *Error correction and repeat resolution.* The goal here is to distinguish between sequencing errors and differences induced by the micro-heterogeneity of different instances of a repeat, and to attempt to reconstruct parts of the repetitive sequence.

(iv) *Layout of contigs using mate-pairs.* The goal here is to use mate-pair information to order and orient contigs relative to each other.

(v) *Computation the consensus sequence.* Finally, the sequence of each contig in the assembly must be determined.

In the following sections we will discuss the main methods for solving these tasks.

3.1 Overlap Comparison of all Reads

The input to an assembly program (or "assembler") is a set $F = \{f_1, \ldots, f_r\}$ of reads, together with mate-pair information and quality values. In order to assemble a set of reads, the assembler must be able to decide whether or not two reads f_i and f_j were sampled from overlapping locations in the source sequence.

Conceptually, this can be done by computing an overlap alignment between each pair of reads or their reverse complements. The detection of an overlap does not necessarily imply adjacency in the target sequence, since an overlap can be *repeat-induced*. In Figure 4 the regions marked R_1 and R_2 indicate two instances of the same repeat with near identical sequences. Hence, reads f_k and f_l form a repeat-induced overlap, whereas reads f_i and f_j form a *true* overlap.

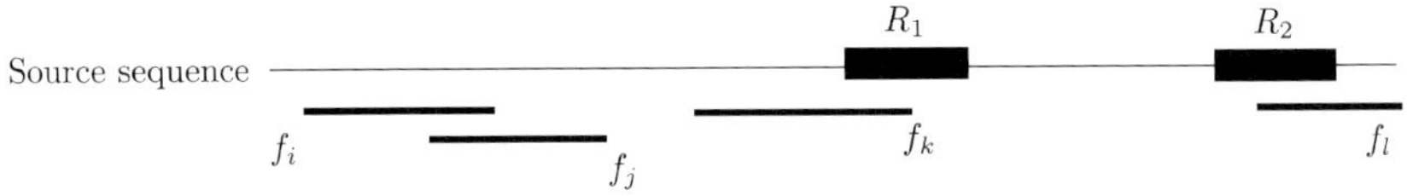

Figure 4 Reads that form true and repeat-induced overlaps. R_1 and R_2 indicate two instances of a repeat.

In a naive approach, one would require $O(r^2)$ sequence comparisons to determine all fragment overlaps. This is not feasible for large genomes where $r \approx 30$–50 million. Since most reads do not actually overlap, this computational expense seems unnecessary. In fact, one can quickly reduce the number of required overlap computations to $O(r)$, by using the "seed-and-extend-and-refine" paradigm. All current assemblers use some version of this idea (Figure 5):

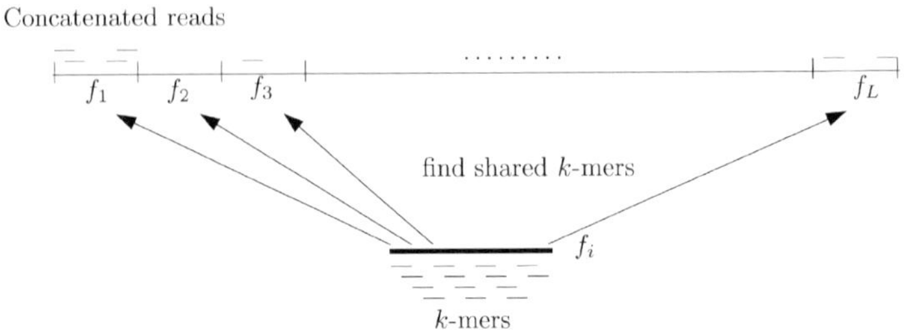

Figure 5 Overlap alignment of reads.

(i) Build a k-mer index H for all reads. This index maps any "k-mer" w (a word of length k) to the set of all occurrences of w in the reads. The value of k should be large enough such that in a random sequence of the same length as the target sequence, the expected number of k-mers is small. In contrast, k should not be too large so as to miss true overlaps due to sequencing errors. This index has two main applications:

   (a) If a pair of reads f_i and f_j do not share at least one k-mer (more sophisticated methods may have more complex requirements), they cannot possibly have a high fidelity overlap alignment and we need not attempt to compute one. If f_i and f_j contain one or more identical k-mers, these k-mers are referred to as *seeds* and the reads are candidates for an *extension*, which entails a more sophisticated and costly overlap computation (see Figure 5).
   (b) If a k-mer w appears significantly more often in the genome than expected, it probably lies in a repeat region of the genome. In this case, to avoid the computation of repeat-induced extensions, the k-mer is not used as a seed.

   A k-mer index can be computed in linear time and space. A first scan over all reads counts the number of k-mers. This allows us to efficiently allocate adjacent memory cells for all positions in the sequences that contain the same k-mer. In a second scan, the positions are written in the allocated positions (see, e.g. Refs. [5,45]).

(ii) The second phase is an extension phase, which makes use of the k-mers computed in the seed phase. Most ideas used here are very similar to BLAST [1,40]. Usually one combines two or more k-mer hits that are near to each other. Then the local alignment is extended in both directions until the quality of the extension starts to deteriorate. The result of this phase is a set of local alignments (they are depicted as longer black diagonals in Figure 6).

(iii) Finally, most algorithms end this stage by refining a set of local alignments with a fast version of the global Needleman–Wunsch algorithm. This can be done by using the shared k-mer information in a number of ways. (i) One can obtain a bound on the quality of the alignment and use it to compute a banded alignment [7]. (ii) One can compute an alignment allowing only k mismatches [6, 39, 40]. (iii) One can use the position of the shared k-mers and compute a chain of the local alignments from the extension phase together with smaller local alignments between their ends (see Figure 6 for an illustration).

The described "overlap" phase of assembly produces a collection of pairwise overlaps between reads which predominantly consists of true positive

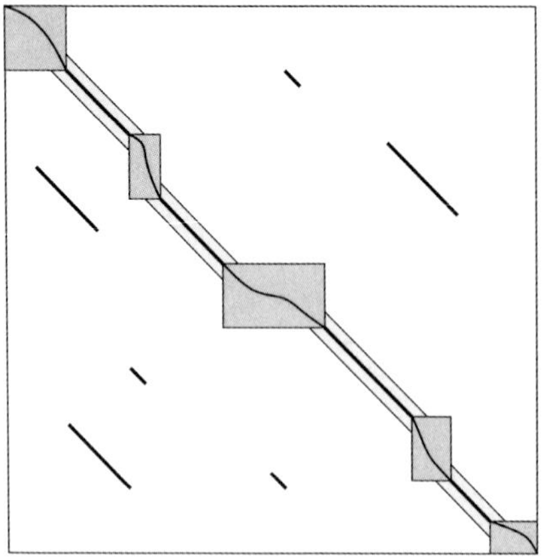

Figure 6 Seed, extend and refine paradigm. First, k-mer seeds being extended, then a banded alignment is computed that explores narrow bands around the extended seeds and, possibly, larger regions between them.

overlaps (Figure 7), i.e. overlaps that result from the fact that the involved reads stem from overlapping positions in the target sequence. However, there will be a number of false positives (repeat-induced overlaps) and false negatives (which may result from the seed-and-extend strategy missing an overlap due to sequencing errors). In Section 3.3 we will discuss how true and repeat-induced overlaps can be used to correct sequencing errors and to classify different repeat instances.

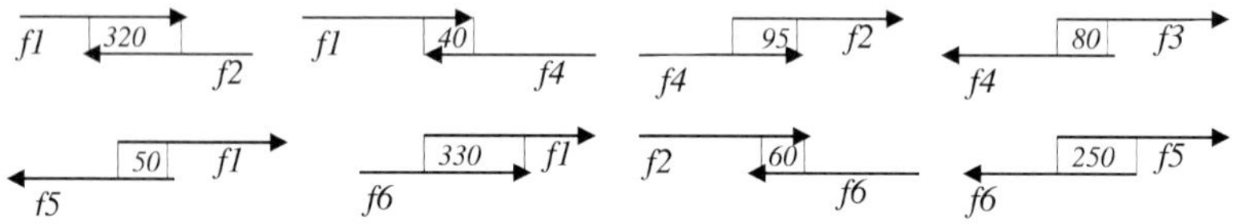

Figure 7 A collection of pairwise overlapping reads

We can view the collection of overlaps in terms of an edge-weighted, semi-directed graph, the *overlap graph* $OG(\mathcal{F})$ (Figure 8). There are two types of edges in this graph. A directed *read-edge* represents a read; the source and target nodes of the edge corresponding to the 5′ and 3′ ends of the read, respectively. The *weight* of a read-edge is simply the length of the corresponding read.

An *overlap-edge* represents an overlap between two reads and joins the two appropriate vertices contained in the corresponding read-edges. The weight of an overlap-edge is set to the negative length of the overlap. If the overlap corresponds to a gapped alignment of the ends of two reads, the amount of

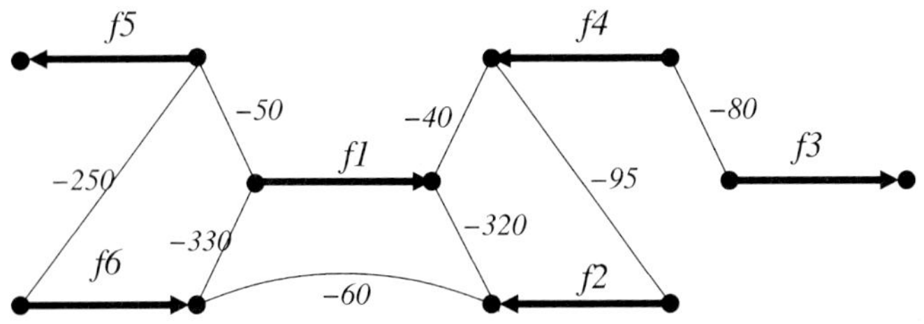

Figure 8 Overlap graph corresponding to the collection of overlaps in Figure 7. Read-edges are shown in bold.

overlap can be more accurately represented by a pair of numbers indicating the length of the two subsequences involved in the alignment.

3.2 Contig Phase: Layout of Reads

Ideally, in the absence of repeat-induced overlaps each connected component C of the overlap graph $OG(\mathcal{F})$ will correspond to a collection of reads sampled from the same local region of the target sequence. However, in practice, due to the abundance of long-range repeats, the overlap graph always consists of one huge, highly connected component.

The goal of the layout phase is to determine sets of reads that possess a consistent layout. Here, a *layout* is defined as an assignment of coordinates to all nodes of C that specify the start position s_i and the end position e_i of each read f_i in C. A layout is called *consistent* if every overlap-edge e is realized in the layout, which is the case when the coordinates assigned by the layout induce the corresponding overlap of the two appropriate reads. A layout is called *correct* if the relative positioning of the reads in the layout corresponds to their relative positioning in the source sequence. Any layout represents the reconstruction of a stretch of contiguous sequence in the target genome (a contig).

A read f_i is said to be *contained* in another read f_j if f_i is equal or highly similar to an internal portion of f_j or the reverse complement of f_j. Since contained reads contribute no additional overlap information, they are usually set aside in the layout phase of assembly. They are brought back into play later to contribute to the computation of arrival statistics and to the scaffolding of contigs using mate-pairs.

The problem of determining a minimal consistent layout of a set of overlapping reads is equivalent to the SCS. As the latter problem is known to be NP-hard [15], assemblers use heuristics to address the problem.

One widely used heuristics greedily "selects" a subset of overlap-edges S such that the union of S and the set of all read-edges F defines an alternating path of reads and overlaps that spans the set of read-edges. Initially, the edge

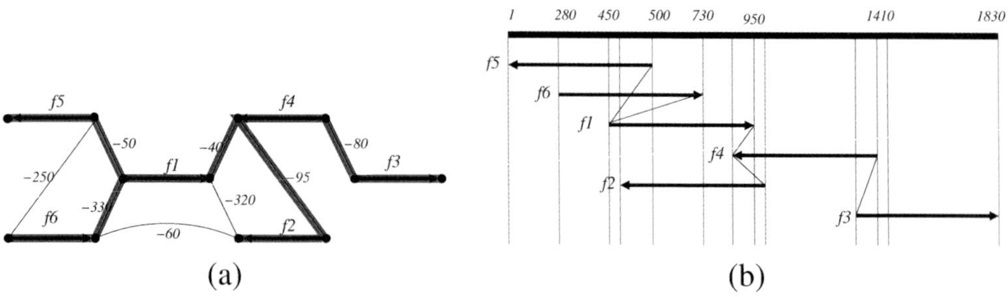

Figure 9 (a) An example of an overlap graph for six reads $\{f_1, \ldots, f_6\}$ that are as assumed to overlap as indicated in Figure 7. The edges of a maximal spanning tree are highlighted. (b) The layout of the reads is defined by the maximal spanning tree.

representing the longest overlap is selected. Then, all overlap-edges in the overlap graph are considered in ascending order of length of overlap. An overlap-edge e is selected if neither of the two nodes of e is already incident to a selected overlap edge.

Another simple heuristics for assigning the coordinates to a component C is to compute a *maximal spanning tree* that includes all read-edges, and maximizes the amount of overlap between reads (Figure 9).

In the presence of repeat-induced overlaps, any read that spans a repeat boundary may potentially overlap with reads from unrelated regions of the genome and thus bring them together in the same component C of the overlap graph. In this case, some of the overlap-edges in C will represent true overlaps, while others will represent repeat-induced overlaps. Both heuristics described above will fail to produce a correct layout whenever they utilize one or more repeat-induced overlap-edges.

As discussed before, many repeat-induced overlap-edges can be avoided in the overlap phase. To alleviate the problem further, one can attempt to distinguish between true overlaps and repeat-induced overlaps by taking a closer look at the overlap alignment. A number of mismatches in the alignment that is significantly higher than expected for the given level of sequencing error indicates that the two reads come from different instances of an inexact repeat, ideally taking the quality values into account.

Once a layout has been computed, a closer study of a multiple alignment of the reads in the layout may yield additional information, provided that sequencing errors will be randomly distributed, whereas repeat-induced discrepancies will occur in a correlated fashion. This is discussed in more detail below.

The most useful combinatorial insight is that if the reads contained in a connected component C of the overlap graph were recruited from different instances of a repeat and if some of the reads span the repeat boundaries, then

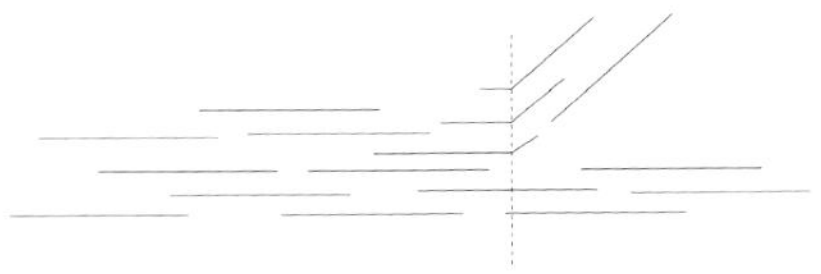

Figure 10 From left to right the reads overlap consistently until we reach the "branch point" at the position indicated by a dotted line. From this position onward, the data is partitioned into two incompatible chains of overlapping reads. Here, the reads on the left of the branch point lie in the interior of a repeat, whereas the reads that span the branch point overlap with a unique flanking sequence.

the latter reads will give rise to inconsistencies in the layout. That means, there will be overlap-edges in C that are not compatible with the overlaps induced by the layout. These incompatible overlaps will involve those reads that span repeat boundaries and a detailed analysis of the pattern of overlaps will uncover potential branch points in the layout (Figure 10).

A *branch point* is a position of a layout within a read at which a single consistent chain of overlapping reads possesses at least two different and mutually exclusive extensions. Whenever a branch point is detected, the adjacent overlaps are removed from the graph and, consequently, the connected component C is partitioned into smaller components, each giving rise to an individual contig.

As mentioned above, a consistent layout of reads defines a contig, which in this case is also called a *unitig* ("*uni*quelly assemble-able con*tig*"), as any given set of reads possesses at most one consistent layout.

Ideally, any unitig u computed in the layout phase will represent a unique stretch of the source sequence and will consist only of reads from that region. We refer to a unitig of this type as a *unique*-unitig or *U-unitig* (Figure 11). Alternatively, and in the absence of inconsistent overlaps, a unitig u may also represent a stretch of sequence that is repeated twice or more in the source sequence and may consist of reads collected from different instances of the repeat.

Methods for distinguishing between U-unitigs and non-unique unitigs make use of the sequencing coverage. For a given level of sequencing coverage, we can work out how many reads we expect to see in a unitig of a given length under the assumption that the unitig represents a unique stretch of the source sequence or that the unitig represents repetitive sequence, respectively.

In other words, a non-unique unitig can often be detected because it contains significantly more reads than expected. Let r be the number of reads and G be the estimated length of the source sequence. It can be shown [31] that for a unitig consisting of r reads and of approximate length ρ, the probability of

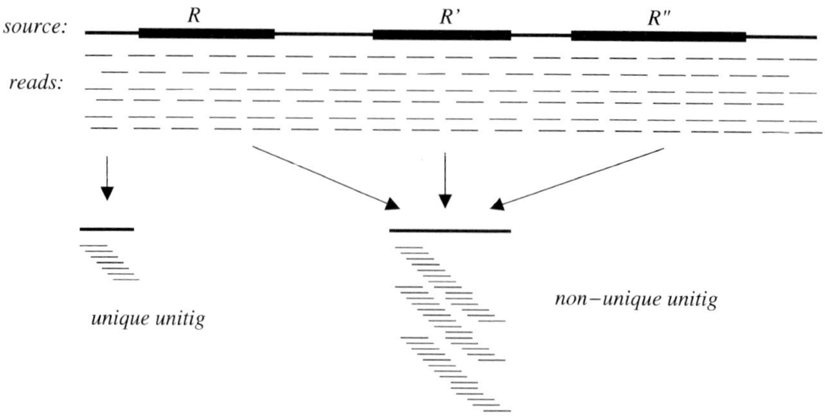

Figure 11 A unitig represents a chain of consistently overlapping reads. However, a unitig does not necessarily represent a segment of unique source sequence. For example, its fragments may come from the interior of the different instances of a long repeat, as shown here. R, R' and R'' represent three instances of the same repeat.

seeing $k-1$ start positions in an interval of length ρ is:

$$\frac{e^{-c}c^k}{k!},$$

with $c := \frac{\rho r}{G}$, if the unitig is not oversampled, and:

$$\frac{e^{-2c}(2c)^k}{k!},$$

if the unitig consists of reads recruited from two instances of a repeat. The *arrival statistic* is the log of the the ratio of these two probabilities:

$$c - (\log 2)k.$$

In practice, a unitig is considered to be unique if its arrival statistic is 10 or above, say.

3.3 Error Correction and Resolving Repeats

In the previous section we discussed how a layout of reads can be collapsed into a contig and how one can detect inconsistencies in the layout that indicate repeat boundaries or how arrival statistics can be used to classify contigs as repetitive.

In this section, we use similar techniques, but with a different goal. Branch-point detection only determines the boundary of a repetitive region to a unique region in the genome and an arrival statistic can merely point to problematic regions. Error correction and repeat resolution approaches take a closer look at the distributions of errors in the layout of a collection of reads and their main task is to determine whether a mismatch in a pairwise alignment is due to a sequencing error, a single nucleotide polymorphism (SNP) or a low copy repeat.

The errors in a repetitive contig and the errors in a nonrepetitive contig are differently distributed. In a nonrepetitive contig errors in overlaps can be explained by sequencing errors which should occur independently from each other in each read. In contrast to this, repetitive contigs by definition consist of reads that are from instances of a repeat from different genomic locations. Depending on the nature of a repeat, two instances differ from each other by a certain amount.

In order to be able to classify sequences as repetitive or nonrepetitive, one needs a suitable null model, i.e. the sequencing error rate in the local genomic region. This error rate was often assumed to be a constant that could be refined using bootstrapping methods [9]. Alternatively, it was estimated using the quality values of bases in the reads. Huang [22] estimated the amount of sequencing errors in a local neighborhood based on the overlaps of an individual read with its overlapping partners (see also Ref. [21]). Developing this idea further, one could obtain an even better estimate of the error rate by constructing a multiple alignment in the layout phase. Such approaches work well if no additional source of error confuses the estimation of the sequencing error. If, however, repetitive overlaps are present, then these approaches cannot be applied directly. Nevertheless, we can assume that we have a rather good idea of the sequencing error for a collection of overlaps.

The fact that the reads are collapsed into a contig means that this difference is small, i.e. in the range of 1–3%. This is still significantly higher than the assumed rate of SNPs and hence this *microheterogeneity* can be used for detecting the different repeat instance (Figure 12).

This simply means that we use the fact that an instance of a repeat differs slightly from other instances. Hence, reads from a certain genomic location *always* differ from the reads in the other location, except in the unlikely event that the corresponding positions are changed by a sequencing error. Some

```
...AGCCGTCAGA...
...AGCCGTCAGA...
...AGCCCTCTGA...
...TGTCGTCTGA...
...AGTCGTCTCA...
...AGTCGTCTGA...
```

Figure 12 Sequencing errors (in red) and micro-heterogeneity of a collapsed repeat (in blue).

assembly programs like Euler [43] and ARACHNE [4] have a built-in, simple error correction phase that corrects numerous mistakes.

However, since the problem is modular, several papers addressed it individually. In Figure 12 differences caused by repeats are shown in blue and differences caused by sequencing errors are shown in red. The blue columns are called *DNPs* (defined nucleotide positions) [52] or *separating columns* [28] and can be used to separate the individual copies of a repeat.

The method of Tammi and coworkers proceeds in a straightforward way. It first prepares multiple alignments which it then refines, using a realignment algorithm [2]. Then, the consensus base in a column is defined as the most frequent base of the column. Whenever we see a certain number of *coinciding* differences from the consensus, the column is a candidate for usage in repeat separation (e.g. the first blue columns in Figure 12). If another candidate column can be found, these candidate columns define a DNP (e.g. the second blue column Figure 12).

Since the above procedure identifies errors that are due to micro-heterogeneity in repeats, we can attribute the remaining errors to the sequencing phase. Hence, the DNPs can also be used for correcting sequencing errors [51].

3.4 Layout of Contigs

In the layout phase, reads are assembled into contigs based on their overlaps, as reported in the overlap graph. Ideally, one may hope that this will give rise to a small number of very large contigs, perhaps one per chromosome arm. However, due to two problems this cannot happen. (i) Shotgun sequencing produces a random sampling of the source sequence, thus the coverage fluctuates along the sequence and some regions will remain unsampled, giving rise to *sequencing gaps* (see Ref. [31] for a mathematical treatment of the statistics for the length and number of such gaps). (ii) Repeats in the source sequence lead to the break-up of potential contigs into smaller ones, as described above.

Hence, a common strategy is to arrange sets of contigs into so-called *scaffolds* or *super-contigs* with the help of mate-pair information. More precisely, a scaffold consists of an ordered list of contigs $(c_1, c_2, \dots, c_t)$, alongside a specification of the orientation of each individual contig (i.e. whether to use c_i or the reverse complement $\bar{c}_i$) and an estimation of the distance between any two consecutive contigs. A scaffold is deemed *correct* if the relative positioning and orientation of its contigs corresponds to the true locations in the source sequence.

As described above, shotgun sequencing projects often use a *paired-end* or *double-barreled* shotgun protocol, in which clones of a given fixed length are sequenced from both ends. This approach produces pairs of reads, called

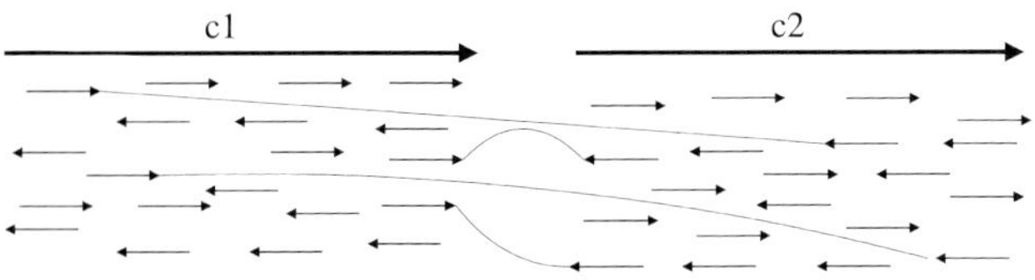

Figure 13 If two assembled contigs c_1 and c_2 correspond to neighboring regions of the source sequence, then we can expect to find mate-pairs that span the gap between them.

mate-pairs, whose relative orientation and mean distance l (with standard deviation σ) are known (Figure 2).

Standard size-selection techniques are used to produce a collection (library), of clones that have approximately the same length. A typical mixture of clone lengths is $l = 2$, 10 and 150 kb. With care, a standard deviation σ of approximately $1/10$ of l can be achieved.

Consider two contigs c_1 and c_2 produced in the layout phase of assembly. If they correspond to neighboring regions in the source sequence, we can expect to find mate-pairs that span the gap between them, as indicated in Figure 13. Such mate-pairs can be used to determine the relative orientation and estimated distance between c_1 and c_2.

Assume that the two contigs c_1 and c_2 are connected by mate-pairs $m_1, m_2, \ldots, m_k$. Each mate-pair provides an estimate of the distance between the two contigs. If these estimates are viewed as independent measurements, then they can be combined into a single estimate using standard statistical calculations.

As the assignment of reads to their mates is error prone, the existence of a single mate-pair linking two different contigs is not deemed significant. It is, however, of great statistical significance if two U-unitigs c_1 and c_2 are linked by two different mate-pairs in a consistent way. Similarly it is very unlikely that two mate-pair specification errors would put together two pairs of reads from the same two local regions of the source genome.

Assume that we are now given a collection of contigs $\{c_1, c_2, \ldots, c_k\}$ and a table of mate-pair information that links pairs of reads that are embedded in the contigs. To discuss the problem in more detail, we introduce the *contig-mate* graph. In this graph, each contig c_i is represented by a directed *contig-edge* having nodes s_i (the start node) and e_i (the end node). So-called *mate-edges* are added between such nodes to indicate that the corresponding contigs contain reads that are mates. For example, the two contigs c_1 and c_2, together with the collection of mates depicted in Figure 14 give rise to the contig-mate graph indicated in Figure 15.

If a set of different mate-pairs link two different contigs c_1 and c_2 in a consistent manner, then the contig-mate graph can be simplified by replacing

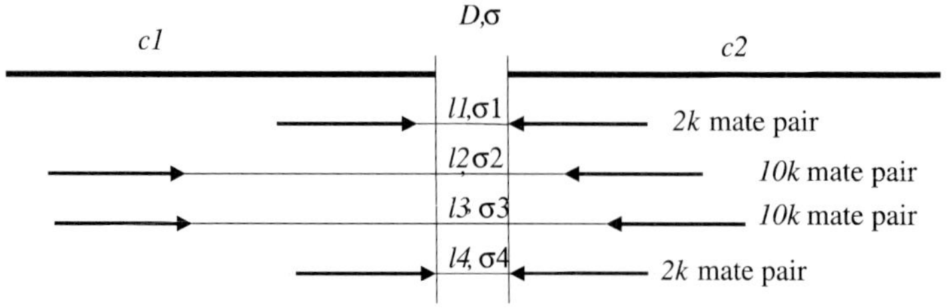

Figure 14 Here we depict two contigs that are linked by four mate-pairs. Each mate-pair provides an estimate (l_i, σ_i) of the gap between the two contigs, and simple statistics can be used to estimate a resulting mean distance D and standard deviation σ.

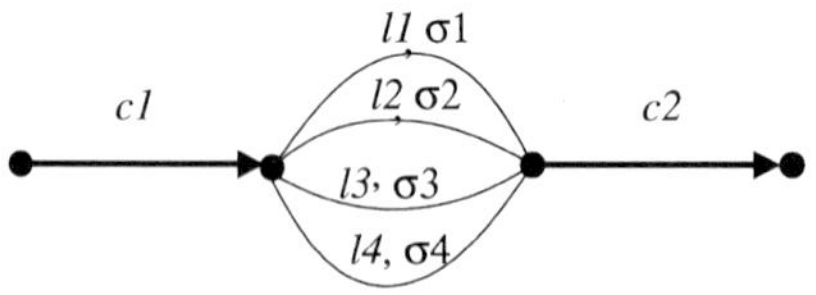

Figure 15 The two contigs and four mate-pairs shown in Figure 14 give rise to two contig-edges and four mate-edges in the contig-mate graph, as shown here.

the set of edges by a single *bundled mate-edge e*, whose mean length μ and standard deviation can be computed from the values for the original mate-edges, using straightforward statistics. Additionally, *e* is assigned a weight to reflect the number of mate-pairs that support it. Further edges can be bundled using so-called *transitive reduction*, which we do not describe here.

The goal of the *scaffolding* phase is to use the contig-mate graph to determine the true relative order and orientation of a set of contigs that are linked by mate-pairs. Most assemblers use different heuristics to address this problem.

We briefly discuss how this problem can be formalized (see Ref. [23] for details). An ordering or scaffolding of a set of contigs can be specified as a path *P* through the corresponding contig-mate graph. To this end, it may be necessary to infer "missing edges" between consecutive contig-edges in the path. To evaluate such a scaffolding one can look at all the mate-edges in the graph. We say that a mate-edge *e* is *satisfied* if the mate-pair layout implied by *e* is compatible with the ordering and orientation of contigs implied by *P*, otherwise *e* is called *unsatisfied*. Thus, the scaffolding phase can be stated as the following optimization problem: for a connected contig-mate graph, find a path *P* through the graph that contains all contig-edges, that possibly uses additional inferred edges and maximizes the number of satisfied mate-edges. This problem has been shown to be NP-hard [23].

Existing assemblers use straightforward heuristics in an attempt to form scaffolds. One heuristics that addresses the stated optimization problem directly is the "greedy path-merging" algorithm [23]. Given a connected contig-mate graph, the algorithm proceeds "bottom-up" as follows, maintaining a valid scaffolding $S \subseteq E$. Initially, all contig-edges $c_1, c_2, \dots c_k$ are selected, but no others. At this stage, the graph consists of k selected paths $P_1 = (c_1), \dots, P_k = (c_k)$. Then, in ordering of decreasing weight, each mate-edge $e = \{v, w\}$ is considered. If v and w lie in the same selected path P_i, then e is a chord of P_i and no action is necessary. If v and w are contained in two different paths P_i and P_j, we attempt to merge the two paths to obtain a new path P_k and accept such a merge, provided the increase of $S(G)$ (the number of satisfied mate-edges) is larger than the increase of $U(G)$ (the number of unsatisfied ones).

3.5 Computation of the Consensus Sequences

In a final step we need to determine the actual sequence for the target molecule. So far, we have discussed how to construct contigs, how to order them in scaffolds and how to address the problem of repeat resolution. The pairwise overlaps between reads provide only an approximate layout of the reads with respect to each other. To obtain a final layout, one needs to solve a special multiple alignment problem, with the following properties:

- The reads of the multiple alignment are almost identical.
- Quality values can be incorporated in the computation of the consensus sequence and be used to assign quality scores to consensus characters.
- The alignment is usually of depth 5–10 and very long (up to millions of base pairs).
- We need to compute the alignment very fast.

The fact that the initial read layout already gives an approximate alignment and the need to compute the alignment quickly results in the application of heuristics to solve the multiple alignment problem, since a generalization of the dynamic programming-based approach would result in a running time of $O(n^k)$ where k is about 5–10.

Most assemblers [4,26,50] implement the idea depicted in Figure 16 in some way. For each contig an alignment is "grown", starting with the pairwise alignment of the two left-most reads or the two reads with the best pairwise similarity. Then, the next read is aligned to the multiple alignment of the previous two reads and so on. Aligning a read to an alignment is usually done by converting the multiple alignment into a profile, and then employing an adaption of the pairwise alignment algorithm to the profile and the read (Figure 16). If quality values are at hand, they can be incorporated in to the alignment computation.

```
CGATAGCTAGG-CTAGCATCGC
      CTAGGGCTAGCATCGGGGCGCC
           GCTAGCAT-GGGGCGCCCTCGATCGTT
                 ATCG--GCGCCCTCG-TCGTTGCTAATAG
```

add next read to box

```
CGCCCTCGTCGTTGCTAATAGCGTTGCGC
```

Figure 16 Computation of the consensus sequence.

Once a multiple alignment has been determined, a consensus character is computed for each column of the multiple alignment. This can be done by simply voting on the majority character or, alternatively, by weighing the vote using the quality values [4]. If prior knowledge of the base composition is at hand, it can even be incorporated in a Bayesian approach [9] which computes the most likely consensus character and also derives a quality value for it. Some assembly programs employ *ad hoc* heuristics to incorporate the quality values [21] or use the approach implemented in the Phrap package [16,18,36]. Phrap avoids computing a multiple alignment altogether. Instead, it chooses a chain of single reads which are chosen such that they provide adjacent intervals of high-quality base calls. In each interval this single high-quality base is chosen for the consensus sequence.

Although this strategy does not use all available information, it avoids some artefacts introduced by the progressive method commonly used for the computation of multiple alignments. For example, Figure 17 shows a typical output of a progressive alignment on the left. Depending on the score function, the last read may result in three different alignments with other reads which are merged into a multi-alignment that introduces two `T`s into the consensus sequence (depicted in blue). However, the multi-alignment on the right is more likely to be correct, since it can be explained with only two sequencing errors in the last read. Such additional characters are to be avoided, since they confound gene prediction algorithms and other

```
...AGCCGT--CTGA...  ...AGCCGT--CTGA.
...AGCCG-T-CTGA...  ...AGCCGT--CTGA.
...AGCCG-T-CTGA...  ...AGCCGT--CTGA.
...AGCCG--TCTGA...  ...AGCCGT--CTGA.
...AGCCGT--CTGA...  ...AGCCGT--CTGA.
...AGCCGTTTCTGA...  ...AGCCGTTTCTGA.
------------------  ----------------
...AGCCGTT-CTGA...  ...AGCCGT--CTGA.
incorrect           correct
```

Figure 17 Common error in consensus computation.

computational sequence analysis tools. A possible strategy to achieve this is to use iterative refinement strategies that correct such mistakes [2].

4 Examples of Existing Assemblers

In Table 1 we give an overview of current assemblers that are able to compute an assembly of large eukaryotic genomes and list some of the genomes that have been applied to. We use the attributes CBC and WGS to indicate whether the assembler follows more the CBC or the WGS paradigm. Assemblers that take clone information and a WGS data set as input are marked as hybrid.

Table 1 Recent assembly programs for eukaryotic genomes.

Name	Year	Strategy	Genomes (examples)
Celera	2000	WGS	*H. sapiens* [54], *M. musculus* [37], *D. melanogaster* [38], *Anopheles gambiae* [19],
GigAssembler	2001	CBC	*H. sapiens* [29]
ARACHNE	2002	WGS	*M. musculus* [4,26]
JAZZ	2002	WGS	*Fugu rubripes, Ciona intestinalis* [3,10]
RePS	2002	WGS	*Oryza sativa* [55,57]
Barnacle	2003	CBC	*H. sapiens* [8]
PCAP	2003	WGS	*Caenorhabdtis briggsae, M. musculus* [20]
Phusion	2003	WGS	*M. musculus* [36]
Atlas	2004	hybrid	*Rattus norvegicus* [18,46]

In the following we give short descriptions of the assemblers listed in Table 1. This will give the flavor of the latest algorithmic approaches and show that all assemblers use similar ideas.

4.1 The Celera Assembler

The Celera assembler was the first WGS assembler to assemble large eukaryotic genomes [38,54]. It screens the reads, removes vector or linker sequence and keeps only the interval with an average sequence identity of 98%. The overlapper module compares all pairs of reads to detect high-fidelity overlaps. To avoid a quadratic number of overlap computations, the overlapper uses a k-mer index to exclude nonrelated pairs from the expensive overlap. This results in a read-overlap graph as described in Section 3.1. Regions of this graph are assembled into contigs whenever the initial arrangement of reads in this region is unique. The Celera assembler incorporates mate-pair information, and orders and orients the contigs. The remaining gaps are closed in a sequence of less and less conservative steps. First contigs are placed if they are "anchored" by two mate-pairs, then if they are anchored by one mate-pair and an overlap path and so on. For each contig a consensus sequence is computed based on a progressive multiple alignment and a local heuristics to remove merging artefacts.

4.2 The GigAssembler

The GigAssembler [29] was designed to assemble the human genome from the CBC data obtained from the HGP [25]. It had to assemble all BACs in a tiling layout of clones. In addition to this input set, it uses mate-pairs, mRNA and EST information to bridge gaps in scaffolds. The GigAssembler screens the input sequences for contaminations and masks known repeats. Additional sequence information (reads of mate-pairs, ESTs, mRNA and BAC end reads) is aligned to the input. Similar to the description in Section 3.1, GigAssembler builds a lookup table of 10-mers and then conducts a detailed alignment in regions where consecutive 10-mers match. The main routine of the GigAssembler builds sequence contigs (called "rafts") from overlapping initial sequence contigs within a clone. Then it builds clone contigs (called "barges") from overlapping clones, and orders and orients the resulting contigs into "super-contigs". These assemblies are combined into full chromosome assemblies.

4.3 The ARACHNE Assembler

The ARACHNE assembler was developed by a group at the MIT that was also a major partner in the HGP. In its first version [4] its functionality was tested by reassembling the genomes of *H. influenza*, *Saccharomyces cervisiae* and *D. melanogaster*, as well as the two smallest human chromosomes, 21 and 22. A later version of ARACHNE [26] was used in the public assembly of the mouse genome. ARACHNE appears to be modeled after the Celera assembler, with a few differences.

As a true WGS assembler its input consists of a set of reads and mate-pairs where the mate-pairs are taken from carefully length selected clone libraries. An overlap phase is conducted as outlined in Section 3.1. In addition, it employs an error correction phase using multiple alignments deduced from the pairwise overlaps.

The contig assembly phase differs from the one employed in the Celera assembler, since it directly incorporates mate-pairs by identifying "paired reads", which are reads of two mate-pairs where the two left and the two right reads overlap, respectively. This is a clever way to form contigs that are consistent in overlap and mate-pair information.

ARACHNE computes repeat boundaries by inspecting the pairwise overlaps. To detect remaining repetitive contigs ARACHNE uses an arrival statistic similar to the one described above, together with the fact that repetitive contigs are likely to have mate-pairs that link them in a contradictory way to other contigs.

ARACHNE uses mate-pairs to build scaffolds using a greedy algorithm that gives priority to merging contigs that are supported by the most links involving the shortest distance. This phase is followed by an attempt to fill gaps using contigs that were previously labeled as repetitive. Since the first labeling was conservative, this will often succeed. A consensus sequence is derived by heuristicly computing a multiple alignment.

4.4 The JAZZ Assembler

The JAZZ assembler is a modular assembler making use of different, sometimes already existing modules. The input reads are trimmed with respect to a window average of the quality values. In addition, they are checked for vector contamination which is then removed. The overlap phase is similar to the description of Section 3.1. An index of 16-mers is constructed. All 16-mers that occur too often are not used for triggering a more expensive alignment step. Then all reads that share more than ten non repetitive 16-mers are aligned using a banded Smith–Waterman algorithm. JAZZ constructs a scaffolded layout of reads. In particular, JAZZ postpones the computation of contigs until the consensus phase, which employs a consensus algorithm similar to Phrap. JAZZ tries to close gaps in scaffolds that are due to repeats in the genome.

4.5 The RePS Sssembler

The RePS uses also Phrap as its main assembly engine. It was primarily used to assemble the rice genome [55]. RePS masks out repeated 20-mers. The masked reads are handed to Phrap. As a post-Phrap step, RePS uses mate-pairs to fill gaps and build scaffolds. The strategies RePS uses are concepts borrowed from the Celera assembler.

4.6 The Barnacle Assembler

Barnacle [8] is an assembler that was used to reassemble the human genome from the public CBC data. In contrast to GigAssembler it does not make use of a physical map, but uses mate-pairs and clone data only (possibly augmented with chromosome assignments). Barnacle computes all pairwise local alignments of the input sequence. This is done using a strategy as described in Section 3.1. Using those overlaps, contigs in the input set are merged whenever possible, thereby reducing the number of contigs by an order of magnitude. The clone overlaps are deduced (two clones overlap if, and only if, at least one contig pair of the corresponding clones overlaps) and conflicts are resolved by heuristically enforcing the layout graph (clones are vertices, overlaps induce edges) to be an interval graph. Barnacle orients the contigs starting from the interval representation of clones and assign coordinates to sub-contigs. Again, possible inconsistencies result in the discarding of the contigs involved.

4.7 The PCAP Assembler

The PCAP assembler is a true WGS assembler that incorporates many aspects of the well known CAP3 assembler [21]. One interesting aspect of PCAP is the way repeats are identified *de novo* during the overlap computation. For this, the set of reads is partitioned into subsets that can be distributed on many computers. Then, in an iterative process, repeats are identified during the overlap computation and used to avoid the computation of repeat induced overlaps. To do this, some overlaps are computed and repetitive regions are identified based on those overlaps. These repeats are used in the next round of overlap computations, and so on.

The overlaps themselves are computed in a manner similar to the approach outlined in Section 3.1. Prior to the construction of contigs, the depth of coverage at every point in the initial layout determined by the overlaps is computed. Using the depth of coverage, overlaps are assigned a score that reflects whether they are repeat induced or not. Only overlaps that are likely to be unique are used in the contigging step. In addition, poor ends of reads are located and clipped and chimeric reads discarded (all based on pairwise overlaps).

Contigs are formed by inspecting read overlaps in decreasing order of their adjusted score. Then the CAP3 algorithm for scaffolding is applied. It consists of finding groups of mate-pairs that indicate a mis-assembly of the contig. If such mate-pairs can be found, the contig is corrected and the mate-pair consistency is checked again.

A simple gap filling strategy based on finding overlap paths is applied, multiple alignments are computed and a consensus sequence is derived as in CAP3. The computation of the consensus sequence involves a heuristic procedure that makes use of the quality values of the reads.

4.8 The Phusion Assembler

The Phusion assembler was primarily designed to assemble the mouse genome from a WGS data set at 7.5$\times$ coverage [36] and was developed in parallel with the ARACHNE assembler [4, 27]. It is a modular assembler in the sense that it incorporates an older program, i.e. Phrap, as an integral part of its operation. It screens the input reads for poor quality reads, which are completely removed, and conducts a screening for vector contamination.

Phusion computes a histogram of all k-mers for a suitable *k*. Similar to the Atlas assembler, it uses this histogram to exclude *k*-mers that occur too seldomly (probably sequencing errors) and too often (probably repeats). The remaining *k*-mers are used to group the reads into contiguous groups, which are then passed to Phrap for assembly. This strategy is quite similar to the Atlas assembler and the compartmentalized assembler used by Celera to assemble a version of the human genome [24].

Phusion uses Phrap as its assembly engine and iteratively computes assemblies of sets of reads. It checks the consistency of the mate-pairs in this set. Whenever an inconsistency is detected, Phusion splits the set and reassembles the parts using Phrap. This results in a number of contigs that might share sequence parts. Phusion tries to join contigs based on the number of shared reads and sequence overlaps, a strategy not unlike that of the GigAssembler [29]. The resulting, larger contigs are scaffolded, using the mate-pair information.

4.9 The Atlas Assembler

The Atlas assembly system is a suite of programs that form a hybrid assembler which uses reads from WGS and from CBC data sets. Thus, it is very similar to the compartmentalized assembler developed at Celera Genomics [24,54].

Atlas trims the input reads based on the error rate in a local window. It builds a *k*-mer index of the WGS reads, since these cover the genome uniformly. Similar to the Phusion assembler, it uses the fact that seldomly occurring *k*-mers are likely to contain sequencing errors, while abundant *k*-mers are likely to be repetitive. Atlas establishes the "rarity" of a *k*-mer in the overlap phase, using such *k*-mers to seed a banded alignment as described in Section 3.1.

The WGS reads are binned by using the localized BAC clone reads to "catch" the corresponding WGS reads. The reads in each BAC bin are assembled using Phrap. Since Phrap does not use mate-pairs during the assembly, the resulting contigs are checked for consistency and, if found to be inconsistent, split using the mate-pair information. The same information is then used to scaffold the resulting contigs. The improved BACs are called *eBACs*.

Atlas performs a meta-assembly of the eBACs. Based on overlap information and independent mapping data, a tiling path of eBACs is computed. The assembly induced by this tiling path is refined using *rolling-Phrap*, which is an iterative procedure calling Phrap in a window that is cleverly moved over the tiling path. The resulting large contigs are linked using mate-pairs and localized BAC reads, and then anchored on the genomic axis using external mapping data.

4.10 Other Assemblers

There are a number of other assemblers that are not described here, either because they have been outdated by more recent developments (e.g. Refs. [14, 50]) or because they have not been used to assemble large eukaryotic genomes. Specifically, we would like to mention the Euler (or Euler-DB) assembler [43], which formulates the assembly problem differently, using a *k*-mer graph.

The last years have seen the development of a host of different assembly programs, which nevertheless share a significant portion of algorithmic ideas. In general, sequence assembly can be seen as a concatenation of algorithmic modules with well-defined interfaces. Hence, we believe that it would be worthwhile to combine the best implementations of these modules, an approach that has been taken by the Amos consortium hosted by The Institute for Genomic Research (TIGR) [53].

5 Conclusion

Assembling whole eukaryotic genomes was deemed impossible only 15 years ago. Yet, an initiative was founded to tackle the seemingly gargantuan task of assembling the human genome. Whole-genome assembly of eukaryotic genomes, once strongly criticized as impractical, has now been successfully applied to a number of large genomes and has become the standard approach. This would not have been possible without bioinformatics support, the development of efficient assembly algorithms and solid engineering to implement those algorithms into robust computer programs that also handle all peculiarities of the data that are not captured in the mathematical models.

Acknowledgements. We would like to thank the Informatics Research group at Celera Genomics which, under the guidance of Gene Myers and Granger Sutton, developed the first large-scale whole genome shotgun assembler, and used it to assemble *D. melanogaster*, *H. sapiens*, and *M. musculus*. Another detailed and careful description of the shotgun assembly problem can be found in [49]. We would like to thank Gene Myers for reviewing an earlier version of the manuscript. His comments helped to greatly improve the chapter.

References

1 ALTSCHUL, S., W. GISH, W. MILLER, E. MYERS AND D. LIPMAN. 1990. A basic local alignment search tool. J. Mol. Biol. **215**: 403–10.

2 ANSON, E. AND E. MYERS. 1997. Realigner: a program for refining DNA sequence multialignments. J. Comput. Biol. **4**: 369–83.

3 APARICIO, S., J. CHAPMAN, E. STUPKA et al. 2002. Whole-genome shotgun assembly and analysis of the genome of *Fugu rubripes*. Science **297**: 1301–10.

4 BATZOGLOU, S., D. B. JAFFE, K. STANLEY, et al. 2002. ARACHNE: a whole genome shotgun assembler. Genome Res. **12**: 177–89.

5 BURKHARDT, S., A. CRAUSER, P. FERRAGINA, H.-P. LENHOF, E. RIVALS AND M. VINGRON. 1999. q-gram based database searching using suffix arrays. Proc. RECOMB **3**: 77–83.

6 CHANG, W. AND E. LAWLER. 1994. Sublinear approximate string matching. Algorithmica **12**: 327–44.

7 CHAO, K., W. PEARSON AND W. MILLER. 1992. Aligning two sequences within a specified diagonal band. Comput. App. Biosci. **8**: 481–7.

8 CHOI, V. AND M. FARACH-COLTON. 2003. Barnacle: an assembly algorithm for clone-based sequences of whole genomes. Gene **320**: 165–76.

9 CHURCHILL, G. AND M. S. WATERMAN. 1992. The accuracy of DNA sequences: estimating sequence quality. Genomics **14**: 89–98.

10 DEHAL, P., Y. SATOU, R. CAMPBELL, et al. 2002/2003. The draft genome of *Ciona intestinalis*: insights into chordate and vertebrate origins. Science **298**: 2157–68.

11 DEWAN, A., A. PARRADO AND T. MATISE. 2002. Map error reduction: using genetic and sequence-based physical maps to order closely linked markers. Human Hered. **54**: 34–44.

12 EWING, B., L. HILLIER, M. WENDL AND P. GREEN. 1998. Base-calling of automated sequencer traces using phred. I. Accuracy assessment. Genome Res. **8**: 175–85.

13 EWING, B. AND P. GREEN. 1998. Base-calling of automated sequencer traces using phred. II. Error probabilities. Genome Res. **8**: 186–94.

14 FLEISCHMANN, R. D., M. ADAMS, O. WHITE, et al. 1995. Whole-genome random sequencing and assembly of Haemophilus influenzae. Science **26**: 496–512.

15 GALLANT, J., D. MAIER AND J. STOERER. 1980. On finding minimal length superstrings. J. Comp. Syst. Sci. **20**: 50–58.

16 GREEN, P. 1994. PHRAP documentation. http: //www.phrap.org.

17 GREEN, P. 1997. Against a whole-genome shotgun. Genome Res. **7**: 410–7.

18 HAVLAK, P., R. CHEN, K. DURBIN, A. EGAN, Y. REN, X.-Z. SONG, G. WEINSTOCK AND R. GIBBS. 2004. The Atlas genome assembly system. Genome Res. **14**: 721–32.

19 HOLT, R. A., G. M. SUBRAMANIAN, A. HALPER, et al. 2002. A whole-genome assembly of *Drosophila*. Science **288**: 129–49.

20 HUANG, X., J. WANG, S. ALURU, S.-P. YANG AND L. HILLIER. 2003. PCAP: a whole-genome assembly program. Genome Res. **13**: 2164–70.

21 HUANG, X. AND A. MADAN. 1999. CAP3: A DNA sequence assembly program. Genome Res. **9**: 868–77.

22 HUANG, X. 1992. A contig assembly program based on sensitive detection of fragment overlaps. Genomics **14**: 18–25.

23 HUSON, D. H., K. REINERT AND E. W. MYERS. 2002. The greedy path-merging algorithm for contig scaffolding. J. ACM **49**: 603–15.

24 HUSON, D. H., K. REINERT, S. A. KRAVITZ, et al. 2001. Design of a compartmentalized shotgun assembler for the human genome. Bioinformatics **17**: 132–9.

25 INTERNATIONAL HUMAN GENOME SEQUENCING CONSORTIUM. 2001. Initial

sequencing and analysis of the human genome. Nature **409**: 860–921.

26 JAFFE, D., J. BUTLER, S. GNERRE, et al. 2003. Whole-genome sequence assembly for mammalian genomes: ARACHNE 2. Genome Res. **13**: 91–6.

27 JAFFE, D., J. BUTLER, S. GNERRE, et al. 2003. Whole genome sequence assembly for mammalian genomes: ARACHNE 2. Genome Res. **13**: 91–6.

28 KECECIOGLU, J. AND J. YU. 2001. Separating repeats in DNA sequence assembly. Proc. RECOMB **5**: 176–83.

29 KENT, W. J. AND D. HAUSSLER. 2001. Assembly of the working draft of the human genome with GigAssembler. Genome Res. **11**: 1541–8.

30 KEVLES, D. J. AND L. HOOD. 1992. *The Code of Codes: Scientific and Social Issues in the Human Genome Project*. Harvard University Press, Cambridge, MA.

31 LANDER, E. S. AND M. S. WATERMAN. 1988. Genomic mapping by fingerprinting random clones: a mathematical analysis. Genomics **2**: 231–9.

32 MARRA, M., L. HILLIER AND R. H. WATERSTON. 1998. Expressed sequence tags – ESTablishing bridges between genomes. Trends Genet. **14**: 4–7.

33 MELDRUM, D. 2000. Automation for genomics, part 1: preparation for sequencing. Genome Res. **10**: 1081–92.

34 MELDRUM, D. 2000. Automation for genomics, part 2: sequencers, microarrays, and future trends. Genome Res. **10**: 1288–303.

35 MEWES, W., K. ALBERMANN, M. BÄHR, et al. 1997. An overview of the yeast genome. Nature **387**: 7–8.

36 MULLIKIN, J. AND Z. NING. 2003. The Phusion assembler. Genome Res. **13**: 81–90.

37 MURAL, R. J., M. D. ADAMS, G. W. MYERS, et al. 2002. A comparison of whole-genome shotgun-derived mouse chromosome 16 and the human genome. Science **296**: 1661–71.

38 MYERS, E. W., G. G. SUTTON, A. L. DELCHER, et al. 2000. A whole-genome assembly of *Drosophila*. Science **287**: 2196–204.

39 MYERS, E. 1990. A fast bit-vector algorithm for approximate string matching based on dynamic programming. J. ACM **46**: 495–515.

40 MYERS, E. 1994. A sublinear algorithm for approximate keyword matching. Algorithmica **12**: 345–74.

41 MYERS, E. 1995. Toward simplifying and accurately formulating fragment assembly. J. Comput. Biol. **2**: 275–90.

42 PALCA, J. 1986. Human genome – Department of Energy on the map. Nature **321**: 371.

43 PEVZNER, P., H. TANG AND M. WATERMAN. 2001. An eulerian path approach to DNA fragment assembly. Proc. Natl Acad. Sci. USA **98**: 9748–53.

44 PEVZNER, P. 2000. *Computational Molecular Biology*. MIT Press, Cambridge, MA.

45 RASMUSSEN, K., J. STOYE AND E. W. MYERS. 2005. Efficient q-gram filters for finding all *epsilon*-matches over a given length. Proc. RECOMB **9**: 189–203.

46 RAT GENOME SEQUENCING CONSORTIUM. 2004. The genome sequence of the brown Norway rat yields insights into mammalian evolution. Nature **428**: 493–521.

47 SCHULER, G., M. BOGUSKI, E. STEWART, et al. 1996. A gene map of the human genome. Science **274**: 540–6.

48 SINSHEIMER, A. L. 1985. The Santa Cruz Workshop – May 1985. Genomics **5**: 954–6.

49 SUTTON, G. AND I. DEW. 2005. Shotgun fragment assembly. In *Series in Systems Biology*. Oxford University Press, Oxford: to appear.

50 SUTTON, G. G., O. WHITE, M. D. ADAMS AND A. R. KERLAVAGE. 1995. TIGR Assembler: A new tool for assembling large shotgun sequencing projects. Genome Sci. Technol. **1**: 9–19.

51 TAMMI, M., E. ARNER, E. KINDLUND AND B. ANDERSSON. 2003. Correcting errors in shotgun sequences. Nucl. Acids Res. **31**: 4663–72.

52 TAMMI, M., E. ARNER, T. BRITTON AND B. ANDERSSON. 2002. Separation of nearly identical repeats in shotgun assemblies using defined nucleotide

positions, DNPs. Bioinformatics **18**: 379–88.

53 TIGR. 2005. The AMOS web page, http://www.tigr.org/software/AMOS.

54 VENTER, J. C., M. D. ADAMS, E. W. MYERS, et al., 2001. The sequence of the human genome. Science **291**: 1145–1434.

55 WANG, J., G. K.-S. WONG, P. NI, et al., 2002. RePS: a sequence assembler that masks exact repeats identified from the shotgun data. Genome Res. **12**: 824–31.

56 WEBBER, J. L. AND E. W. MYERS. 1997. Human whole-genome shotgun sequencing. Genome Res. **7**: 401–9.

57 YU, J., S. HU, J. WANG, et al. 2002. A draft sequence of the rice genome (*Oryza sative* L. ssp. *indica*). Science **296**: 79–92.

Part 3 Sequence Analysis

3
Sequence Alignment and Sequence Database Search

Martin Vingron

1 Introduction

In evolutionary studies two characters are called homologous when they share common evolutionary ancestry. Genes may also be homologous, which usually is reflected by similarity among their DNA or amino acid sequences. Furthermore, homology among genes frequently implies that they are functionally similar. Thus, there are two good reasons to compare the sequences of genes or proteins, i.e. the unraveling of evolutionary relationships and extrapolating function from one gene to another.

The basis for the study of sequence similarity is the comparison of two sequences which will be dealt with in Section 2. Sequence comparisons are performed in large numbers when searching sequence databases for sequences that are similar to a query sequence. Algorithms for this purpose need to be fast, even at the expense of sensitivity. Section 3 discusses the widely used heuristic approaches to database searching. However, the algorithms we are designing for the purpose of quantifying sequence similarity can only be as good as our understanding of evolutionary processes and thus they are far from perfect. Therefore, results of algorithms need to be subjected to a critical test using statistics. Methods for the assessment of the statistical significance of a finding are introduced in Section 4.

Genes do not come in pairs, but rather in large families. Consequently, the need arises to align more than two sequences at a time, which is done by multiple alignment programs. Computationally a very hard problem, it has attracted considerable attention from the area of algorithm development. Section 5 presents the basic approaches to multiple sequence alignment.

Bioinformatics - From Genomes to Therapies Vol. 1. Edited by Thomas Lengauer

ISBN: 978-3-527-31278-8

Section 6 builds on the knowledge of a multiple alignment and introduces how to exploit the information contained in several related sequences for the purpose of identifying additional related sequences in a database. The last section covers methods and resources to structure the entire space of protein sequences.

2 Pairwise Sequence Comparison

2.1 Dot plots

Dot plots are probably the simplest way of comparing sequences [55]. A dot plot is a visual representation of the similarities between two sequences. Each axis of a rectangular array represents one of the two sequences to be compared. A window length is fixed, together with a criterion under which two sequence windows are deemed to be similar. A typical choice for this similarity criterion would be a certain fraction of matching residues within a window. Whenever one window in one sequence resembles another window in the other sequence, a dot or short diagonal is drawn at the corresponding position of the array. Thus, when two sequences share similarity over their entire length a diagonal line extends from one corner of the dot plot to the diagonally opposite corner. If two sequences only share patches of similarity this is revealed by diagonal stretches.

Figure 1 shows an example of a dot plot. There, the coding DNA sequences of the α- and β-chains of human hemoglobin are compared to each other. For this computation the window length was set to 31. The program adds up the matches within a window and the gray value at the position corresponding to the center of the window is set according to the quality of the match at that position. One can clearly discern a diagonal trace along the entire length of the two sequences. Note the jumps where this trace changes to another diagonal of the array. These jumps correspond to the position where one sequence has more (or fewer) letters than the other one. Figure 1 was produced using the program "dotter" [71].

Dot plots are a powerful method of comparing two sequences. They do not predispose the analysis in any way such that they constitute the ideal first-pass analysis method. Based on the dot plot the user can decide whether they deal with a case of global, i.e. beginning-to-end, similarity or local similarity. "Local similarity" denotes the existence of similar regions between two sequences that are embedded in the overall sequences which lack similarity. Sequences may contain regions of self-similarity which are frequently termed internal repeats. A dot plot comparison of the sequence itself will reveal internal repeats by displaying several parallel diagonals (see also Chapter 7).

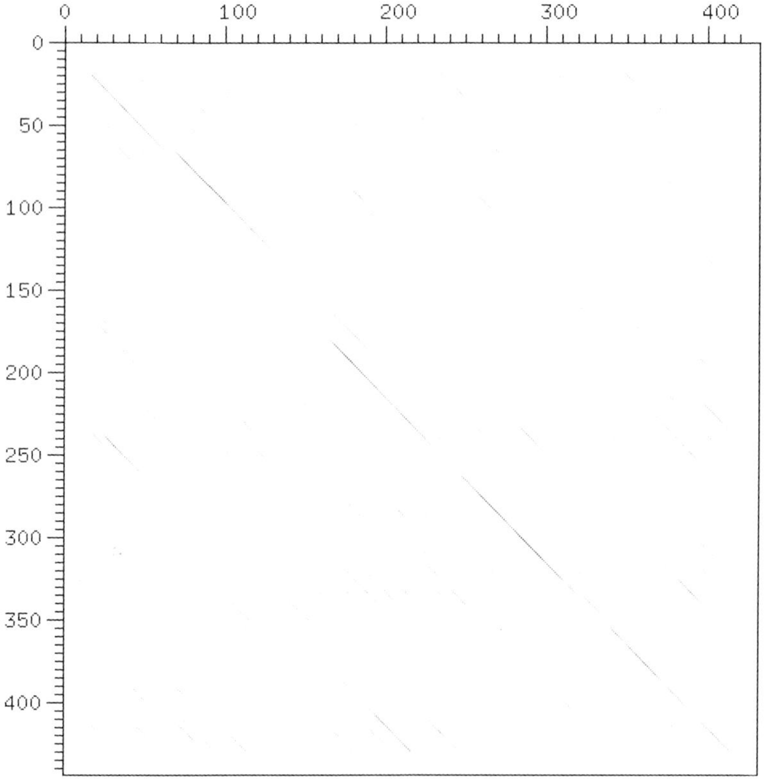

Figure 1 Dot plot comparing two hemoglobin sequences. The horizontal axis corresponds to the sequence of the human β-hemoglobin chain; the vertical sequence (numbered from top to bottom) represents the human α-hemoglobin chain.

Instead of simply deciding if two windows are similar, a quality function may be defined. In the simplest case, this could be the number of matches in the window. For amino acid sequences the physical relatedness between amino acids may give rise to a quantification of the similarity of two windows. For example, when a similarity matrix on the amino acids (like the Dayhoff matrix, see below) is used one might sum up these values along the window. However, when this similarity matrix contains different values for exact matches this leads to exactly matching windows of different quality. The dot plot method of Argos [5] is an intricate design that reflects the physical relatedness of amino acids. The program dotter [71] is an X-windows-based program that allows for displaying dot plots for DNA, for proteins and for comparison of DNA to protein.

```
Hemoglobin alpha-1  1  MVLSPADKTNVKAAWGKVGAHAGEYGAEALERMFLSFPTTKTYFPHF-D  48
                       :.|:|.:|:.|.|.||||  :.:|.|:|||.|:::.:|.|:.:|..| |
Hemoglobin beta     1 MVHLTPEEKSAVTALWGKV--NVDEVGGEALGRLLVVYPWTQRFFESFGD  48

Hemoglobin alpha-1 49 LSH-----GSAQVKGHGKKVADALTNAVAHVDDMPNALSALSDLHAHKLR  93
                      ||.     |:::||:|||||.:|:::::||:|::..::::||:||.:||:
Hemoglobin beta    49 LSTPDAVMGNPKVKAHGKKVLGAFSDGLAHLDNLKGTFATLSELHCDKLH  98

Hemoglobin alpha-1 94 VDPVNFKLLSHCLLVTLAAHLPAEFTPAVHASLDKFLASVSTVLTSKYR 142
                      |||.||:||::.|:..||.|:..||||:|:|:.:|.:|:|:..|:.||:
Hemoglobin beta    99 VDPENFRLLGNVLVCVLAHHFGKEFTPPVQAAYQKVVAGVANALAHKYH 147
```

Figure 2 Sequence alignment between the amino acid sequences of human hemoglobin α- and β-chains. Note that these are the same genes for which the dot plot of the corresponding coding DNA sequences is shown in Figure 1.

2.2 Sequence Alignment

A sequence alignment [81] is a scheme of writing one sequence on top of another such that the residues in the same position are deemed to have a common evolutionary origin. If the same letter occurs at the same position in both sequences, then this position has been conserved in evolution (or, coincidentally, mutations from another ancestral residue have given rise to the same letter twice). If the letters differ it is assumed that both derive from the same ancestral letter, which could be one of the two or neither. Homologous sequences may have different length, though, which is generally explained through insertions or deletions in sequences. Thus, a letter or a stretch of letters may be paired up with dashes in the other sequence to signify such an insertion or deletion. Since an insertion in one sequence can always be seen as a deletion in the other one sometimes uses the term "indel" (or, simply, "gap"). Figure 2 depicts an example of an alignment. The sequences aligned there are the proteins derived from the coding sequences compared in Figure 1. Note that the first stretch of contiguously aligned amino acids (up to the WGKV match) corresponds to the first diagonal stretch in the dot plot of Figure 1. The subsequent insertion of 2 amino acids in the α-chain corresponds to linking this first diagonal to the second one, which is located around position 100. Likewise, the next five-letter gap in the alignment corresponds to the join from the second diagonal to the third, starting around position 200 in the dot plot.

In such a simple evolutionarily motivated scheme, an alignment mediates the definition of a distance for two sequences. One generally assigns a score of zero to a match, some positive number to a mismatch and a larger positive number to an indel. By adding these values along an alignment one obtains a score for this alignment. A distance function for two sequences can be defined by looking for the alignment which yields the minimum score.

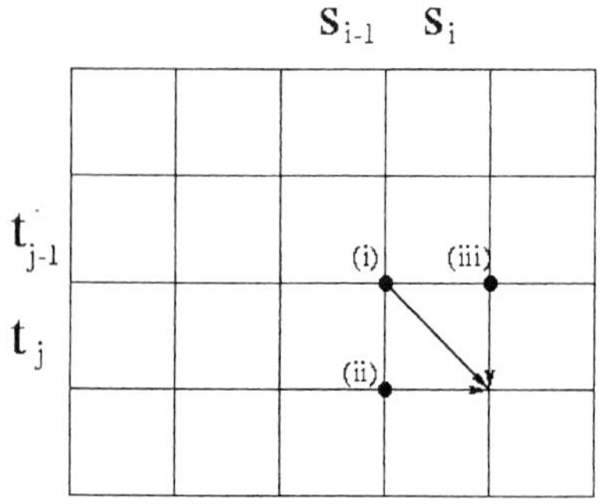

Figure 3 Schematic representation of the edit matrix comparing two sequences. The arrows indicate how an alignment may end according to the three cases described in the text.

Naively, the alignment that realizes the minimal distance between two sequences could be identified by testing all possible alignments. This number, however, is prohibitively large; luckily, using dynamic programming, the minimization can be effected without explicitly enumerating all possible alignments of two sequences. To describe this algorithm [64] denote the two sequences by $s = s_1, \ldots, s_n$ and $t = t_1, \ldots, t_m$. The key to the dynamic programming algorithm is the realization that for the construction of an optimal alignment between two stretches of sequence $s_1, \ldots, s_i$ and $t_1, \ldots, t_j$ it suffices to inspect the following three alternatives:

(i) The optimal alignment of $s_1, \ldots, s_{i-1}$ with $t_1, \ldots, t_{j-1}$, extended by the match between s_i and t_j;

(ii) The optimal alignment of $s_1, \ldots, s_{i-1}$ with $t_1, \ldots, t_j$, extended by matching s_i with a gap character "–";

(iii) The optimal alignment of $s_1, \ldots, s_i$ with $t_1, \ldots, t_{j-1}$, extended by matching a gap character "–" with t_j.

Each of these cases also defines a score for the resulting alignment. This score is made up of the score of the alignment of the so far unaligned sequences that used plus the cost of extending this alignment. In case (i), this cost is determined by whether or not the two letters are identical; in cases (ii) and (iii), the cost of extension is the penalty assigned to a gap. The winning alternative will be the one with the best score (Figure 3).

To implement this computation one fills in a matrix the axes of which are annotated with the two sequences s and t. It is helpful to use north, south, west and east to denote the sides of the matrix. Let the first sequence extend from west to east on the north side of the matrix. The second sequence extends from north to south on the west side of the matrix. We want to fill the matrix starting in the north-western corner, working our way southward row by row, filling each row from west to east. To start, one initializes the northern and western margin of the matrix, typically with gap penalty values. After this initialization the above rules can be applied. A cell (i, j) that is already filled contains the score of the optimal alignment of the sequence $s_1, \ldots, s_i$ with $t_1, \ldots, t_j$. The score of each such cell can be determined by inspecting the cell immediately north-west of it [case (i)], the cell west [case (ii)] and the one north [case (iii)] of it, and deciding for the best scoring option. When the procedure reaches the south-eastern corner, that last cell contains the score of the best alignment. The alignment itself can be recovered as one backtracks from this cell to the beginning, each time selecting the path that had given rise to the best option.

The idea of assigning a score to an alignment and then minimizing or maximizing over all alignments is at the heart of all biological sequence alignment. However, many more considerations have influenced the definition of the scores and made sequence alignment applicable to a wide range of biological settings. First, note that one may either define a distance or a similarity function of an alignment. The difference lies in the interpretation of the values. A distance function defines positive values for mismatches or gaps and then aims at minimizing this distance. A similarity function assigns high values to matches and low values to gaps, and then maximizes the resulting score. The basic structure of the algorithm is the same for both cases. In 1981, Smith and Waterman [69] showed that for global alignment, i.e. when a score is computed over the entire length of both sequences, the two concepts are in fact equivalent. Thus, it is now customary to choose the setting that gives more freedom for appropriately modeling the biological question of interest.

In the similarity framework one can easily distinguish among the different possible mismatches and also among different kinds of matches. For example, a match between two tryptophans is usually regarded to be more important than a match between two alanines. Likewise, the pairing of two hydrophobic amino acids like leucine and isoleucine is preferable to the pairing of a hydrophobic with a hydrophilic residue. Scores are used to describe these similarities and are usually represented in the form of a symmetric 20×20 matrix, assigning a similarity score to each pair of amino acids. Although easy to understand from the physical characteristics of the amino acids, the values in such a matrix are usually derived based on an evolutionary model that enables one to estimate whether particular substitutions are preferred or avoided. To be more precise, the similarity score for 2 amino acids is defined as the logarithm of the likelihood ratio of the two residues being homologous versus finding them at their corresponding positions due to chance. This approach has been pioneered by Dayhoff [17] who computed a series of amino acid similarity matrices. Each matrix in this series corresponds to a particular evolutionary distance among sequences. This distance is measured in a unit called 1 PAM, for 1 Accepted Point Mutation (in 100 positions). The matrices carry names like PAM120 or PAM250, and are supposed to be characteristic for evolutionary distances of 120 or 250 PAM, respectively. Other more recent series of matrices are the BLOSUM matrices [27] or the VT series of matrices [57]. For every matrix one needs to find appropriate penalties for gaps.

The treatment of gaps deserves special care. The famous algorithm by Needleman and Wunsch [60] did not impose any restrictions on the penalty assigned to a gap of a certain length. For reasons of computational speed, later gap penalties were restricted to a cost function linear in the number of deleted (inserted) residues [64]. This amounts to penalizing every single indel. However, since a single indel tends to be penalized such that it is considerably inferior to a mismatch, this choice resulted in longer gaps being quite expensive and thus unrealistically rare. As a remedy, one mostly uses a gap penalty function which charges a *gap open penalty* for every gap that is introduced and penalizes the length with a *gap extension penalty* which is charged for every inserted or deleted letter in that gap. Clearly, this results in an affine linear function in the gap length, frequently written as $g(k) = a + b * k$ [80].

With the variant of the dynamic programming algorithm first published by Gotoh [23] it became possible to compute optimal alignments with affine linear gap penalties in time proportional to the product of the lengths of the two sequences to be aligned. This afforded a speed-up by an order of magnitude compared to a naive algorithm using the more general gap function. A further breakthrough in alignment algorithms development was provided by an algorithm that could compute an optimal alignment using computer memory only proportional to the length of one sequence instead of their product. This algorithm by Myers and Miller [59] is based on work by Hirshberg [29].

Depending on the biological setting, several kinds of alignment are in use. When sequences are expected to share similarity extending from the beginning of the sequences to their ends, they are aligned globally. This means that each residue of either sequence is part either of a residue pair or a gap. In particular, it implies that gaps at the ends are charged like any other gap. This, however, is a particularly unrealistic feature of a global alignment. While sequences may very well share similarity over their entire length (see the example dot plot of two hemoglobin chains in Figure 1), their respective N- and C-termini usually are difficult to match up, and differences in length at the ends are more of a rule than an exception. Consequently, one prefers to leave gaps at the ends of the sequences unpenalized. This variant is easy to implement in the dynamic programming algorithm. Two modifications are required. First, the initialization of the matrix needs to reflect the gap cost of zero in the margin of the matrix. Second, upon backtracking, one does not necessarily start in the corner of the matrix, but rather searches the margins for the maximum from which to start. Variants of this algorithm that penalize only particular end-gaps are easy to derive and can be used, for example, to fit one sequence into another or to overlap the end of one sequence with the start of another.

In many cases, however, sequences share only a limited region of similarity. This may be a common domain or simply a short region of recognizable similarity. This case is dealt with by so-called local alignment in an algorithm due to Smith and Waterman [69]. Local alignment aims at identifying the best pair of regions, one from each sequence, such that the optimal (global) alignment of these two regions is the best possible. This relies on a scoring scheme that maximizes a similarity score because otherwise an empty alignment would always yield the smallest distance. Naively, the algorithm to compute a local alignment would need to inspect every pair of regions and apply a global alignment algorithm to it. The critical idea of Smith and Waterman was to offer the maximization in each cell of the matrix a fourth alternative: a zero to signify the beginning of a new alignment. After filling the dynamic programming matrix according to this scheme, backtracking starts from the cell in the matrix that contains the largest value.

Upon comparing a dot plot and a local alignment one might notice regions of similarity visible in the dot plot, but missing in the alignment. While in many cases there exist gap penalty settings that would include all interesting matching regions in the alignment, generally it requires the comparison with the dot plot to notice possible misses. This problem is remedied by an algorithm due to Waterman and Eggert [82] which computes suboptimal, local and nonoverlapping alignments. It starts with the application of the Smith–Waterman algorithm, i.e. a dynamic programming matrix is filled and backtracking from the matrix cell with the largest entry yields the best local alignment. Then the algorithm proceeds to delineate a second-best local alignment. Note that this cannot be obtained by backtracking from the second-best matrix cell. Such an approach would yield an alignment largely overlapping the first one and thus containing little new information. Instead, those cells in the dynamic programming matrix are set to zero from where backtracking would lead into the prior alignment. This can be regarded as "resetting" the dynamic programming matrix after having deleted the first alignment. Then the second best alignment is identified by looking for the maximal cell in the new matrix and starting backtracking from there. Iteration of this procedure yields one alternative, nonoverlapping alignment after the other in order of descending quality. Application of this algorithm avoids possibly missing matching regions because even under strong gap penalties the procedure will eventually show all matching regions.

There is an interesting interplay between parameters, particularly the gap penalty, and the algorithmic variant used. Consider a pair of sequences whose similar regions can in principle be strung together into an alignment. Under a weak gap penalty the Smith–Waterman algorithm has a chance to identify this entire alignment. On the other hand, not knowing about the similarity between the sequences ahead of time, a weak gap penalty might also yield all kinds of spurious aligned regions. The Waterman–Eggert algorithm is a valid alternative. The gap penalty can be chosen fairly stringently. The first (i.e. the Smith–Waterman) alignment will then identify only the best-matching region out of all the similar regions. By iterating the procedure, though, this algorithm will successively identify the other similar regions as well. For a detailed discussion of these issues, see Vingron and Waterman [79].

3 Database Searching I: Single-sequence Heuristic Algorithms

This section takes a first look at the problem of identifying those sequences in a sequence database that are similar to a given sequence. This task arises, for example, when a gene has been newly sequenced and one wants to determine whether a related sequence already exists in a database. Generally, two settings can be distinguished. The starting point for the search may either be a single sequence, with the goal of identifying its relatives, or a family of sequences, with the goal of identifying further members of that family. Searching through a database needs to be fast and sensitive, but the two objectives contradict each other. Fast methods have been developed primarily for searching with a single sequence and this will be the topic of this section.

When searching a database with a newly determined DNA or amino acid sequence – the so-called query sequence – the user typically lacks knowledge of whether an expected similarity might span the entire query or just part of it. Likewise, they will be ignorant of whether the match will extend along the full length of some database sequence or only part of it. Therefore, one needs to look for a local alignment between the query and any sequence in the database. This immediately suggests the application of the Smith–Waterman algorithm to each database sequence. One should take care, though, to apply a fairly stringent gap penalty such that the algorithm focuses on the regions that really match. After sorting the resulting scores the top scoring database sequences are the candidates of interest.

Several implementations of this procedure are available, most prominently the SSEARCH program from the FASTA package [63]. There exist implementations of the Smith–Waterman algorithm that are tuned for speed like one using special processor instructions [85] and, among others, one by Barton [9].

Depending on implementation, computer and database size, a search with such a program takes on the order of 1 min.

The motivation behind the development of other database search programs has been to emulate the Smith–Waterman algorithm's ability to discern related sequences while at the same time performing the job in much less time. To this end, one usually makes the assumption that any good alignment that one wishes to identify contains, in particular, some stretch of ungapped similarity. Furthermore, this stretch will tend to contain a certain number of identically matching residues and not only conservative replacements. Based on these assumptions, most heuristic programs rely on identifying a well-matching core and then extending it or combining several of these. With hindsight, the different developments in this area can further be classified according to a traditional distinction in computer science by which one either preprocesses the query or the text (i.e. the database). Preprocessing means that the string is represented in a different form that allows for faster answers to particular questions, e.g. whether the string contains a certain subword.

The FASTA program (part of a package [63] that usually goes by the same name) sets a size k for k-tuple subwords. For DNA sequences, the parameter k might typically be set to 7, while for amino acid sequences 2 would be a reasonable choice. The program then looks for diagonals in the comparison matrix between the query and search sequence along which many k-tuples match. This can be done very quickly based on a preprocessed list of k-tuples contained in the query sequence. The set of k-tuples can be identified with an array whose length corresponds to the number of possible tuples of size k. This array is linked to the indices of the positions at which the particular k-tuples occur in the query sequence. Note that a matching k-tuple at index i in the query and at index j in the database sequence can be attributed to a diagonal by subtracting one index from the other. Therefore, when inspecting a new sequence for similarity one walks along this sequence inspecting each k-tuple. For each of them one looks up the indices of the positions at which it occurs in the query, computes the index-difference to identify the diagonal and increases a counter for this diagonal. After inspecting the search sequence in this way a diagonal with a high count is likely to contain a well-matching region. In terms of the execution time, this procedure is only linear in the length of the database sequence and can easily be iterated for a whole database. Of course this rough outline needs to be adapted to focus on regions where the match density is high and link nearby, good diagonals into alignments.

The other widely used program to search a database is called BLAST [1,3]. BLAST follows a similar scheme in that it relies on a core similarity, although with less emphasis on the occurrence of exact matches. This program also aims at identifying core similarities for later extension. The core similarity is defined by a window with a certain match density on DNA or with an amino acid similarity score above some threshold for proteins. Independent of the exact definition of the core similarity, BLAST rests on the precomputation of all strings which are similar in the given sense to any position in the query. The resulting list may contain on the order of 1000 or more words, each of which if detected in a database gives rise to a core similarity. In BLAST nomenclature this set of strings is called the neighborhood of the query. In fact, the code to generate this neighborhood is exceedingly fast.

Given the neighborhood, a finite automaton is used to detect occurrences in the database of any string from the neighborhood. This automaton is a program constructed "on the fly" and specifically for the particular word neighborhood that has been computed for a query. Upon reading through a database of sequences, the automaton is given an additional letter at a time and decides whether the string that ends in this letter is part of the neighborhood. If so, BLAST attempts to extend the similarity around the neighborhood and if this is successful reports a match.

As with FASTA, BLAST has also been adapted to connect good diagonals and report local alignments with gaps. BLAST converts the database file into its own format to allow for faster reading. This makes it somewhat unwieldy to use in a local installation unless someone takes care of the installation. FASTA, however, is slower, but easier to use. There exist excellent web servers that offer these programs, in particular at the National Center for Biotechnology Information [43] and at the European Bioinformatics Institute [41] where BLAST or FASTA can be used on up-to-date DNA and protein databases.

According to the above-mentioned distinction among search methods into those that preprocess the pattern and those that preprocess the text, there also is the option of transforming a DNA or amino acid database such that it becomes easier to search. This route was taken, for example, by a group from IBM developing the FLASH [14] program. They devised an intricate, although supposedly very space-consuming technique of transforming the database into an index for storing the offsets of gapped k-tuples. The QUASAR program by Burkhard and coworkers [13] preprocesses the database into a so-called suffix array, similar to a suffix tree, yet simple to keep on disk. Programs in practical use for quickly searching entire genomes are BLAT [50] and SSAHA [61].

With the availability of expressed sequence tags (ESTs) it has become very important to match DNA sequence with protein sequence in such a way that a possible translation can be maintained throughout the alignment. Both FASTA and BLAST packages contain programs for this and related tasks. When coding DNA is compared to proteins, gaps are inserted in such a way as to maintain a reading frame. Likewise, a protein sequence can be searched versus a DNA sequence database. The search of DNA versus DNA with an emphasis on matching regions that allow for a contiguous translation is not so well supported. Although a dynamic programming algorithm for this task is feasible, the existing implementation in BLAST compares all reading frames.

4 Alignment and Search Statistics

Alignment score is the product of an optimization, mostly a maximization procedure. As such it tends to be a large number sometimes suggesting biological relatedness where there is none. In pairwise comparisons the user still has a chance to study an alignment by eye in order to judge its validity; however, upon searching an entire database automatic methods are necessary to attribute a statistical significance to an alignment score.

In the early days of sequence alignment, the statistical significance of the score of a given pairwise alignment was assessed using the following procedure. The letters of the sequences are permuted randomly and a new alignment score is calculated. This is repeated roughly 100 times, and the mean and standard deviation of this sample are calculated. The significance of the given alignment score is reported in "number of standard deviations above the mean", also called the Z-score. Studying large numbers of random alignments is correct, in principle. However, the significance of the alignment should then be reported as the fraction of random alignments that score better than the given alignment. The procedure described assumes that these scores are distributed normally. Since the random variable under study – the score of an optimal alignment – is the maximum over a large number of values, this is not a reasonable assumption. In fact, the lack of fit quickly becomes obvious when trying to fit a normal distribution to the data. The second argument against this way of calculating significance is a pragmatic one: the procedure needs to be repeated for every alignment under study because the effect of the sequence length cannot be accounted for.

Based on the work of several researchers [48, 70], it has meanwhile become apparent that alignment score as well as scores from database searches obey a so-called extreme-value distribution. This is not surprising given that extreme-value distributions typically describe random variables that are the result of maximization. In sequence alignment, there are analytical results confirming the asymptotic convergence to an extreme-value distribution for the case of local alignment without gaps, i.e. the score of the best-matching contiguous diagonal in a comparison [18]. This is also a valid approximation to the type of matching effected in the database search program BLAST. Thus, this approach has become widely used and, in fact, has contributed significantly to the popularity of database search programs because significance measures have made the results of the search much easier to interpret.

The statistical significance of an event like observing a sequence alignment of a certain quality is the probability to observe a better value as a result of chance alone. This quantity is refereed to as the *p*-value. For example, a *p*-value of 10^{-3} is interpreted as expecting to see an excess of the given threshold in one in a 1000 experiments. To compute this one needs to model chance alignments, which is precisely what the statistician means by deriving the distribution of a random variable. The probability that a chance result would exceed an actually obtained threshold S is 1 minus the value of the cumulative distribution function evaluated at that threshold. In sequence alignment, this cumulative distribution function is generally expressed as [48]:

$$\exp(mnKe^{\lambda S})$$

where m and n are the lengths of the sequences compared, and K and λ are parameters which need to be computed (where possible) or derived by simulation. K and λ depend on the scoring matrix used (e.g. the PAM120 matrix) and the distribution of residues. Hence, for any scoring system these parameters are computed beforehand and the statistical significance of an alignment score S is then computed by evaluating the formula with the length of the two sequences compared.

The most prominent case for which the parameters K and λ can be defined analytically is local alignment without gaps. Algorithmically this amounts to computing a Smith–Waterman alignment under very high gap penalties such that the resulting alignment will simply not contain any gaps. Since this notion of alignment also guides the heuristic used by the BLAST database search program, the resulting statistical estimates are primarily used in database searching. In this application, one of the lengths is the length of the input sequence and the other length can be chosen on the order of the length of the concatenated sequences from the database that is being searched. Alternatively, one can think of the database search as a repetition of many individual pairwise comparisons, which amounts to repeating the experiment "sequence comparison" many times. In this setting, the number of false positives one expects to find can be determined as the product of the *p*-value of the individual comparison and the number of times the experiment is repeated, i.e. the number of sequences in the database. This expected number of false positives is referred to as the *E*-value. A typical *E*-value threshold for a database search would be, for example, 1, indicating that the score cutoff is chosen such that among the sequences faring better than the cutoff one expects to find one false-positive hit.

When gaps are allowed, the determination of K and λ is more complex because an approximation of the distribution function of alignment score by an extreme-value distribution as above is not always valid. Generally speaking, it is allowed only for sufficiently strong gap penalties where alignments remain compact as opposed to spanning the entire sequences. Under sufficiently strong gap penalties, though, it has been demonstrated that the approximation is indeed valid just like for infinite gap penalties [79]. However, it is not possible any more to compute the values of the parameters K and λ analytically. As a remedy one applies simulations in which many alignments of randomly generated sequences are computed and the parameters are determined based on fitting the empirical distribution function with an extreme-value distribution [83]. As in the case above, this procedure allows for determining parameters beforehand and computing significance by putting the lengths of the sequences into the formula.

The question remains of how to determine whether approximation by an extreme-value distribution is admissible for a certain scoring scheme and gap penalty setting one is using. This can be tested on randomly generated (or, simply, unrelated) sequences by computing a global alignment between sequences under that particular parameter setting. If the result has a negative sign (averaged over many trials or on very long random sequences), then the approximation is admissible. This is based on a theorem due to Arratia and Waterman [6], and subsequent simulation results reported by Waterman and Vingron [84]. In particular, a gap open penalty of 12 with an extension penalty of 2 or 3 for the case of the PAM250 matrix, as well as any stronger combination, allows for approximation by the extreme-value distribution.

In database searching the fitting need not be done on randomly generated sequences. Under the assumption that the large majority of sequences in a database are not related to the query, the bulk of the scores generated upon searching can be used for fitting. This approach is taken by Pearson in the FASTA package. It has the advantage that the implicit random model is more realistic since it is taken directly from the data actually searched. Along a similar line of thought, Spang and Vingron [72] tested significance calculations in database searching by evaluating a large number of search results. Their study showed that one should not simply use the sum of the lengths of all the sequences in the database as the length parameter in the formula for the extreme-value distribution. This would overestimate the length that actually governs the statistics. Instead, a considerably shorter effective length can determined for a particular database using simulations. This effect is probably due to the fact that alignments cannot start in one sequence and end in the next one, which makes the number of feasible starting points for random alignments smaller than the actual length of the database.

5 Multiple Sequence Alignment

For many genes a database search reveals a whole number of homologous sequences. Then, one wishes to learn about the evolution and the sequence conservation in such a group. This question surpasses what can reasonably be achieved by the sequence comparison methods described in Section 3. Pairwise comparisons do not readily exhibit positions that are conserved among a whole set of sequences and tend to miss subtle similarities that become visible when observed simultaneously among many sequences. Thus, one wants to simultaneously compare several sequences.

A multiple alignment arranges a set of sequences in a scheme such that positions believed to be homologous are written in a common column (Figure 4). As in a pairwise alignment, when a sequence does not possess an amino acid in a particular position, this is denoted by a dash. There also are conventions similar to the ones for pairwise alignment regarding the scoring of a multiple alignment. The so-called sum-of-pairs (SOP) [2] score adds the scores of all the induced pairwise alignments contained in a multiple alignment. For a linear

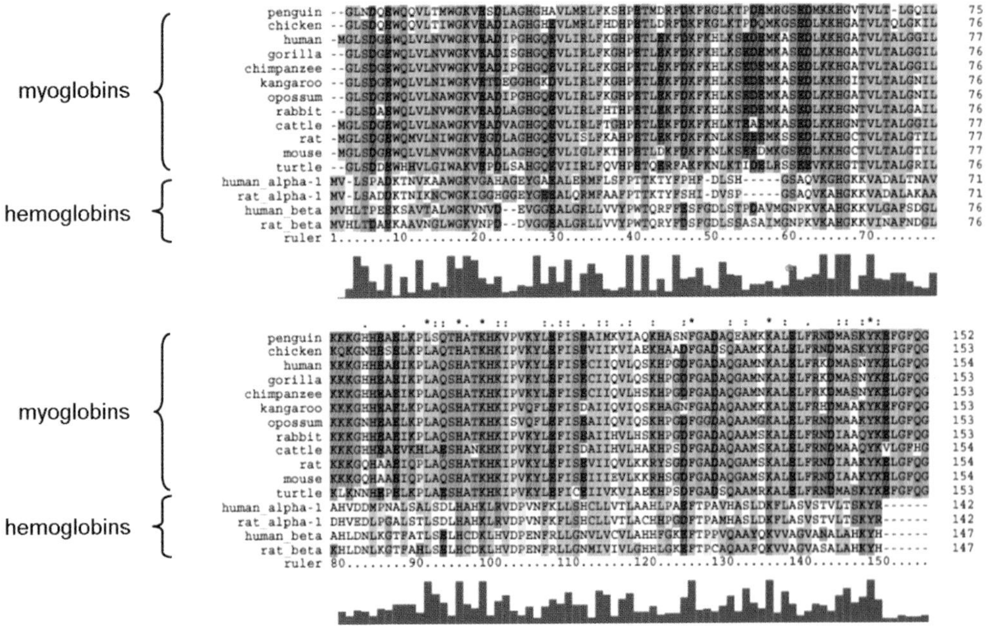

Figure 4 Example of a multiple sequence alignment: an alignment of amino acid sequences of myoglobins and hemoglobins from a number of species. Each sequence begins in the top block and continues in the bottom block. The color code indicates physicochemical attributes of amino acids. The bar diagram below the alignment quantifies the degree of conservation in the column above.

gap penalty this amounts to scoring each column of the alignment by the sum of the amino acid pair scores or gap penalties in this column. Although it would be biologically meaningful, the distinctions between global, local and other forms of alignment are rarely made in a multiple alignment. The reason for this will become apparent below when we describe the computational difficulties in computing multiple alignments.

In general, the columns of a multiple alignment cannot be determined based on the set of all pairwise alignments. Quite the contrary, pairwise alignments may contradict each other in that one set of alignments opts to place, say, residue a from sequence i in one column with residue b from sequence j, while from another set of pairwise alignments it may follow that a should be in one column with another letter c from sequence j. If one wishes to assemble a multiple alignment from pairwise alignments one has to avoid "closing loops", i.e. one can put together pairwise alignments as long as no new pairwise alignment is included involving a sequence which is already part of the multiple alignment. In particular, pairwise alignments can be merged when they align one sequence to all others, when a linear order of the given sequences is maintained or when the sequence pairs for which pairwise alignments are given form a tree. While all these schemes allow for the ready definition of algorithms that output multiply aligned sequences, they do not include any information stemming from the simultaneous analysis of several sequences.

An alternative approach is to generalize the dynamic programming optimization procedure applied for pairwise alignment to the delineation of a multiple alignment that maximizes, for example, the SOP score. The algorithm used [80] is a straightforward generalization of the global alignment algorithm. This is easy to see, in particular, for the case of the column-oriented SOP scoring function avoiding an affine gap penalty in favor of the simpler linear one. With this scoring, the arrangement of gaps and letters in a column can be represented by a Boolean vector indicating which sequences contain a gap in a particular column. Given the letters that are being compared, one needs to evaluate the scores for all these arrangements. However conceptually simple this algorithm may be, its computational complexity is rather forbidding. For n sequences it is proportional to 2^n times the product of the lengths of all sequences. The space requirement of this algorithm is on the order of the product over the lengths of the n sequences, which constitutes an even greater obstacle to its practical application.

There exists software to compare three sequences with this algorithm that additionally implements a space-saving technique [46]. For more than three sequences, algorithms have been developed that aim at reducing the search space while still optimizing the given scoring function. The most prominent program of this kind is MSA2 [25,44]. An alternative approach is used by DCA

[36, 73], which implements a "divide-and-conquer" philosophy. The search space is repeatedly subdivided by identifying anchor points through which the alignment is highly likely to pass.

However, none of these approaches scales well to large numbers of sequences to be aligned. The most common remedy is reducing the multiple alignment problem to an iterated application of the pairwise alignment algorithm. However, in doing so, one also aims at drawing on the increased amount of information contained in a set of sequences. Instead of simply merging pairwise alignments of sequences, the notion of a profile [24] has been introduced in order to grasp the conservation patterns within subgroups of sequences. A profile is essentially a representation of an already computed multiple alignment of a subgroup of sequences. This alignment is "frozen" for the remaining computation. Other sequences or other profiles can be compared to a given profile based on a generalized scoring scheme defined for this purpose. The advantage of scoring a sequence versus a profile over scoring individual sequences lies in the fact that the scoring schemes for profile matching reflect the conservation patterns among the already aligned sequences. (Profiles are discussed in more detail in Chapter 11.)

Given a profile and a single sequence, the two can be aligned using the basic dynamic programming algorithm together with the accompanying scoring scheme. The result will be an alignment between sequence and profile that can readily be converted into a multiple alignment now comprising the sequences underlying the profile plus the new one. Likewise, two profiles can be aligned with each other, resulting in a multiple alignment containing all sequences from both profiles. Various multiple alignment strategies can be implemented with these tools. Most commonly, a hierarchical tree is generated for the given sequences, which is then used as a guide for iterative profile construction and alignment. This alignment strategy is called "progressive", and was introduced in papers by Taylor [76], Corpet [16] and Higgins [28]. Higgins' program Clustal [42] and, in particular, its latest version ClustalW are probably the most widely used programs for multiple sequence alignment [47]. Two recent variants of progressive alignment are MUSCLE [21] and PROBCONS [19]. Other programs in practical use are the MSA2 program and DCA. Lee and coworkers [54] developed a program that focuses on fast alignment of highly similar sequences, e.g. ESTs, using an algorithm termed partial order alignment.

Progress has been made also on the problem of local multiple alignment. The algorithm behind the Dialign [37, 56] program relies on collecting local similarities among all pairs of sequences and then assembles those into multiply aligned regions. Similarly, T-Coffee [62] allows for inclusion of both local and global alignments, as well as other possible information like structural similarity, and merges those consistently into a multiple alignment.

Since iterative profile alignment tends to be guided by a hierarchical tree, this step of the computation also influences the final result. Usually the hierarchical tree is computed based on pairwise comparisons and their resulting alignment scores. Subsequently, this score matrix is used as input to a clustering procedure like single linkage clustering or UPGMA (unweighted pair group method with arithmetic mean) [74]. However, it is well understood that in an evolutionary sense such a hierarchical clustering does not necessarily result in a biologically valid tree. Thus, when allowing this tree to determine the multiple alignment there is the danger of pointing further evolutionary analysis of this alignment in the wrong direction. Consequently, the question has arisen of a common formulation of evolutionary reconstruction and multiple sequence alignment. The cleanest, although biologically somewhat simplistic, model attempts to reconstruct ancestral sequences to attribute to the inner nodes of a tree [65]. Such reconstructed sequences at the same time determine the multiple alignment among the sequences. In this "generalized tree alignment" one aims at minimizing the sum of the edge lengths of this tree, where the length of an edge is determined by the alignment distance between the sequences at its incident nodes. As to be expected, the computational complexity of this problem again makes its solution unpractical. The practical efforts in this direction go back to the work of Sankoff [65, 66]. Hein [26] and Schwikowski and Vingron [68] produced software [38, 40] relying on these ideas.

With the increased interest in analysis of regulatory regions in DNA, the problem of finding subtle local similarities, in particular in DNA sequences, has received much interest. Many programs for the detection of common sequence motifs use probabilistic modeling and/or machine learning approaches. In particular, the mathematical technique of the Gibbs sampler has lent its name also to a motif-finding program, the Gibbs Motif Sampler [31,53]. Bailey and Elkan [7] designed the MEME [33] program which relies on an expectation maximization algorithm. A number of pattern-finding programs have been compared by Tompa and coworkers [78].

6 Multiple Alignments, Hidden Markov Models (HMMs) and Database Searching II

Information about which residues are conserved and thus important for a particular family is crucial not only for the purpose of multiply aligning a set of sequences, but is also very valuable in the context of identifying related sequences in a database. A multitude of methods has been developed that aim at identifying sequences in a database which are related to a given family. The first one was the notion of a profile that was described above and was actually introduced in the context of database searching. As in multiple alignment, profiles help in emphasizing conserved regions in a database search. Thus, a sequence that matches the query profile in a conserved region will receive a higher score than a database sequence matching only in a divergent part of an alignment. This feature is of enormous help in distinguishing truly related sequences.

Algorithmically, profile searching simply uses the dynamic programming alignment algorithm for aligning a sequence to a profile on each sequence in the database. Of course, this is computationally quite demanding and much slower than the heuristic database search algorithms like BLAST or FASTA. Typically, the multiple alignment underlying the profile describes a conserved domain which one expects to find within a database sequence. Therefore, in this context, it is important that end gaps should not be penalized. Furthermore, gap penalties for profile matching frequently vary along the profile in order to reflect the existence of gaps within the underlying multiple alignment. Through this mechanism one attempts to allow new gaps preferentially in regions where gaps have been observed already. However, different suggestions exist as to the choice and derivation method for these gap penalties [77].

In 1994, Haussler and coworkers [52] and Baldi and coworkers [8] introduced HMMs for the purpose of identifying family members in a database. An HMM is a generative probabilistic model in the sense that we can think of it as a machine that generates strings of symbols; in biological applications, typically the letters of a biological sequence. It has "states" and each state will output a symbol according to a distribution associated to this state. After a state has output a symbol, a transition to one of its successor states occurs according to a specified transition probability. These transitions are Markovian, meaning that the transitions leading out of a state are governed only by this state's transition probabilities and not by how the machine got to arrive in this state. The "hidden" element in the HMM comes from the image that an observer gets to see the generated symbol series and then needs to infer which series of states gave rise to it or what the underlying distributions might look like. HMMs and related algorithms are discussed in depth by Durbin and coworkers [20].

The structure of a *profile HMM* mimics a multiple alignment. We think of it as a machine that emits a sequence which would typically be randomly drawn based on a given multiple alignment, according to the distribution of letters in its columns. If gaps were forbidden, the emitted sequence would essentially draw one letter from each column of the alignment. Insertions and deletions, however, imply that the generated sequence may differ in length from the multiple alignment, with some columns possibly skipped or new letters inserted in the emitted sequence. Figure 5 schematically shows the

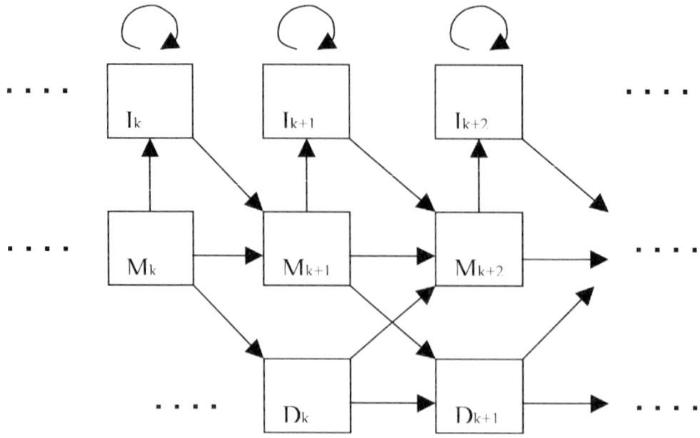

Figure 5 Sketch of the structure of a profile HMM.

states and transitions that realize this structure. The middle row represents a series of match states (M). These represent the columns of the given alignment and emit letters according to a distribution that is supposed to fit the corresponding column of the alignment. A transition into an insert state (denoted I, arranged in the top row) lets the machine emit an additional letter, with the possibility of remaining in this insert state and emitting more additional letters as indicated by the self-loop at the insert nodes. The transition from a match state into a delete state (D, bottom row) leads to the emitted sequence skipping one or more of the following columns of the alignment, which corresponds to a deletion in the emitted sequence with respect to the alignment.

In this manner, the profile HMM can output sequences which, by way of their generating state sequence, are aligned relative to the given multiple alignment. The task of aligning a sequence to a profile HMM can therefore be phrased in the probabilistic setting of "What is the most likely sequence of states to have given rise to this sequence?". This is solved by the so-called Viterbi algorithm, which largely resembles the classical dynamic programming sequence alignment algorithm in its structure. Alternatively, one can ask for the probability of the observed sequence as such, independent of which path generated it. This is computed by summing over the different state sequences that could have produced the sequence. Here, the fully probabilistic formalization is superior to an *ad hoc* score definition which would not allow for posing and answering this question. Algorithmically, this summation can be computed efficiently by the so-called forward algorithm.

There is a standard learning algorithm, the Baum–Welch algorithm, to determine emission and transition probabilities of an HMM given a set of learning data. When training a profile HMM, one has the sequences of the multiple alignment at hand, which may be too small a set for parameter determination in many cases. The problem becomes manageable, though, when one uses the residue distributions in the columns as a guideline for the emission probabilities and chooses the transition probabilities to reflect, in essence, the way gaps should be handled. Adding a possible correction (the "pseudocounts") for sampling artifacts, this choice of parameters can either be used directly or as a starting configuration for a subsequent application of the Baum–Welch algorithm in order to refine parameter estimation. Nevertheless, training an HMM is a very difficult problem and the Baum–Welch algorithm may only find a local optimum.

The first application of HMMs in sequence analysis seems to be due to Churchill [15], who applied the technique to the segmentation of sequences based on their composition. Profile HMMs followed later, addressing the same problem as multiple alignment profiles. A widely used implementation of profile HMMs is the HMMER package [32]. The two concepts of HMMs and profiles are formally very similar, although set in a different language. Bucher and Karplus [12] introduced generalized profiles, and showed that the two concepts are equally powerful in their abilities to model sequence families and detect related sequences. Nevertheless, due to the coherent probabilistic description language and a broad spectrum of good software implementations, HMMs have found widespread acceptance. Many other areas in computational molecular biology, e.g. gene finding, have also profited greatly from the introduction of HMMs.

The fact that a profile or HMM can pick out new sequences also related to the given family suggests that these should be used to update the profile or HMM used as search pattern. This idea leads to iterative search algorithms which search the database repeatedly, each time updating the query pattern with some or all of the newly identified sequences. PSI-BLAST [3] is a very successful implementation of this idea. It starts with a single sequence, and after the first search constructs profiles from conserved regions among the query and newly identified sequences. Without allowing for gaps (to increase search speed) these new profiles are used to repeat the search. Generally, PSI-BLAST quickly converges after updating these profiles again and generally is very successful in delineating all the conserved regions a sequence may share with other sequences in a database. In the realm of HMMs, SAM is a very careful implementation of the idea of iterated searches [39, 49].

It is the generally held view that searching a database with a profile or HMM produces extreme-value distributed random scores just like single-sequence database searching. The quality of the fit to the extreme-value distribution may, however, depend on the particular given alignment. This has been substantiated with mathematical arguments only for the case of ungapped profile matching [22]. Nevertheless, this basic understanding of the statistical behavior of database-matching methods is a crucial element of iterative search programs. Without clear and reliable cutoff values one could not decide which sequences to integrate into the next search pattern and would run the danger of including false positives, thus blurring the information in the pattern.

Both single-sequence search methods and profile/HMM-based methods have been thoroughly validated during recent years [11]. Databases of structurally derived families, e.g. SCOP [34, 58], have made it possible to search a sequence database with a query, and exactly determine the number of false positives and false negatives. For every search one determines how many sequences one misses (false negatives) in dependence of the number of false positive matches. If the sequence statistics is accurate, the number of false positives correlates well with the E-value, i.e. the number of false positives expected by chance. This way of validating search methods allows for making objective comparisons and for determining how much quality one actually gains with slower methods over faster, less accurate methods.

7 Protein Families and Protein Domains

The companion question to the one that assigns related sequences from a database to a given query sequence or family is the question that tries to assign to a query sequence the family of which it is a member or the domains that it contains. One resource for this purpose is the InterPro database [4], which contains amino acid patterns that are descriptive for particular domains, families or functions. The InterPro database summarizes information from several other motif databases including, among others, Prosite [30] and Pfam [10]. One can either scan a sequence against this database [86] or rely on precomputed information that is stored along with the sequences in the databases. The Pfam database contains precomputed HMMs for protein domains. A query sequence can be matched against this library of HMMs in order to identify known domains in the query sequence. Here, too, match statistics plays a crucial role in order to determine the significantly matching domains. A server that allows one to scan a sequence versus all Pfam domains can be found at the Sanger center [45]. Software has also been developed to recognize the Pfam HMMs in either coding DNA or in genomic DNA. In the latter case, the program combines the HMM matching with the distinction between coding and noncoding DNA.

Apart from finding and cataloguing domains of proteins, efforts have also been made to structure the space of all protein sequences into homologous groups or orthologous families. Linial and coworkers have developed the Protonet [67] system, hierarchically structuring the set of all proteins. Krause and coworkers [51] developed SYSTERS [35] to delineate protein families and supply consensus sequences of these families to be searched with a DNA or protein query sequence. Koonin and coworkers put special emphasis on the delineation of orthologous genes, and collect this information in the COG and KOG databases [75].

8 Conclusions

The problems and methods introduced above have been instrumental in the advance in our understanding of genome function, organization and structure. While some years ago human experts would check every program output, nowadays sequence analysis routines are being applied in an automatic fashion creating annotation that is included in various databases. This holds true for similarity relationships among sequences and extends all the way to the prediction of genomic structure or to function prediction based on similarity. Although the quality of the tools has increased dramatically, the possibility of error and, in particular, its perpetuation by further automatic methods exists. Thus, it is apparent that the availability of these high-throughput computational analysis tools is a blessing and a problem at the same time.

References

1 Altschul, S. F., W. Gish, W. Miller, E. W. Myers and D. J. Lipman. 1990. Basic local alignment search tool. J. Mol. Biol. **215**, 403–10.

2 Altschul, S. F. and D. J. Lipman. 1989. Trees, stars, and multiple biological sequence alignment. SIAM J. Appl. Math. **49**, 197–209.

3 Altschul, S. F., T. L. Madden, A. A. Schaffer, J. H. Zhang, Z. Zhang, W. Miller and D. J. Lipman. 1997. Gapped BLAST and PSI-BLAST: a new generation of protein database search programs. Nucleic Acids Res. **25**, 3389–402.

4 Appweiler, R., T. K. Attwood, A. Bairoch, et al. 2001. The InterPro database, an integrated documentation resource for protein families, domains and functional sites. Nucleic Acids Res. **29**, 37–40.

5 Argos, P. 1987. A Sensitive procedure to compare amino-acid-sequences. J. Mol. Biol. **193**, 385–96.

6 Arratia, R. and M. S. Waterman. 1994. A phase transition for the score in matching random sequences allowing deletions. Ann. Appl. Prob. **4**, 200–25.

7 Bailey, T. L. and C. Elkna. 1995. The value of prior knowledge in discovering motifs with MEME. Proc. ISMB **3**, 21–9.

8 Baldi, P., Y. Chauvin, T. Hunkapiller and M. A. Mcclure. 1994. Hidden Markov-models of biological primary sequence information. Proc. Natl Acad. Sci. USA **91**, 1059–63.

9 Barton, G. J. 1993. An efficient algorithm to locate all locally optimal alignments between two sequences allowing for gaps. Comput. Appl. Biosci. **9**, 729–34.

10 Bateman, A., E. Birney, L. Cerruti, et al. 2002. The Pfam protein families database. Nucleic Acids Res. **30**: 276–80.

11 Brenner, S. E., C. Chothia and T. J. P. Hubbard. 1998. Assessing sequence comparison methods with

reliable structurally identified distant evolutionary relationships. Proc. Natl Acad. Sci. USA **95**: 6073–6078.

12 Bucher, P., K. Karplus, N. Moeri and K. Hofmann. 1996. A flexible motif search technique based on generalized profiles. Comput. Chem. **20**: 3–23.

13 Burkhardt, S., A. Crauser, P. Ferragina, H.-P. Lenhof, E. Rivals and M. Vingron. 1999. q-gram based database searching using a suffix array (QUASAR). Proceedings of the Third International Conference on Computational Molecular Biology (RECOMB), ACM Press, New York: 77–83.

14 Califano, A. and I. Rigoutsos. 1993. FLASH: a fast look-up algorithm for string homology. Proc. ISMB **1**: 56–64.

15 Churchill, G. A. 1989. Stochastic-models for heterogeneous DNA-sequences. Bull. Math. Biol. **51**: 79–94.

16 Corpet, F. 1988. Multiple sequence alignment with hierarchical-clustering. Nucleic Acids Res. **16**: 10881–90.

17 Dayhoff, M. O., W. C. Barker and L. T. Hunt. 1978. Establishing homologies in protein sequences. Atlas of Protein Sequences and Structure **5**: 345–52.

18 Dembo, A., S. Karlin and O. Zeitouni. 1994. Critical phenomena for sequence matching with scoring. Ann. Prob. **22**: 2022–39.

19 Do, C. B., M. S. P. Mahabhashyam, M. Brudno and S. Batzoglou. 2005. ProbCons: probabilistic consistency-based multiple sequence alignment. Genome Res. **15**: 330–40.

20 Durbin, R., S. Eddy, A. Krogh and G. Mitchison. 1998. *Biological Sequence Analysis*. Cambridge University Press, Cambridge.

21 Edgar, R. C. 2004. MUSCLE: multiple sequence alignment with high accuracy and high throughput. Nucleic Acids Res. **32**: 1792–7.

22 Goldstein, L. and M. S. Waterman. 1994. Approximations to profile score distributions. J. Comput. Biol. **1**: 93–104.

23 Gotoh, O. 1982. An improved algorithm for matching biological sequences. J. Mol. Biol. **162**: 705–8.

24 Gribskov, M., A. D. McLachlan and D. Eisenberg. 1987. Profile analysis – detection of distantly related proteins. Proc. Natl Acad. Sci. USA **84**: 4355–8.

25 Gupta, S. K., J. D. Kececiogly and A. A. Schaffer. 1995. Improving the practical space and time efficiency of the shortest-path approach to sum-of-pairs multiple sequence alignment. J. Comp. Biol. **2**: 459–72.

26 Hein, J. 1990. Unified approach to alignment and phylogenies. Methods Enzymol. **183**: 626–45.

27 Henikoff, S. and J. G. Henikoff. 1992. Amino-acid substitution matrices from protein blocks. Proc. Natl Acad. Sci. USA **89**: 10915–9.

28 Higgins, D. G., A. J. Bleasby and R. Fuchs. 1992. Clustal-V – Improved software for multiple sequence alignment. Comput. Appl. Biosci. **8**: 189–91.

29 Hirschberg, D. S. 1977. Algorithms for longest common subsequence problem. J. ACM **24**: 664–75.

30 Hofmann, K., P. Bucher, L. Falquet and A. Bairoch. 1999. The PROSITE database, its status in 1999. Nucleic Acids Res. **27**: 215–9.

31 http://bayesweb.wadsworth.org/gibbs/gibbs.

32 http://hmmer.wustl.edu.

33 http://meme.sdsc.edu.

34 http://scop.mrc-lmb.cam.ac.uk/scop.

35 http://systers.molgen.mpg.de.

36 http://www.bibiserv.techfak.uni-bielefeld.de/dca.

37 http://www.bibiserv.techfak.uni-bielefeld.de/dialign.

38 http://www.bioweb.pasteur.fr/Seqanal/interfaces/treealign-simple.

39 http://www.cse.ucsc.edu/research/compbio/sam.

40 http://www.dkfz.de/tbi/services/3w/start.

41 http://www.ebi.ac.uk.

42 http://www.ebi.ac.uk/clustalw.

43 http://www.ncbi.nlm.nih.gov.

44 http://www.ncbi.nlm.nih.gov/CBBresearch/Schaffer/msa.

45 http://www.sanger.ac.uk.

46 HUANG, X. Q. 1993. Alignment of three sequences in quadratic space. Appl. Comput. Rev. **1**: 7–11.

47 JEANMOUGIN, F., J. D. THOMPSON, M. GOUY AND D. G. HIGGINS. 1998. Multiple sequence alignment with Clustal X. Trends Biochem. Sci. **23**: 403–5.

48 KARLIN, S. AND S. F. ALTSCHUL. 1990. Methods for assessing the statistical significance of molecular sequence features by using general scoring schemes. Proc. Natl Acad. Sci. USA **87**: 2264–8.

49 KARPLUS, K., C. BARRETT AND R. HUGHEY. 1998. Hidden Markov models for detecting remote protein homologies. Bioinformatics **14**: 846–56.

50 KENT, W. J. 2002. BLAT – The BLAST-like alignment tool. Genome Res. **12**: 656–64.

51 KRAUSE, A., J. STOYE AND M. VINGRON. 2005. Large scale hierarchical clustering of protein sequences. BMC Bioinformatics **6**: 15.

52 KROGH, A., M. BROWN, I. S. MIAN, SJÖLANDER AND D. HAUSSLER. 1994. Hidden Markov models in computational biology – applications to protein modeling. J. Mol. Biol. **235**: 1501–31.

53 LAWRENCE, C. E., S. F. ALTSCHUL, M. S. BOGUSKI, J. S. LIU, A. F. NEUWALD AND J. C. WOOTTON. 1993. Detecting subtle sequence signals – a Gibbs sampling strategy for multiple alignment. Science **262**: 208–14.

54 LEE, C., C. GRASSO AND M. F. SHARLOW. 2002. Multiple sequence alignment using partial order graphs. Bioinformatics **18**: 452–64.

55 MAIZEL, J. V. AND R. P. LENK. 1981. Enhanced graphic matrix analysis of nucleic acid and protein sequences. Proc. Natl Acad. Sci. USA **78**: 7665–9.

56 MORGENSTERN, B. 1999. DIALIGN 2: improvement of the segment-to-segment approach to multiple sequence alignment. Bioinformatics **15**: 211–8.

57 MULLER, T. AND M. VINGRON. 2000. Modeling amino acid replacement. J. Comput. Biol. **7**: 761–776.

58 MURZIN, A. G., S. E. BRENNER, T. HUBBARD AND C. CHOTHIA. 1995. SCOP – a structural classification of proteins database for the investigation of sequences and structures. J. Mol. Biol. **247**: 536–40.

59 MYERS, E. W. AND W. MILLER. 1988. Optimal alignments in linear-space. Comput. Appl. Biosci. **4**: 11–17.

60 NEEDLEMAN, S. B. AND C. D. WUNSCH. 1970. A general method applicable to search for similarities in amino acid sequence of 2 proteins. J. Mol. Biol. **48**: 443–53.

61 NING, Z., A. J. COX AND J. C. MULLIKIN. 2001. SSAHA: a fast search method for large DNA databases. Genome Res. **11**: 1725–9.

62 NOTREDAME, C., D. G. HIGGINS AND J. HERINGA. 2000. T-Coffee: a novel method for fast and accurate multiple sequence alignment. J. Mol. Biol. **302**: 205–17.

63 PEARSON, W. R. AND D. J. LIPMAN. 1988. Improved tools for biological sequence comparison. Proc. Natl Acad. Sci. USA **85**: 2444–8.

64 SANKOFF, D. 1972. Matching sequences under deletion/insertion constraints. Proc. Natl Acad. Sci. USA **69**: 4–6.

65 SANKOFF, D. 1975. Minimal mutation trees of sequences. SIAM J. Appl. Math. **28**: 35–42.

66 SANKOFF, D., R. J. CEDERGREN AND G. LAPALME. 1976. Frequency of insertion–deletion, transversion, and transition in evolution of 5S ribosomal-RNA. J. Mol. Evol. **7**: 133–49.

67 SASSON, O., A. VAAKNIN, H. FLEISCHER, E. PORTUGALY, Y. BILU, N. LINIAL AND M. LINIAL. 2003. ProtoNet: hierarchical classification of the protein space. Nucleic Acids Res. **31**: 348–52.

68 SCHWIKOWSKI, B. AND M. VINGRON. 1997. The deferred path heuristic for the generalized tree alignment problem. J. Comput. Biol. **4**: 415–31.

69 SMITH, T. F. AND M. S. WATERMAN. 1981. Identification of common molecular subsequences. J. Mol. Biol. **147**: 195–7.

70 SMITH, T. F., M. S. WATERMAN AND C. BURKS. 1985. The statistical distribution of nucleic acid similarities. Nucleic Acids Res. **13**: 645–56.

71 SONNHAMMER, E. L. L. AND R. DURBIN. 1995. A dot-matrix program with dynamic threshold control suited for genomic DNA

and protein-sequence analysis. Gene-Combis **167**: 1–10.

72 SPANG, R. AND M. VINGRON. 1998. Statistics of large-scale sequence searching. Bioinformatics **14**: 279–84.

73 STOYE, J. 1998. Multiple sequence alignment with the divide-and-conquer method. Gene **211**: GC45–56.

74 SWOFFORD, D. L. AND G. J. OLSEN (eds.). 1990. *Phylogeny Reconstruction*. Sinauer Associates, Sunderland, MA.

75 TATUSOV, R. L., N. D. FEDOROVA, J. D. JACKSON, et al. 2003. The COG database: an updated version includes eukaryotes. BMC Bioinformatics **4**: 41.

76 TAYLOR, W. R. 1987. Multiple sequence alignment by a pairwise algorithm. Comp. Appl. Biosci. **3**: 81–7.

77 TAYLOR, W. R. 1996. A non-local gap-penalty for profile alignment. Bull. Math. Biol. **58**: 1–18.

78 TOMPA, M., N. LI, T. L. BAILEY, et al. 2005. Assessing computational tools for the discovery of transcription factor binding sites. Nat. Biotechnol. **23**: 137–44.

79 VINGRON, M. AND M. S. WATERMAN. 1994. Sequence alignment and penalty choice – review of concepts, case-studies and implications. J. Mol. Biol. **235**: 1–12.

80 WATERMAN, M. S. 1984. Efficient sequence alignment algorithms. J. Theor. Biol. **108**: 333–7.

81 WATERMAN, M. S. 1995. *Introduction to Computational Molecular Biology*. Chapman & Hall, London.

82 WATERMAN, M. S. AND M. EGGERT. 1987. A new algorithm for best subsequence alignments with application to transfer RNA–ribosomal-RNA comparisons. J. Mol. Biol. **197**: 723–8.

83 WATERMAN, M. S. AND M. VINGRON. 1994. Rapid and accurate estimates of statistical significance for sequence data-base searches. Proc. Natl Acad. Sci. USA **91**: 4625–8.

84 WATERMAN, M. S. AND M. VINGRON. 1994. Sequence comparison significance and Poisson approximation. Stat. Sci. **9**: 367–81.

85 WOZNIAK, A. 1997. Using video-oriented instructions to speed up sequence comparison. Comput. Appl. Biosci. **13**: 145–50.

86 ZDOBNOV, E. M. AND R. APWEILER. 2001. InterProScan – an integration platform for the signature-recognition methods in InterPro. Bioinformatics **17**: 847–8.

4 Phylogeny Reconstruction

Ingo Ebersberger, Arndt von Haeseler, and Heiko A. Schmidt

1 Introduction

In 1973 Theodosius Dobzhansky said "Nothing in biology makes sense except in the light of evolution" [27]. Although more than 30 years old, the citation still remains valid. Biologists nowadays use the massive amount of sequence data to infer the phylogenetic relationship of contemporary organisms. DNA, long words over a finite alphabet of four nucleotides, is transmitted from one generation to the next. The copying process and environmental factors lead to an accumulation of mutations in the sequence. Such mutations manifest themselves as slight changes in the DNA sequence, so-called *substitutions*. The vertical transmission (in time) of DNA together with the discrete nature of the mutations makes the molecule an ideal target to study phylogenetic relationship of organisms. Consequently, sequence-based phylogenies of organisms have been determined from many different genes. Such gene-trees have provided surprisingly new and sometimes controversial insights into the evolutionary relationships of organisms. However, research and debates still focus on the best methodology. That is, how do we measure similarity or dissimilarity, how can we model the process of substitution, how can we accurately infer the tree? Despite this ongoing discussion, molecular phylogenies are nowadays a routine tool for biologists interested in the evolution of organisms.

Moreover, the application of molecular phylogenies goes beyond the reconstruction of phylogenetic trees for organisms. Gene trees or, more general, sequence trees serve as an important source of information to understand how sequences are related. From this relatedness it is then possible to infer the function of an unknown sequence (see also Chapter 32). Not only function can be inferred, but also structure can be deduced from trees (Chapter 11). From sequence trees we can deduce the evolutionary history of the sequences themselves. We can determine regions that are conserved or highly variable and we can detect sequences that show a highly aberrant substitution pattern. Moreover, we can detect duplications of genes or parts of the genome; thus,

Bioinformatics - From Genomes to Therapies Vol. 1. Edited by Thomas Lengauer

ISBN: 978-3-527-31278-8

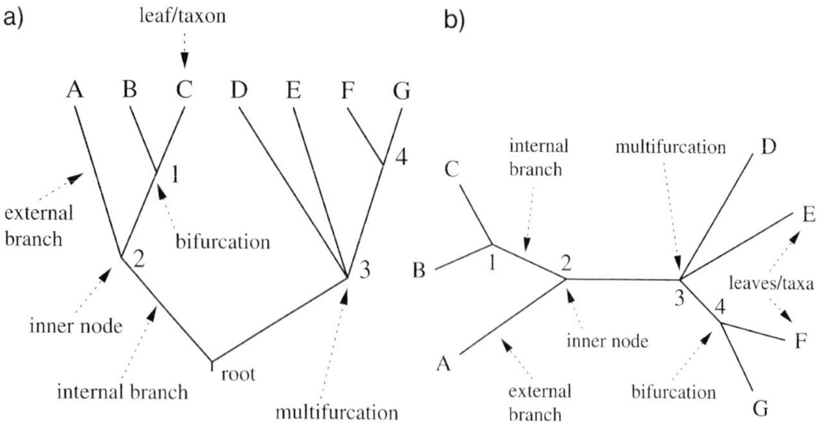

Figure 1 Phylogenetic relationships among a set of seven taxa represented by a rooted (a) and an unrooted (b) tree.

trees serve as analytical tools in comparative genomics (Chapter 37). Even the coevolution of host–pathogen interactions (Chapter 41) is mirrored in the similarity of trees from both groups.

Again, the conclusions strongly depend on the accuracy of the reconstruction method. To critically evaluate the result a basic understanding of the different approaches to phylogenetic inference is required. In this chapter we present a very basic introduction. We set the stage by a brief introduction in the terminology, followed by a summary about current approaches to model evolutionary changes (Section 2). Section 3 describes the three fundamental principles of phylogenetic inference, i.e. maximum parsimony, distance-based and maximum likelihood (ML) inference. The subsequent section deals with the optimization problem of finding the "best" tree(s) with respect to some objective function. With the advent of phylogenomics one wants to reconstruct a species tree from a collection of multiple genetic loci. This question leads naturally to supertree methods introduced in Section 5. Finally, Section 6 summarizes attempts to infer evolutionary relationships if the data do not evolve according to a tree. Processes like horizontal gene transfer or recombination destroy the tree-likeness of the data. In such instances it is better to reconstruct networks rather then trees.

1.1 Reconstructing a Tree from its Leaves

The fundamental axiom in evolutionary biology is the assumption that any two taxa share a common ancestor at some time point in their history. Thus, following backward in time the lineages along which these taxa have evolved, they will eventually coalesce. Considering a large set of taxa, consecutive coalescent events result in an ever-decreasing number of predecessors until only two lineages remain. The ancient taxon in which these lineages eventually merge then represents the most recent ancestor common to all taxa in the dataset. The correspondence to a tree is apparent and it is therefore of little surprise that trees play the key role in phylogenetic research. The dominant role of trees in the area of phylogeny reconstruction is already manifested in the phylogeneticist's vocabulary (Figure 1). Contemporary taxa are dubbed *leaves*, leaves are connected via *external branches* to *internal nodes* (their common ancestors) and internal nodes themselves are connected via *internal branches*. Nodes that give rise to two descendants are termed *bifurcations*, nodes with a larger number of descendants are referred to as *multifurcations* (Biologists usually judge it as unlikely that three or more lineages emerged at precisely the same time from a shared ancestor; thus, we will concentrate on bifurcating trees.) Eventually, if the direction of the evolutionary process is known the ancestor of all leaves in the tree is identified. To stay in the picture, this node is termed the *root*. If directional information is not available, the relationships of the taxa is represented in an *unrooted tree*. However, in this case the temporal succession of ancestors remains undetermined. The reconstruction of the phylogenetic relationships for a set of taxa and their representation by a tree can be separated into two subproblems. (i) What is the order individual taxa split from their shared ancestors, i.e. what is the topology of the tree? (ii) What is the evolutionary time that has passed since any two taxa last shared a common ancestor, i.e. how long are the corresponding branches of the tree? In most cases no hard evidence (such as a comprehensive fossil record) exists to directly reconstruct the evolutionary steps transforming one ancestral taxon into its descendants. Rather, we get hold only of the end-points of this process and are more or less ignorant about anything that has happened in the past. Thus, we are facing the problem of reconstructing the phylogenetic tree just by looking at its leaves.

1.2 Phylogenetic Relationships of Taxa and their Characters

Although one is typically interested in the relationships of the taxa, the reconstruction procedure is usually not based on the taxon as a whole. For practical reasons one vicariously concentrates on individual characteristics of these taxa, usually either morphological or molecular features. We will refer to such representative characteristics as *characters* and to their peculiar expression in the individual taxa as the *character state*.

To collect the raw data for phylogenetic analyses, the variety of states for a particular character in a set of taxa has to be assessed first. Next, the possible transformations of the character states during evolution has to be reconstructed, which can then be used for phylogenetic inference. Irrespective of what type of character has been chosen, two general approaches can be followed to trace the phylogenetic signal in the data. One can identify those (evolutionary novel) character states that are shared among a subset of the taxa. Such a congruency is interpreted as a result of shared descent [68]. Based on the pattern and extent of congruent character states, the degrees of relationships among the taxa can then be inferred. Alternatively, the extent of evolutionary change for the particular character between any pair of taxa can be assessed. From the resulting pairwise evolutionary distances again a phylogenetic tree can be reconstructed. We will outline both approaches in greater detail below (Section 3). Eventually, the evolutionary history of the particular character is extrapolated to the taxon level.

Inherent to any character-based strategy for phylogeny reconstruction is the assumption that comparisons are performed only between homologous characters, i.e. characters related by descent. Although the assignment of homology constitutes one of the major issues in evolutionary studies [68, 114, 134], we will take for granted that this postulation is met.

1.2.1 The Problem of Character Inconsistencies

To date, tree reconstruction is frequently based on more than only a single character. This adds the advantage that different characters can complement each other by adding resolution to different parts of the tree. However, the reverse effect occurs as well: trees reconstructed from different characters can disagree. Given that incompatible groupings of taxa are supported significantly by the respective data, two alternative explanations are possible. First, the incompatible groupings are based on a misinterpretation of the data. For example, taxa can share the same character state not due to a shared ancestry, but rather because the particular state arose independently at least twice during evolution. Phylogeny reconstruction methods that model the evolutionary process (see Sections 3.2 and 3.3) usually account for this problem. However, if such parallel, convergent or back mutations remain unrecognized (or are neglected) an erroneous tree reconstruction is possible. In the second explanation, the evolution of each character state is correctly reconstructed. Such genuine discrepancies between inferred trees have various causes. Among the most frequently stated are processes like the random sorting of ancestral polymorphisms (e.g. Ref. [130]) and horizontal gene transfer [25, 28, 91] (Figure 2). If one is suspicious that either of these scenarios could apply, several independently evolving characters should be analyzed. The most frequently observed tree is then usually the tree reflecting the evolutionary relationships of the taxa as a whole. Alternatively, if it seems appropriate to visualize such discrepant phylogenetic signals in the data, a network rather than a tree can be chosen to represent the phylogeny of the taxa (see Section 6).

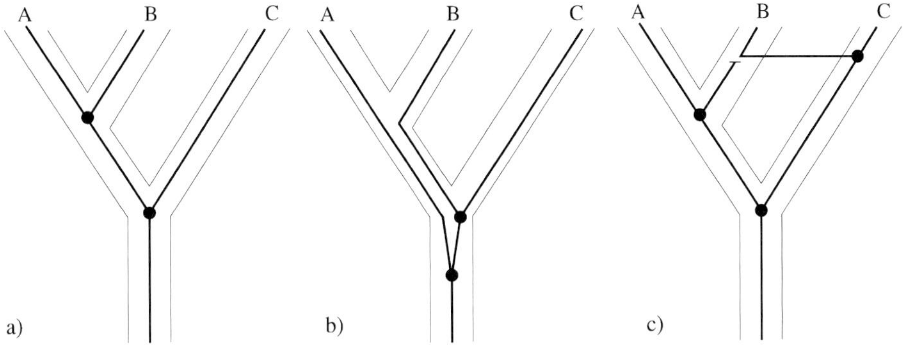

Figure 2 Trees for individual characters (inner trees) can differ from the species tree (outer trees). (a) The phylogenetic history of the character follows that of the species. (b) The random sorting of ancestral polymorphisms at subsequent speciation events. (c) Horizontal gene transfer, i.e. the lateral transfer of individual genes or DNA sequences between species. Both random sorting of ancestral polymorphisms and horizontal gene transfer can result in phylogenies that are incongruent to the species tree or to trees reconstructed from other characters.

1.2.2 **Finding the Appropriate Character Set**

In theory, phylogenetic relationships can be reconstructed from any set of homologous characters subject to evolutionary change. Depending on the scope of a study and the collection of taxa, however, certain types of characters might be more suitable than others. Changes in shape, or morphology in general, are the most conspicuous effects of evolution. Therefore, the field of phylogeny reconstruction was dominated by the analysis of morphological characters for a long time. However, with the expansion of molecular biology the focus has shifted considerably. Initially, the immunological and electrophoretic analysis of structural and electrical properties of proteins [85], presence or absence of genomic features such as restriction enzyme recognition sites [47], or DNA–DNA hybridization [45, 137] were used to measure the extent of character change on the molecular level. As time proceeded, the presence and absence or linear order of regulatory elements and genes (see also Chapter 8), and recently even expression data [83] have been employed for phylogenetic inference. However, the most dominant role is still taken by the direct comparison of biological sequences. Sequences change in the course of time and any two sequences derived from a common ancestor will diverge. The pattern and extent of differences between two related sequences is then used to reconstruct their evolutionary history. Initially, due to experimental constraints, the comparison of protein sequences was prevalent. Nowadays, analyses rely almost entirely on DNA sequences and even those studies comparing amino acid sequences usually derive these from the corresponding DNA sequences. The advantages of DNA sequence data are apparent. DNA sequences can be obtained with considerable ease from any taxon and even the comparison of entire genomes has now become feasible. Allowing for some simplifications, each nucleotide position in the DNA sequence can be regarded as an independently evolving character and the number of possible states is strictly limited to the four bases: adenine (A), guanine (G), cytosine (C) and thymine (T). Eventually, different DNA sequences in a genome can evolve with different rates. This allows for easy adaption of the dataset to the evolutionary time scale for which phylogenetic resolution is required.

2 Modeling DNA Sequence Evolution

The substitution of nucleotides in a DNA sequence, i.e. the replacement of one nucleotide by a different one, is usually considered a random event. As a consequence, an important prerequisite for the reconstruction of phylogenetic relationships among species is the prior specification of a *model of substitution*, which provides a statistical description of DNA sequence evolution [97]. If we consider the substitution of one nucleotide by another one at any given site in a sequence as a random event and, furthermore, assume that a series of such random events occurs during some time interval, then theses events form a homogeneous Poisson process [37], if two very mild assumptions are met:

(i) The occurrence of a substitution in the time interval (t_1, t_2) is independent of a substitution in another time interval (t_3, t_4), where (t_1, t_2) and (t_3, t_4) do not overlap.

(ii) There is a constant $\mu > 0$, such that for any time interval $(t, t+h)$, $h > 0$ and h small, the probability that one event occurs is independent of t and is proportional to μh. The probability that more than one substitution occurs during $(t, t+h)$ becomes vanishing small as $h \to 0$.

The latter condition implies the so-called time homogeneity and, moreover, it implies that the probability of one substitution is proportional to the length of the time interval, i.e. the size of h. As substitutions are assumed to occur spontaneously and independently from past or future substitutions, homogenous Poisson processes are a simple approach to model the evolution of DNA. Moreover, under conditions (1) and (2) the number of substitutions $X(t)$ that occur up to any arbitrary time t is Poisson distributed with parameter μt [37]. Thus:

$$P_i(t) = [(\mu t)^i \exp(-\mu t)]/i!\,, \tag{1}$$

is the probability that $i = 0, 1, 2, \dots$ substitutions occur in the time interval $(0, t)$. On average, μt substitutions with variance μt are expected. Note that the parameters μ (nucleotide substitutions per site per unit time) and t (the time) cannot be estimated separately, but only through their product μt (number of substitutions per site up to time t).

The nucleotide substitution process of DNA sequences described by the Poisson process can be generalized to a so-called Markov process that uses a rate matrix (typically called Q with elements Q_{xy}), which specifies the relative rates of change for each nucleotide. The most general form of the Q-matrix is shown in Figure 3. Rows follow the order A, C, G and T so that, for example, the second term of the first row is the instantaneous rate of change from base A to base C. This rate is given by the frequency of base C (π_C) times a relative rate parameter, describing (in this case) how often the substitution A to C occurs during evolution with respect to all other possible substitutions. Thus, each nondiagonal entry in the matrix represents the flow from nucleotide x to nucleotide y, while the diagonal elements are chosen to make the sum of each row equal to zero. They represent the total flow that leaves nucleotide x. Accordingly, we can write the total number of substitutions per unit time (i.e. the total substitution rate μ) as:

$$\mu = -\sum_{x} Q_{xx}\pi_x, x \in \{A, C, G, T\}. \quad (2)$$

Models like the one summarized in Figure 3 belong to the general class of time-homogenous time-continuous Markov models. When applied to modeling nucleotide substitutions, they share the following set of assumptions:

- The rate of change from x to y at any nucleotide position in a sequence is independent of the nucleotide that occupied this position prior to x (Markov property).
- Substitution rates do not change over time (homogeneity).
- The waiting time until the first substitution occurs follows a continuous distribution (time continuity).

$$Q = \begin{pmatrix} -(a\pi_C + b\pi_G + c\pi_T) & a\pi_C & b\pi_G & c\pi_T \\ g\pi_A & -(g\pi_A + d\pi_G + e\pi_T) & d\pi_G & e\pi_T \\ h\pi_A & i\pi_C & -(h\pi_A + i\pi_C + f\pi_T) & f\pi_T \\ j\pi_A & k\pi_C & l\pi_G & -(j\pi_A + k\pi_C + l\pi_G) \end{pmatrix}$$

Figure 3 Instantaneous rate matrix Q. Each entry in the matrix represents the instantaneous substitution rate from nucleotide x to nucleotide y (rows and columns follow the order A, C, G and T). a to l are rate parameters describing the relative rate one nucleotide is substituted by any other nucleotide. $\pi_A, \pi_C, \pi_G, \pi_T$ correspond to the nucleotide frequencies. Diagonal elements are chosen such that each row sums up to zero.

Once the evolutionary model (and thus the Q-matrix) is specified, the probabilities $P(t)$ of change from one nucleotide to any other during evolutionary time t is computed as follows:

$$P(t) = \exp(Qt)\,. \tag{3}$$

Each entry $P_{xy}(t)$ of the resulting probability matrix $P(t)$ specifies the probability to observe nucleotide y at time point t if the original nucleotide at this site was x.

2.1 Nucleotide Substitution Models

From the instantaneous substitution rate matrix Q in Figure 3 various submodels can be derived. Among these, the so-called stationary time-reversible models are the ones most commonly used. These models introduce the constraint that for any two nucleotides i and j the rate of change from i to j is the same as from j to i (thus, $a = g, b = h, c = j, d = i, e = l, f = l$ in Figure 3). Under these conditions the values of π_N (N = A, G, C, T) correspond to the stationary frequencies of the four nucleotides, respectively (i.e. $\pi \cdot Q = 0$). If all eight parameters of a reversible Q-matrix are specified separately, the general time reversible model [92] is derived. The most simplest (fewest number of parameter) model assumes that the equilibrium frequencies of the four nucleotides are 0.25 each and that any nucleotide has the same rate to be replaced by any other. These assumptions correspond to a Q-matrix with $\pi_A = \pi_C = \pi_G = \pi_T = 1/4$, and $a = b = c = d = e = f = 1$. This resembles the well-known Jukes-Cantor model [82]. An overview of the hierarchy of the most common substitution models is shown in Figure 4.

2.2 Modeling Rate Heterogeneity

The nucleotide substitution models described so far implicitly assume that the rate of nucleotide substitution is the same for any position in the DNA sequence. However, it is well known that this is an oversimplification. For example, substitutions occur at an about 10 times higher frequency at C and G nucleotides when the C is followed by a G along the sequence [71]. Similarly, selective constraints maintaining functional DNA sequences result in varying substitution rates along a DNA sequence. To account for such site-dependent rate variations, a plausible model for the distribution of rates over sites is required. Most commonly, a continuous probability distribution, the Γ-distribution with expectation 1 and variance $1/\alpha$, is used [61]. By adjusting the shape parameter α, the Γ-distribution allows varying degrees of rate heterogeneity (Figure 5). For $\alpha > 1$, the distribution is bell-shaped and models weak rate heterogeneity among sites. For $\alpha < 1$, the Γ-distribution takes on

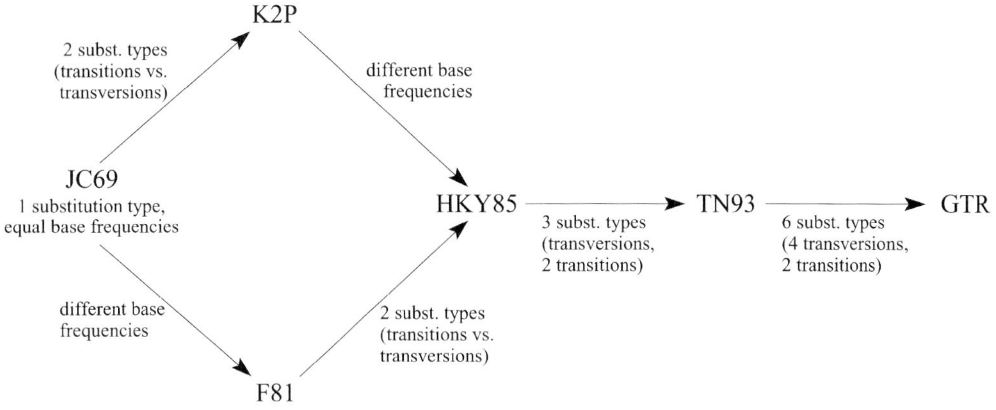

Figure 4 A hierarchy of the most commonly used nucleotide substitution models. JC69: Jukes and Cantor (1969) [82]; F81: Felsenstein (1981) [42]; K2P: Kimura two-parameter model (1980) [84]; HKY85: Hasegawa, Kishino and Yano (1985) [65]; TN93: Tamura and Nei (1993) [149]; GTR: general time reversible model [92]. Many more models are possible and an extensive overview is given in Ref. [74].

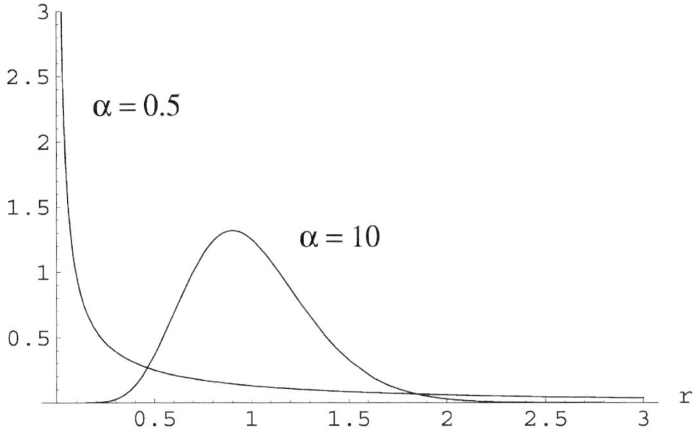

Figure 5 Probability density functions of the Γ-distribution for different values of the shape parameter α. The x-axis represents the relative substitution rate r of a site.

a characteristic L-shape, which describes strong rate heterogeneity, i.e. some sites have very high substitution rates, while the other sites are practically invariable.

2.3 Codon Models

Heterogeneous substitution rates become a particular issue for DNA that codes for proteins. Amino acid sites in a protein sequence are under different selective constraints, depending on their relevance for the protein func-

$$q_{xy} = \begin{cases} 0 & \text{if } x \text{ and } y \text{ differ at two or three nucleotide positions} \\ \pi_y & \text{if } x \text{ and } y \text{ differ by one synonymous transversion} \\ \kappa\pi_y & \text{if } x \text{ and } y \text{ differ by one synonymous transition} \\ \omega^{(h)}\pi_y & \text{if } x \text{ and } y \text{ differ by one nonsynonymous transversion} \\ \omega^{(h)}\kappa\pi_y & \text{if } x \text{ and } y \text{ differ by one nonsynonymous transition} \end{cases}$$

Figure 6 Instantaneous rate that codon x at site h is replaced by codon y. π_y represents the stationary marginal frequency of codon y, κ denotes the transition/transversion rate ratio and $\omega^{(h)}$ the nonsynonymous/synonymous substitution rate ratio at site h.

tion. Accordingly, nucleotide substitutions causing the encoded amino acid to change (replacement substitutions) will have fixation probabilities that depend on the selective constraint imposed on the encoded amino acid. In contrast, silent substitutions, i.e. a change in the DNA sequence has no effect on the encoded protein, are invisible to selective forces acting on the protein sequence. As a result, the ratio of nonsynonymous and synonymous substitution rates (ω) will vary among sites in a DNA sequence, with $\omega = 1$ indicating no selection, $\omega < 1$ representing purifying selection by removing replacement mutations and $\omega > 1$ representing diversifying positive selection/adaptive evolution. Codon models have been specifically designed to model the evolution of protein-coding DNA sequences [59]. An example is shown in Figure 6 based on an extension of the HKY85 model [65] (see Figure 4). Note that, in contrast to the conventional substitution models, codon models consider the replacement of one nucleotide triplet (codon) by another. Thus, we obtain $4^3 - 3 = 61$ possible character states at a site, the codon (the three stop codons are not taken into account). Obviously, the assignment of a distinct substitution rate ratio ω to each codon position would lead to a vast over-parameterization of the model. Therefore, either a set of predefined ω-values or statistical distributions, both discrete and continuous, are used to account for varying ω-values among sites [104, 156, 157].

3 Tracing the Evolutionary Signal

Given a set of homologous DNA sequences whose phylogeny is known, inferences can be made about the evolutionary forces molding the contemporary DNA sequences from their shared ancestral sequence. Conversely, with a concept or a model at hand of how DNA sequences evolve one can aim to reconstruct the phylogenetic tree based on the DNA sequences. In either case, however, a meaningful sequence alignment is required. Thus, the sequences need to be aligned such that homologous nucleotides in different sequences form a column. To account for the insertion and deletion of nucleotides during evolution, gaps are introduced to achieve this positional homology. Chapter

3 deals with methods to compute sequence alignments. We will not dwell on further methodological details and take it for granted that an alignment is available. Based on this alignment, several criteria exist to compute the tree that best reflects the evolutionary relationships of the data. We will explain four principles for tree generation.

3.1 The Parsimony Principle of Evolution

Parsimony methods share as an optimality criterion that among various alternative hypotheses the one that requires the minimal number of assumptions should be chosen. In the context of DNA sequence evolution one searches for the tree(s) that explain the observed diversity in the contemporary sequences with the minimal number of nucleotide substitutions (Figure 7). Usually only a fraction of the differences in a sequence alignment determine the so-called parsimonious tree(s). They are called phylogenetically informative in a parsimony analysis. For instance, position 8 in the alignment (Figure 7) best supports a tree grouping sequences *W* and *X*, and *Y* and *Z*, respectively. In

	Position								
	1	2	3	4	5	6	7	8	9
Sequence W:	C	G	C	A	C	T	G	T	T
Sequence X:	C	G	C	A	C	T	G	T	T
Sequence Y:	T	G	A	A	C	T	G	C	T
Sequence Z:	C	G	G	A	C	T	G	C	T
	*		*					*	

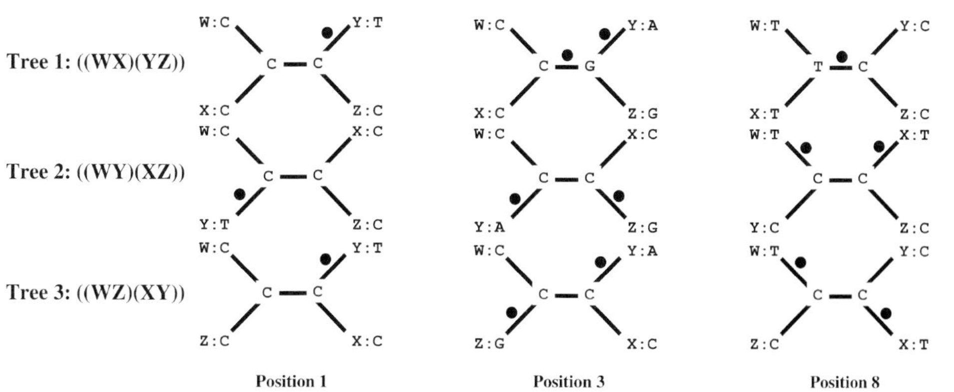

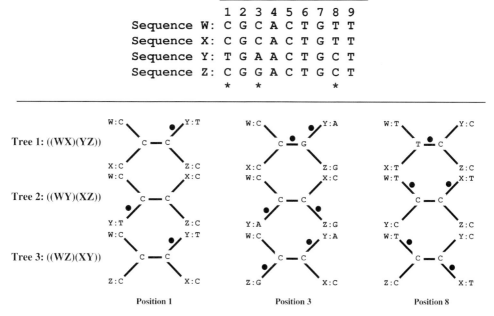

Figure 7 Maximum parsimony tree reconstruction from an alignment of four DNA sequences. For the alignment columns labeled with a "*" all three possible unrooted tree topologies are shown. Labels at the leafs denote the taxon and the represented nucleotide. Nucleotides at the inner nodes represent one parsimonious reconstruction. Nucleotide substitutions are represented by black dots. Position 8 is the only position that distinguishes the three trees with respect to the number of substitutions. Tree 1 requires only a single substitution compared to trees 2 and 3, which require two substitutions each.

contrast, to explain the sequence diversity at positions 1 and 3, two substitutions are necessary regardless of the tree structure. Such positions are called phylogenetically non-informative in a parsimony analysis. The tree that is supported best by the phylogenetically informative sites is then the maximum parsimony tree. Note, however, that no unique solution is guaranteed since more than one most parsimonious tree might exist.

3.1.1 Generalized Parsimony

To date, a vast number of modifications of the initial criterion of maximum parsimony exist [39, 40, 50, 88]. Instead of referring to each and every modification separately, we would like to present the generalized idea of parsimony [128, 129] from which the individual modifications can be easily derived. In a mathematical terminology, one aims to identify those trees in the space of all possible trees which minimize the following equation:

$$L(\tau) = \sum_{k=1}^{B} \sum_{j=1}^{L} \omega_j \cdot \text{diff}(x_{k'j}, x_{k''j}) \,, \tag{4}$$

where $L(\tau)$ is called the length of the tree τ, B is the total number of branches in the tree, L is the number of nucleotide positions analyzed (alignment length), and k' and k'' are the two nodes connected by branch k displaying the nucleotides $x_{k'j}$ and $x_{k''j}$. These can be either the observed nucleotides present in the alignment or, in the case of internal nodes, the optimal nucleotide assignments. Finally, $\text{diff}(x, y)$ represents the cost-matrix that specifies the cost of the transformation from nucleotide x to nucleotide y and ω_j is a specific weight for each alignment position. Thus, diff and $\omega = (\omega_1, \ldots, \omega_L)$ allow for specifying *a priori* some beliefs about the importance of positions and substitutions for the tree reconstruction, e.g. cost matrix **A** in Figure 8 reflects a Jukes–Cantor type of evolution, whereas cost matrix **B** down-weights transitions relative to transversions.

A =

	A	C	G	T
A	–	1	1	1
C	1	–	1	1
G	1	1	–	1
T	1	1	1	–

B =

	A	C	G	T
A	–	5	1	5
C	5	–	5	1
G	1	5	–	5
T	5	1	5	–

Figure 8 Cost matrices for generalized parsimony. In matrix **A** substitutions between all four nucleotides invoke the same cost. Matrix **B** represents a slightly more sophisticated model. More weight is put on transversions than on transitions.

3.1.2 **Multiple/Parallel Hits**

Parsimony principles rely on the assumption that a group of related sequences share a certain nucleotide by descent. However, this only approximates the true evolutionary events if the overall amount of sequence changes is low. Thus, multiple changes at the same site in one taxon or parallel independent changes at the same site in different taxa are sufficiently infrequent not to be an issue. However, when considerably diverged sequences are used for tree reconstruction or marked substitution rate heterogeneity among sites exists, multiple/parallel hits can cause severe problems in both assessing the correct number of nucleotide substitutions along the phylogenetic tree and in inferring the correct tree topology.

3.2 Distance-based Methods

In contrast to parsimony methods with a biologically motivated approach to tree reconstruction, distance-based methods choose a mathematical route [43]. A phylogenetic tree is reconstructed for a set of taxa from their pairwise evolutionary distances. To this end, a distance matrix D is calculated from all possible pairwise sequence comparisons. Entry D_{ij} of this matrix represents the distance between sequence i and sequence j. In a simple approach D_{ij} is computed as the edit distance (Hamming distance), i.e. the minimum number of substitutions required to transform sequence i into j. However, multiple changes at the same position cannot be accounted for and therefore the Hamming distance will sometimes underestimate the true number of substitutions. To rectify this, models of sequence evolution are invoked that correct for multiple changes (see Section 2). Various methods were suggested for inferring a tree from a distance matrix. Common, although in fact not especially designed for phylogenetic tree reconstruction, are *clustering methods*. Clustering methods do not have an explicit objective function to be optimized. UPGMA, the most widespread clustering method, will serve as an example.

3.2.1 **UPGMA**

The "Unweighted Pair Group Method using Arithmetic means" groups those two taxa first whose evolutionary distance is minimal. Consider taxa A, B, C, and D with evolutionary distances as shown in Figure 9a. The taxa A and B with distance 6 are clustered first. Subsequently, A and B are treated as one compound taxon AB, and pairwise distances to the remaining taxa C and D are computed. $D_{(AB)C}$ is calculated as the arithmetic mean of the individual distances D_{AC} and D_{BC}, thus $D_{(AB)C} = (7+8)/2 = 7.5$. Likewise, we compute $D_{(AB)D} = (13+14)/2 = 13.5$. Now the cycle is repeated for the new 3×3 distance matrix. We obtain $((AB)C)$ and D as the two remaining

	A	B	C
B	6		
C	7	8	
D	13	14	11

a)

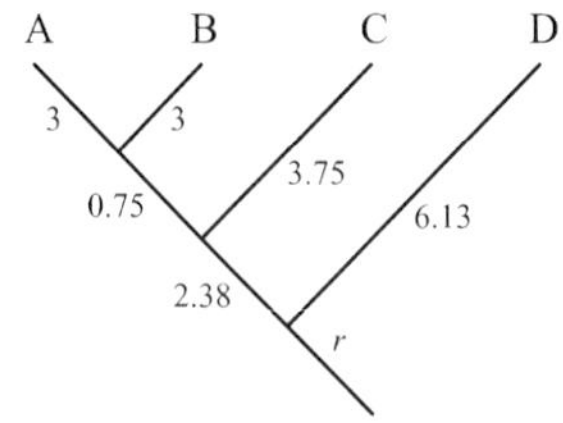

b)

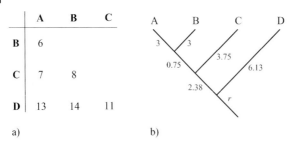

Figure 9 Reconstruction of a phylogenetic tree with the UPGMA method. From the matrix of pairwise sequence distances (a) the phylogenetic tree shown in (b) is reconstructed. Numbers in (b) represent the branch lengths inferred under the assumption of a molecular clock. r identifies the root of the tree.

taxa with $D_{((AB)C)D} = (13 + 14 + 11)/3 = 12.7$. Finally, D and $((AB)C)$ are merged to conclude the procedure. The full tree $(((AB)C)D)$ with branch lengths is displayed in Figure 9b. Thus, UPGMA reconstructs a rooted tree, where branch lengths are computed such that the distances from root r to the leaves A, B, C, and D are identical (6.13 in our example). More generally, such trees fulfill the so-called ultrametric inequality, i.e. for each triple of taxa X, Y and Z:

$$D_{XY} \leq \max[D_{XZ}, D_{YZ}]. \tag{5}$$

Equation 5 is equivalent to the statement that two of the three distances are the same and at least as large as the third distance. More interestingly, the reverse is also true. If for a distance matrix the ultrametric inequality is fulfilled, then the distance matrix is representable by a rooted tree such that the distances D_{ij} are identical to the sum of the branch lengths connecting the two taxa X and Y in the tree. In biological parlance, if the distances computed from a set of aligned sequences obey the ultrametric inequality then the sequences evolve according to a molecular clock, i.e. they accumulate substitutions at the same rate (see Section 2). Therefore, UPGMA can give misleading trees if the distances reflect a substantial departure from the molecular clock. To arrive at a correct tree topology nevertheless, the distance matrix can be corrected for unequal rates of evolution among the lineages under study (*transformed distance method* [87]). The such modified distance matrix can then be used to infer the tree topology using UPGMA.

3.2.2 **Neighbors-relation Methods**

To overcome the restriction of the molecular clock, the characterization of unrooted trees with branch lengths is helpful. If it is required, alternative routes can be taken at a later step in the tree reconstruction to located the

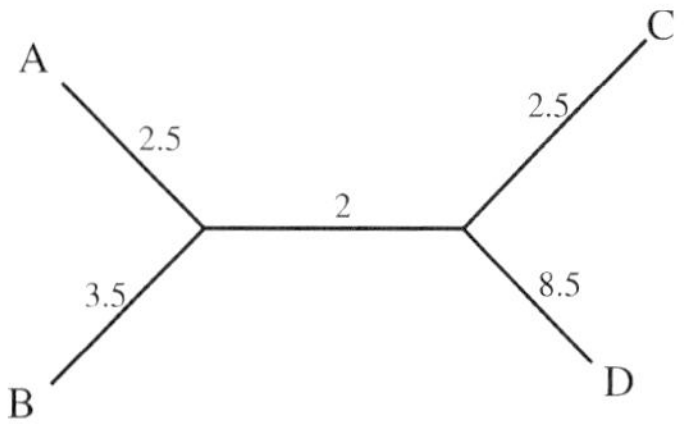

Figure 10 An unrooted quartet tree and its branch lengths reconstructed from the distance matrix in Figure 9(a).

position of the root (see Section 3.6). The celebrated *four-point condition* and its relaxations [5, 20] state conditions when a distance matrix is representable as a tree. A distance matrix is representable as a tree if and only if for all quartets W, X, Y and Z in a taxon set the following holds:

$$D_{WX} + D_{YZ} \leq \max[D_{WY} + D_{XZ}, D_{WZ} + D_{XY}] \,. \tag{6}$$

The distance matrix in Figure 9(a) fulfills this criterion and the corresponding unrooted quartet tree is displayed in Figure 10. However, for real data the four-point condition is rarely met. Thus, one relaxes this condition by introducing the concept of *neighbors*. Two taxa are called neighbors in an unrooted tree if they are connected through a single internal node. For instance the taxa A and B in Figure 10 are neighbors, while the taxa A and C or B and D are not. This concept of neighborhood was generalized to distance matrices [5] and resulted in a series of tree reconstruction methods [5, 51, 131].

3.2.3 **Neighbor-joining Method**

A widely used method based on the neighbors-relation concept is the NJ method by Saitou and Nei [123]. NJ is a clustering algorithm. During each clustering step, two taxa or clusters of taxa are identified as neighbors in the tree, if their grouping results in a tree whose overall length is minimal, i.e. the sum of the lengths of all branches is minimal (minimum-evolution criterion [21]). To this end, one starts with a star-like tree. Subsequently, two taxa X and Y are identified that minimize:

$$S_{XY} = \frac{1}{2(N-2)} \sum_{k=3}^{N} (D_{Xk} + D_{Yk}) + \frac{1}{2} D_{XY} + \frac{1}{N-2} \sum_{3 \leq i \leq j \leq N} D_{ij} \,. \tag{7}$$

The cycle of calculating a new distance matrix and identifying the next neighbors is continued until the initially star-like tree is fully resolved (see also Section 12). For details of the NJ algorithm, see Ref. [147]. Since then, several weighted and improved versions of the NJ algorithm have been published [16, 53].

3.2.4 **Least-squares Methods**

We have described the application of cluster methods to phylogeny reconstruction. However, another view of the reconstruction problem based on a distance matrix is the specification of an objective function we want to optimize. From a mathematical view, we want to find a tree together with its branch lengths such that the distance of two taxa X, Y in the tree, i.e. the sum of the branch lengths connecting the two taxa in the tree, is close to D_{XY}.

The least-squares method provides such a measure for the goodness of fit of the tree and its branch lengths to the data. The best tree (τ_{LS}) under this criterion minimizes the following equation:

$$R(\tau) = \sum_{XY} (T_{XY} - D_{XY})^2 , \qquad (8)$$

where T_{XY} is the sum of the branch lengths along the unique path connecting sequences X and Y. Cavalli-Sforza and Edwards [21] and Fitch and Margoliash [49] were among the first to apply the least-squares theory to the tree-reconstruction problem. However, the big challenge is the determination of the tree topology.

3.3 The Criterion of Likelihood

The third method of tree reconstruction is based on the principle of ML [48] which was made popular in the field by Felsenstein in 1981 [42]. The general idea of ML is as simple as it is appealing: for a given model M and its parameters θ_M the probability or likelihood of observing data D can be calculated. Those parameters are chosen that maximize the likelihood of observing the data. For the particular problem of inferring a phylogenetic tree from biological sequence data the tree topology τ is introduced such that:

$$\tau_{ML} = \underset{(\tau,\theta_M)}{\operatorname{argmax}} P(D|\tau, \theta_M). \qquad (9)$$

Note the subtle, but far-reaching, difference to the principle of maximum parsimony. The general concept of sequence evolution inherent to maximum parsimony, i.e. that one sequence is transformed into another via the least number of changes, is replaced by an explicit model of sequence evolution to describe the substitution process. From this the most significant advantage of ML becomes apparent: it allows us to incorporate any model of biological sequence evolution into the tree reconstruction process. In this way, it opens access to the full use of statistical approaches to compare alternative phylogenetic hypotheses, as well as to test fit and robustness of individual models of sequence evolution. A further advantage compared to the previous two approaches is the possibility to make full use of the sequence information.

In a likelihood framework also constant alignment sites provide information about the tree topology and its branch lengths.

3.4 Calculating the Likelihood of a Tree

We have described in Section 2 how to calculate the probability of observing a difference at a given site in two sequences. We now extend this to compute the probability to find a certain nucleotide pattern (A_s) in column s of N aligned DNA sequences, e.g. the pattern `CCTC` at position 1 of the alignment shown in Figure 7. This probability depends on the model of DNA sequence evolution and on the tree relating the N nucleotides in the alignment column: $P(A_s|\tau,\theta_M)$. We assume that all positions in the alignment of length L evolve according to the same evolutionary model M and evolve independently from each other. Then, the probability of the alignment given a tree and a model is a function of the tree τ and θ_M. Thus:

$$P(A|\tau,\theta_M) = \prod_{s=1}^{L} P(A_s|\tau,\theta_M). \tag{10}$$

To avoid numerical problems caused by underflows and rounding errors during the calculation the likelihood of the data is usually calculated in log-scale, such that:

$$\log[P(A|\tau,\theta_M)] = \sum_{s=1}^{L} \log[P(A_s|\tau,\theta_M)]. \tag{11}$$

Equation (11) facilitates computation of the likelihood of an alignment, if θ_M, τ and its branch lengths are specified. In reality, however, we face the reverse situation. Starting from a given alignment, we aim to infer the underlying phylogenetic tree together with its branch lengths. In order to do so we regard these parameters as variables. Once we have decided on an evolutionary model and have specified its parameter values, we can adjust the tree topology and the branch lengths such that Eq. (11) is maximized. While straightforward and efficient ways exist to obtain ML branch lengths for a specific tree topology (e.g. Ref. [42]), it is a computationally demanding problem to obtain an optimal tree topology. Section 4 explains the details.

3.5 Bayesian Statistics in Phylogenetic Analysis

The likelihood approach outlined so far determines the quality of a tree by calculating the probability of observing the alignment A given the tree τ and the model of sequence evolution specified by θ_M (see Eq. 11). If we consider a particular combination of τ and θ_M as an evolutionary hypothesis, H, we have inferred $P(A|H)$, the probability of A given H. However, it might be interesting to address the reverse question: what is the probability that the evolutionary hypothesis H is correct given the data, i.e. $P(H|A)$ (see [37, p. 106])? Applying Bayes' theorem, we can calculate this posterior probability as:

$$P(H|A) = \frac{P(A|H)P(H)}{P(A)} \,. \tag{12}$$

Rewriting the equation for the problem of tree reconstruction we obtain:

$$P(\tau, \theta_M|A) = \frac{P(A|\tau, \theta_M)P(\tau, \theta_M)}{P(A)} \,, \tag{13}$$

where $P(\tau, \theta_M)$ is the prior probability to choose the tree τ and the model with its parameters. $P(A)$ is the total probability of the alignment A.

Equation (13) can be used in two ways for making phylogenetic inferences from a set of DNA sequences. If one is only interested in identifying the tree that is best supported by the data, one simply determines the tree and the θ_M that maximize Eq. 13. Because $P(A)$ is a constant it can be ignored during optimization. Alternatively, the posterior probability for every possible realizations of τ and θ_M (H_i) can be calculated. This identifies not only the H_i that is supported best by the data, but allows us also to assess how much better the support is compared to the alternative hypotheses [139]. However, it is easy to see that this is feasible for only a very limited number of sequences (see Section 4.1). Thus, Markov chain Monte Carlo (MCMC) simulations are used to estimate the posterior probabilities [105].

Imagine that the individual H_i comprise points in a landscape and $P(H_i|A)$ corresponds to their respective (unknown) altitude. A MCMC simulation is similar to a walk through this landscape that visits the individual points. This walk, however, is not totally random, but guided in a way that higher points are visited more often than lower ones. Thus, when the walk is finished an altitude profile of the landscape is generated from the number of times a particular point was visited. In practice, MCMC simulations work the following way. Starting from any H_i the transition to a new hypothesis H_j, e.g. a new tree topology, is proposed with a probability $q(H_i, H_j)$. This proposal is then accepted with probability:

$$\min\left(1, \frac{P(A|H_j)P(H_j)q(H_i, H_j)}{P(A|H_i)P(H_i)q(H_j, H_i)}\right) , \tag{14}$$

otherwise remaining at H_i. If H_j is supported better by the data than H_i, then H_j is always accepted. Otherwise, H_j is accepted with a probability that depends on how much worse the support of H_j is compared to H_i. The latter option ensures to escape from local optima (see Section 4). If the transition is accepted, H_j will be sampled and the chain moves on. Given that the chain could run for a sufficiently long time, the number of times H_i has been sampled reflects its posterior probability. However, in most cases it is not clear how long a chain should be run. To reduce potential biases of the Monte Carlo estimates, initial sample points are generally discarded [9]. This *burn-in* procedure has the effect that the chain samples only near-optimal hypotheses. Moreover, one samples only every 1000th hypothesis to generate more or less independent samples [46]

Today, more sophisticated MCMC simulations are performed that use several Markov chains whose "temperatures" differ [57, 77]. So-called "hot" chains have a high acceptance probability of inferior transition proposals. To stay in the above picture, they are used for a more global exploration of the landscape since their affinity to areas of high altitude is low. In contrast, "cold" chains with their low acceptance probability of inferior hypotheses are used for a thorough exploration of local areas in the landscape. Hypothesis sampling is done only from the cold chain. However, from time to time, the temperatures of the chains are swapped such that a hot chain is turned into a cold chain and *vice versa*. However, many more variants of MCMC sampling of phylogenies exist and the field is quickly evolving [33,74–76,95].

3.6 Rooting Trees/Molecular Clock

So far we have introduced various methods of inferring the relationships between sequences (or taxa). Unfortunately, most of the methods described above lack an inherent criterion for assigning directionality to the evolutionary process. As a consequence, they are unable to identify the root of a phylogenetic tree. To obtain a rooted tree, nevertheless, it is required (and possible) to add supplementary information into the tree reconstruction procedure.

3.6.1 Outgroup Rooting

Among the various methods for rooting a tree, it is most intuitive to divide the taxa into two subgroups: a monophyletic ingroup, i.e. taxa that share a common ancestor to the exclusion of all other taxa in the dataset, and an outgroup, whose more distant relationship to any member of the ingroup is either known or at least reasonable to assume (Figure 11a). It is then straightforward to conclude that the node that joins the outgroup to the ingroup represents the root of the ingroup subtree (r_{subtree} in Figure 11) [110, 140]. Though simple, this approach requires some considerations. Despite their clear position outside the ingroup, outgroup taxa should be as closely related to the ingroup taxa as possible. This will increase the probability of

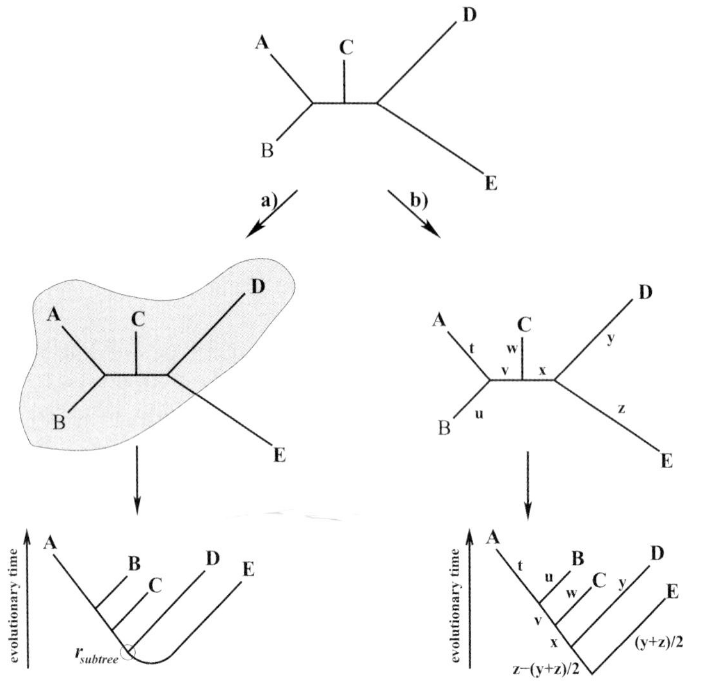

Figure 11 Two alternative principles for rooting phylogenetic trees. (a) Outgroup rooting. The set of taxa A–E is divided into an ingroup (shaded in grey) and the outgroup, taxon E. The node that joins the outgroup to the ingroup represents the root of the ingroup-subtree (r_{subtree}). The root of the entire tree, i.e. the common ancestor of all taxa, must be located somewhere on the outgroup branch. However, its exact position remains unknown. (b) Distance-based rooting. t to z denote the lengths of the individual branches in the tree. The root of the entire tree is identified as the midpoint of the path connecting the two taxa with the largest evolutionary distance.

reliably identifying homologous sequence positions using standard alignment procedures. Furthermore, it minimizes the risk of misplacing the outgroup due to its large evolutionary distance from the ingroup [99, 127, 154]. In addition to these more general requirements, some additional guidelines exist for rooting phylogenetic trees by an outgroup. First, more than one taxon should be included into the outgroup [100]. Furthermore, different outgroup taxa should be used to check whether the root placement depends on the choice of the outgroup [150].

3.6.2 **Midpoint Rooting and Molecular Clock**

As we have seen, the choice of a meaningful outgroup for rooting a phylogenetic tree can become a considerable problem. This is especially relevant when groups are analyzed whose phylogenetic relationships are unclear. In such cases additional assumptions about the evolutionary process are imposed that help rooting the tree.

Given that per unit of time any lineage accumulates the same amount of sequence changes (molecular clock) the point in the tree that is equally distant from all terminal taxa can be assigned as the root (see Section 3.2). In reality, however, the assumption of a molecular clock is frequently violated. If this is neglected rooting under the clock assumption tends to place the root in a part of the tree that is evolving at a high evolutionary rate. Midpoint rooting slightly relaxes the constraints imposed by the molecular clock assumption. It places the root on the midpoint of the path connecting the two most distantly related taxa in the phylogenetic tree (Figure 11b). Compared to the molecular clock scenario this retains only the postulation that the evolutionary rate has to be the same along the two most divergent lineages in the dataset. Midpoint rooting identifies the localization of the root correctly when this criterion is met [148].

4 Finding the Optimal Tree

So far we have outlined the principles to construct a phylogenetic tree from a set of aligned sequences. However, it still is unclear how to find the tree that reflects the relationships between the taxa best. We can differentiate between two general concepts of searching the tree space comprised by all possible tree topologies for the desired optimal tree: (i) *exhaustive* searches, which guarantee the identification of the optimal tree, and (ii) the computationally less-demanding *heuristic* searches that, however, do not necessarily obtain the globally optimal tree.

4.1 Exhaustive Search Methods

In the conceptually simplest approach, the exhaustive search, each and every possible bifurcating tree in the tree space is evaluated under the selected optimality criterion. The identification of the optimal tree(s) is then straightforward and the computational challenge is limited to exploring all of the tree space. To accomplish this, one starts with the (unique) unrooted tree that connects three randomly chosen taxa from the dataset. Subsequently, the remaining taxa are added in a step-wise fashion, such that the ith taxon is added separately to each of the $2i - 5$ branches of every possible tree for the $i - 1$ previous taxa. Obviously, the addition of every taxon increases the number of possible trees by the number of branches to which the new taxon can be connected [41]. Thus, the total number of unrooted trees for a set of n

taxa is:

$$B(n) = \prod_{k=3}^{n}(2k-5) = \frac{(2n-5)!}{2^{n-3}(n-3)!}. \tag{15}$$

The limitations of the exhaustive search are evident. Already a compilation of 20 taxa, a dataset that is nowadays easily exceeded, requires the evaluation of over 2×10^{20} different trees. This is, and presumably will remain, computationally infeasible.

Branch-and-bound methods [93] provide an alternative approach to finding a globally optimal solution without the need of evaluating all tree topologies. Instead, a guided search in the tree space is performed omitting those subspaces that cannot contain an optimal tree [67]. The rationale is simple and has the only prerequisite that the criterion of tree evaluation, i.e. the objective function F, is nondecreasing when new taxa are added to a particular subtree. If we want to minimize F, then we start with the computation of an upper bound F_{upper}, e.g., we evaluate any arbitrary n-taxon tree. Subsequently, using a three-taxon tree as a primer we recursively reconstruct the possible n-taxon trees. However, as we move along in our reconstruction procedure, i.e. with the addition of more and more taxa into the trees, we compare F of the resulting subtrees with F_{upper}. As soon as F_{subtree} exceeds F_{upper} we know that the search path leads to a subspace which contains trees where F is always larger that F_{upper}. Thus, no further reconstruction is required and another search path is evaluated. Alternatively, if we end up with a n-taxon tree, we store the new tree as candidate and update F_{upper} to the new value. The estimation of F_{upper} is crucial for the computational efficiency. Therefore, a number of improvements have been added to this basic scheme [67, 148]. These refinements are mainly designed to further reduce the exploration of tree space. They include methods for obtaining a near-optimal tree for an assessment of the initial upper bound, as well as schemes for generating a suitable order in which the taxa are added to the subtrees. For instance, by adding divergent taxa first, the length of the initial subtrees is increased, allowing for a quicker identification of subtrees that exceed the upper bound for the tree length.

Despite these improvements, exact searches eventually run into computational problems when data sets become large. For these cases, the considerably faster heuristic methods for tree reconstruction are required.

4.2 Heuristic Search Methods

Heuristic methods for tree reconstruction earn a substantial speed-up in computation time by jettisoning a guaranteed globally optimal solution to the tree search problem. With contemporary software it is possible to reconstruct trees from datasets of more than 1000 taxa (e.g. [63, 107, 141]). Thus, nowadays biological datasets hardly ever reach the computational limits of tree reconstruction software, provided that one is willing to abandon the guaranteed optimality.

4.2.1 Hill Climbing and the Problem of Local Optimization

The problem to find an optimal tree for a set of taxa can again be illustrated by the metaphor of exploring a landscape. A hiker aims to visit the point with the highest altitude in a hilly area. Due to the poor visibility, the highest peak cannot be identified *a priori*. Thus, the hiker remains with the only option to climb any slope he encounters first until he has reached its top. Up there he checks his altimeter and is either confident to have reached one of the highest points in this area and finishes his search or he invests more effort and climbs another hill. This kind of search strategy is called a local search.

Applying this approach to the tree search problem, we start off with any tree and modify it in a stepwise fashion, usually accepting only such modifications that result in an improved tree according to the chosen objective function. At a certain point no further improvement is possible and thus we have reached the top of the hill. At this point we have no means of deciding whether we have found the globally optimal tree or merely a local optimum and an optimization with a different initial tree would obtain better results. Consequently, tree searches based on a local search have to cope with three challenges: (i) the identification of a reasonable tree to start the search with, (ii) the implementation of a stepwise hill-climbing algorithm for the tree search and (iii) the avoidance of getting stuck in local optima that are highly suboptimal in terms of cost.

4.2.1.1 Identification of the Starting Tree

Reasonable starting trees are quickly obtained via so-called "greedy" strategies. The tree reconstruction is divided into several subproblems which are then sequentially solved by always choosing the solution that looks best given the current situation. In star decomposition methods (Figure 12), we begin with an assignment of all taxa to the terminal nodes of a star-like tree. Subsequently, all trees are evaluated that can be obtained by joining any two of the terminal taxa into a new group. The tree that scores best under the chosen optimality criterion forms the basis for the next step. The iteration of pairwise joining and tree evaluation continues until the tree is fully resolved.

Alternatively, we can directly construct a binary tree from scratch by inserting the taxa into a tree in a stepwise fashion (Figure 13) [38]. First, a set of three taxa is used to form a unique binary tree. Next, a fourth taxon is chosen for insertion into the initial tree. Since the taxon can be attached on any of the three branches of the initial tree, we have three possible topologies for the

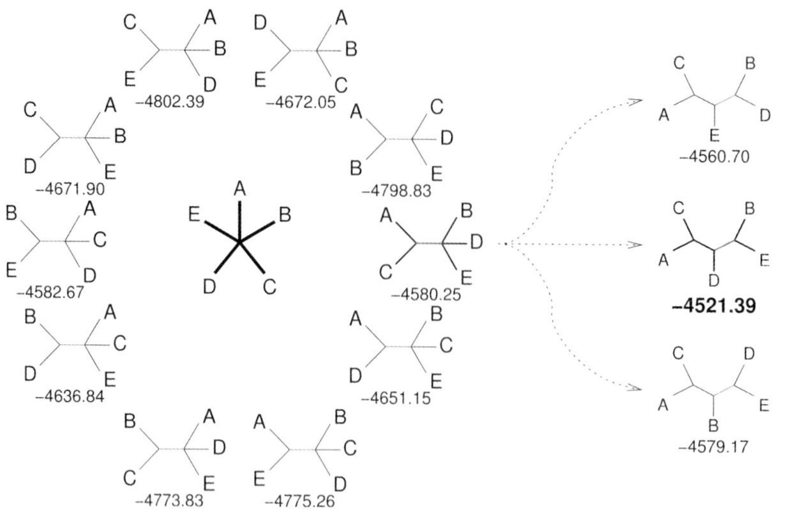

Figure 12 Star decomposition.

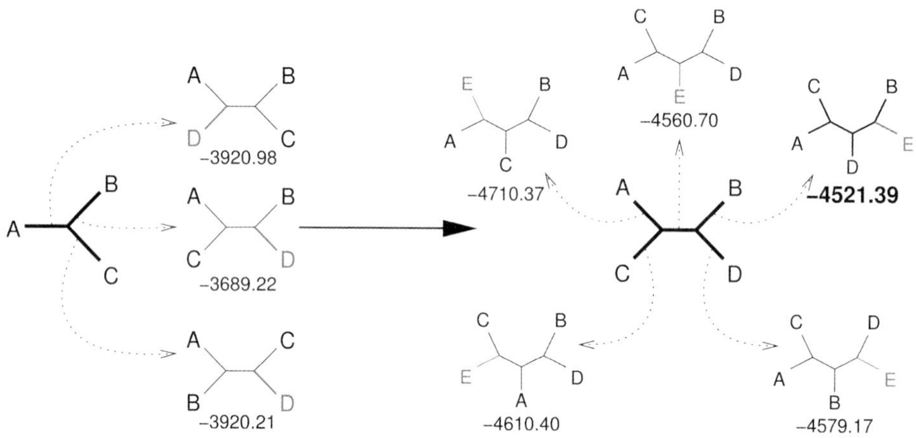

Figure 13 Stepwise insertion.

four-species tree. All of these will be evaluated and the best tree will be stored for insertion of the fifth taxon. The iteration continues until the tree includes all taxa in the dataset.

It is straightforward to see why both star decomposition methods as well as the stepwise insertion procedure are prone to obtaining only locally optimal trees. Any decision concerning the position of a taxon in the tree is fixed for the remaining part of the reconstruction procedure.

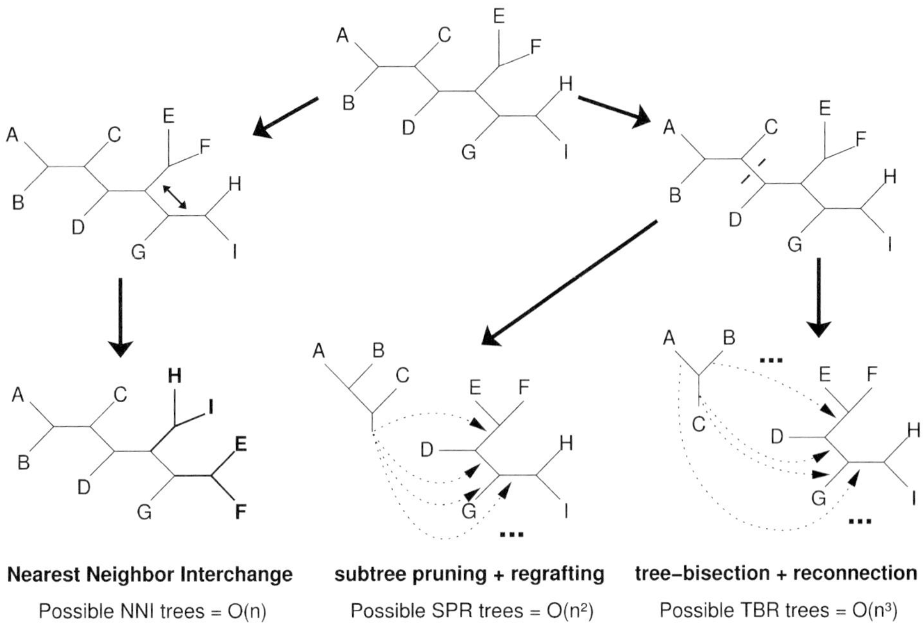

Figure 14 Three methods that accomplish branch swapping.

4.2.1.2 **Optimization Procedure and Avoidance of Local Optima**

In order to escape local optima, tree-rearrangement methods were suggested that override previous decisions concerning the placement of taxa in the tree. In brief, the initial "optimal" tree is modified such that a part of the tree is excised and re-inserted elsewhere. The trees resulting from such "branch swaps" are evaluated and subjected to one or more acceptance criteria. While a better tree is always accepted, trees inferior to the one already obtained can be accepted under certain conditions [77]. This deviation from the strict hill-climbing approach facilitates the transition to better trees that require more than one rearrangement of the current best tree. Note, the similarity to the MCMC approach (see Section 3.5). Currently, three branch-swapping methods are in use (Figure 14). Nearest neighbor interchange, the simplest approach, takes any internal branch of the tree and swaps two of the four connected subtrees. In this way, a total of $O(n)$ alternative trees are evaluated. (Note that only swapping two subtrees located on the opposite sides of the internal branch leads to the formation of a new tree!) Subtree pruning and regrafting ($O(n^2)$) excises a subtree and regrafts it with the cut surface at any branch on the tree. Tree bisection and reconnection is the most exhaustive way of swapping branches ($O(n^3)$). The tree is bisected along an internal branch and the resulting subtrees are rejoined at any pair of branches.

As noted, any of the branch-swapping methods is capable of guiding the tree-reconstruction procedure out of a local optimum. However, no guarantee is given that this does not simply lead into the next local optimum. Apparently, if the branch swapping is continued for a sufficient amount of time it becomes likely that sooner or later the global optimum will be found. However, how shall we recognize the globally optimal tree once we found it and for how long should we continue the tree search?

4.2.2 Modeling Tree Quality

It is inherent in the heuristic approach that, no matter how long we search, we can never be sure that we have found the globally optimal tree. Thus, we need a concept of tree quality, as we are continuing the search. In most cases it is essentially up to the user how long the search is continued. Either a predefined number of optimization steps or a lower limit by which new trees have to improve can be used as stopping criteria. However, both criteria are arbitrary and a well-founded basis for deciding when to end the search would be desirable. Recently, a method was suggested that is based on the rate of finding better trees during the search [152]. Let $F_1, F_2, \ldots, F_j$ denote the values of the objective function F for the trees found at iteration $1, 2, \ldots, j$. Then the sequence $r(k)$ of record times (i.e. number of iteration at which a better tree is found) is defined by:

$$r(1) = 1,\ r(k+1) = \min\{j | F_j > F_{r(k)}\}. \tag{16}$$

This sequence is used to estimate the point in time, r_{stop}, when to stop the search based on the probability of yet finding a better tree. Using the theory detailed in Refs. [23,120], one can estimate on the fly an upper 95% confidence limit $r_{95\%}$ of r_{stop}. Once $r_{95\%}$ iterations have been carried out and a better tree has not been detected the program will stop. It can then be concluded that with a probability of 95% no better tree will be found during this search. On the other hand, if a better tree is found before $r_{95\%}$ is hit, the $r_{95\%}$ is re-estimated on the basis of the new record time added to the sequence $r(k)$ and the search continues.

4.2.3 Heuristics for Large Datasets

The considerable ease with which DNA sequences are obtained nowadays results in ever-increasing datasets available for phylogeny reconstruction. As a consequence there is a demand for increasing the capacity of tree reconstruction software. One way to satisfy the needs is the development of parallelized versions of tree reconstruction programs, e.g. fastDNAml-based programs [111,141,142], TREE-PUZZLE [132], GAML [14] and MRBAYES [2].

The objectives for further improvements on the computational basis can be quickly summarized. (i) Finding in a shorter time a better starting tree for subsequent optimization. IQPNNI [152] accomplishes this by limiting the number of computation steps to place a new taxon during tree reconstruction. PhyNav [151], on the other hand, reduces the initial tree space by choosing for each group of closely related taxa one representative. From the resulting representative leaf set a scaffold is reconstructed, to which the initially deferred taxa are subsequently added such that an optimal tree is obtained. (ii) Improving the algorithms for tree optimization. For instance, PHYML [63] has implemented a fast algorithm for nearest-neighbor interchange, and RAxML [141] provides an improved version for subtree pruning and regrafting. (iii) The utilization of alternative approaches for tree reconstruction, such as a metapopulation genetic algorithm [96]. (iv) The dissection of the tree-reconstruction problem into a set of subproblems that can be solved on several CPUs in parallel. Some of these improvements are recent developments and it is not clear yet which combination will be optimal for tree reconstruction. In a sense an all-embracing optimal solution might be elusive since it is likely that different combinations will perform optimally on different data sets. Thus, it seems impossible to provide guidelines for when to use what program.

5 The Advent of Phylogenomics

A common problem for the accurate reconstruction of evolutionary relationships among taxa is the limited amount of phylogenetic signal in the data which, in addition, is frequently blanketed by noise. In view of the various genome sequencing efforts it seems trivial to enhance the signal-to-noise-ratio by the simple addition of more data [69]. However, even with the availability of whole-genome sequences, alignments remain limited to the level of individual genes in many cases. Both the rearrangement of genetic information in different taxa and the in part substantial sequence divergence of nonfunctional parts of the genome prevent the generation of meaningful longer sequence alignments. To extend the amount of information, nonetheless, disjoint datasets derived from multiple genomic loci can be combined for the analysis. This intersection of phylogenetics and genomics is referred to as phylogenomics.

5.1 Multilocus Datasets

Two approaches have been suggested for combining multilocus datasets from the same set of taxa for phylogenetics analysis. In supermatrix approaches [126] (also referred to as "total evidence" [89]) all individual sequence alignments are concatenated to form one large superalignment. The tree reconstruction is then based on this superalignment using standard methods. In

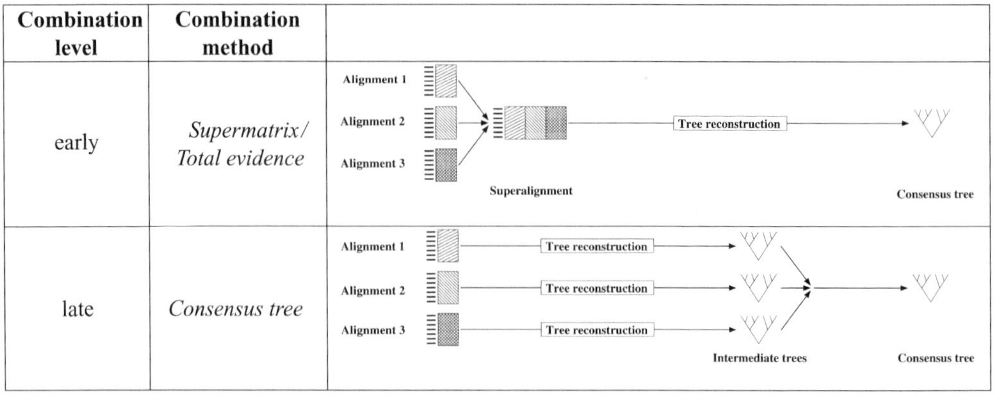

Figure 15 Two alternative methods to reconstruct a single phylogenetic tree from a set of disjoint alignments. In the early-level combination, the individual alignments represented by the patterned boxes are concatenated first to form a single superalignment. Standard phylogeny reconstruction programs can then be applied to reconstruct a tree from the superalignment. In the late-level combination a phylogenetic tree is reconstructed first for each alignment separately. The individual trees are later combined into a single consensus tree.

this approach, the phylogenetic information present in the individual alignments is combined early in the phylogenetic analysis. Hence, we refer to them as *early-level combination* methods (Figure 15, "early"). Alternatively, the information present in the individual alignments can be combined late in the phylogenetic analysis. Trees are reconstructed first for each alignment separately. These individual trees are then combined at a later step to form a so-called *consensus tree* (Figure 15 "late"). However, in contrast to the concatenation of individual alignments, which is simple text-editing, the combination of trees requires some further considerations.

A frequently used method for computing a consensus from a compilation of trees is based on the principle of identifying the set of compatible splits among these trees. To this end, splits comprise bipartitions of the taxon set that are induced by cutting a tree at any edge. More formally, splits are represented by the symbol "|" (Figure 16). Note that cutting at an external edge creates only trivial splits present in all trees. These are usually discarded from the analysis. Thus, we can induce for any tree with taxon set $\mathcal{N}$ a split $\mathcal{A}|\mathcal{B}$, such that $\mathcal{A} \cup \mathcal{B} = \mathcal{N}$ and $\mathcal{A} \cap \mathcal{B} = \emptyset$.

From the tree in Figure 16 we deduce the splits $\{A, B\}|\{C, D, E\}$ and $\{A, B, C\}|\{D, E\}$ (or shorter $AB|CDE$ and $ABC|DE$) We note that taxon C has changed sides. Thus, if we compute all four possible intersections between the splits only one will be empty. More formally, two splits $\mathcal{A}|\mathcal{B}$ and $\mathcal{C}|\mathcal{D}$ are said to be compatible if one of the four possible intersections $\mathcal{A} \cap \mathcal{C}$, $\mathcal{A} \cap \mathcal{D}$, $\mathcal{B} \cap \mathcal{C}$, $\mathcal{B} \cap \mathcal{D}$ is empty. If two intersections are empty the splits are identical. It is

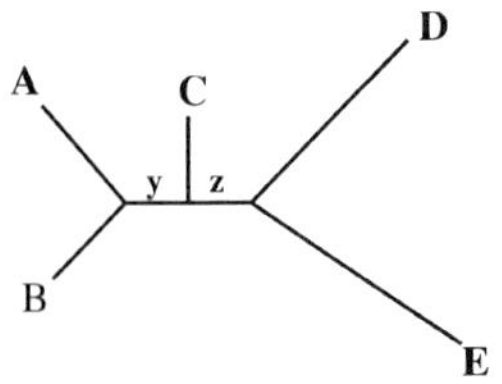

Figure 16 Two nontrivial splits can be derived from this tree. Cutting at the edge y induces the split $AB|CDE$. Cutting at edge z moves taxon C from the right-hand side of the split to the left-hand side and results in $ABC|DE$.

easy to see that splits derived from a tree are always pairwise compatible. On the other hand, a collection of splits that are pairwise compatible fit on a tree. Hence, collections of pairwise compatible splits are another way of encoding trees. For multilocus data the resulting trees are not necessarily the same and, thus, one needs approaches to summarize the results. The easiest form to summarize the result is simply counting the fraction ℓ at which a certain split occurs in a set of trees. If we collect only splits with $\ell > 50\%$, then the resulting system of splits is pairwise compatible and therefore representable as a tree [136] which we call M_{50} or majority rule consensus tree [102] (Figure 17). The cutoff value ℓ can of course be raised to construct more stringent majority consensus trees M_ℓ [102].

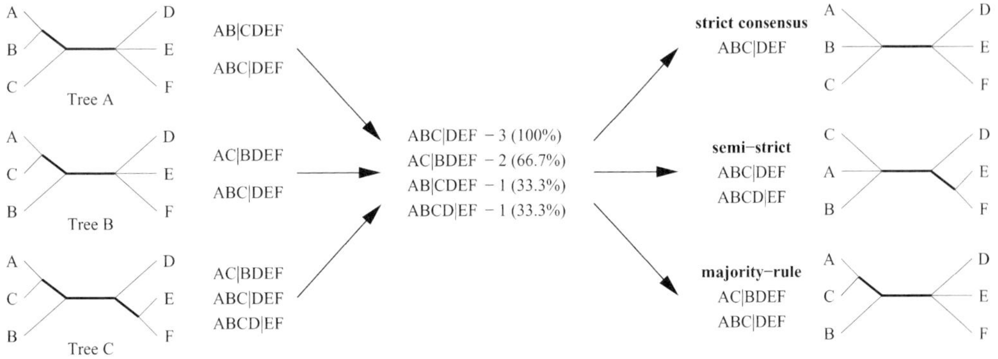

Figure 17 Examples for consensus methods to summarize a set of trees with identical taxon sets: strict (M_{strict}), semi-strict ($M_{\text{semi-strict}}$) and 50% majority rule consensus (M_{50}).

More restrictive cases of majority consensus are the strict consensus M_{strict} [122] that incorporates only splits present in all trees, and the semi-strict consensus $M_{\text{semi-strict}}$ [15] that contains all splits which are not contradicted by any split from the input trees (Figure 17). Many further methods exist for generating a consensus tree (e.g. Refs. [1,19,81,133]).

The application of consensus methods extends beyond the combination of trees from multilocus datasets. In principle, they can be used to summarize any set of trees, e.g., derived from Jackknife [106] or Bootstrap analysis [35, 44], sampled from MCMC simulations [77], or obtained by randomized input orders [144] to assess the reliability (or uncertainty) in the reconstructed trees.

5.2 Combining Incomplete Multilocus Datasets: Supertrees and their Methods

Consensus methods have one serious limitation – they are restricted to trees of equal size and taxon sets. The mutual coverage of currently available gene sequences and taxa is far from being satisfactory [32]. Thus, consensus methods are only applicable to very special and restricted multilocus data. This results in a trade-off between the number of taxa and the number of loci used in an analysis. Thus, one can study either many loci with only few taxa or vice versa. This situation will improve as more sequence data accumulate, especially in the wake of completely sequenced genomes. However, incomplete data will still remain simply because not all genes are present in the genomes of all taxa. Consequently, the question emerges of how to incorporate multiple incomplete datasets into phylogenetic analysis.

In principle, similar strategies are applicable as outlined in Section 5.1. Supermatrix methods use concatenated alignments. However, this requires that tree reconstruction methods must be able to handle the missing data. Simply discarding alignment positions with gaps would leave the user with only completely sampled loci or even no data at all.

When the data is combined at a late level in the analysis, several strategies are feasible. Methods have been proposed to combine separately reconstructed overlapping (typically rooted) trees of the different loci into one so-called supertree [11, 60] (Figure 18). Supertree approaches can be divided into two classes: agreement supertrees [12] and optimization supertrees [155].

5.2.1 Agreement Supertrees

Agreement supertree methods reconstruct a supertree based on those groupings that are shared or at least are uncontested among the set of rooted source trees [12]. This reflects the assumption that all source trees can in principle be obtained simply by pruning different sets of branches from one large tree, the parent tree, i.e. the source trees are compatible. It should, therefore, be straightforward to reconstruct the topology of the parent tree from the topologies of the source trees. Unfortunately, different parent trees may frequently lead to the same set of partial trees. In other words, agreement supertree reconstruction may result in different parent trees. The first supertree method available [60] was designed to find all possible parent trees

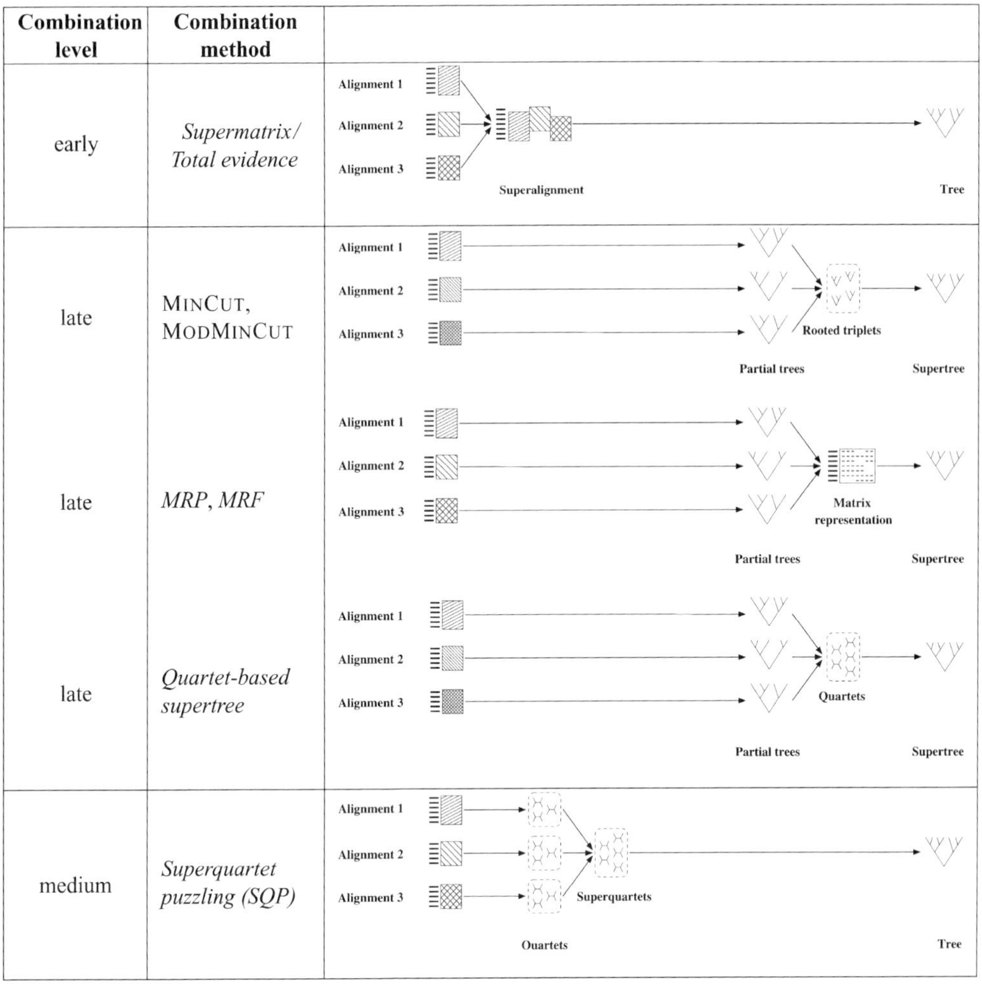

Figure 18 A number of different methods to construct a single phylogenetic tree from a set of alignments with incomplete, but overlapping taxon sets. In early-level combination all alignments are concatenated into one large (super)alignment (or supermatrix, missing sequences are filled with gaps) from which the tree is reconstructed. In late-level combination the (typically rooted) trees are decomposed into sub-structures like rooted triplets (to obtain common nestings) or quartets, or are re-coded into a binary matrix representation (see Figure 19). These are then used to reconstruct a supertree. In medium-level combination with SQP the data is combined via quartets computed from each alignment. The resulting superquartets are then amalgamated into an overall tree.

for a set of partial trees and compute the strict consensus from the different parent (see Section 5.1). The resulting supertree, however, displays only those bipartitions that are supported in all parent trees, but some information about the structure present only in a fraction of the parent trees might be concealed.

Thus, subsequent approaches like OneTree [17] returned only a single possible parent tree as the supertree. Obviously, the strict requirement of the source trees being compatible severely limits the applicability of these supertree methods. Any real application leads to source trees that are incompatible for a variety of reasons, one of which is just by chance. Thus, subsequent agreement supertree methods, such as the MinCut Supertree algorithm [135] or the ModMinCut Supertree algorithm [112] aimed to overcome the requirement of compatibility. In brief, they introduced a weighting of the links between taxa (i.e. common occurrence in the same subtrees) during the reconstruction of supertrees, such that the weight of a link increases the more source trees display this link. Subsequently, if a subtree cannot be resolved further due to incompatible source trees, the subtree is resolved by pruning those links with the lowest weight (MinCut) greedily. Furthermore, the ModMinCut Supertree algorithm [112] aims to keep links that are uncontradicted, even if they are established only by a single input tree, which would cause the MinCut Supertree algorithm to discard it. Further agreement supertree methods have been suggested recently and a comprehensive overview is given in Refs. [10,12].

5.2.2 **Optimization Supertrees**

Here, the set of input trees is decomposed into smaller entities. These entities serve as input to reconstruct an overall tree based on an objective function.

Matrix representation methods are one example. Prior to constructing the supertree, the rooted input trees are encoded into a binary matrix. Typically each internal node in the (rooted) input trees is encoded either by its adjacent subtree (Ragan/Baum scheme [7, 119], Figure 19a) by assigning "1" to taxa within the subtree and "0" otherwise, or its adjacent sister groups (Purvis' scheme [118], Figure 19b) assigning "0" to the taxa in one sister group and "1" to the other. All other and missing taxa of the tree are assigned "?". The obtained matrix representation of the input trees is then used as input alignment to reconstruct a supertree (Figure 18). For this purpose, various optimizing algorithms can be applied, such as (i) parsimony (MRP [7, 119]), which is to date the by far most common method, (ii) distance-based methods (MRD or average consensus method [94]), and (iii) finding the optimal tree which requires the least changes between ones and zeros (flips) to be congruent with the matrix (MRF [22]).

As an alternative to the matrix representation method, quartet-based supertree methods make direct use of the topological information in the input trees. To this end, the source trees are decomposed into quartet trees, which then serve as building blocks to reconstruct of the supertree [115, 121] (Figure 18).

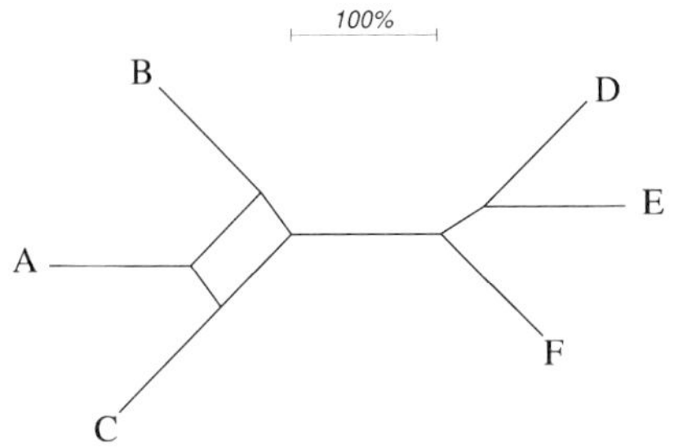

Figure 20 Consensus network representing all four splits collected from the three input trees in Figure 17. Branch lengths are drawn relative to the percent occurrence in the input trees. The compatible splits $ABC|DEF$ and $ABCF|DE$ form the tree-like branches in the right part of the graph, while the pairwise incompatible splits $AB|CDEF$ and $AC|BDEF$ form the net structure on the left. Note that the contraction of one set of parallel branches each obtains the corresponding tree responsible for the incompatible split.

One way to visualize incompatible splits present in the data goes along the lines with these consensus methods. Instead of stopping at the 50% cutoff which guarantees that the outcome is, in fact, a tree, one keeps adding less frequent splits obtained from a set of input trees to the split system, i.e. the set of splits. More generally, the application of a cutoff value r (analogous to ℓ in consensus trees) allows the selection of any split system S_r that is present in at least a portion r of all input trees. Pushing r below 50% ($r \leq 0.5$) may lead to a splits system that no longer conforms to a tree. Visualizing such incompatible splits systems provides insights into the extent and pattern of heterogeneity of the phylogenetic signal in the data. In such a network, one split is represented either by a single branch or by parallel branches, indicating incompatible splits as in Figure 20, where $AC|BDEF$ and $AB|CDEF$ are incompatible.

It has been shown (see Ref. [73]), that a split system S_r with cutoff fraction r does contain any subset larger than $\lfloor 1/r \rfloor$ splits which are all pairwise incompatible. S_r is said to be $(\lfloor 1/r \rfloor)$-compatible. The split system in Figure 20, for example, is 2-compatible, containing the subset of two pairwise incompatible splits $AC|BDEF$ and $AB|CDEF$. All split systems S_r with $r > 0.5$ are 1-compatible, which means that there are no incompatible splits and, hence, the resulting topology would again be a tree.

The amount of pairwise incompatible splits obviously determines the complexity of the network containing them. Median networks [4, 72] can contain cubes of dimension up to $\lfloor 1/r \rfloor$, and might thus be utterly complex. For example, a split system $S_{0.25}$ can be 4-compatible and, hence, needs four dimensions to be visualized.

Median networks are a very general type of network which can be reconstructed from the binary encoding of a split system. To this end, for each split taxa on one side of the split are assigned ones, those on the other side zeros. Then, intermediate states (representing the inner nodes in the median network) are computed from the binary sequences in a parsimonious fashion (see Section 3.1). It has been shown [4,72] that by pruning branches from a median network one can extract all the most parsimonious trees for the split system. Due to the fact that median networks can grow arbitrarily incomprehensible, less-complex approximations are often applied. Split graphs [6, 30, 31], for example, attempt to filter the splits and branches drawn, to derive a planar graph, i.e. a graph without intersecting edges. Refer to Ref. [72] for a more detailed overview.

6.1.2 **Constructing Split Systems from Trees**

Commonly, split systems are collected from a set of input trees with equal taxon sets as they are obtained, for example, from Bootstrap analysis or MCMC sampling (see Section 3.5). Similar to consensus trees (see Section 5.1), such split systems are visualized as so-called *consensus networks* [70]. Such consensus networks (see Figure 20) visualize the area and the extent of contradiction of the phylogenetic signal found in the input trees. However, like supertrees (Section 5.2), network reconstruction is not restricted to trees with equal taxon set, but can also be done from overlapping trees using the Z-closure method [78]. In accordance to the amalgamation of trees to supertrees, such networks are then called "super-networks" (Note, that the "super" prefix in super-networks does not follow the same notation as in supertrees, super-alignments, or superquartets, since is not network constructed from networks, but from trees.)

6.1.3 **Constructing Split Systems from Sequence Data**

Although applications such as consensus networks and super-networks were suggested quite recently, one should note that the basic idea of representing evolutionary processes by networks rather than trees is not new.

In contrast to approaches for network reconstruction based on collections of splits derived from trees, distance-based methods including split decomposition [6] or Neighbor-Net [18] have been suggested for constructing split systems directly from the data.

Applied to viral sequences from Ref. [124], methods like split decomposition and Neighbor-Net can easily identify the presence of the three recombinant HIV strains SE7812_2, UG266 and VI1310-1.7 (Figure 21b and c). These would have been wrongly classified based on the result of a tree reconstruction method like BioNJ (Figure 21a). To avoid such missinterpretation of the data, phylogenetic networks should be taken into account at early stages of the analysis.

Programs like SplitsTree [80], T-REX [101] NETWORK [3] and Spectronet [72] provide easy to use means that can be readily applied to a phylogenetic

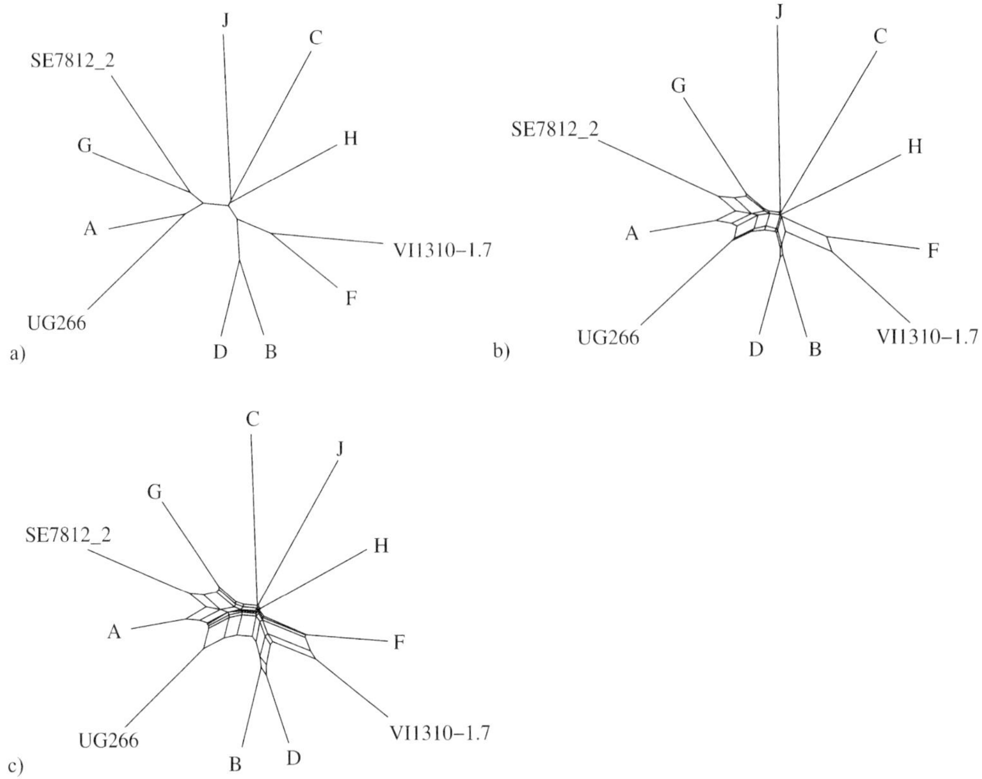

Figure 21 Results for a set of HIV dataset from Ref. [124] containing reference strains (A, B, C, D, F, G, H and J) together with three recombinants (SE7812_2, UG266 and VI1310-1.7). (a) BioNJ tree (eight splits), (b) split decomposition (14 splits) and (c) Neighbor-Net (19 splits).

analysis. These packages were used to analyze viral data [29, 124], hybridization events [98] and intra-specific data [4], respectively.

It should be noted that such split-based networks provide only a visualization of ambiguities in the data and do not qualify as methods to infer the reasons for the net-like structure [18].

6.2 Reconstructing Reticulate Evolution and Further Analyses

As mentioned above, in the case that different methods, different loci or even just different parts of the very same gene show conflicting phylogenetic signals, various causes might account for the observed conflicts.

One would certainly first check whether the conflicting evolutionary hypotheses are really significantly different [58, 66, 138]. If so, we can envi-

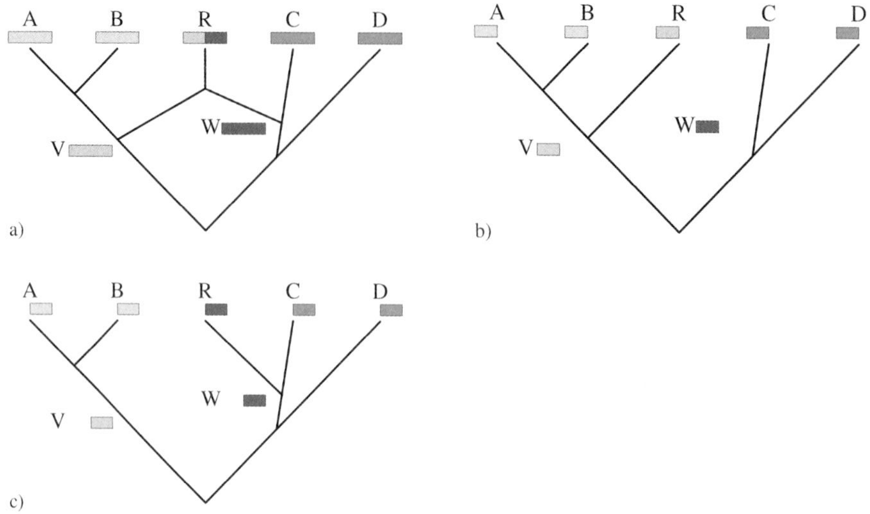

Figure 22 A case of reticulate evolution. (a) The recombination between strain V and W forms the recombinant strain R. The two parts of the sequence reflect different evolutionary histories (b) and (c) of the reticulation.

sion several biological mechanisms to produce reticulations, i.e. network-like evolution. In the particular example of virus evolution, reassortment, i.e. the mixture of viral chromosomes within a cell co-infected by different viral lineages [86], and recombination, i.e. the reciprocal exchange of genomic regions among chromosomes [113, 117] both of which are highly abundant in viruses [52, 125], are two examples. Horizontal gene transfer [8, 25, 62] and genome hybridization or fusion [28, 103, 158] constitute other possibilities. Sometimes reticulations may be simply due to parallel substitutions in different organisms. In such a situation the loops in the network indicate the occurrence of reverse or parallel mutations.

In general, purely split-based networks are not sufficient to illustrate an event of reticulate evolution like the recombination shown in Figure 22, because one cannot represent this evolutionary history as a set of splits, although one could map the separate gene trees. Hence, recently network methods like hybridization networks, recombination networks or galled trees [64, 79] have been suggested to reconstruct reticulate evolution.

The pros and cons of the different algorithms still need to be evaluated. However, the comparison of different phylogenetic networking strategies is not easy. From Figure 21 it is apparent that Neighbor-Net is more liberal in introducing noncompatible splits (19 splits in the Neighbor-Net compared to 14 splits in the split decomposition network and eight splits in a fully resolved tree.) Simulation studies, successfully applied in phylogenetic inference, might be one way to evaluate accuracy and robustness of phylogenetic network methods. To this end a carefully designed experimental setup is required. An alternative approach is the introduction of optimality criteria (least-squares, likelihood) in the network construction process. Unfortunately, the development of phylogenetic networks in a likelihood framework is still in its infancy [143, 146, 153]. Only with such methods is one able to decide whether the nontree-like signals are biologically plausible or are merely an artifact of the reconstruction procedure.

In this chapter, we have only touched upon some phylogenetic network methods. Space limitations do not allow a full account of all existing methods. The pyramidal clustering technique [26] is, like Neighbor-Net, an agglomerative approach. Statistical geometry in sequence space [36] and its descendants [109, 145] are alternative approaches to summarize the extent of tree-likeness in the data without reconstructing phylogenetic networks.

The field of phylogenetic networks will certainly profit from the data produced in whole-genome projects. As independently evolving DNA segments of eukaryotic genomes may display different evolutionary histories, the correct history of the taxa, as carriers of these segments, is probably more adequate. However, before we can reliably do this, we need to distinguish true loops in a network from artificial loops generated by too simple assumptions about the evolutionary process.

References

1 ADAMS III, E. N. 1986. *N*–trees as nestings: complexity, similarity, and consensus. J. Classif. **3**: 299–317.

2 ALTEKAR, G., S. DWARKADAS, J. P. HUELSENBECK AND F. RONQUIST. 2004. Parallel Metropolis coupled Markov chain Monte Carlo for Bayesian phylogenetic inference. Bioinformatics **20**: 407–15.

3 BANDELT, H.-J., P. FORSTER AND A. RÖHL. 1999. Median-joining networks for inferring intraspecific phylogenies. Mol. Biol. Evol. **16**: 37–48.

4 BANDELT, H.-J., P. FORSTER, B. C. SYKES AND M. B. RICHARDS. 1995. Mitochondrial portraits of human populations using median networks. Genetics **141**: 743–53.

5 BANDELT, H.-J. AND A. W. M. DRESS. 1986. Reconstructing the shape of a tree from observed dissimilarity data. Adv. Appl. Math. **7**: 309–43.

6 BANDELT, H.-J. AND A. W. M. DRESS. 1992. Split decomposition: a new and useful approach to phylogenetic analysis of distance data. Mol. Phylogenet. Evol. **1**: 242–52.

7 BAUM, B. R. 1992. Combining trees as a way of combining data sets for phylogenetic inference, and the desirability of combining gene trees. Taxon **41**: 3–10.

8 BERGTHORSSON, U., K. L. ADAMS, B. THOMASON AND J. D. PALMER. 2003. Widespread horizontal transfer of mitochondrial genes in flowering plants. Nature **424**: 197–201.

9 BESAG, J. AND P. J. GREEN. 1993. Spatial statistics and bayesian computation (with discussion). J. R. Statist. Soc. B **55**: 25–37.

10 BININDA-EMONDS, O. R. P., K. E. JONES, S. A. PRICE, M. CARDILLO, R. GRENYER AND A. PURVIS. 2004. Garbage in,

garbage out: data issues in supertree construction. In BININDA-EMONDS O. R. P. (ed.), *Phylogenetic Supertrees: Combining Information to Reveal the Tree of Life*. Kluwer, Dordrecht: 267–80.

11 BININDA-EMONDS, O. R. P. 2004. *Phylogenetic Supertrees: Combining Information to Reveal the Tree of Life*. Kluwer, Dordrecht.

12 BININDA-EMONDS, O. R. P. 2004. The evolution of supertrees. TREE **19**: 315–22.

13 BININDA-EMONDS, O. R. P. 2004. Trees versus characters and the supertree/supermatrix "paradox". Syst. Biol. **53**: 360–1.

14 BRAUER, M. J., M. T. HOLDER, L. A. DRIES, D. J. ZWICKL, P. O. LEWIS AND D. M. HILLIS. 2002. Genetic algorithms and parallel processing in maximum-likelihood phylogeny inference. Mol. Biol. Evol. **19**: 1717–26.

15 BREMER, K. 1990. Combinable component consensus. Cladistics **6**: 369–72.

16 BRUNO, W. J., N. D. SOCCI AND A. L. HALPERN. 2000. Weighted neighbor joining: a likelihood-based approach to distance-based phylogeny reconstruction. J. Mol. Evol. **17**: 189–97.

17 BRYANT, D. AND M. STEEL. 1995. Extension operations on sets of leaf-labeled trees. Adv. Appl. Math. **16**: 425–53.

18 BRYANT, D. AND V. MOULTON. 2004. Neighbor-Net: an agglomerative method for the construction of phylogenetic networks. Mol. Biol. Evol. **21**: 255–65.

19 BRYANT, D. 2003. A classification of consensus methods for phylogenetics. Bioconsensus: Proc. of Tutorial and Workshop on Bioconsensus II DIMACS-AMS, Piscataway, NJ: 55–66.

20 BUNEMAN, P. 1971. The recovery of trees from measurements of dissimilarity. In HODSON F. R., D. G. KENDALL AND P. TAUTU (eds.), *Mathematics in the Archeological and Historical Sciences*. Edinburgh University Press, Edinburg: 387–95.

21 CAVALLI-SFORZA, L. L. AND A. W. F. EDWARDS. 1967. Phylogenetic analysis: models and estimation procedures. Evolution **21**: 550–70

22 CHEN, D., L. DIAO, O. EULENSTEIN, D. FERNÁNDEZ-BACA AND M. J. SANDERSON. 2003. Flipping: a supertree construction method. In JANOWITZ M. F., F.-J. LAPOINTE, F. R. MCMORRIS, B. MIRKIN AND F. S. ROBERTS (eds.), *DIMACS Series in Discrete Mathematics and Theoretical Computer Science* American Mathematical Society, Providence, RI: 135–60.

23 COOKE, P. 1980. Optimal linear estimation of bounds of random variables. Biometrika **67**: 257–8.

24 DE QUEIROZ, A., M. J. DONOGHUE AND J. KIM. 1995. Separate versus combined analysis of phylogenetic evidence. Annu. Rev. Ecol. Syst. **26**: 657–81.

25 DELWICHE, C. F. AND J. D. PALMER. 1996. Rampant horizontal transfer and duplication of rubisco genes in eubacteria and plastids. Mol. Biol. Evol. **13**: 873–82.

26 DIDAY, E. AND P. BERTRAND. 1986. An extension of hierarchical clustering: the pyramidal representation. In GELSEMA, E. S. AND KANAL, L. N. (eds.), *Pattern recognition in Practice*. North-Holland, Amsterdam: 411–24.

27 DOBZHANSKY, T. 1973. Nothing in biology makes sense except in the light of evolution. Am. Biol. Teach. **35**: 125–29.

28 DOOLITTLE, W. F. 1999. Phylogenetic classification and the universal tree. Science **284**: 2124–8.

29 DOPAZO, J., A. DRESS AND A. VON HAESELER. 1993. Split decomposition: a technique to analyze viral evolution. Proc. Natl. Acad. Sci. USA **90**: 10320–4.

30 DRESS, A., D. HUSON AND V. MOULTON. 1996. Analyzing and visualizing sequence and distance data using SplitsTree. Discr. Appl. Math. **71**: 95–109.

31 DRESS, A. W. M. AND D. H. HUSON. 2004. Constructing splits graphs. IEEE/ACM Trans. Comput. Biol. Bioinform. **1**: 109–15.

32 DRISKELL, A. C., C. ANÉ, J. G. BURLEIGH, M. M. MCMAHON, B. C. O'MEARA AND M. J. SANDERSON. 2004. Prospects for building the tree of life from large sequence databases. Science **306**: 1172–4.

33 DRUMMOND, A., A. RAMBAUT, B. SHAPIRO AND O. G. PYBUS. 2005.

Bayesian coalescent inference of past population dynamics from molecular sequences. Mol. Biol. Evol. **22**: 1185–92.

34 EERNISSE, D. AND A. G. KLUGE. 1993. Taxonomic congruence versus total evidence, and amniote phylogeny inferred from fossils, molecules, and morphology. Mol. Biol. Evol. **10**: 1170–95.

35 EFRON, B. 1979. Bootstrap methods: another look at the Jackknife. Ann. Stat. **7**: 1–26.

36 EIGEN, M., R. WINKLER-OSWATITSCH AND A. DRESS. 1988. Statistical geometry in sequence space: a method of quantitative comparative sequence analysis. Proc. Natl. Acad. Sci. USA **85**: 5913–7.

37 EWENS, W. J. AND G. R. GRANT. 2001. *Statistical Methods in Bioinformatics: An Introduction*. Springer, New York, N.Y.

38 FARRIS, J. S., A. G. KLUGE AND M. J. ECKHARDT. 1970. A numerical approach to phylogenetic systematics. Syst. Zool. **19**: 172–89.

39 FARRIS, J. S. 1977. Phylogenetic analysis under Dollo's Law. Syst. Zool. **26**: 77–88.

40 FARRIS, J. S. 1996. Parsimony jackknifing outperforms neighbor-joining. Cladistics **12**: 99–124.

41 FELSENSTEIN, J. 1978. The number of evolutionary trees. Syst. Zool. **27**: 27–33.

42 FELSENSTEIN, J. 1981. Evolutionary trees from DNA sequences: a maximum likelihood approach. J. Mol. Evol. **17**: 368–76.

43 FELSENSTEIN, J. 1984. Distance methods for inferring phylogenies: a justification. Evolution **38**: 16–24.

44 FELSENSTEIN, J. 1985. Confidence limits on phylogenies: an approach using the bootstrap. Evolution **39**: 783–91.

45 FELSENSTEIN, J. 1987. Estimation of hominoid phylogeny from a DNA hybridization data set. J. Mol. Evol. **26**: 123–31.

46 FELSENSTEIN, J. 2004. *Infering Phylogenies*. Sinauer, Sunderland, MA.

47 FERRIS, S. D., A. C. WILSON AND W. M. BROWN. 1981. Evolutionary tree for apes and humans based on cleavage maps of mitochondrial DNA. Proc. Natl. Acad. Sci. USA **78**: 2432–6.

48 FISHER, R. A. 1912. On an absolute criterion for fitting frequency curves. Mess. Math. **41**: 155–60.

49 FITCH, W. M. AND E. MARGOLIASH. 1967. Construction of phylogenetic trees. Science **155**: 279–84.

50 FITCH, W. M. 1971. Toward defining the course of evolution: Minimum change for a specific tree topology. Syst. Zool. **20**: 406–16.

51 FITCH, W. M. 1981. A non-sequential method for constructing trees and hierarchical classifications. J. Mol. Evol. **18**: 30–7.

52 FITCH, W. M. 1996. The variety of human virus evolution. Mol. Phylogenet. Evol. **5**: 247–58.

53 GASCUEL, O. 1997. BIONJ: an improved version of the NJ algorithm based on a simple model of sequence data. Mol. Biol. Evol. **14**: 685–95.

54 GATESY, J., C. MATTHEE, R. DESALLE AND C. HAYASHI. 2002. Resolution of a supertree/supermatrix paradox. Syst. Biol. **51**: 652–64.

55 GATESY, J., R. H. BAKER AND C. HAYASHI. 2004. Inconsistencies in arguments for the supertree approach: supermatrices versus supertrees of *Crocodylia*. Syst. Biol. **53**: 342–55.

56 GATESY, J. AND M. S. SPRINGER. 2004. A critique of matrix representation with parsimony supertrees. In BININDA-EMONDS, O. R. P. (ed.), *Phylogenetic Supertrees: Combining Information to Reveal the Tree of Life*. Kluwer, Dordrecht: 369–88.

57 GEYER, C. J. 1991. Markov chain Monte Carlo maximum likelihood. Proc. 23rd Symp. on the Interface Interface Foundation, Fairfax Station: 156–63.

58 GOLDMAN, N., J. P. ANDERSON AND A. G. RODRIGO. 2000. Likelihood-based tests of topologies in phylogenetics. Syst. Biol. **49**: 652–70.

59 GOLDMAN, N. AND Z. YANG. 1994. A codon-based model of nucleotide substitution for protein-coding DNA sequences. Mol. Biol. Evol. **11**: 725–36.

60 GORDON, A. D. 1986. Consensus supertrees: the synthesis of rooted trees containing overlapping sets of labelled leaves. J. Classif. **3**: 335–48.

61 GU, X., Y.-X. FU AND W.-H. LI. 1995. Maximum likelihood estimation of the heterogeneity of substitution rate among nucleotide sites. Mol. Biol. Evol. **12**: 546–57.

62 GUINDON, S. AND G. PERRIÈRE. 2005. Intragenomic base content variation is a potential source of biases when searching for horizontally transferred genes. Mol. Biol. Evol. **18**: 1838–40.

63 GUINDON, S. AND O. GASCUEL. 2003. A simple, fast, and accurate algorithm to estimate large phylogenies by maximum likelihood. Syst. Biol. **52**: 696–704.

64 GUSFIELD, D. AND V. BANSAL. 2005. A fundamental decomposition theory for phylogenetic networks and incompatible characters. Proc. RECOMB **9**: 217–32.

65 HASEGAWA, M., H. KISHINO AND T.-A. YANO. 1985. Dating of the human–ape splitting by a molecular clock of mitochondrial DNA. J. Mol. Evol. **22**: 160–74.

66 HASEGAWA, M. AND H. KISHINO. 1994. Accuracies of the simple methods for estimating the bootstrap propability of a maximum-likelihood tree. Mol. Biol. Evol. **11**: 142–5.

67 HENDY, M. D. AND D. PENNY. 1982. Branch and bound algorithms to determine minimal evolutionary trees. Math. Biosci. **60**: 133–42.

68 HENNIG, W. 1966. *Phylogenetic systematics*. University of Illinois Press, Urbana, IL.

69 HILLIS, D. M. 1996. Inferring complex phylogenies. Nature **383**: 130–1.

70 HOLLAND, B. R., D. PENNY AND M. D. HENDY. 2003. Outgroup misplacement and phylogenetic inaccuracy under a molecular clock – a simulation study. Syst. Biol. **52**: 229–38.

71 HOLLIDAY, R. AND G. W. GRIGG. 1993. DNA methylation and mutation. Mutat. Res. **285**: 61–7.

72 HUBER, K. T., M. LANGTON, D. PENNY, V. MOULTON AND M. HENDY. 2002. Spectronet: a package for computing spectra and median networks. Appl. Bioinform. **20**: 159–61.

73 HUBER, K. T. AND V. MOULTON. 2005. Phylogenetic Networks. In GASCUEL. (ed.), *Mathematics of Evolution and Phylogeny*. Oxford University Press, Oxford: 178–204.

74 HUELSENBECK, J. P., B. LARGET AND M. E. ALFARO. 2004. Bayesian phylogenetic model selection using reversible jump Markov chain Monte Carlo. Mol. Biol. Evol. **21**: 1123–33.

75 HUELSENBECK, J. P., B. LARGET, R. MILLER AND F. RONQUIST. 2002. Potential applications and pitfalls of Bayesian inference of phylogeny. Syst. Biol. **51**: 673–88.

76 HUELSENBECK, J. P., F. RONQUIST, R. NIELSEN AND J. P. BOLLBACK. 2001. Bayesian inference of phylogeny and its impact on evolutionary biology. Science **294**: 2310–4.

77 HUELSENBECK, J. P. AND F. RONQUIST. 2001. MRBAYES: Bayesian inference of phylogenetic trees. Bioinformatics **17**: 754–5.

78 HUSON, D. H., T. DEZULIAN, T. KLÖPPER AND M. A. STEEL. 2004. Phylogenetic super-networks from partial trees. In Proc. 4th Workshop on Algorithms in Bioinformatics, Bergen: 388–99.

79 HUSON, D. H., T. KLOEPPER, P. J. LOCKHART AND M. STEEL. 2005. Reconstruction of reticulate networks from gene trees. Proc. RECOMB **9**: 233–49.

80 HUSON, D. H. 1998. SplitsTree: analyzing and visualizing evolutionary data. Bioinformatics **14**: 68–73.

81 JERMIIN, L. S., G. J. OLSEN, K. L. MENGERSEN AND S. EASTEAL. 1997. Majority-rule consensus of phylogenetic trees obtained by maximum-likelihood analysis. Mol. Biol. Evol. **14**: 1296–302.

82 JUKES, T. H. AND C. R. CANTOR. 1969. Evolution of protein molecules. In MUNRO, H. N. (ed.), *Mammalian Protein Metabolism*. Academic Press, New York: 21–123.

83 KHAITOVICH, P., S. PÄÄBO AND G. WEISS. 2005. Toward a neutral evolutionary model of gene expression. Genetics **170**: 929–39.

84 KIMURA, M. 1980. A simple method for estimating evolutionary rates of base substitutions through comparative studies of nucleotide sequences. J. Mol. Evol. **16**: 111–20.

85 KING, M.-C. AND A. C. WILSON. 1975. Evolution at two levels in humans and chimpanzees. Science **188**: 107–16.

86 KLEMPA, B., H. A. SCHMIDT, R. ULRICH, S. KALUZ, M. LABUDA, H. MEISEL, B. HJELLE AND D. H. KRÜGER. 2003. Genetic interaction between distinct Dobrava Hantavirus subtypes in *Apodemus agrarius* and *A. flavicollis* in nature. J. Virol. **77**: 804–9.

87 KLOTZ, L. C., N. KOMAR, R. L. BLANKEN AND R. M. MITCHELL. 1979. Calculation of evolutionary trees from sequence data. Proc. Natl. Acad. Sci. USA **76**: 4516–20.

88 KLUGE, A. G. AND J. S. FARRIS. 1969. Quantitative phyletics and the evolution of anurans. Syst. Zool. **18**: 1–32.

89 KLUGE, A. G. 1989. A concern for evidence and a phylogenetic hypothesis of relationships among Epicrates (Boidae, Serpentes). Syst. Zool. **38**: 7–25.

90 KLUGE, A. G. 1998. Total evidence or taxonomic congruence: cladistics or consensus classification. Cladistics **14**: 151–8.

91 KURLAND, C. G., B. CANBACK AND O. G. BERG. 2005. Horizontal gene transfer: a critical view. Proc. Natl. Acad. Sci. USA **100**: 9658–62.

92 LANAVE, C., G. PREPARATA, C. SACCONE AND G. SERIO. 1984. A new method for calculating evolutionary substitution rates. J. Mol. Evol. **20**: 86–93.

93 LAND, A. AND A. DOIG. 1960. An automatic method for solving discrete programming problems. Econometrica **28**: 497–520.

94 LAPOINTE, F.-J. AND G. CUCUMEL. 1997. The average consensus procedure: combining of weighted trees containing identical or overlapping sets of taxa. Syst. Biol. **46**: 306–12.

95 LARGET, B. AND D. L. SIMON. 1999. Markov chain Monte Carlo algorithms for Bayesian analysis of phylogenetic trees. Mol. Biol. Evol. **16**: 750–9.

96 LEMMON, A. R. AND M. C. MILINKOVITCH. 2002. The metapopulation genetic algorithm: an efficient solution for the problem of large phylogeny estimation. Proc. Natl. Acad. Sci. USA **99**: 10516–21.

97 LIÒ, P. AND N. GOLDMAN. 1998. Models of molecular evolution and phylogeny. Genome Res. **8**: 1233–44.

98 LOCKHART, P. J., P. A. MCLENACHAN, D. HAVELL, D. GLENNY, D. HUSON AND U. JENSEN. 2001. Phylogeny, radiation, and transoceanic dispersal of New Zealand alpine buttercups: molecular evidence under split decomposition. Ann. Missouri Bot. Gard. **88**: 458–77.

99 MADDISON, D. R., M. RUVOLO AND D. L. SWOFFORD. 1992. Geographic origins of human mitochondrial DNA: phylogenetic evidence from control region sequences. Syst. Biol. **41**: 111–24.

100 MADDISON, W. P., M. J. DONOGHUE AND D. R. MADDISON. 2005. Outgroup analysis and parsimony. Syst. Zool. **33**: 83–103.

101 MAKARENKO, V. 2001. T-REX: reconstructing and visualizing phylogenetic trees and reticulation networks. Bioinformatics **17**: 664–8.

102 MARGUSH, T. AND F. R. MCMORRIS. 1981. Consensus *n*-trees. Bull. Math. Biol. **43**: 239–44.

103 MARTIN, W. 1999. Mosaic bacterial chromosomes: a challenge en route to a tree of genomes. BioEssays **21**: 99–104.

104 MASSINGHAM, T. AND N. GOLDMAN. 2005. Detecting amino acid sites under positive selection and purifying selection. Genetics **169**: 1753–62.

105 MAU, B., M. A. NEWTON AND B. LARGET. 1999. Bayesian phylogenetic inference via Markov chain Monte Carlo methods. Biometrics **55**: 1–12.

106 MILLER, R. G. 1974. The Jackknife – a review. Biometrika **61**: 1–15.

107 MINH, B. Q., L. S. VINH, A. VON HAESELER AND H. A. SCHMIDT. 2005. pIQPNNI – parallel reconstruction of large maximum likelihood phylogenies. Bioinformatics **21**: 3794–6.

108 MIYAMOTO, M. M. AND W. M. FITCH. 1995. Testing species phylogenies and phylogenetic methods with congruence. Syst. Biol. **44**: 64–76.

109 NIESELT-STRUWE, K. AND A. VON HAESELER. 2001. Quartet-mapping, a generalization of the likelihood-mapping procedure. Mol. Biol. Evol. **18**: 1204–19.

110 NIXON, K. C. AND J. M. CARPENTER. 1993. On outgroups. Cladistics **9**: 413–26.

111 OLSEN, G. J., H. MATSUDA, R. HAGSTROM AND R. OVERBEEK. 1994. fastDNAml: a tool for construction of phylogenetic trees of DNA sequences using maximum likelihood. Comput. Appl. Biosci. **10**: 41–8.

112 PAGE, R. D. M. 2002. Modified Mincut supertrees. In Proc. 2nd Workshop on Algorithms in Bioinformatics, Rome: 537–51.

113 PARASKEVIS, D., K. DEFORCHE, P. LEMEY, G. MAGIORKINIS, A. HATZAKIS AND A.-M. VANDAMME. 2005. SlidingBayes: exploring recombination using a sliding window approach based on Bayesian phylogenetic inference. Bioinformatics **21**: 1274–5.

114 PATTERSON, C. 1988. Homology in classical and molecular biology. Mol. Biol. Evol. **5**: 603–25.

115 PIAGGIO-TALICE, R., G. BURLEIGH AND O. EULENSTEIN. 2004. Quartet supertrees. In BININDA-EMONDS, O. R. P. (ed.), *Phylogenetic Supertrees: Combining Information to Reveal the Tree of Life*. Kluwer, Dordrecht: 173–91.

116 POSADA, D. AND K. A. CRANDALL. 2001. Intraspecific gene genealogies: trees grafting into networks. TREE **16**: 37–45.

117 POSADA, D. AND K. A. CRANDALL. 2002. The effect of recombination on the accuracy of phylogeny estimation. J. Mol. Evol. **54**: 396–402.

118 PURVIS, A. 1995. A composite estimate of primate phylogeny. Philos. Trans. R. Soc. Lond. B **348**: 405–21.

119 RAGAN, M. A. 1992. Phylogenetic inference based on matrix representation of trees. Mol. Phylogenet. Evol. **1**: 53–8.

120 ROBERTS, D. L. AND A. R. SOLOW. 2003. Flightless birds: when did the dodo become extinct? Nature **426**: 245.

121 ROBINSON-RECHAVI, M. AND D. GRAUR. 2001. Usage optimization of unevenly sampled data through the combination of quartet trees: an eutherian draft phylogeny based on 640 nuclear and mitochondrial proteins. Isr. J. Zool. **47**: 259–70.

122 ROHLF, F. J. 1982. Consensus indices for comparing classifications. Math. Biosci. **59**: 131–44.

123 SAITOU, N. AND M. NEI. 1987. The neighbor-joining method: a new method for reconstructing phylogenetic trees. Mol. Biol. Evol. **4**: 406–25.

124 SALEMI, M. AND A.-M. VANDAMME. 2003. *The Phylogenetic Handbook: A Practical Approach to DNA and Protein Phylogeny*. Cambridge University Press, Cambridge.

125 SALMINEN, M. O., J. K. CARR, D. S. BURKE AND F. E. MCCUTCHAN. 1995. Genotyping of HIV-1. *Human Retroviruses and AIDS Compendium*. Los Alamos National Laboratory III-30–III-34, Los Alamos, NM.

126 SANDERSON, M. J., A. PURVIS AND C. HENZE. 1998. Phylogenetic supertrees: assembling the trees of life. TREE **13**: 105–9.

127 SANDERSON, M. J. AND H. B. SHAFFER. 2002. Troubleshooting molecular phylogenetic analyses. Annu. Rev. Ecol. Syst. **33**: 49–72.

128 SANKOFF, D. AND J. B. KRUSKAL. 1983. Time warps, string edits, and macromolecules. Addison-Wesley, Reading, MA.

129 SANKOFF, D. 1975. Minimal mutation trees of sequences. SIAM J. Appl. Math. **28**: 35–42.

130 SATTA, Y., J. KLEIN AND N. TAKAHATA. 2000. DNA archives and our nearest relative: the trichotomy problem revisited. Mol. Phylogenet. Evol. **32**: 259–75.

131 SATTATH, S. AND A. TVERSKY. 1977. Additive similarity trees. Psychometrika **42**: 319–45.

132 SCHMIDT, H. A., K. STRIMMER, M. VINGRON AND A. VON HAESELER. 2002. TREE-PUZZLE: maximum likelihood phylogenetic analysis using quartets and parallel computing. Bioinformatics **18**: 502–4.

133 SCHMIDT, H. A. 2003. *Phylogenetic Trees from Large Datasets.* PhD Thesis. Universität Düsseldorf.

134 SCHUH, R. T. 2000. *Biological Systematics: Principles and Applications.* Cornell University Press, Ithaca, NY.

135 SEMPLE, C. AND M. STEEL. 2000. A supertree method for rooted trees. Discr. Appl. Math. **105**: 147–58.

136 SEMPLE, C. AND M. STEEL. 2003. *Phylogenetics.* Oxford University Press, Oxford.

137 SHELDON, F. H. 1987. Phylogeny of herons estimated from DNA–DNA hybridization data. Auk **104**: 97–108.

138 SHIMODAIRA, H. AND M. HASEGAWA. 1999. Multiple comparisons of log-likelihoods with applications to phylogenetic inference. Mol. Biol. Evol. **16**: 1114–6.

139 SINSHEIMER, J. S., J. A. LAKE AND R. J. A. LITTLE. 1996. Bayesian hypothesis testing of four-taxon topologies using molecular sequence data. Biometrics **52**: 715–28.

140 SMITH, A. B. 1994. Rooting molecular trees: problems and strategies. Biol. J. Linn. Soc. **51**: 279–92.

141 STAMATAKIS, A. P., T. LUDWIG AND H. MEIER. 2005. RAxML-III: a fast program for maximum likelihood-based inference of large phylogenetic trees. Bioinformatics **21**: 456–63.

142 STEWART, C. A., D. HART, D. K. BERRY, G. J. OLSEN, E. A. WERNERT AND W. FISCHER. 2001. Parallel implementation and performance of fastDNAml – a program for maximum likelihood phylogenetic inference. In Proc. Int. Conf. on High Performance Computing and Communications, Denver, CO: 191–201.

143 STRIMMER, K., C. WIUF AND V. MOULTON. 2001. Recombination analysis using directed graphical models. Mol. Biol. Evol. **18**: 97–9.

144 STRIMMER, K. AND A. VON HAESELER. 1996. Quartet puzzling: a quartet maximum-likelihood method for reconstructing tree topologies. Mol. Biol. Evol. **13**: 964–9.

145 STRIMMER, K. AND A. VON HAESELER. 1997. Likelihood-mapping: a simple method to visualize phylogenetic content of a sequence alignment. Proc. Natl Acad. Sci. USA **94**: 6815–9.

146 STRIMMER, K. AND V. MOULTON. 2000. Likelihood analysis of phylogenetic networks using directed graphical models. Mol. Biol. Evol. **17**: 875–81.

147 STUDIER, J. A. AND K. J. KEPPLER. 1988. A note on the neighbor-joining algorithm of Saitou and Nei. Mol. Biol. Evol. **5**: 729–31.

148 SWOFFORD, D. L., G. J. OLSEN, P. J. WADDELL AND D. M. HILLIS. 1996. Phylogeny reconstruction. In HILLIS, D. M., C. MORITZ, AND B. K. MABLE (eds.), *Molecular Systematics*, Sinauer, Sunderland, MA: 407–514.

149 TAMURA, K. AND M. NEI. 1993. Estimation of the number of nucleotide substitutions in the control region of mitochondrial DNA in humans and chimpanzees. Mol. Biol. Evol. **10**: 512–26.

150 TARRÍO, R., F. RODRÍGUEZ-TRELLES AND F. J. AYALA. 2001. Shared nucleotide composition biases among species and their impact on phylogenetic reconstructions of the drosophilidae. Mol. Biol. Evol. **18**: 1464–73.

151 VINH, L. S., H. A. SCHMIDT AND A. VON HAESELER. 2005. PhyNav: a novel approach to reconstruct large phylogenies. In WEIHS, C. AND W. GAUL (eds.), *Classification – The Ubiquitous Challenge.* Springer, Berlin: 386–93.

152 VINH, L. S. AND A. VON HAESELER. 2004. IQPNNI: moving fast through tree space and stopping in time. Mol. Biol. Evol. **21**: 1565–71.

153 VON HAESELER, A. AND G. A. CHURCHILL. 1993. Network models for sequence evolution. J. Mol. Evol. **37**: 77–85.

154 WHEELER, W. C. 2005. Nucleic acid sequence phylogeny and random outgroups. Cladistics **6**: 363–8.

155 WILKINSON, M., J. L. THORLEY, D. PISANI, F.-J. LAPOINTE AND J. O. MCINERNEY. 2004. Some desiderata for liberal supertrees. In BININDA-EMONDS, O. R. P. (ed.), *Phylogenetic*

Supertrees: Combining Information to Reveal the Tree of Life. Kluwer, Dordrecht: 227–46.

156 YANG, Z. AND R. NIELSEN. 1998. Synonymous and nonsynonymous rate variation in nuclear genes of mammals. J. Mol. Evol. **46**: 409–18.

157 YANG, Z. AND W. J. SWANSON. 2002. Codon-substitution models to detect adaptive evolution that account for heterogeneous selective pressures among site classes. Mol. Biol. Evol. **19**: 49–57.

158 ZILLIG, W. 1991. Comparative biochemistry of Archaea and Bacteria. Curr. Opin. Genet. Dev. **1**: 544–51.

5
Finding Protein-coding Genes

David C. Kulp

1 Introduction

Gene finding in genomic DNA sequences is a critical step in the functional annotation of genomes. Over the past approximately quarter century increasingly sophisticated methods have been developed to better understand and catalog the mechanisms of transcription, splicing and translation, and to predict the gene products, be they peptide sequences or RNA genes. With the advent of large-scale sequencing, software programs were developed to automate gene prediction.

In this chapter the common techniques for computational gene finding are introduced. Basic concepts and terminology are given in Section 2. Sections 3–5 discuss feature prediction for both content and signal features, and Section 6 introduces the standard dynamic programming formalism for incorporating multiple features into complete gene structure predictions. Some performance results for *ab initio* gene finding are given in Section 7. Practical gene finding must also consider available experimental mRNA, protein and genomic sequence data. Some of these homology methods as well as other integrative approaches are described in Section 8. Finally, the chapter concludes with some caveats about the practical limitations of automated gene prediction.

2 Basic DNA Terminology

Since one DNA strand is the complement of the other, in DNA analysis only one strand is stored in the databases. Which strand is represented is generally arbitrary and unimportant, but in this chapter the represented sequence is called the forward strand and the implicit complement is the reverse. A DNA sequence is always represented, by convention, in the direction of DNA replication. The left end of the sequence is referred to as upstream or $5'$ and the right end is downstream or $3'$.

Bioinformatics - From Genomes to Therapies Vol. 1. Edited by Thomas Lengauer

ISBN: 978-3-527-31278-8

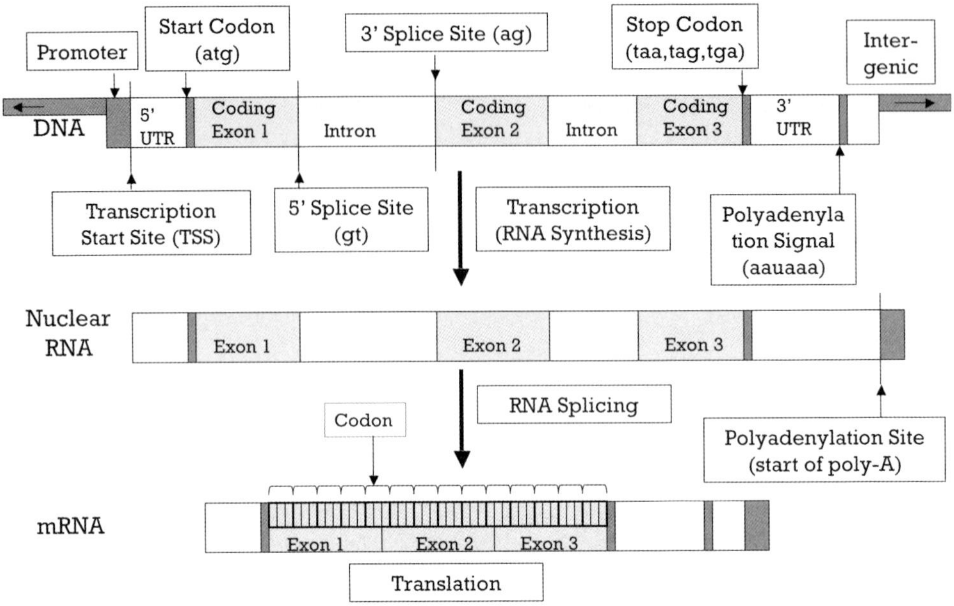

Figure 1 The central dogma of molecular biology in eukaryotes. A primary transcript region starting at the promoter is copied into pre-mRNA. The transcript is then spliced in eukaryotes to produce the mature cytoplasmic message. The message is translated into peptides. Note that codons may span splice boundaries. Although not shown in this diagram, splicing is possible in the untranslated regions.

For the purposes of this discussion, a gene is defined as the subsequence of genomic DNA that is transcribed by RNA polymerase – usually Pol II when referring to eukaryotic transcription. The gene structure further includes those features on the mRNA involved with splicing and translation, i.e. the splice and translation start and stop sites. For convenience, all of these features are usually annotated with respect to the original DNA sequence as shown in Figure 1. Transcription occurs on single-stranded DNA, on either the forward or reverse strand. By convention the gene structure is annotated on the informational or sense strand (not the template or antisense strand).

Although transcription has been observed to occur at the same physical genomic position on both strands, we usually assume for simplicity that there are no overlapping transcripts. Automatic gene finders must evaluate both the explicitly represented sequence and the implicit reverse complement – this is usually performed simultaneously.

Predicting genes in eukaryotes is considerably more challenging than in prokaryotes because of splicing. Most transcribed mRNA (pre-mRNA) in eukaryotes is spliced into smaller sequences called processed or mature mRNA through the excision of introns by the spliceosome complex, leaving a set of concatenated exons to be passed to the ribosome. The introns are located between the 5′ and 3′ splice sites, also called donor and acceptor sites. At least 99% of 5′ splice sites begin with "GT" and 3′ splice sites end with "AG", called the consensus dinucleotides.

The spliceosome concatenates exons separated by often long introns. Each exon can be as short as a few nucleotides. Thus, while gene finding in prokaryotes involves indentifying a single contiguous coding sequence, gene finding in eukaryotes requires a combinatorial search of many different possible exons.

3 Detecting Coding Sequences

Identification of the protein-coding domain sequence (CDS) between the start and stop codons is of great interest because the translated peptide product can be directly inferred from the CDS. The ribosome, after binding to the mRNA, begins translation at an AUG triplet (ATG in DNA). The ribosome matches these triplets, called codons, consecutively with tRNAs adding deterministically one of the 20 amino acids to a polypeptide sequence for each codon according to the genetic code (Table 1). The translation process is terminated when one of three stop codons (UAA, UAG or UGA) is encountered. Thus, the CDS on the DNA sequence is composed of a sequence of codons that code for

Table 1 The standard genetic code showing codon and amino acid single- and three-letter abbreviations

TTT	F	Phe	TCT	S	Ser	TAT	Y	Tyr	TGT	C	Cys
TTC	F	Phe	TCC	S	Ser	TAC	Y	Tyr	TGC	C	Cys
TTA	L	Leu	TCA	S	Ser	TAA	*	Ter	TGA	*	Ter
TTG	L	Leu	TCG	S	Ser	TAG	*	Ter	TGG	W	Trp
CTT	L	Leu	CCT	P	Pro	CAT	H	His	CGT	R	Arg
CTC	L	Leu	CCC	P	Pro	CAC	H	His	CGC	R	Arg
CTA	L	Leu	CCA	P	Pro	CAA	Q	Gln	CGA	R	Arg
CTG	L	Leu	CCG	P	Pro	CAG	Q	Gln	CGG	R	Arg
ATT	I	Ile	ACT	T	Thr	AAT	N	Asn	AGT	S	Ser
ATC	I	Ile	ACC	T	Thr	AAC	N	Asn	AGC	S	Ser
ATA	I	Ile	ACA	T	Thr	AAA	K	Lys	AGA	R	Arg
ATG	M	Met	ACG	T	Thr	AAG	K	Lys	AGG	R	Arg
GTT	V	Val	GCT	A	Ala	GAT	D	Asp	GGT	G	Gly
GTC	V	Val	GCC	A	Ala	GAC	D	Asp	GGC	G	Gly
GTA	V	Val	GCA	A	Ala	GAA	E	Glu	GGA	G	Gly
GTG	V	Val	GCG	A	Ala	GAG	E	Glu	GGG	G	Gly

Prokaryotes also use an additional GTG initiation codon. There are other rare genetic codes as well. See http://www.ncbi.nlm.nih.gov/Taxonomy/Utils/wprintgc.cgi.

the corresponding protein, beginning with the start codon and ending with a stop codon.

Note that the notions CDS and exon are not synonymous, although frequently exchanged in gene-finding literature. Exons refer to those DNA segments that are not excised by splicing, i.e. all of the sequence corresponding to the mature mRNA. Exonic sequences can be either coding (CDS) or untranslated regions (UTRs). A CDS typically begins and ends in the middle of an exon. Introns are possible upstream and (rarely) downstream of a CDS. Splicing need not occur on codon boundaries.

3.1 Reading Frames

There are six possible reading frames along double-stranded DNA – three on each strand. A CDS beginning translation at position i is in reading frame $f = i$ modulo 3. Reading frames on the opposite strand are conventionally labeled as $-f$. We say that a codon is in frame if its position modulo 3 is the same as the CDS in question. In particular, an in-frame stop codon terminates a CDS, but out-of-frame stop codons have no effect. A sequence of consecutive codons between a start and stop codon is called an open reading frame (ORF).

Genes in prokaryotes are relatively easy to identify by searching for ORFs of a minimum length, say 300 nucleotides. In random DNA, an in-frame stop codon is expected about every 21 codons (63 nucleotides) and the chance of an ORF longer than this becomes increasingly unlikely. Small ORFs can truly be coding, especially in eukaryotes, due to splicing and asymmetric nucleotide distributions can easily allow for long ORFs, requiring more sophisticated pattern recognition methods, as described below.

3.2 Coding Potential

The distribution of codons is subject to evolutionary and biophysical constraints. The G + C content (fraction of G and C nucleotides) among genomes varies, which affects codon frequencies for different organisms. Amino acid frequencies are not uniform and arrangements of amino acids in polypeptides are, of course, also not random. These effects, as well as other DNA and mRNA structural and processing constraints, lead to biases in the frequency and ordering of codons, called codon bias. Moreover, basal expression levels have been observed to relate to the levels of available tRNAs, so codons in higher expressed genes are more significantly biased towards the abundant tRNAs [27]. Synonymous codon bias, closely related to codon bias, describes the differing frequencies of codons coding for the same amino acid.

Codon usage tables that list codon frequencies have been compiled for many organisms (http://www.kazusa.or.jp/codon). To test for coding potential is to assess whether the frequencies of codons in a candidate ORF are statistically similar to the codon usage for the organism and these measures are typically the heart of all gene-finding programs. For example, about 2.1% of human codons are "ATC" and among its class of synonymous codons that encode isoleucine ("ATC", "ATT" and "ATA"), "ATC" is used about 47% of the time.

A representative method of this class of coding potential measures is the Gribskov codon preference statistic [26]. The relative frequency of a codon, C, among its class of synonymous codons, $f_{\text{class}(C)}(C)$, is computed using a codon-usage table derived from highly expressed genes and compared to the relative frequency of the codon, $g_{\text{class}(C)}(C)$, in a background model according to position-independent nucleotide composition:

$$S(C) = \log\left[f_{\text{class}(C)}(C)/g_{\text{class}(C)}(C)\right] .$$

For example, suppose the codon, C, is "ATC", and its codon usage is as above and the frequencies of A, T, C and G in the query sequence are (0.21, 0.21, 0.29, 0.29). Then $S(C = \text{ATC}) = \log(0.47/0.41) = 0.14$. The normalized sum of log-likelihood ratios over a window w (of, say, 25 codons) provides an indication of relative coding potential and expression. By convention, the normalized sum is exponentiated to generate the final Gribskov statistic. Such methods are often used in plots for preliminary visualization of a novel genomic sequence and are sometimes sufficient to support manual gene prediction in prokaryotes (Figure 2).

Measures such as the Gribskov codon preference statistic lack consideration for the positions of codons relative to each other. Observed dependencies among adjacent codons led to the proposal of several measures based on pairs of codons (dicodons). In an important benchmark paper, Fickett and Tung [22], assessed most of the extant methods and their conclusion was that dicodon (or hexamer, more generally) measurements were superior to all other methods.

The three-periodic fifth-order Markov model is a particularly appealing formulation of hexamer statistics that is widely used in modern gene finders. Proposed by Borodovsky [7], such Markov models are used to represent the probability distributions of the four possible nucleotides at each of the three base positions in a codon. Suppose we are interested in a codon beginning at position i composed of individual nucleotides b_i, b_{i+1} and b_{i+2}. Three separate Markov models are defined: $P_0(b_i|b_{i-5}\ldots b_{i-1})$, $P_1(b_{i+1}|b_{i-4}\ldots b_i)$ and $P_2(b_{i+2}|b_{i-3}\ldots b_{i+1})$. Each probability distribution is generated from simple frequency counts of each possible nucleotide in the context of the previous five nucleotides. Training sets of millions of codons are available from annotated

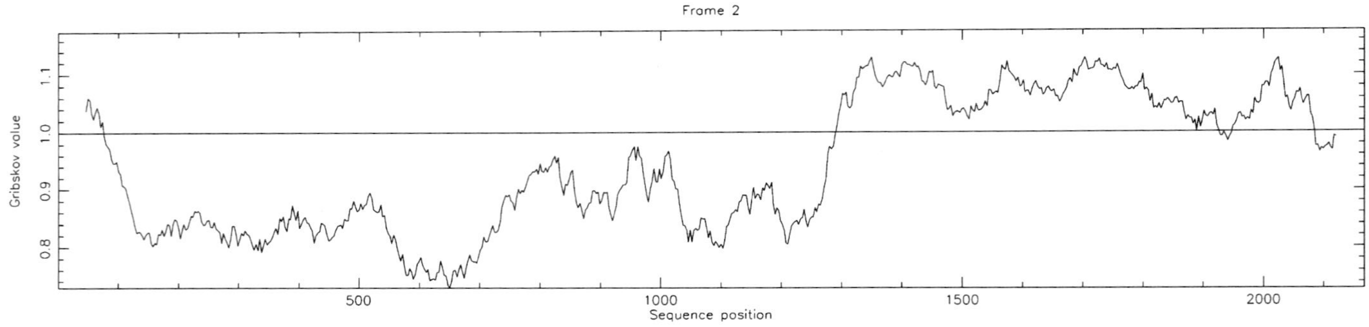

Figure 2 Codon bias computed by the Gribskov codon preference statistic. Frame 2 of the *Pseudomonas amiC* gene as generated by the program syco from the EMBOSS software collection. The region starting around position 1300 has consistently high coding potential, suggesting a coding region.

GenBank sequences. Assuming conditional independence, the probability of a codon is just the product of the probabilities of the individual bases b_i, b_{i+1}, and b_{i+2}. A separate null model is defined for noncoding bases, $P_{\mathrm{nc}}(b_i|b_{i-5}\ldots b_{i-1})$. A log-likelihood ratio score for a codon starting at i is then:

$$\sum_{f=0..2}\left[\log P_f(b_{i+f}|b_{i+f-5}\ldots b_{i+f-1}) - \log P_{\mathrm{nc}}(b_{i+f}|b_{i+f-5}\ldots b_{i+f-1})\right] .$$

These scores can be accumulated over a window as with the Gribskov measure.

To obtain accurate estimates, a fifth-order Markov model requires sufficiently large numbers of observations in each context. This is not always available and, in some cases, even longer contexts may be plentiful. Salzberg describes a variant that uses different context length depending on the available data [66].

The fact that statistics related to codon usage aid in the identification of CDS and also in the estimation of expression levels indicates an inherent classification weakness in the use of codon statistics, i.e. that genes with low expression levels are more difficult to find because their statistics are weaker. Low expressors are more difficult to detect experimentally as well, further biasing the codon statistics gathered regarding known genes.

4 Gene Contents

In addition to the statistical regularities in CDS, other discriminating properties of coding exons and introns have been observed. These features of variable-length DNA sequences are sometimes referred to as content.

For example, noncoding DNA is expected to have a relatively neutral distribution of nucleotides with exceptions such as physical–chemical constraints and the presence of repeats. Thus, models of noncoding DNA have been devised similar to coding potential measures. A simple and common implementation is the use of Markov models for intron and intergenic DNA analagous to the three fifth-order Markov models for coding DNA. The log probabilities of different models can then be compared. Figure 3 shows the distribution of probabilities from fifth-order coding and intron models, and the distribution of the difference in log probabilities.

Guigo and Fickett [29] showed that all content measures are highly correlated with G + C bias. Thus, it is common to compute Markov distributions by partitioning the training data into discrete isochores (extended regions of G + C bias in the genome) based on windowed-G + C content [9].

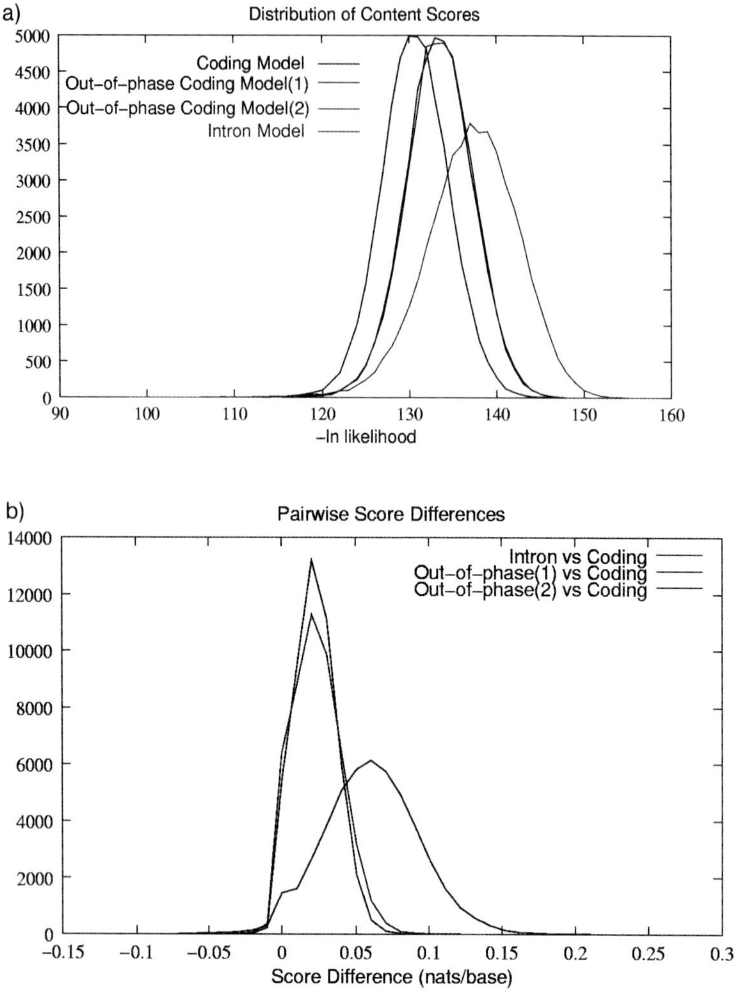

Figure 3 Score distributions for Markov chain models of coding and intronic DNA. Three Markov chains were trained on 53 183 460 coding bases and one Markov chain was trained on 16 149 264 intronic bases from the well-annotated protein-coding exons of *Caenorhabdtis elegans*. For each of 48 124 exons, a 99-nucleotide in-frame coding region was scored using the coding model, two out-of-phase coding models and the intron model. (a) The distribution of log-likelihoods for fifth-order Markov chains. (b) The distribution of differences in log-likelihoods per base of the fifth-order coding model versus the intron and out-of-phase fifth-order models for each coding exon. (Pairwise comparisons with likelihoods from out-of-phase models were only made if there were no in-frame stop codons in the alternative frame.) The implication here is that although the overall distributions for the coding and noncoding models are very similar, a comparison of scores for individual exons shows good separation (i.e. most model score differences are greater than zero). This simply shows that the fifth-order Markov models are reasonably good at classifying coding regions.

The lengths of exons and introns differ – often significantly [32]. Exons follow an approximately log-normal distribution with a mean length of about 140 bases in most eukaryotes, but the typical length of introns varies significantly by organism. Many of the model organisms such as fly and worm have intron lengths within a relatively tight range of about 70 bases – the minimal required intron length for efficient splicing. Mammalian introns are typically much longer than exons due to prolific insertions of repetitive elements; they are rarely less than 100 bases and have a long exponential distribution to 10^6 bases.

5 Gene Signals

Gene structure is defined by the start and stop positions in DNA of exons and CDS. Through laboratory experimentation, molecular biologists have shown that for each of these sites there are necessary, conserved motifs that govern the transcription and translational machinery. With respect to gene finding, Staden [75] distinguished these control sites as signals as compared to variable length content. Signal features loosely correspond to binding sites or special functional patterns recognized by the polymerase, spliceosome or ribosome.

If it were possible to automatically detect all signals independently, then the gene-finding solution would be complete. For the most part, however, no one signal can reliably be detected on its own. Later in this chapter we will learn how to combine these measures along with coding potential to achieve superior gene finding performance. First, we explore a few of the methods for independent signal detection.

5.1 Splice Sites

Degenerate matches to a motif can be detected using a position-specific weight matrix [74]. Weight matrices are commonly used in many biosequence approximate matching problems. A weight matrix is a (2-D) array $W(1 \leq i \leq m, 1 \leq j \leq 4)$ such that $W(i, j)$ is the probability of nucleotide j at position i in a motif of length m. Frequencies can be used to generate these probabilities, priors can be introduced when data is sparse or more sophisticated contexts can be represented such as dinucleotide frequencies (e.g. an order-1 Markov weight matrix at each position resulting in a $m \times 16$ matrix). For example, almost all introns begin with the consensus dinucleotide "GT" and end with "AG", but the regions around the beginning and end of the intron (the splice sites) have less specific nucleotide patterns. Figure 4 shows examples of weight matrices for splice sites in *C. elegans*.

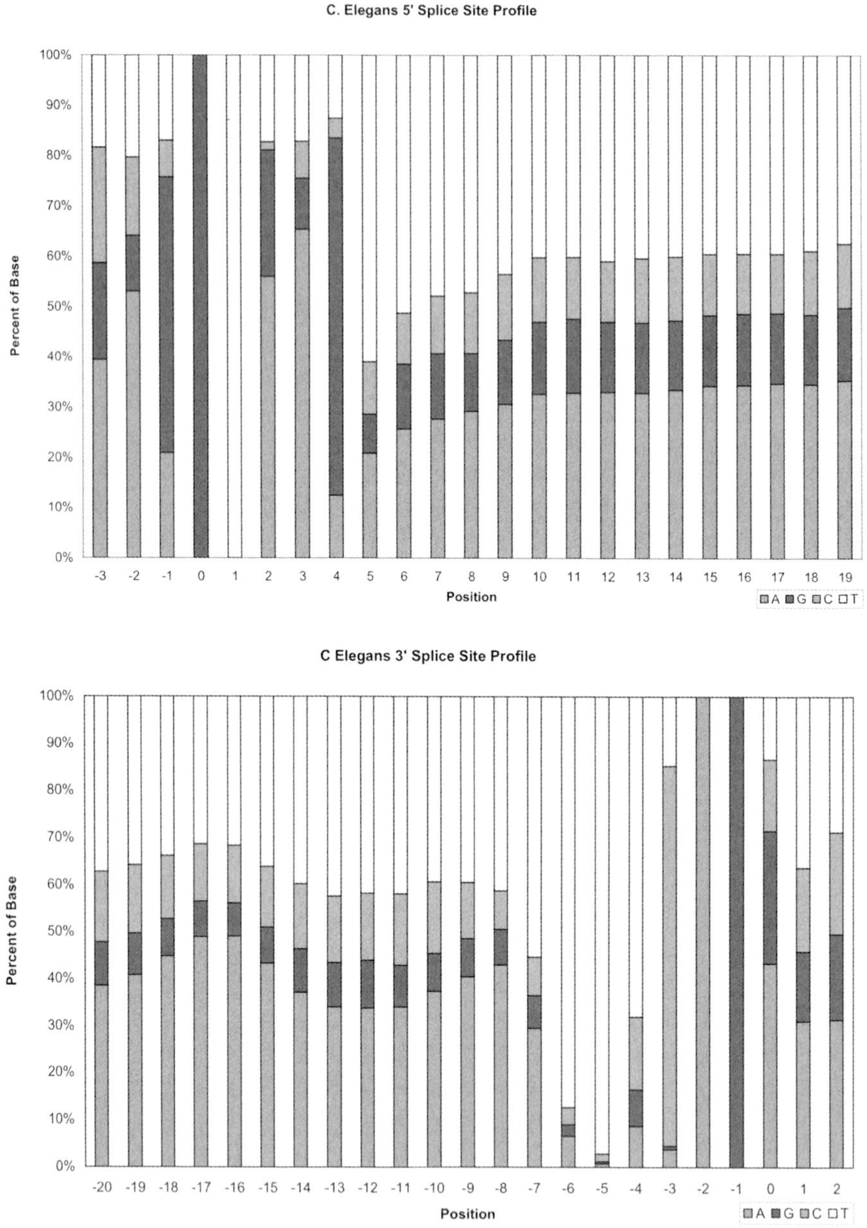

Figure 4 *C. elegans* splice site weight matrices. Windows of 20 bases downstream of the $5'$ exon junction ("GT", the beginning of the intron, is positions 0 and 1) and 20 bases upstream of the $3'$ exon junction ("AG", the end of the intron, is positions -1 and -2) were selected from curated gene sequences [16].

Given any test sequence represented as a 2-D matrix $S(1 \leq i \leq m, 1 \leq j \leq 4)$, where $S_{i,j} = 1$ for the nucleotide j found at position i and 0 elsewhere, then the likelihood of the feature can be defined simple as $\prod_{i=1 \ldots m} \prod_{j=1 \ldots 4} W_{i,j}^{S_{i,j}}$. Stormo presented thermodynamic, likelihood, and information theoretic justifications for the use of weight matrices [76].

Moving beyond the simple weight matrix is the maximal dependence decomposition decision tree (MDD) method that captures local, but nonadjacent dependencies [9]. The MDD method evaluates a set of rules to determine which weight matrix to use to score the sequence. The rules are based on the dependencies between positions in the motif. For example, in eukaryotic $5'$ splice sites, it can be shown that the distribution of nucleotides in the conserved columns $-3 \cdots + 6$ around the consensus GT are most correlated to the nucleotide G in position $+5$. Thus, the training set is partitioned according to the "+5" value into two sets such that the score function for each set is conditionally independent of that nonadjacent position. This leaves only adjacent dependencies or independent positions, which can be easily modeled using conventional weight matrices as above. (Other approaches have been proposed to detect dependencies among nonadjacent bases with similar performance characteristics, e.g. a Bayesian Network structure inference method was proposed by Cai and coworkers [13].)

The primary limitation of weight matrices is the inability to model insertions and deletions. To handle more complex motifs, profile hidden Markov models (HMMs) (see also Chapter 3) and related probabilistic state machine models can be employed in a similar manner as for protein sequences [20,42].

The $3'$ splice site is slightly more complicated because the upstream pyrimidine-rich branch site contributes to its recognition. However, the branch site is variable and so not amenible to fixed-width matrix methods. As a result, many recognition methods have been proposed for splice site recognition that allow for the incorporation of multiple distinct sequence features as inputs (branch site, splice site, intron content, exon content). One of the more effective techniques for combining different sequence features is discriminant analysis in which weights for different features are fitted to maximize the discrimination between true and decoy sites [72,82].

Probably the current leading method for splice site prediction is GeneSplicer – an extension of the MDD metric [59]. Other techniques include neural networks, boolean logic rules, decision trees, support vector machines (SVMs) and many others (e.g. Refs. [8,9,61,73]).

In addition to the conventional splice site recognition, about 1% of splice sites have non-canonical dinucleotides. This is largely ignored in gene finding, but is addressed by Burset and coworkers [11].

In general, recent methods achieve reasonable performance in splice site detection, but are unavoidably burdened by large numbers of decoy sites, resulting in false-positive rates around 5% when recognizing about 90% true splice sites.

5.2 Translation Initiation

Identifying the beginning of translation is perhaps surprisingly challenging. Part of the difficulty is that the database is rife with experimentally unconfirmed start sites. In addition, the signal for start sites tends to be rather weak. In prokaryotes, the Shine–Delgarno motif serves as a binding site for the ribosome preceding the first codon. The consensus motif is AGGAGG, but it can take on short and degenerate forms. Kozak observed that in higher eukaryotes translation usually begins at the first start codon after the transcription start site [41]. However, for the purposes of *ab initio* gene finding, this is usually of little help since the beginning of the transcript is also unknown and cannot be reliably predicted in large DNA sequences without a high false-positive rate [21] (and even the annotated transcription start sites are often wrong due to truncated mRNAs). In vertebrates, a consensus of `gccaccATGg` (start codon in caps) is observed and weight matrices have been developed from reliable start sites using first and second order models similar to approaches for splice sites [2, 40], but these methods are subject to high false-positive rates.

Like with splice site detection, many of the conventional machine learning techniques have been successfully applied including neural networks, linear and quadratic discriminant analysis and SVMs [58, 64, 67, 85]. The most successful independent predictor of translation initiation is an SVM classifier that remarkably identifies almost 100% of start sites with well less than 1% false positives on a standard test set [48].

5.3 Translation and Transcription Termination

Recognizing one of the stop codons (TAA, TAG or TGA) is trivial assuming that the reading frame is known. Conversely, more probable stops are those with high coding potential upstream of the site and low coding potential downstream. The transcript following the stop codon is typically one long exon. Splicing after the stop codon is rare.

Finally, transcription is terminated by a polyadenylation signal with a consensus of `AATAAA` although there are many variants. The motif is small, is not located predictably near other contextual features, is frequently unannotated in the databases and may not even be present. Moreover, it is estimated that in human about half of all transcripts have multiple 3′ termination sites and these are often imprecisely cleaved [80]. Graber [25] describes a pseudo-probabilistic model for detecting termination in yeast. Again, standard weight matrix and discriminant analysis methods have shown moderate success for detecting the termination site [47,77]. In practice, transcription termination is largely ignored in *ab initio* prediction. Instead, it is often considered sufficient to detect just the gene structure from the start to stop codons.

6 Integrating Gene Features

So far, we have learned that there are different gene features (signals and contents) each with statistically significant discriminative power. There are numerous scoring methods for different features assessed independently, yet we intuit that combining these features is likely to yield better results. However, there is an exponential combination of possible labelings of exons, introns and intergenic regions (i.e. any segment can begin or end at any position). Thus, the gene finder is faced with two major problems: how to effectively combine features and how to efficiently explore the possible gene structures. The solution for both of these problems is using dynamic programming. In some implementations, logical adjacent features are combined into a single score and then a dynamic program is applied.

6.1 Combining Local Features

In the same way that multiple features were used as input to the long list of machine learning classifiers in splice site recognition (Section 5), so too can multiple features be combined to recognize larger functional units. Zhang is a major proponent of this strategy of recognizing exons based on combined information from the flanking signals and content, noting that the *in vivo* recognition of exons in transcription is believed to largely be driven by the interactions of DNA-binding complexes that straddle the exons, according to the exon definition model [5,30]. Zhang has produced a suite of methods for recognizing the $5'$ and $3'$ UTR exons, initial coding exon, internal coding exons and last coding exon using quadratic discriminant analysis, each recognition module combining multiple, different features, with excellent performance [77,83].

The choice of the fundamental functional units of gene recognition differ among gene finding programs. For example, the nucleotide is the basic unit in HMMGene, Genie treats splice sites and coding exons separately, and MZEF combines these local features into a single functional unit. However, in all cases, the same dynamic programming technique can be used to combine these functional units into complete gene structure predictions.

6.2 Dynamic Programming

Snyder and Stormo [71] showed how the optimal combination of features could be obtained using dynamic programming. Let us define the states of our gene finder as the different types of functional units in our gene model, $Q = q_1 \ldots q_m$. We say that the sequence, $X = x_1 \ldots x_n$ is labeled by a corresponding sequence of states, $\Phi = \phi_1 \ldots \phi_n$, called the "parse", where $\phi_i \in Q$. Quite simply, a parse formally describes the gene structure, e.g. intergenic DNA from position 0 to 100, 5′ UTR from 101 to 150, initial CDS exon from 151 to 200, etc.

The key idea behind dynamic programming is the assumption that the score of a parse can be decomposed into independent segments or, at least, segments that are only locally dependent. This independence assumption is clearly violated in some cases. For example, tertiary protein structure obviously implies specific long-range interactions among codons. Nevertheless, this is a reasonable approximation for gene finding that offers significant computational advantage.

If every possible segment can be scored independently, then the parse with the best score can be computed recursively. Given an input DNA sequence X and possible states Q, then define a score matrix $S(j,k)$ that holds the score of the best parse of the subsequence $x_1 \ldots x_j$ ending with a segment at x_j in state q_k. Define $s(i,j,k)$ as the independent score of a segment from $x_i \ldots x_j$ of state k. [These $s(i,j,k)$ terms are based on feature scores from one of the many methods discussed and alluded to in the previous sections, such as coding potential, splice site scores, etc.] If we assume for simplicity in this formulation that any segment of any length can follow any other segment (we will improve on this momentarily), then $S(j,k)$ is defined as the best score from all positions $i < j$ in any possible state in Q plus the score of a segment in state k from $i \ldots j$. For example, the best score for labeling a DNA sequence such that that an initial exon ends at position 200 is computed by considering the score for an initial exon starting from every position less than 200 following any of the other possible states (5′ UTR, intron, intergenic, etc.):

$$S(j,k) = \max_{i<j, l \in Q} \left(S(i,l) + s(i,j,k)\right) . \qquad (1)$$

The general form of this dynamic program is usually called the Viterbi algorithm and the process of predicting a parse is often called decoding [23]. While $S(;)$ holds only the best score, it is straightforward to also simultaneously compute a trace-back describing the parse that achieved the best score.

The formulation here is different from the conventional presentation of dynamic programming for biosequence analysis (Chapter 3) because segments can take on arbitrary length. As a result, running time is quadratic in the length of the sequence and number of states – prohibitively expensive except for very small sequences, but at least not exponential.

6.3 Gene Grammars

In order to ensure that the evaluation of all reasonable gene structures for long DNA sequences can be achieved in acceptable running time we add grammatical constraints that ensure only legal and sensible parses are considered. Dong and Searls [18, 68] were the first major proponents for describing gene structure in linguistic terms. (For a thorough treatment in a modern gene finding system, see also Ref. [46].) The key idea behind grammatical constraints is that different segment states can only appear within specific contexts. For example, an intron can only following an exon. The grammatical constraints for genes can be expressed in a so-called regular grammar and can be visualized as a finite state machine. Figure 5 shows such a state diagram for a simplified gene model.

We call neighboring pairs of states transitions (e.g. intron following exon). In addition to strict contextual constraints on state transitions, we observe that some state transitions are allowed, but are less likely than others. For example, it is less likely that an intron will be followed by a terminal exon than an internal exon. We define $t(l,k)$ as the score for a transition from state l to state k. These are usually assigned based on frequencies of observed transitions in training data. In addition, we define a function $T(k)$ that returns a set of allowable previous states for k.

One specific type of transition constraint in $T(\cdot)$ is especially important, i.e. the frame constraint. In order to ensure that the total number of bases in the CDS is a multiple of 3, states must be created to ensure that introns split codons in a frame consistent manner. For example, if one base precedes an intron then two bases must follow it before the next full codon (see Figure 5.)

In Section 4 we observed that exons and introns had predictable length distributions as well as maximum and minimum lengths that are rarely or never exceeded, e.g. coding exons are rarely larger than a few thousand bases and introns are almost never smaller than about 50 bases. Therefore, we restrict the allowable length segments considered in our dynamic program by introducing $\min(k)$ and $\max(k)$ values for each state k.

From this, we have an improved method for scoring possible parses that extends Eq. (1) as:

$$S(j,k) = \max_{\substack{i<j,\ l \in T(k) \\ j-i > \min(k) \\ j-i < \max(k)}} S(i,l) + s(i,j,k) + t(l,k) \tag{2}$$

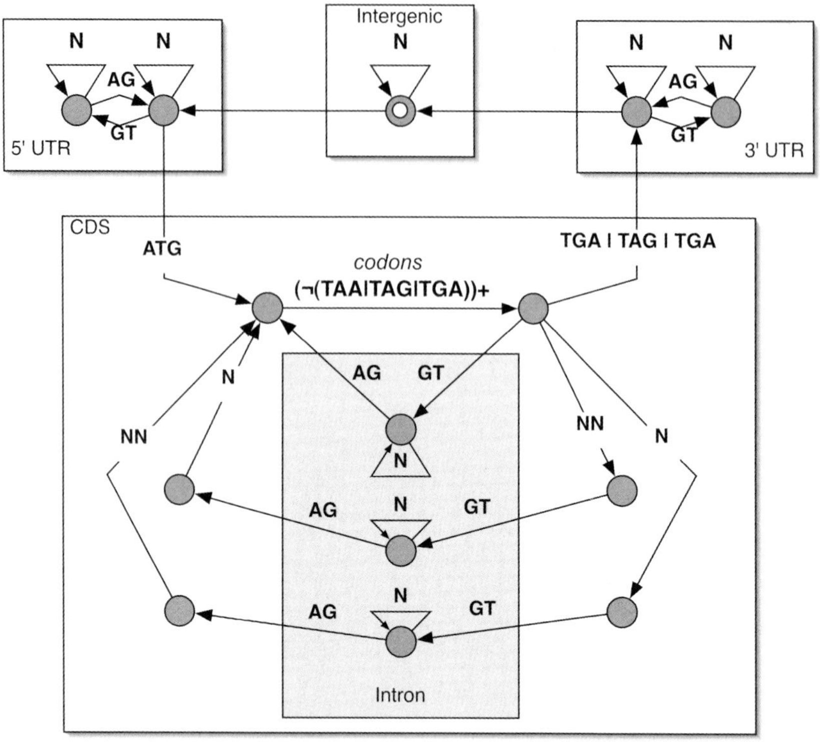

Figure 5 A finite state automaton (FSA) that recognizes gene structures. This simplified FSA recognizes legal protein-coding gene structures such that the total CDS length is a multiple of 3. Nucleotides are matched along the arcs and states are associated with nodes. N represents any base. "¬" indicates negation and "+" indicates one or more repeated times. The score functions $s(i,j,k)$ provide scores for sequences along the arcs. $l \in T(k)$ if there is a directed edge from k to l. $t(l,k)$ map to the possible outward arcs for node k. In this model, a single variable length codon state is used. In more sophisticated models there might be elaborate splice site states, states for initial, internal and final coding exons, promoter and polyadenylation sites, and reverse strand genes. The double circle is the start and end state.

The addition of length and state transition restrictions significantly improves running time, while ensuring that only meaningful parses are considered. In addition, software engineers for different systems have employed other tricks to improve the speed of gene prediction to approximately linear in the length of the input DNA sequence [9, 46, 55].

When the segment scores and transition scores are defined as log probabilities, which is easily derived from feature scoring methods such as weight matrices and Markov models, then we say that such a model of gene structure is a stochastic regular grammar or equivalently a HMM. (Note that sometimes the inclusion of variable length segments in the model is called a generalized HMM (GHMM) [44] or a state-duration HMM [60].) The dynamic program is then a maximum likelihood optimization:

$$\arg\max_{\Phi}(-\log P(X,\Phi)) \; .$$

Such models are called generative models because the score functions are decomposed into conditional probability terms of the form $-\log P(x_i \ldots x_j|q_k)$, corresponding to the $s(i,j,k)$ score function, and $-\log P(q_l|q_k)$, the $t(l,k)$ transition score. It is sometimes convenient to describe HMMs as generating the sequence X via a random walk through the finite state machine. The decoding step is, then, the prediction of the most likely random path, Φ, that generated the observed data (see also Chapter 3).

Almost all of the successful, modern gene finders are based on this HMM framework including the most widely used *ab initio* gene finders GENSCAN [9] and FGENES [65]. Furthermore, we will shortly see that improvements to gene finders with respect to the inclusion of homologous protein, aligned cDNA and orthologous DNA are all extensions of this basic HMM framework.

An additional advantage of the probabilistic framework is that the score functions can be learned systematically using standard learning procedures, i.e. a maximum likelihood optimization using the forward–backward algorithm [60]. In practice, the parameters for the different score functions are trained independently in most gene-finding programs, but HMMGene [43] is a notable exception that achieves good performance. In addition, using the same algorithm for a test sequence it is possible to obtain the score of any single feature (such as an internal coding exon) in the context of all possible parses that might contain it. Studies have shown that these scores are meaningful metrics for ranking the confidence of different segments of a prediction [9,46,63].

7 Performance Comparisons

Performance of different gene finders has been assessed by several researchers including studies by Reese and coworkers [62] on *Drosophila* and Rogic and coworkers [63] on mammalian sequences. First, in Reese and coworkers, a 2.9-Mb contiguous DNA sequence was subjected to automated analysis by a battery of gene-finding programs and compared with the gene structures from a careful manual curation. Assessing false-positive rate (over-prediction) in this test was problematic because full-length gene structures were known with certainty for only a fraction of genes, full-length cDNAs were rejected if they did not meet certain automatic criteria such as having a good splice site score and for those uncertain genes there was a serious bias because automated gene finder predictions had been used by the manual curators to guide their annotations. Five *ab initio* gene finders were tested and standard evaluation statistics were collected (using the same metrics as in Ref. [12]). Table 2 presents the performance predicting individual exons and the entire CDS from start to stop codon. The only strong conclusion that can be made is that HMM-based gene finders (FGENES, Genie and HMMGene) perform comparably and superior to the other methods. We also know from these tests and others that gene finders naturally perform better when trained with examples from the organism being tested or related species [78].

Table 2 *Ab initio* performance for the *Adh* locus in *Drosophila*

	FGENES (v1/v2/v3)			GeneID (v1/v2)		Genie	HMMGene	Grail
Exon								
Sn	0.65	0.44	0.75	0.27	0.58	0.70	0.68	0.42
Sp	0.49	0.68	0.24	0.29	0.34	0.57	0.53	0.41
Missing	0.11	0.46	0.06	0.54	0.21	0.08	0.05	0.24
Wrong	0.32	0.17	0.53	0.48	0.47	0.17	0.20	0.29
CDS								
Sn	0.30	0.09	0.37	0.02	0.26	0.40	0.35	0.14
Sp	0.27	0.18	0.10	0.05	0.10	0.29	0.30	0.12
Missing	0.09	0.35	0.09	0.44	0.14	0.05	0.07	0.16
Wrong	0.24	0.25	0.52	0.22	0.30	0.11	0.15	0.24

Sn refers to the fraction of known genes that were predicted exactly correct. Sp is the fraction of predicted genes that were exactly correct. Missing is the fraction of known genes with no overlapping prediction. Wrong is the fraction of predictions that do not overlap annotated genes. High Sn and Sp and low Missing and Wrong values are better. Sn and Missing were determined from a different test set than Sp and Wrong. FGENES and GeneID were run under multiple parameter settings to produce different sensitivity/specificity trade-offs. See Ref. [62] for details. Importantly, different versions of these programs have typically been developed since these tests, so conclusions should be qualitative regarding methodology only.

Recognizing that biases could exist in gene prediction programs if testing data included gene structures used in training, Rogic and coworkers assessed the performance of seven *ab initio* gene finders on 195 mammalian gene structures that were submitted to GenBank after the programs were released. In the overall results shown in Table 3, the ranges of performance measures is comparable to that for invertebrates and the gene finder HMMGene shows a

Table 3 *Ab initio* performance for mammalian genes [63]

	FGENES	GeneMark	Genie	GENSCAN	HMMGene	Morgan	MZEF
Exon							
Sn	0.67	0.53	0.71	0.70	0.76	0.46	0.58
Sp	0.67	0.54	0.70	0.70	0.77	0.41	0.59
Missing	0.12	0.13	0.19	0.08	0.12	0.20	0.32
Wrong	0.09	0.11	0.11	0.09	0.07	0.28	0.23

Measures are interpreted as in Table 2.

significant advantage over other methods. However, to complicate matters, Rogic and coworkers also found that gene finder performance was frequently highly dependent on gene or genome characteristics such as type of CDS exon (initial, internal or terminal) and G + C content.

A third study, by Guigo and coworkers, has shown that gene finders perform notably worse on long DNA sequences than for the short test sequences that contain only one gene found in assessment studies [28], so mammalian performance shown here is probably an upper bound and should be evaluated only relatively.

8 Using Homology

A second class of gene finders is those that take advantage of homologous sequences from databases of cDNA, DNA and protein sequences. Alignments of cDNA indicate the exon–intron structure. Conserved sequences between orthologous chromosomes indicates functional DNA, i.e. regulatory and protein-coding sequences, and protein–DNA similarity identifies putative CDS.

8.1 cDNA Clustering and Alignments

The gold standard for gene structures are derived from the alignment of cDNAs (complementary DNA from mRNA) to DNA. A full-length cDNA requires only the identification of the CDS, which is typically assumed to be the largest ORF. However, full-length cDNAs are rare. Instead, tens of millions of cDNA fragments called expressed sequence tags (ESTs) with lengths of several hundred bases have been deposited in GenBank. These sequences are typically random sequencing reads from the $3'$ or $5'$ ends of libraries of cloned full-length or partial mRNAs. The primary difficulties with ESTs are the relatively short size, the frequent sequencing errors and the sheer number of such sequences.

Occasionally ESTs are analyzed individually in the hope of identifying fragments of coding regions. For this purpose, specialized HMMs similar to profile HMMs (see Chapter 3) have been developed to identify the reading frame with the highest coding potential while allowing for frame shifts that interrupt the CDS [35]. The methods employ similar, but simpler, Markov model scoring metrics and state machines than those used for gene finders in DNA (Figure 6).

Most EST analysis is based on assembling multiple EST sequences to form longer cDNA sequences. If no genome sequence is available, then rapid clustering methods are often employed to generate groups of homologous

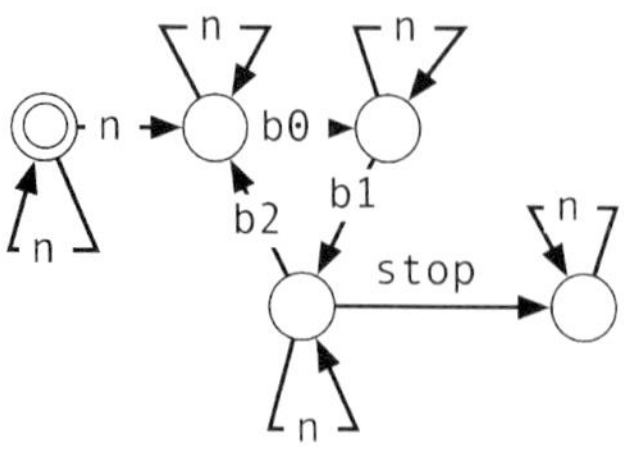

Figure 6 A finite state automaton for labeling the CDS (codons of "b0 b1 b2") in EST sequences allowing for frameshifts. A function such as the fifth-order Markov model is used for the scoring of the b0, b1 and b2 arcs.

ESTs [10]. These groups are then input to a conventional fragment assembler (see Chapter 2). This approach typically generates undesirable chimeric or partial assemblies and alternative isoforms can cause havoc.

In the conventional assembly approach, each EST contributes to just one assembled cDNA, but if multiple isoforms exist, then an EST should naturally be a part of multiple cDNA assemblies. A splice graph captures all of the possible isoforms implied by a set of ESTs [33]. In the splice graph, a virtual genome sequence is deduced, nodes are positions in the genome, and arcs connect positions that are adjacent in aligned ESTs. Figure 7 shows an example visualization of a splice graph. The splice graph makes clear that the number of possible isoforms can, in the worst case, be exponential to the number of exons.

With a completed genome sequence, cDNAs can be aligned to the DNA, which serves as a template, and the gene structure can be delineated by the connected set of overlapping, aligned ESTs. This is superior to the previous cluster-based transcript assembly because (i) errors in ESTs can be corrected by comparison with DNA, (ii) chimeras are less likely and (iii) the genome is directly annotated providing exon–intron structure.

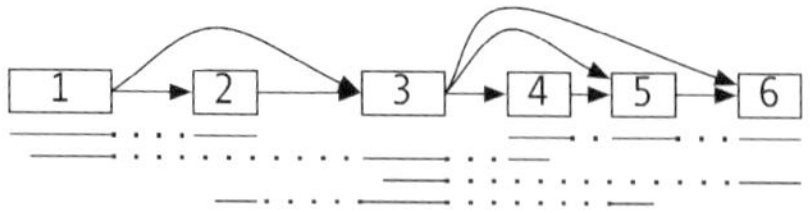

Figure 7 Splice graph. Nodes are positions along the horizontal axis, representing a DNA sequence. (When no genome sequence is available, the DNA sequence is virtual and ambiguous, i.e. implied exclusively by the differences between homologous ESTs.) Nodes are connected based on EST evidence. Those nodes adjacent on the DNA are merged into exons (numbered boxes). Arcs between boxes are introns. Assembled ESTs are shown below the splice graph. Dotted lines show where an EST spans across a DNA gap. Note that the genomic DNA is not necessary to build a splice graph. However, errors in the EST sequences can easily introduce false variants. To address this, multiple ESTs are usually required to confirm alternative splicings.

Highly accurate programs for cDNA–DNA alignment that include special handling for small exons, accurate splice site definition and EST errors are now available, most notably GMAP [79] or BLAT [38]. Imposing strict alignment criteria is usually advisable such as requiring that at least 90% of the EST be aligned with 95% identity in the aligned region. Even stricter alignment criteria are common, such as requiring that all intron gaps in an alignment conform to consensus dinucleotide splice sites and span a minimal number of genomic bases.

EST sequences are obtained from the 3′ and 5′ ends of cDNA clone inserts (and sometimes random internal positions). Due to the construction of the vector sequence containing the cDNA insert, sequencing of the two ends occurs on opposite strands of the insert. By convention these sequences are deposited in GenBank without reverse complementing and so usually a 5′ EST sequence is of the sense strand and a 3′ EST is of the anti-sense strand. Therefore, the orientation of a gene on the DNA can be inferred by comparing the EST read direction and the orientation of the sequence in the EST–DNA alignment. However, the labeling of the read direction and the strand that is submitted to the database is only a convention and there are frequent errors in the EST database, mostly among older database entries due to lane shifts from gel-based sequencing machines [1]. Other characteristics can be used to infer orientation of the EST including comparisons to other aligned sequences, presence of a poly-A tail (or poly-T prefix), presence of a polyadenylation signal and, most effectively, the consensus dinucleotides in the splice sites, if the EST splices. Shendure and Church [69] describe one laboratory's orientation procedure although no single software program currently exists to perform this orientation. (A lingering problem remains that EST–DNA alignments may indicate anti-sense transcription – a phenomenon that has been increasingly documented [81]; however, conventional gene modeling and genome annotation prohibit overlapping transcripts.)

Once individual ESTs are oriented and aligned, longer transcripts can be derived by merging gene structures implied by overlapping ESTs. As with EST analysis without a genome sequence, conflicting alignments can imply a large number of putative alternative splice forms. The PASA program is one of several programs that can be used to generate a a set of such putative transcripts from EST–DNA alignments [31,37]. In PASA, a dynamic program, is employed to assemble a minimal set of unique transcripts from compatible EST alignments, i.e. those ESTs that agree in all of their inferred splicings.

Not all genes will be present in EST libraries because differentially expressed genes may be absent or expressed at low levels in the sampled tissues. Furthermore, due to limitations in full-length cDNA cloning and the long lengths of some transcripts, many genes are not fully covered by EST sequences. Thus, assembly of ESTs tends to result in fragmented, partial transcripts.

One method that has been used to specifically deal with the incomplete EST information is Genie [46]. The method assumes that assembled transcripts from EST–DNA alignments define true, but incomplete, gene structures and so the *ab initio* gene-finding algorithm described in Section 6.3 is employed only in the breaks between alignments. This can be achieved in a straightforward manner by modifying the $t(;)$ transcription score function to prohibit transitioning into some states at specific positions within the sequence depending on the EST–DNA alignment evidence (e.g. transitioning into an exon state in the middle of an EST-defined intron is prohibited in the dynamic program).

Lastly, it is possible to leverage the EST data to infer gene bounds even when only incomplete EST alignments exist using EST mate pairs. For some cDNA inserts both the $5'$ and $3'$ ends have been sequenced. When both ESTs are aligned to the same chromosome, within a reasonable genomic distance, and compatibly ordered and oriented, then one can infer that the entire region between the mate pairs corresponds to a single primary transcript. In a similar way as above, an *ab initio* gene finder can be constrained to predict exactly one primary transcript in the defined region.

8.2 Orthologous DNA

When two organisms are sufficiently similar to identify and align orthologous genomic sequence, but sufficiently distant so that nonfunctional DNA has mutated, then the comparative analysis of the two genomes can be directly applied to gene finding (see also Chapter 37). Organisms such as chicken and mouse are of a reasonable evolutionary distance from human to support this sort of comparison. The key assumptions are that the number and approximate content of CDS regions between the two species are well conserved, while introns, UTRs and intergenic DNA have drifted significantly from the common ancestor. Thus, if the conserved regions can be identified, then they are most likely coding regions and so the gene-finding problem is to combine these conserved CDS segments into a more complete gene structure.

There are two main approaches to the problem: *ad hoc* weighting schemes and principled pair HMMs. Twinscan [39] and SGP2 [56] are examples of the former method. For example, in Twinscan, the dynamic program corresponding to the gene grammar described in Section 6.3 is augmented with scores from BLAST sequence similarity matches between the two genomes. In other words, for any candidate region $x_i \dots x_j$ in a CDS state, q_k, the score function $s(i, j, k)$ is improved according to the quality of the genome–genome

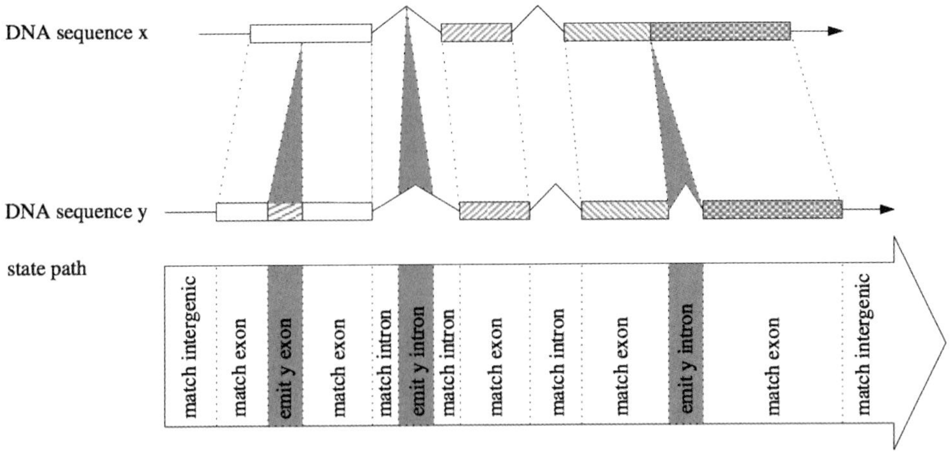

Figure 8 Pair HMM. Two DNA sequences are generated simultaneously from left to right. At each step, a subsequence (possibly of length zero) is emitted for each genome. The score for the pairs of subsequences is based on the local statistics like a conventional *ab initio* gene finder as well as the similarity of the two subsequences. (From Meyer and Durbin [51].)

alignment in that region. Thus, the method performs gene finding on one genome sequence using the second genome as evidence.

The second, more elegant, approach is to simultaneously align and label both genome sequences according to a probabilistic model. DoubleScan [51] and SLAM [3] embody this class of gene finders. The approach, called a pair HMM, is a generalization of the *ab initio* method and is best understood by considering an HMM-based gene finder as a generative model that produces labeled sequences of DNA, as previously described in Section 6.3. In a conventional HMM, one or more nucleotides are emitted for each state; in a pair HMM two sequences of nucleotides are emitted for each state. To address asymmetries such as inserts in one genome additional states must be added to allow for null string emissions in one genome. Figure 8 shows a diagram of the generative process.

Our score function $s(i,j,k)$ is extended to consider the scores of pairs of segments, i.e. $s(i,j,m,n,k)$ is the score for simultaneously emitting DNA sequences $x_i \ldots x_j$ and $y_m \ldots y_n$ in state q_k. The dynamic program for a pair HMM is not much different from the conventional single sequence HMM although the pair HMM must consider, theoretically, all possible segments $x_i \ldots x_j$ and $y_m \ldots y_n$ in every possible state, which adds a very significant computational burden. In practice this computational burden may not be worth the investment since the *ad hoc* score enhancement methods are fast and have been shown to perform well.

Conserved noncoding sequences (CNS), which are usually regulatory sequences, are often a problem for comparative modeling methods because they are interpreted as coding. In some implementations special CNS states are introduced and, in theory, if the CNS sequences lack sufficient coding potential, then they will be labeled CNS instead of CDS. In practice, mispredicting CNS remains a challenge, particularly for highly conserved genes.

The comparative method has also been extended to model the proper phylogenetic distance among three or more genomes in a phylo-HMM [70] and also to successfully annotate very closely related species such as among primates [50].

8.3 Protein Homology

The simplest application of protein homology is the identification of putative CDS subsequences from sequence alignment. For example, BLASTX [24] performs six-frame translation of a DNA query sequence and rapidly identifies those regions that are similar to known proteins. The user must assemble the fragmentary evidence.

The most elegant and specific use of protein homology is employed by the GeneWise [6] program, which merges the profile HMM used in protein remote homology searching (see Chapter 11) and the gene finding HMM model described in Section 6.3 into a unified DNA–protein alignment. The method is a pair HMM, similar to those used in comparative genomic analysis, but in this case the model is more complex because the two generated sequences use different alphabets, and additional constraints must be included to ensure proper pairing of amino acids and codons according to the genetic code. The model also includes a basic set of splice site recognition states to allow for the alignment of the protein sequence across introns (similar to cDNA–DNA alignment) as well as nucleotide insertion and deletion states to allow for errors and frame shifts.

GeneWise produces only partial gene structures corresponding to the region of protein alignment on the DNA. However, importantly, such alignments provide highly accurate predictions when a sufficiently close homolog is available. Moreover, the prediction of splice sites is particularly good due to the constraint of the alignment of the protein across introns.

HMM gene-finding programs such as FGENES++ [65] and Genie [45] employ a more *ad hoc* approach in which scores for coding features are artificially inflated when database similarities are found, in a similar manner as the comparative genome program Twinscan [39]. As a protein–DNA alignment improves, the score for labeling the DNA region as coding improves. In this way, a complete gene structure is predicted with protein homology evidence contributing, but it is not used exclusively nor is it required. Such an approach requires careful tuning of the contribution of protein homology to avoid over-prediction or over extension of coding exons.

8.4 Integrative Methods

There has been much work on integrative methods of combining multiple gene finders and homology evidence, both principled and *ad hoc*, for whole genome analysis, but we do not review them in detail here. Several programs have attempted to integrate the predictions of multiple gene finders within a probabilistic model (e.g. Refs. [4, 53, 57]). Other programs provide an abstract framework of an HMM gene finder that allows a software developer to incorporate arbitrary feature scoring methods [34]. Most genome centers and informatics sites maintain "pipelines" for automated annotation. Many of these are not portable or are tightly bound to other institutional software infrastructure. Two noteworthy examples are the annotation pipelines of the NCBI (http://www.ncbi.nlm.nih.gov/genome/guide/build.html) and Ensembl [17]. Both include sophisticated and comprehensive methods for reliable whole genome gene prediction.

9 Pitfalls: Pseudogenes, Splice Variants and the Cruel Biological Reality

As this chapter concludes its tour of gene-finding methods, it is worth a brief mention of some of the unfortunate difficulties that make gene prediction a hard problem that is unlikely to be satisfactorily solved in the near term. The challenges almost all lie in the complexity of genome organization that is not (and often cannot be) modeled by the various gene-finding techniques [52]. Here are several issues:

- Exons can be extremely small – only a few nucleotides, which is insufficient to detect coding potential. Worse, re-splicing has been observed in which an exon is entirely removed.
- Noncanonical splice sites are expected to occur, on average, about once every 10 genes in the genome; however, with few exceptions, methods assume GT/AG splice sites.
- Pseudogenes are numerous in many organisms including the human. There are specialized programs to detect retrotranscribed and nonfunctional genes (e.g. Ref. [15]), but young pseudogenes are often quite difficult to distinguish by sequence statistics.

- Alternative splicing is prolific in higher eukaryotes. It is estimated that 60% of human genes have multiple isoforms. We saw in Section 8.1 that cDNA methods can be used to enumerate possible splice variants. HMM-based methods like those we learned here can be used to generate suboptimal parses that sometimes represent alternative isoforms [14]. Recent predictive methods have been developed for detecting alternative splice sites [19]. Nevertheless, alternative splicing remains a serious impediment to automated genome annotation.
- Untranslated exons make up a large fraction of the typical gene, yet there is very little, if any, signal differentiating UTRs from intergenic DNA.
- The size of most genomes implies that gene-like patterns are likely to occur frequently in intergenic DNA. In order to control false-positive rates, gene-finding programs must also sacrifice true positives. Studies of gene finders in large DNA sequences reveal frequent overprediction [28].
- Anti-sense transcription [81], high rates of transcription outside of known protein-coding genes [36] and large classes of small noncoding RNAs have been observed [54]. Besides causing difficulties in the use of cDNA–DNA alignment evidence for gene finding, these findings emphasize that important, functional, nonprotein-coding genes are being transcribed at high rates and that new classes of genes may yet be discovered. Alternatively, the high rates of transcription observed by Kampa and coworkers [36] also suggests that the genome is less of a programmed machine and more of a stochastic process in which nonfunctional transcription may be occurring at high levels. Such "noise" in the system is insurmountable using sequence analysis alone.

As a result of these and other complications, manual genome annotation is expected to remain the definitive source for gene structures for a long time. An excellent source of manual curations is the VEGA system [49].

10 Further Reading

Chapter 3 of this book introduces HMMs. Further background on HMMs and probabilistic modeling of gene sequences – the techniques that dominate gene finding – is best found in the book *Biological Sequence Analysis* [20] and Rabiner's oft-cited tutorial [60]. An excellent gene-finding bibliography is maintained by Wentian Li (http://www.nslij-genetics.org/gene/). The primary literature for generalized HMMs (e.g. GENSCAN [9] and Genie [44,46]) and comparative gene finders (e.g. DoubleScan [51], SLAM [3], and Shadower

[50]) is particularly good for new readers. There are many reviews of gene-finding techniques and Zhang's [84] is a relatively recent good one. The website www.genefinding.org is also a useful resource for developers.

References

1 Aaronson, J., B. Eckman, R. Blevins, J. Borkowski, J. Myerson, S. Imran and K. Elliston. 1996. Toward the development of a gene index to the human genome: an assessment of the nature of high-throughput EST sequence data. Genome Res. **6**: 829–45.

2 Agarwal, P. and V. Bafna. 1998. The ribosome scanning model for translation initiation: implications for gene prediction and full-length cDNA detection. Proc. ISMB **6**: 2–7.

3 Alexandersson, M., S. Cawley and L. Pachter. 2003. SLAM: cross-species gene finding and alignment with a generalized pair hidden Markov model. Genome Res. **13**: 496–502.

4 Allen, J., M. Pertea and S. Salzberg. 2004. Computational gene prediction using multiple sources of evidence. Genome Res. **14**: 142–8.

5 Berget, S. 1995. Exon recognition in vertebrate splicing. J. Biol. Chem. **270**: 2411–4.

6 Birney, E., M. Clamp and R. Durbin. 2004. GeneWise and Genomewise. Genome Res. **14**: 988–95.

7 Borodovsky, M. and J. McIninch. 1993. GeneMark: parallel gene recognition for both DNA strands. Comput. Chem. **17**: 123–33.

8 Brunak, S., J. Engelbrecht and S. Knudsen. 1991. Prediction of human mRNA donor and acceptor sites from the DNA sequence. J. Mol. Biol. **220**: 49–65.

9 Burge, C. and S. Karlin. 1997. Prediction of complete gene structures in human genomic DNA. J. Mol. Biol. **268**: 78–94.

10 Burke, J., D. Davison and W. Hide. 1999. d2_cluster: a validated method for clustering EST and full-length cDNA sequences. Genome Res. **9**: 1135–42.

11 Burset, M., I. Seledtsov and V. Solovyev. 2000. Analysis of canonical and non–canonical splice sites in mammalian genomes. Nucleic Acids Res. **28**: 4364–75.

12 Burset, M. and R. Guigo. 1996. Evaluation of gene structure prediction programs. Genomics **34**: 353–67.

13 Cai, D., A. Delcher, B. Kao and S. Kasif. 2000. Modeling splice sites with Bayes networks. Bioinformatics **16**: 152–8.

14 Cawley, S. and L. Pachter. 2003. HMM sampling and applications to gene finding and alternative splicing. Bioinformatics **19** (Suppl. 2): II36–41.

15 Coin, L. and R. Durbin. 2004. Improved techniques for the identification of pseudogenes. Bioinformatics **20** (Suppl. 1): I94–I100.

16 Consortium, T. 1998. Genome sequence of the nematode *C. elegans*: a platform for investigating biology. Science **282**: 2012–8.

17 Curwen, V., E. Eyras, T. Andrews, L. Clarke, E. Mongin, S. Searle and M. Clamp. 2004. The Ensembl automatic gene annotation system. Genome Res. **14**: 942–50.

18 Dong, S. and D. Searls. 1994. Gene structure prediction by linguistic methods. Genomics **23**: 540–51.

19 Dror, G., R. Sorek and R. Shamir. 2005. Accurate identification of alternatively spliced exons using support vector machine. Bioinformatics **21**: 897–901.

20 Durbin, R., S. Eddy, A. Krogh and G. Mitchison. 1998. *Biological Sequence Analysis*. Cambridge University Press, Cambridge.

21 Fickett, J. and A. Hatzigeorgiou. 1997. Eukaryotic promoter recognition. Genome Res. **7**: 861–78.

22 Fickett, J. and C. Tung. 1992. Assessment of protein coding measures. Nucleic Acids Res **20**: 6441–50.

23 Forney, G. 1973. The Viterbi algorithm. Proc. IEEE **61**: 268–78.

24 Gish, W. and D. States. 1993. Identification of protein coding regions by database similarity search. Nat. Genet. **3**: 266–72.

25 Graber, J., G. McAllister and T. Smith. 2002. Probabilistic prediction of *Saccharomyces cerevisiae* mRNA 3′-processing sites. Nucleic Acids Res. **30**: 1851–8.

26 Gribskov, M., J. Devereux and R. Burgess. 1984. The codon preference plot: graphic analysis of protein coding sequences and prediction of gene expression. Nucleic Acids Res. **12**: 539–49.

27 Grosjean, H. and W. Fiers. 1982. Preferential codon usage in prokaryotic genes: the optimal codon-anticodon interaction energy and the selective codon usage in efficiently expressed genes. Gene **18**: 199–209.

28 Guigo, R., P. Agarwal, J. Abril, M. Burset and J. Fickett. 2000. An assessment of gene prediction accuracy in large DNA sequences. Genome Res. **10**: 1631–42.

29 Guigo, R. and J. Fickett. 1995. Distinctive sequence features in protein coding genic non-coding, and intergenic human DNA. J. Mol. Biol. **253**: 51–60.

30 Guo, M. and S. Mount. 1995. Localization of sequences required for size-specific splicing of a small *Drosophila* intron *in vitro*. J. Mol. Biol. **253**: 426–37.

31 Haas, B., A. Delcher, S. Mount, J. Wortman et al. 2003. Improving the Arabidopsis genome annotation using maximal transcript alignment assemblies. Nucleic Acids Res. **31**: 5654–66.

32 Hawkins, J. 1988. A survey on intron and exon lengths. Nucleic Acids Res. **16**: 9893–908.

33 Heber, S., M. Alekseyev, S. Sze, H. Tang and P. Pevzner. 2002. Splicing graphs and EST assembly problem. Bioinformatics **18** (Suppl. 1): S181–8.

34 Howe, K., T. Chothia and R. Durbin. 2002. GAZE: a generic framework for the integration of gene-prediction data by dynamic programming. Genome Res. **12**: 1418–27.

35 Iseli, C., C. Jongeneel and P. Bucher. 1999. ESTScan: a program for detecting, evaluating, and reconstructing potential coding regions in EST sequences. Proc. ISMB **7**: 138–48.

36 Kampa, D., J. Cheng, P. Kapranov, et al. 2004. Novel RNAs identified from an in-depth analysis of the transcriptome of human chromosomes 21 and 22. Genome Res. **14**: 331–42.

37 Kan, Z., E. Rouchka, W. Gish and D. States. 2001. Gene structure prediction and alternative splicing analysis using genomically aligned ESTs. Genome Res. **11**: 889–900.

38 Kent, W. 2002. BLAT – the BLAST-like alignment tool. Genome Res. **12**: 656–64.

39 Korf, I., P. Flicek, D. Duan and M. Brent. 2001. Integrating genomic homology into gene structure prediction. Bioinformatics **17** (Suppl. 1): S140–8.

40 Kozak, M. 1987. An analysis of 5′-noncoding sequences from 699 vertebrate messenger RNAs. Nucleic Acids Res. **15**: 8125–48.

41 Kozak, M. 1991. Structural features in eukaryotic mRNAs that modulate the initiation of translation. J. Biol. Chem. **266**: 19867–70.

42 Krogh, A., M. Brown, I. Mian, K. Sjolander and D. Haussler. 1994. Hidden Markov models in computational biology. Applications to protein modeling. J. Mol. Biol. **235**: 1501–31.

43 Krogh, A. 1997. Two methods for improving performance of an HMM and their application for gene finding. Proc. ISMB **5**: 179–86.

44 Kulp, D., D. Haussler, M. Reese and F. Eeckman. 1996. A generalized hidden Markov model for the recognition of human genes in DNA. Proc. ISMB **4**: 134–42.

45 Kulp, D., D. Haussler, M. Reese and F. Eeckman. 1997. Integrating database homology in a probabilistic gene structure model. Pac. Symp. Biocomput. **2**: 232–44.

46 KULP, D. 2003. Protein-coding gene structure prediction using generalized hidden Markov models. University of California. PhD Dissertation.

47 LEGENDRE, M. AND D. GAUTHERET. 2003. Sequence determinants in human polyadenylation site selection. BMC Genomics **4**: 7.

48 LI, H. AND T. JIANG. 2004. A class of edit kernels for SVMs to predict translation initiation sites in eukaryotic mRNAs. Proc. RECOMB **8**: 262–71.

49 LOVELAND, J. 2005. VEGA, the genome browser with a difference. Brief. Bioinform. **6**: 189–93.

50 MCAULIFFE, J., L. PACHTER AND M. JORDAN. 2004. Multiple-sequence functional annotation and the generalized hidden Markov phylogeny. Bioinformatics.

51 MEYER, I. AND R. DURBIN. 2002. Comparative *ab initio* prediction of gene structures using pair HMMs. Bioinformatics **18**: 1309–18.

52 MOUNT, S. 2000. Genomic sequence, splicing, and gene annotation. Am. J. Hum. Genet. **67**: 788–92.

53 MURAKAMI, K. AND T. TAKAGI. 1998. Gene recognition by combination of several gene-finding programs. Bioinformatics **14**: 665–75.

54 OTA, T., Y. SUZUKI, T. NISHIKAWA, ET AL. 2004. Complete sequencing and characterization of 21,243 full-length human cDNAs. Nat. Genet. **36**: 40–5.

55 PARRA, G., E. BLANCO AND R. GUIGO. 2000. GeneID in *Drosophila*. Genome Res. **10**: 511–5.

56 PARRA, G., P. AGARWAL, J. ABRIL, T. WIEHE, J. FICKETT AND R. GUIGO. 2003. Comparative gene prediction in human and mouse. Genome Res. **13**: 108–17.

57 PAVLOVIC, V., A. GARG AND S. KASIF. 2002. A Bayesian framework for combining gene predictions. Bioinformatics **18**: 19–27.

58 PEDERSEN, A. AND H. NIELSEN. 1997. Neural network prediction of translation initiation sites in eukaryotes: perspectives for EST and genome analysis. Proc. ISMB **5**: 226–33.

59 PERTEA, M., X. LIN AND S. SALZBERG. 2001. GeneSplicer: a new computational method for splice site prediction. Nucleic Acids Res. **29**: 1185–90.

60 RABINER, L. 1989. A tutorial on hidden Markov models and selected applications in speech recognition. Proc. IEEE **77**: 257–86.

61 REESE, M., F. EECKMAN, D. KULP AND D. HAUSSLER. 1997. Improved splice site detection in Genie. J. Comput. Biol. **4**: 311–23.

62 REESE, M., G. HARTZELL, N. HARRIS, U. OHLER, J. ABRIL AND S. LEWIS. 2000. Genome annotation assessment in *Drosophila melanogaster*. Genome Res. **10**: 483–501.

63 ROGIC, S., B. OUELLETTE AND A. MACKWORTH. 2002. Improving gene recognition accuracy by combining predictions from two gene-finding programs. Bioinformatics **18**: 1034–45.

64 SALAMOV, A., T. NISHIKAWA AND M. SWINDELLS. 1998. Assessing protein coding region integrity in cDNA sequencing projects. Bioinformatics **14**: 384–90.

65 SALAMOV, A. AND V. SOLOVYEV. 2000. *Ab initio* gene finding in *Drosophila* genomic DNA. Genome Res. **10**: 516–22.

66 SALZBERG, S., M. PERTEA, A. DELCHER, M. GARDNER AND H. TETTELIN. 1999. Interpolated Markov models for eukaryotic gene finding. Genomics **59**: 24–31.

67 SALZBERG, S. 1997. A method for identifying splice sites and translational start sites in eukaryotic mRNA. Comput. Appl. Biosci. **13**: 365–76.

68 SEARLS, D. 1992. The Linguistics of DNA. Am. Sci. **80**: 579–91.

69 SHENDURE, J. AND G. CHURCH. 2002. Computational discovery of sense–antisense transcription in the human and mouse genomes. Genome Biol. **3**: 1–14.

70 SIEPEL, A. AND D. HAUSSLER. 2004. Combining phylogenetic and hidden Markov models in biosequence analysis. J. Comput. Biol. **11**: 413–28.

71 SNYDER, E. AND G. STORMO. 1993. Identification of coding regions in genomic DNA sequences: an application of dynamic programming and neural networks. Nucleic Acids Res. **21**: 607–13.

72 SOLOVYEV, V., A. SALAMOV AND C. LAWRENCE. 1994. Predicting internal exons by oligonucleotide composition and discriminant analysis of spliceable open reading frames. Nucleic Acids Res. **22**: 5156–63.

73 SONNENBURG, S., G. RATSCH, A. JAGOTA AND M. KLAUS-ROBERT. 2002. New methods for splice site recognition. In Proc. Int. Conf. on Artificial Neural Networks, Madrid: 329–36.

74 STADEN, R. 1984. Computer methods to locate signals in nucleic acid sequences. Nucleic Acids Res. **12**: 505–19.

75 STADEN, R. 1990. Finding protein coding regions in genomic sequences. Methods Enzymol. **183**: 163–80.

76 STORMO, G. 1990. Consensus patterns in DNA. Methods Enzymol. **183**: 211–21.

77 TABASKA, J. AND M. ZHANG. 1999. Detection of polyadenylation signals in human DNA sequences. Gene **231**: 77–86.

78 WATERSTON, R., K. LINDBLAD-TOH, E. BIRNEY, ET AL. 2002. Initial sequencing and comparative analysis of the mouse genome. Nature **420**: 520–62.

79 WU, T. AND C. WATANABE. 2005. GMAP: a genomic mapping and alignment program for mRNA and EST sequences. Bioinformatics **21**: 1859–75.

80 YAN, J. AND T. MARR. 2005. Computational analysis of 3′-ends of ESTs shows four classes of alternative polyadenylation in human, mouse, and rat. Genome Res. **15**: 369–75.

81 YELIN, R., D. DAHARY, R. SOREK, ET AL. 2003. Widespread occurrence of antisense transcription in the human genome. Nat. Biotechnol. **21**: 379–86.

82 ZHANG, L. AND L. LUO. 2003. Splice site prediction with quadratic discriminant analysis using diversity measure. Nucleic Acids Res. **31**: 6214–20.

83 ZHANG, M. 1997. Identification of protein coding regions in the human genome by quadratic discriminant analysis. Proc. Natl. Acad. Sci. USA **94**: 565–8.

84 ZHANG, M. 2002. Computational prediction of eukaryotic protein–coding genes. Nat. Rev. Genet. **3**: 698–709.

85 ZIEN, A., G. RATSCH, S. MIKA, B. SCHOLKOPF, T. LENGAUER AND K. MULLER. 2000. Engineering support vector machine kernels that recognize translation initiation sites. Bioinformatics **16**: 799–807.

6
Analyzing Regulatory Regions in Genomes

Thomas Werner

1 General Features of Regulatory Regions in Eukaryotic Genomes

Regulatory regions share several common features despite their obvious divergence in sequence. Most of these common features are not evident directly from the nucleotide sequence, but result from the restraints imposed by functional requirements. Therefore, understanding of the major components and events during the formation of regulatory DNA–protein complexes is crucial for the design and evaluation of algorithms for the analysis of regulatory regions. Transcription initiation from polymerase II (Pol II) is the best understood example so far and will be a major focus of this chapter. However, the mechanisms and principles revealed from promoters are mostly valid for other regulatory regions as well.

Algorithms for the analysis and recognition of regulatory regions draw from the underlying biological principles, to some extent, in order to generate suitable computational models. Therefore, a brief overview over the biological requirements and mechanisms is necessary to understand what are the strengths and weaknesses of the individual algorithms. The choice of parameters and implementation of the algorithms largely control the sensitivity and speed of a program. The specificity of software recognizing regulatory regions in DNA is determined, to a large extent, by how closely the algorithm follows what will be called the biological model from hereon. Several overviews of this topic have been published [27,96].

1.1 General Functions of Regulatory Regions

The biological functionality of regulatory regions is generally not a property evenly spread over the regulatory region in total. Functional units usually are defined by a combination of defined stretches that can be delimited and possess an intrinsic functional property (e.g. binding of a protein or a curved DNA structure). Several functionally similar types of these stretches of DNA are already known and will be referred to as *elements*. Those elements are

Bioinformatics - From Genomes to Therapies Vol. 1. Edited by Thomas Lengauer

ISBN: 978-3-527-31278-8

neither restricted to regulatory regions nor individually sufficient for the regulatory function of a promoter or enhancer. The function of the complete regulatory region is composed from the functions of the individual elements either in an additive manner (independent elements) or by synergistic effects (modules).

1.2 Most Important Elements in Regulatory Regions

Transcriptional regulation depends on sequence elements that are directly accessible from the genomic DNA sequence such as transcription factor (TF)-binding sites (TFBSs), repeats and hairpins (repeats that can form hairpin-like structures by self-complementarity). In addition, various elements not easily detectable in the sequence are important. Most of these affect chromatin structure and accessibility such as histone acetylation and methylation status as well as DNA methylation status. Such phenomena not directly linked to the local DNA sequence are usually summarized under epigenetic effects.

1.3 TFBSs

Binding sites for specific proteins are most important among the sequence elements. They consist of about 10–30 nucleotides, not all of which are equally important for protein binding. As a consequence, individual protein-binding sites vary in sequence, even if they bind to the same protein. There are nucleotides contacted by the protein in a sequence-specific manner, which are usually the best-conserved parts of a binding site. Different nucleotides are involved in DNA backbone contacts, i.e. contacting the sugar-phosphate framework of the DNA helix (not sequence specific as they do not involve the bases A, G, C or T). There are also internal "spacers" not contacted by the protein at all. In general, protein-binding sites exhibit enough sequence conservation to allow for the detection of candidates by a variety of sequence similarity-based approaches. Potential binding sites can be found almost all over the genome and are not restricted to regulatory regions. Quite a number of binding sites outside regulatory regions are also known to bind their respective binding proteins [57]. Therefore, the abundance of predicted binding sites is not just a shortcoming of the detection algorithms, but reflects biological reality. Often it is not possible either to identify individual binding proteins as they might bind as part of multi-protein complexes [68]. This illustrates another important point: TF binding *in vivo* is usually context dependent. The isolated TF will bind to a cognate site quite differently if brought together in a reaction tube as naked protein and oligonucleotide probe than *in vivo* where adaptive DNA structure and a host of other proteins are present. As became evident from several chromatin immunoprecipitation

(ChIP) studies, even *in vivo* binding of a TF does not automatically imply a function in transcription control as was found in a genome-wide study which identified many more cAMP response element-binding protein (CREB)-binding sites than CREB-regulated genes [45].

1.4 Sequence Features

Regulatory DNA also contains several features not directly resulting in recognizable sequence conservation. For example, two copies of a direct repeat (approximate or exact) are conserved in sequence with respect to each other, but different direct repeats are not similar in sequence at all. Nevertheless, direct repeats are quite common within regulatory DNA regions. They consist either of short sequences, which are repeated twice or more frequently within a short region, or they can be complex repeats, which repeat a pattern of two or more elements. (More details on sequence repeats and how to detect them can be found in Chapter 7.) Repeat structures are often associated with enhancers. Enhancers are DNA structures that enhance transcription over a distance without being promoters themselves. One example of a highly structured enhancer is the interleukin-2 enhancer [74]. Other sequence features that are hard to detect by computer methods include the relatively weak nucleosomal positioning signals [46], DNA stretches with intrinsic three-dimensional (3-D) structures (like curved DNA, e.g. Ref. Ref. [81]), methylation signals (if there are definite signals for methylation at all) and other structural elements.

1.5 Structural Elements

Currently, secondary structures are the most useful structural elements with respect to computer analysis. Secondary structures are mostly known for RNAs (see Chapter 14) and proteins (see Chapter 9), but they also play important roles in DNA. DNA can form double hairpins called cruciform DNA representing the hairpin structures of RNA and can be important for transcriptional regulation [59]. Potential secondary structures can be easily determined and even scored via the negative enthalpy that should be associated with the actual formation of the hairpin (single-strand) or cruciform (double-strand) structure. Secondary structures are also not necessarily conserved in the primary nucleotide sequence, but are subject to strong positional correlation within the 3-D structure, i.e. the orientation of the double helix in space. Without any doubt 3-D aspects of DNA sequences are very important for the functionality of such regions. However, existing attempts to calculate such structures in reasonable time have met with mixed success and cannot be used for a routine sequence analysis at present. Part of that difficulty is that DNA structure can be quite flexible and structural changes are readily induced by interacting proteins [68].

1.6 Organizational Principles of Regulatory Regions

This section will mainly concentrate on eukaryotic polymerase II promoters, as they are currently the best-studied regulatory regions.

1.6.1 Overall Structure of Pol II Promoters

Promoters are DNA regions capable of specific initiation of transcription (start of RNA synthesis) and consist of three basic regions (see Section 1.6.3). The part determining the exact nucleotide for transcriptional initiation is called the core promoter, and is the stretch of DNA sequence where the RNA polymerase and its cofactors assemble on the promoter.

The region immediately upstream of the core promoter is called the proximal promoter and usually contains a number of TFBSs responsible for the assembly of an activation complex. This complex in turn recruits the polymerase complex. It is generally accepted that most proximal promoter elements are located within a stretch of about 250–500 nucleotides upstream of the actual transcription start site (TSS).

The third part of the promoter is located even further upstream and is called the distal promoter. This region usually regulates the activity of the core and the proximal promoter, and also contains TFBSs. However, distal promoter regions and enhancers exhibit no principal differences. If a distal promoter region acts position and orientation independent it is called an enhancer.

1.6.2 TFBS in Promoters

The TFBSs within promoters (and likewise most other regulatory sequences) do not show any general patterns with respect to location and orientation within the promoter sequences, although particular functionality may be associated with a specific location or association within the promoter [89].

Even functionally important binding sites for a specific TF may occur almost anywhere within a promoter. For example, functional activating protein 1 (AP-1, a complex of two TFs: one from the Fos and one from the Jun family)-binding sites can be located far upstream, as in the rat bone sialoprotein gene where an AP-1 site located about 900 nucleotides upstream of the TSS inhibits expression [97]. An AP-1 site located close to the TSS is important for the expression of Moloney murine leukemia virus [75]. Moreover, functional AP-1 sites have also been found inside exon 1 (downstream of the TSS) of the proopiomelanocortin gene [11] as well as within the first intron of the *fra-1* gene [6], both locations outside the promoter. Similar examples can be found for several other TF sites, illustrating why no general correlation of TF sites within specific promoter regions can be defined. TFBSs can be found virtually everywhere in promoters, but in individual promoters possible locations are much more restricted. A closer look reveals that the function of an AP-1-binding site often depends on the relative location and, especially, on the sequence context of the binding site. The AP-1 site in the above-mentioned rat bone sialoprotein gene overlaps with a set of glucocorticoid-responsive element (GRE, the DNA sequence that is bound by the glucocorticoid receptor which is a TF) half-sites (nuclear factor-binding sites are often composed of two almost identical half-sites separated by a spacer of a few nucleotides), which are crucial for the suppressive function.

The context of a TF site is one of the major determinants of its role in transcription control. As a consequence of context requirements, often TF sites are grouped together and such functional groups have been described in many cases. A systematic attempt at collecting synergistic or antagonistic pairs of TFBSs has been made with the COMPEL database [51]. In many cases, a specific promoter function (e.g. a tissue-specific silencer) will require more than two sites. Promoter subunits consisting of groups of TFBSs that carry a specific function independent of the promoter will be referred to as *promoter modules*. Arnone and Davidson originally gave a more detailed definition of promoter modules [1]. In summary, promoter modules contain several TFBSs which act together to convey a common function like tissue-specific expression. The organization of binding sites (and probably also of other elements) of a promoter module appears to be much more restricted than the apparent variety of TF sites and their distribution in the whole promoter suggests. Within a promoter module both sequential order and distance can be crucial for function, indicating that these modules may be the critical determinants of a promoter rather than individual binding sites. Promoter modules are always constituted by more than one binding site. Since promoters can contain several modules that may use overlapping sets of binding sites, the conserved context of a particular binding site cannot be determined from the primary sequence. The corresponding modules must be detectable separately before the functional modular structure of a promoter or any other regulatory DNA region can be revealed by computer analysis. One well-known general promoter module is the core promoter, which will be discussed in more detail below. However, the basic principles of modular organization are also true for most, if not all, other regulatory regions and are neither peculiar nor restricted to promoters.

1.6.3 Module Properties of the Core Promoter

The core promoter module can be defined functionally by its capability to assemble the transcription initiation complex and orient it specifically towards the TSS of the promoter [100], defining the exact location of the TSS. Various combinations of about four distinguishable core promoter elements that constitute a general core promoter can achieve this. This module includes the TATA box, the initiator region (INR), an upstream activating element and a downstream element (Figure 1). The TATA box is a basic transcription element, which is located about 20–30 nucleotides upstream of the actual TSS and is known to bind to the TATA box-binding protein (TBP). However, this is also where the straightforward definition of a core promoter module ends because not all four elements are required or some elements can be too variable to be recognizable by current computer tools.

The first group is made up of TATA box-containing promoters without a known initiator. Successful positioning of the initiation complex can start at the TATA box-containing promoters by the TFIID complex, which contains the TBP as well as several other factors. Together with another complex of general TFs, termed TFIIB, this leads to the assembly of an initiation complex [22]. If an appropriate upstream TFBS cooperates with the TATA box, no special initiator or downstream sequences might be required, which allows for the assembly of a functional core promoter module from just two of the four elements. This represents one type of a distinct core promoter that contains a TATA box, common among cellular genes in general.

The second group is TATA-less promoters with a functional initiator. As is known from a host of TATA-less promoters, however, the TATA box is by no means an essential element of a functional core promoter. An INR combined with a single upstream element has also been shown to be capable of specifically initiating transcription [41], although initiators cannot be clearly defined at the sequence level so far. Generally, a region of 10–20 nucleotides around the TSS is thought to represent the initiator. A remarkable array of four different upstream TF sites (SP1, AP-1, ATF or TEF1) was shown to confer inducibility by T-antigen to this very simple promoter, i.e. mediated transcriptional activation upon binding of T-antigen. T-antigen is a potent activating protein from a (simian) virus called SV40. This is an example of a TATA-less distinct promoter that can be found in several genes from the hematopoietic lineage (generating blood cells).

The third group is made up of a composite promoter consisting of both a TATA box and an initiator. This combination can be found in several viral promoters and it has been shown that an additional upstream TFBS can influence whether the TATA box or the initiator element will determine the promoter properties [21]. The authors showed that upstream elements can significantly increase the efficiency of the INR in this combination; in particular, SP-1 sites made the TATA box almost obsolete in their example. The combination of TATA box with an INR had the general effect of inducing resistance against the detrimental effects of a TFIIB mutant, which interfered

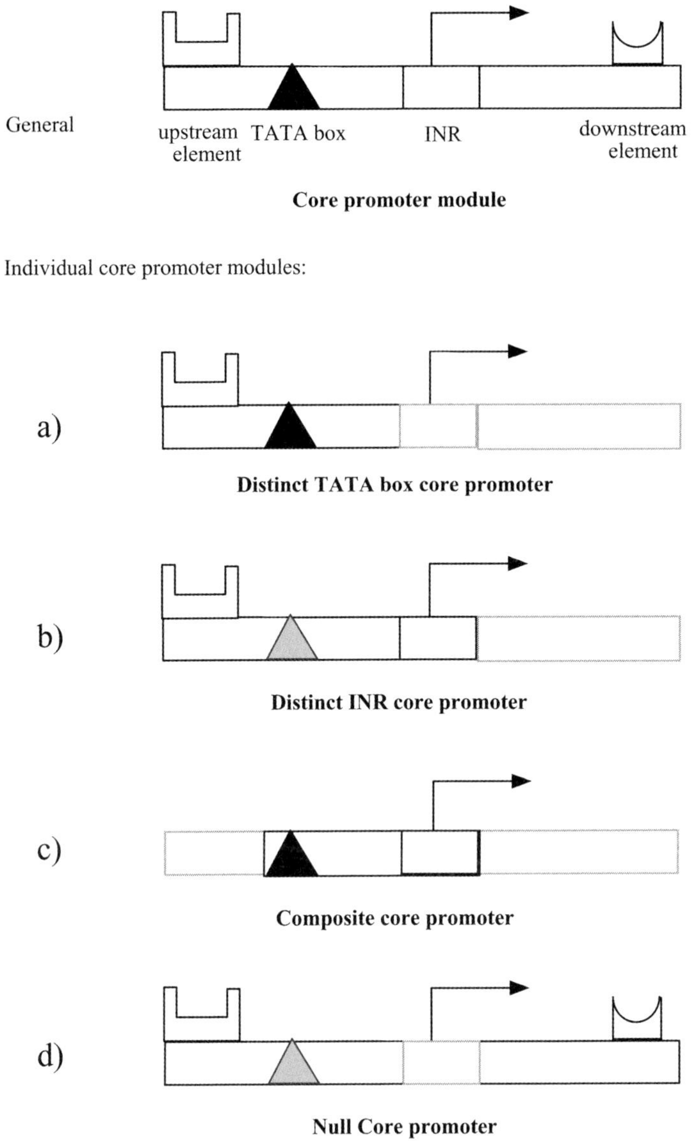

Figure 1 General structure of a Pol II core promoter and four different setups (a–d) of a Pol II core promoter. Simultaneous presence of all four elements is not always essential. The shapes above the bar symbolize additional protein-binding sites and the arrow indicates the TSS.

with expression from TATA-only promoters. This is also an example for of more indirect effects of specific arrangements in promoters that may not be apparent unless special conditions occur.

The last group consists of so-called null-promoters, which have neither a TATA box nor an initiator, and rely exclusively on upstream and downstream elements [66].

Basically, at least the four different core promoter types detailed above have been identified so far, all of which represent valid combinations of core promoter sites (reviewed in Ref. [66]). If the combinations involving upstream and downstream elements are also considered, seven core promoter modules are possible (most of which can be actually found in genes and consist of the four variants in Figure 1(a–d) adding upstream or downstream elements or both).

The only apparent common denominator of transcription initiation within a promoter would be that there must be at least one core promoter element anywhere within a certain region. This assumption is wrong. Both the spacing and/or sequential order of elements within the core promoter module are of utmost importance regardless of the presence or absence of individual elements (as a rule; however, there appear to be some exceptions). Moreover, many distinct promoters have requirements for specific upstream or downstream elements and will only function with their specific TF. Moving around the initiator, the TATA box and, to some extent, also upstream elements can have profound effects on promoter functions. For example, insertion of just a few nucleotides between the TATA box and an upstream TFBS (TF MyoD) in the desmin gene promoter cuts the expression levels by more than half [62]. Moreover, the promoter structure can affect later stages of gene expression like splicing [23]. It was also shown for the rat β-actin promoter that a few mutations around the TSS (i.e. within the initiator) could render that gene subject to translational control [7].

As a final note, the mere concept of one general TATA box and one general INR is an oversimplification. There are several clearly distinguishable TATA boxes in different promoter classes [35] and the same is true for the INR region, which also has several functionally distinct implementations as the glucocorticoid-responsive INR in the murine thymidine kinase gene [73], the C/EBP-binding INR in the hepatic growth factor gene promoter [48] or the YY-1-binding INR [90].

Most of the principles of variability and restrictions detailed above for the core promoter modules are also true for other promoter modules that modify transcriptional efficiency rather that determining the start point of transcription as the core promoter does. The bottom line is that the vast majority of alternative combinatorial arrangements of the elements that can be derived from a particular promoter might not contribute to the function of the promoter. Module-induced restrictions are not necessarily obvious from the primary sequences. Figure 2 shows a schematic Pol II promoter with the initiation complex assembled that illustrates that it matters where a specific

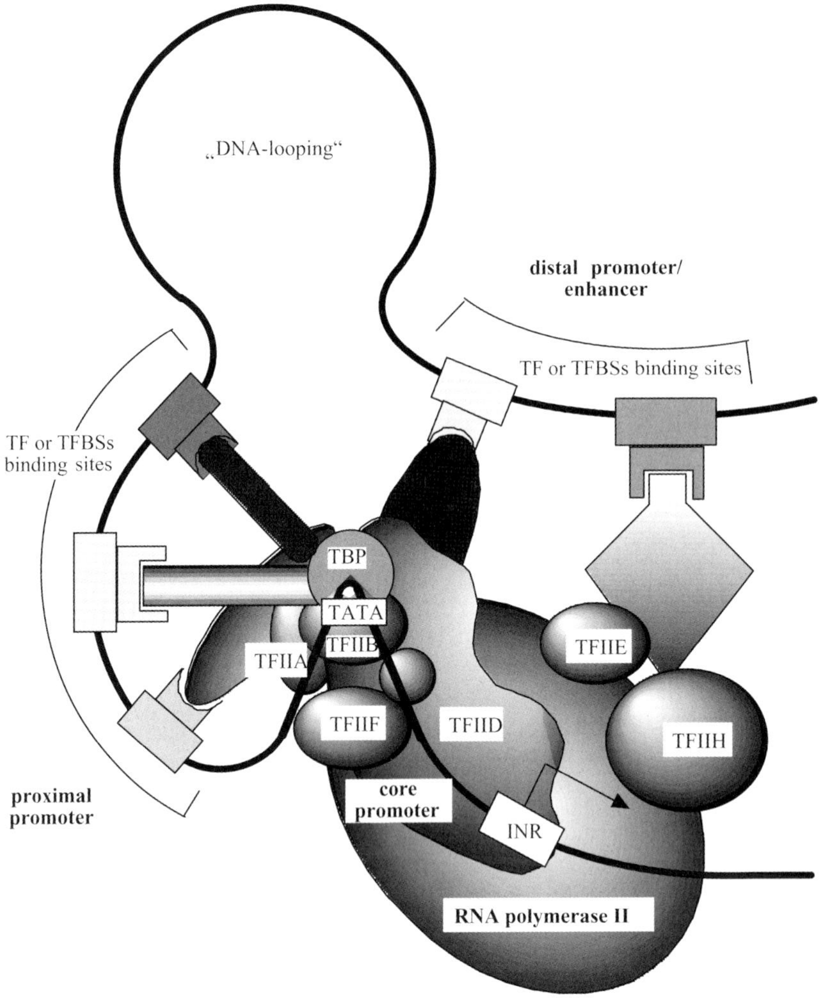

Figure 2 Transcription initiation complex bound to a schematic promoter.

protein is bound to the DNA in order to allow for proper assembly of the molecular jigsaw puzzle of the initiation complex. This is not immediately obvious from inspection of promoter sequences because there exist several (but a strictly limited set of) alternative solutions to the assembly problem. As complicated as Figure 2 may appear, it still ignores all aspects of chromatin rearrangements and nucleosomal positions, which also play an important role in transcription regulation. Stein and coworkers, initially in 1995 and in a 2001 follow-up paper, have detailed an example of the profound influence of these effects on promoter–protein complex assembly and function for the osteocalcin promoter [63, 84]. However, chromatin-related effects are not yet considered in any of the promoter prediction methods. Therefore, we do not go into any more details here.

1.7 Bioinformatics Models for the Analysis and Detection of Regulatory Regions

Algorithms used to analyze and detect regulatory regions are necessarily based on some kind of usually simplified model of what a regulatory region should look like. All of these models inevitably compromise between accuracy with respect to the biological model (the standard of truth) and computational feasibility of the model. For example, a computational model based on *a priori* 3-D structure prediction derived from molecular dynamics using sophisticated force fields may be the most accurate model for a region, but cannot be used for the analysis of real data due to excessive demand on computational resources. On the other hand, a model based on simple sequence similarities detected by IUPAC consensus (see also Section 2.1) sequences can be easily used on a PC, but results will usually not match the biological truth in an acceptable manner.

1.8 Statistical Models

It was noted several years ago that promoters and most likely also other regulatory regions like enhancers contain more TFBSs that nonregulatory sequences. Therefore, an analysis of the relative frequencies of such sites within a sliding window can yield some information on the potential regulatory character of a stretch of DNA, which is the prototype of simple statistical models. Several programs exist that rely to some extent on this type of statistics. Another set of statistical models calculates local GC content bias and uses this feature to discriminate potential promoters from other sequences. Such nucleotide bias statistics are only used in combination with other features (see Section 1.8.1) as they do not exhibit sufficient discriminatory power on their own.

A new breed of statistical models has been successfully introduced into promoter analysis more recently. These models focus on statistical analysis based on identification of promoter-associated words (not predefined such as TFBSs) using methods coming from other fields such as speech recognition. These methods are currently the best performers in promoter finding.

1.8.1 Mixed Models

It is clear from Section 1.6 that a binding site description-based pure statistical model is an oversimplification that will adversely affect the accuracy of prediction despite its attractive ease of implementation. Therefore, mixed models are also used that take at least some regional information into consideration and can be seen as statistical models split into compartments. Within the compartments solely statistical features are considered, but promoter organization is somewhat reflected by the arrangement of the compartments, which represent different promoter regions.

1.8.2 Organizational Models

The last category consists of models that try to closely follow the organizational principles of real regulatory regions. In order to accomplish this, individual promoter elements like TFBSs as well as their relative order and distances are encoded in a formal model, which reflects the setup of a single promoter or a small group of functionally similar promoters. Although they match the biological situation best, their widespread application requires an enormous amount of automation and background logistics such as high-quality promoter databases, automatic methods to derive the computational models as well as means of evaluating the resulting models. In the meantime, most of the basic requirements have been met, but real-life application is just picking up as this book is written.

However, such approaches are well suited for elucidating the molecular basis of coregulation of genes in a particular coexpressed cluster of genes from microarray experiments. So far this has been shown mainly for a simple eukaryote, *Saccharomyces cerevisiae* (yeast) [69]. Considerable progress has been made already in applying combinatorial TFBS models to higher eukaryotes such as mammalian systems, mostly based on experimental evidence [14,40].

2 Methods for Element Detection

2.1 Detection of TFBSs

TFBSs are the most important elements within regulatory DNA regions like promoters or enhancers. The majority of the known TFs recognize short DNA stretches of about 10–15 nucleotides in length that show different degrees of internal variation. Successful detection of protein-binding sites in DNA sequences always relies on precompiled descriptions of individual binding sites. Such descriptions are usually derived from a training set of four or more authentic binding sites. However, the criteria applied for the decision whether a site is authentic or not vary considerably among authors of different publications. One of the first approaches to define protein-binding sites used IUPAC consensus sequences, which indicate the predominant nucleotide or nucleotide combination at each position in a set of example sequences (e.g.

SIGNAL SCAN [70]). The IUPAC string `TGASTCA` indicates that the first three positions are most frequently T, G and A, while the fourth position may be C or G, followed by T, C and A in most cases. IUPAC consensus sequences became very popular as they are extremely easy to define from even a small set of sequences, and their definition does not require more than a pencil and a sheet of paper.

However, IUPAC consensus sequences strongly depend on the sequence set used for definition because IUPAC consensus findings are based on majority rules. Adding or removing a single sequence can change the assigned nucleotide at a position while it would have little effect in a corresponding weight matrix. Cavener defined some rules that we have used for several years now and, in our experience, IUPAC consensus sequences defined that way can be useful [16]. However, IUPAC consensus sequences may reject biologically functional binding sites due to a single mismatch (or an ill-defined IUPAC sequence).

The concept of nucleotide weight matrix (NWM) descriptions was developed in the 1980s as an alternative to IUPAC strings (e.g. Refs. [83, 86]). Basically, weight matrices use an alignment of sequences to first generate a nucleotide distribution matrix representing the complete nucleotide distribution at each position of the alignment. Then some sort of weighting algorithm is used to adjust the matrix to the biological situation (also detailed in Section 5.3.1). However, although weight matrices proved to be generally superior to IUPAC strings, their greatest disadvantage is the absolute requirement for predefined matrices, which are more complicated to construct than IUPAC strings and require specific software. This delayed widespread use of weight matrices for almost a decade, although the methods were principally available. They remained mostly unused because only a few special matrices had been defined (e.g. Ref. [12]). The situation changed when in 1995 two (overlapping) matrix libraries for TF sites were compiled and became widely available almost simultaneously [17,72]. MATRIX SEARCH [17] transformed the TRANSFAC database as completely as possible (starting at two binding sites for one factor) into matrices using a log-odds scoring approach. The MatInspector library [72] was originally largely based on a stringent selection from the matrix table of the TRANSFAC database, including the matrices derived from the ConsInspector library [32, 33] and several genuine matrices. The Information Matrix Database was compiled from the TRANSFAC matrix table and the TFD. In the meantime, the MatInspector library became independent from TRANSFAC and is updated regularly by Genomatix Software (Munich; currently more than 600 matrices), whereas IMD (another weight matrix database) has not been updated recently.

2.2 Detection of Novel TFBS Motifs

All the above covers the various approaches used to describe and find known TFBS motifs, i.e. there is always evidence that a known TF binds to such regions. There is another group of methods that join knowledge about evolutionary relationship of promoters with pattern-finding algorithms to detect phylogenetically conserved TFBSs. Examples of publications in this field include comparison of conserved human mouse patterns with [64] or without [61] direct sequence alignment, as well as approaches no longer restricted to two sequences such as FootPrinter [8] or PhyME, which includes overrepresentation into the probabilistic score of its findings [80].

A completely different set of methods deals with the detection of potential TFBS patterns solely based on their occurrence in a set of sequences without any biological knowledge about the particular TF binding to such regions. I separate such methods from the TFBS recognition methods as an unknown proportion of significant motifs detected this way may in fact not be TFBSs at all, but may be conserved for other reasons. Nevertheless, these methods do contribute to the generation of hypotheses about hitherto unknown TFBS patterns. Available matrix detection programs were reviewed some time ago [34] and a comparison of these methods by application to a test set of sequences has been published [36] (see Ref. [85] for a more recent review of the topic). A very recent study focused on matrix generation programs with no real emphasis on search programs [88]. For convenience, Table 1 summarizes some methods for the detection of TFBSs that are available in the internet with emphasis on programs featuring a WWW interface.

Various newer approaches have been published in the meantime, ranging from excellent purely mathematically motivated pattern detection (e.g. from Pevzner's group [50] or using self-organizing maps [65]) to strong connection

Table 1 Internet-accessible methods to detect promoter elements (TFBSs)

Program	Availability	Comments
MatInspector	http://www.genomatix.de	Genomatix matrices; free of charge use for academics (limited) after registration
SIGNAL SCAN	http://bimas.dcrt.nih.gov/molbio/ signal	IUPAC consensus library
MATRIX SEARCH	http://bimas.dcrt.nih.gov/molbio/ matrixs	IMD matrix library (TRANSFAC + TFD)
TFSearch	http://www.cbrc.jp/research/ db/TFSEARCH	TRANSFAC matrices
TESS	http://www.cbil.upenn.edu/tess/	TRANSFAC matrices
MATCH	http://www.gene-regulation.com/cgi-bin/ pub/programs/match/bin/match	TRANSFAC matrices; free of charge use for academics (restricted public version)

between biological [71] and experimental data with pattern detection [13]. This list only represents an arbitrary collection of very few papers in the field and the selection was purely driven by the desire to cite at least one method for each basic approach. I will not discuss *de novo* detection methods in any more detail here, as the major scope of this chapter is not *de novo* detection of patterns, but regulatory sequences analysis, which is usually based on precompiled pattern collections.

2.3 Detection of Structural Elements

Regulatory sequences are associated with a couple of other individual elements or sequence properties in addition to the factor-binding sites. Among these are secondary structure elements like the HIV-1 TAR region (*trans*-activating region, which constitutes an RNA enhancer, e.g. Ref. [10]), cruciform DNA structures (symmetric double hairpins of both strands in DNA, e.g. Ref. [92]) or simple direct repeats (e.g. Ref. [5]). Three-dimensional structures like curved DNA [54] also influence promoter function. Most of these elements can be detected by computer-assisted sequence analysis [20, 43], but none of them is really promoter specific and all such elements can be found frequently outside of promoters. The promoter or enhancer function arises from the combination of several elements that need to cooperate to exert transcription control which none of them can achieve alone. This also illustrates the main problem of promoter recognition. It is necessary to compile several individually weak signals into a composite signal, which then indicates a potential promoter without being overwhelmed by the combinatorial complexity of potential element combinations.

2.4 Assessment of Other Elements

Several methods employ statistical measures of sequence composition to include features of regulatory sequences, which cannot be described by the three types discussed above. These includes frequencies of oligonucleotides (dinucleotides, trinucleotides and hexamers are used most frequently), CpG islands (CG dinucleotides are usually underrepresented in mammalian genomes except in part of coding and regulatory sequences; CpG islands are regions where the dinucleotide is NOT underrepresented [38]) and periodicity of weak sequence patterns (`AA`, `TT`, etc.). Definitions of such elements are usually too weak to make any significant contribution to current prediction programs. However, this situation might well change due to the unprecedented amounts of continuous genomic sequences that become available in the course of the current genome-sequencing projects.

3 Analysis of Regulatory Regions

Basically, two different tasks can be distinguished in the analysis of regulatory regions. The first task is analysis aimed at the definition of common features based on sets of known regulatory sequences. This is a prerequisite for the definition of descriptions suitable for large-scale application for prediction of potential regulatory regions within new anonymous sequences, which can be seen as the second task.

3.1 Comparative Sequence Analysis

Comparative sequence analysis is one of the most powerful methods to deduce regulatory features and organization. Two main types of comparative analysis can be distinguished. The first approach compares regulatory regions, e.g. promoters within one species such as promoters coexpressed under particular conditions, or simply all (known) promoters within a genome to deduce general features. The second approach compares only orthologous regulatory sequences (again promoters are the most prominent representatives) in order to elucidate which features and elements have remained conserved in evolution. Such features should be closely associated with conserved functions of the corresponding regulatory regions. While comparative analysis within species affords no distinction between pure statistical findings and functional conservation, phylogenetic analysis of orthologous regulatory sequences should indicate predominantly functionally conserved features. However, intragenomic comparison may differentiate between individual functions, whereas phylogenetic analysis will always yield a summary over all conserved functions. Thus, very often a combination of both approaches is the best way to go [25].

3.2 Training Set Selection

One of the most important steps in comparative sequence analysis is the selection of suitable training sets of sequences. If a training set of promoters consists only of constitutively expressed sequences (constant level of expression, no or little regulation), little can be learned about any kind of tissue-specific expression regardless of the methods applied. Inclusion of too many wrong sequences (e.g. that are not promoters at all or promoters not involved in the regulation under investigation, see alternative promoters below) may also prevent any meaningful analysis. Although this observation appears trivial at first, it becomes a real issue when data are scarce and less well-characterized sequences have to be used.

Control sets known not to be functionally similar to the training sets are about as important as the training sets. However, true negative regions are even scarcer than known regulatory regions. Negative often means just "no positive functions found", which can also be due to failures or simply means that the sequences have not been tested at all. Therefore, statistical negative control sequences are often required. Random sequences can be generated easily, but often are of limited use, as they do not represent several important features of natural DNA correctly. This includes underrepresented features (e.g. CpG islands), asymmetric features (e.g. strand specificity), local changes in GC content or repetitive DNA elements. Selection of appropriate control sequences can be a major effort, but is also crucial for the validity of the evaluation of any method. Common problems with controls are either known or unknown biases in the control set or circularity problems, i.e. the training and the test sets of sequences are related or overlap. The availability of large continuous stretches of genomic DNA from the genome-sequencing projects constantly improves this situation. Genomic sequences should always be the first choice for controls as they reflect the natural situation.

3.3 Statistical and Biological Significance

The quality of sequence pattern recognition is often optimized to improve the correlation of the methods with the data (positive and negative training sets). However, in most cases it is not possible to collect sufficient data to perform a rigorous correlation analysis. Therefore, bioinformatics methods often rely on statistical analysis of their training sequences and optimize for the statistically most significant features. Unfortunately, this kind of optimization does not always reflect the evolutionary optimization of regulatory sequences that is always optimizing several features at once. This problem is different from overfitting of data as it is more about optimization criteria than parameter fitting *per se*.

The dynamics of biological function often necessitates suboptimal solutions. For example, real sequences usually do not contain binding sites with the highest affinity for their cognate protein because binding *and* dissociation of the protein is required for proper function. The perfect binding site with the highest binding affinity would interfere with the dissociation and is therefore strongly selected against.

3.4 Context Dependency

The biological significance of any sequence element is defined by the regulatory function it can elicit. This is usually dependent on a functional context rather than being a property of individual elements. Therefore, statistical significance of the features or scores of individual elements is neither necessary nor sufficient to indicate biological significance. Recognition of the functional context in an essentially linear molecule like DNA can be achieved by correlation analysis of individual elements, which became an important part of all semi-statistical or specific modeling approaches discussed below. The context is also an important parameter in statistical analysis. For example, an element frequently found all over the genome could become even statistically significant if only the immediate vicinity of a binding partner's binding sites is analyzed such as in case of transcriptional modules. Therefore, lack of statistical significance may just indicate that the wrong context was chosen for the analysis.

4 Methods for Detection of Regulatory Regions

There are several methods available for the prediction of regulatory DNA regions in new sequence data. Table 2 lists methods available with a special focus on programs that provide a WWW interface. Unfortunately, there is no "one-does-it-all" method, and all methods have their individual strong and weak points. There was a fairly recent review on the subject including most relevant programs, with one exception [4]. The program PromoterInspector

Table 2 Internet-accessible promoter/promoter region prediction tools

Program	Availability	Comments
Promoter prediction		
Ab initio promoter finding (large-scale sequences)		
PromoterInspector	http://genomatix.de	free of charge use for academics (limited) after registration
Dragon PromoterFinder	http://research.i2r.a-star.edu.sg/promoter/promoter1_5/DPF	free for academics
Promoter finding in preselected sequence ranges		
Eponine	http://servlet.sanger.ac.uk:8080/ eponine	free for academics
FirstEF	http://rulai.cshl.org/tools/FirstEF	free for academics
Promoter module/region recognition		
ModelInspector	http://www.genomatix.de	free of charge use for academics (limited) after registration; modules of two TF sites (MatInspector library)

[77] was not included as it predicts promoter regions and neither strand orientation nor the TSS.

A program doing an excellent job in one case might be a complete failure in another case in which other methods are successful. Therefore, we will describe a number of methods without intending any rank by order of discussion. We will rather follow the functional hierarchy that appears to apply to the different regulatory regions. However, the apparent best application range will be indicated.

4.1 Scaffold/Matrix Attachment Regions (S/MARs)

A chromatin loop is the region of chromosomal DNA located between two contact points of the DNA with the nuclear matrix marked by so-called S/MARs. The nuclear matrix is a mesh of proteins filling the interdomain space inside the nucleus where S/MARs form highly flexible structures that are necessary, but not sufficient, for anchoring at chromosomal DNA to the matrix [42].

Transcriptional regulation requires the association of DNA with this nuclear matrix, which retains a variety of regulatory proteins. S/MARs are composed of several elements, including TFBSs, AT-rich stretches, potential cruciform DNA and DNA-unwinding regions, to name a few of the most important S/MAR elements. There is an excellent recent review on chromatin domains and S/MAR functions [9]. Singh and coworkers published a method to detect potential S/MAR elements in sequences and made the method available via WWW (http://www.ncgr.org/MAR-search/) [79]. Their method is based on a statistical compilation of the occurrence of a variety of S/MAR features (called rules). Accumulation of sufficient matches to these rules will be predicted as potential S/MAR regions. The specificity of the method depends critically on the sequence context of the potential S/MAR sequences. Another approach utilizes a single S/MAR associated sequence element to locate potential S/MARs [91]. Therefore, results are difficult to evaluate by comparisons. We developed another approach to define especially AT-rich MARs called SMARTest, which is available on the web at http://www.genomatix.de. SMARTest is based on a library of MAR-associated nucleotide weight matrices and determines S/MARs independent of any larger sequence context [37]. Therefore, the method is suitable for testing isolated S/MAR fragments. MARFinder and SMARTest are complementary, and should be seen in combination rather than as alternatives.

4.2 Enhancers/Silencers

Enhancers are regulatory regions that can significantly boost the level of transcription from a responsive promoter regardless of their orientation and distance with respect to the promoter as long as they are located within the same chromatin loop. Silencers are basically identical to enhancers and follow the same requirements, but exert a negative effect on promoter activities. Both regulatory regions are also relevant in disease processes, as detailed in a recent review [55]. At present there are no specific programs to detect enhancers and silencers. However, programs designed to detect the internal organization of promoters are probably also suitable to detect at least some enhancers and silencers since these regions often also show a similar internal organization as promoters.

4.3 Promoters

Promoters were described in detail above – they are just mentioned here again to place them into context.

4.4 Programs for Recognition of Regulatory Sequences

There are several ways promoter recognition tools can be categorized. We will focus on the main principles and intended usage of the programs rather than technical details. Two generally distinct approaches have been used so far in order to achieve *in silico* promoter recognition. The majority of programs focus on *general promoter recognition*, which represents the first category.

The second category of tools aims at *specific promoter recognition* relying on more detailed features of promoter subsets like combinations of individual elements. The beauty of this approach is its excellent specificity, which is extremely helpful if only promoters of a certain class are of interest or megabases of sequences have to be analyzed. The bad news here is limited applicability, i.e. each promoter group or class requires a specifically predefined model before sequences can be analyzed for these promoters. This may result in a huge number of false negatives in large-scale analysis.

We will briefly discuss individual methods in these two categories with emphasis on the implementation of the biological principles of promoter features. Recently, a practical comparison of the majority of available tools based on general promoter models has been carried out [4], which was the first large-scale update since the original comparison carried out by Fickett and Hatzigeorgiou in 1997 [29]. There was another review in between those two studies by Ohler and Niemann [67]. Therefore, we will not go into details on the performance of the methods.

4.4.1 Programs Based on Statistical Models (General Promoter Prediction)

These programs aim at the detection of Pol II promoters by a precompiled general promoter model that is part of the method. Learning methods range from supervised artificial neuronal networks over statistical analyses to simple counting of features to a threshold. One group of programs in this category (see below) concentrates on recognition of core promoter properties and infers promoter location solely on that basis, whereas the other group consists of programs that take into account also the proximal promoter region of about 250–300 nucleotides upstream of the TSS. General recognition models were usually based on training sets derived from the Eukaryotic Promoter Database (EPD) and various sets of sequences without known promoter activities. The EPD originally was an excellent collection of DNA sequences that fulfill two conditions: they have been shown experimentally to function as promoters and the TSS is known. Recently, EPD also started to incorporate promoters not fulfilling these stringent conditions [78].

The beauty of the above approaches is their generality, which does not require any specific knowledge about a particular promoter in order to make a prediction. This appears ideal for the analysis of anonymous sequences for which no *a priori* knowledge is available. The bad news was that the specificity of all such general approaches implemented was very limited for quite some time. However, the development of PromoterInspector [77] heralded a new era of promoter prediction, combining acceptable sensitivity with high specificity. Other programs that followed performed comparably [3]. These general promoter prediction approaches were the first to provide acceptable *a priori* promoter prediction on a whole chromosome and now genome scale [76]. Specificities were originally reported just below 50%, but in the meantime many of the orphan predictions (in the middle of unannotated sequence) have found their genes and transcripts boosting specificity to between 80 and 90%. Only a really complete annotation of the genomes will tell the true specificity of those methods. Nevertheless, it is clear that the goal of highly specific promoter prediction in whole mammalian genomes has been achieved.

Some general promoter model-based programs employ methods already described for identification of individual promoter elements (usually TBFBS IUPAC or weight matrix descriptions), but try to derive more general features from a collection of such elements rather than emphasizing individual elements. These methods may be called *statistical element analyses* and treat the proximal promoter as a purely statistical problem of TFBS accumulations, sometimes fine-tuned by some sort of weighting based on occurrence frequencies of TFBSs in promoters as compared to a negative sequence set. Despite the complicated modular structure of promoters outlined above there is a solid rational basis for this general model. All promoters must have a functional core promoter module often containing a TATA box, which is the prime target of the majority of the general promoter prediction tools. This is also one of the reasons that some programs confine their analysis to the core promoter region, which avoids problems with the much more diverse proximal regions. Biological knowledge is solely used to select the training sets and a variety of methods is used to learn the distinctive patterns.

Without exception, TFBS-based statistical element analysis suffers from a huge number of false-positive predictions (typically about one prediction in 10 000–30 000 nucleotides).

4.4.2 **Programs Utilizing Mixed Models**

These programs also rely on statistical promoter models, but include directly or indirectly some organizational features of promoters, placing them in between the pure statistical models and attempts to approximate the biologically important structured organization of promoters. Again, the first-generation methods will only be summarized. FunSiteP [53] as well as the approach taken by Audic and Claverie [2] fall into this category.

4.4.3 **Programs Based on Specific Promoter Recognition**

The second, more recent and far more successful concept should be called *functional element analysis*, as it relies heavily on biological knowledge about the relative importance of individual elements and derives discriminative features on that basis. These methods carry out a sophisticated compositional analysis of the proximal promoter analysis to detect unique features within that region that can be used to distinguish promoters from nonpromoters without understanding the details, but using any pre-existing knowledge for feature selection.

This category of methods introduces the functional context in the form of heuristic rules or tries to learn the context from comparative sequence analysis. These methods emphasize specific modeling of promoters or promoter substructures rather than general recognition. Therefore, it is not possible to directly assess the promoter prediction capabilities of these methods. However, in many cases recognizing a common substructure between promoters can be very helpful, especially for experimental design. Although these programs were also published during the time the first-generation general promoter prediction programs appeared, they are still useful in whole-genome scans due to their very high specificity, warranting a more detailed discussion here.

The method FastM was derived from the program ModelGenerator [31] and takes advantage of the existence of NWM libraries. It can be accessed via a WWW interface (http://genomatix.gsf.de part of GEMS launcher) and allows for a straightforward definition of any modules of two TFBSs by simple selection from the MatInspector Library [72]. This now enables definition and detection of wide variety of synergistic TFBS pairs. These pairs are often functional promoter modules conferring a specific transcriptional function to a promoter as shown in Refs. [52, 56]. FastM models of two binding sites can successfully identify promoters sharing such composite elements, but are not promoter specific. Composite elements can also be located in enhancers or similar structures. The latest version of FastM enables definition of complete, highly specific promoter class models including up to 10 individual elements, also including IUPAC strings, repeats and hairpin structures.

The program FrameWorker [15] automates several of the steps taken manually in FastM in order to ground specific promoter modeling on as much an algorithmic basis as possible. FastM requires crucial parameters such as strand orientation, distance ranges, order of elements, as well as the individual nature of the elements (e.g. which weight matrix to use) to be determined by the user. FrameWorker, in contrast, automatically determines theses parameters from a comparison of an (still manually selected) set of input sequences within user-defined ranges. However, determination of the individual weight matrices to be used, as well as their number, distances and relative order, does not require previous knowledge.

Another approach aimed at modeling promoter substructures consisting of two distinct elements is TargetFinder [58]. This method combines TFBSs with features extracted from the annotation of a database sequence to afford selective identification of sequences containing both features within a defined length. The advantage is that TargetFinder basically also follows the module-based philosophy, but allows inclusion of features that have been annotated by experimental work for which no search algorithm exists. Naturally, this excludes analysis of new anonymous sequences. The program is accessible via a WWW interface (http://gcg.tigem.it/TargetFinder.html).

It should be mentioned here that Fickett also employed the idea of a two-TFBS module to successfully detect a subclass of muscle-specific regulatory sequences governed by a combination of MEF2 and MyoD [28]. However, this was also a very specific approach and no general tool resulted from that work. The MEF2/MyoD model can be used to define a corresponding module with FastM. Wasserman and Fickett also published a modeling approach based on clustering of a preselected set of NWM (defined in the same study) correlated with muscle-specific gene expression [93]. They were able to detect about 25% of the muscle-specific regulatory regions in sequences outside their training set and more than 60% in their training set. They classify their method as regulatory module detection. However, their results suggest that they probably detect a collection of different, more specific modules with respect to the definition given above. Although the method is not promoter specific and the specificity is moderate, it is a very interesting approach that has potential

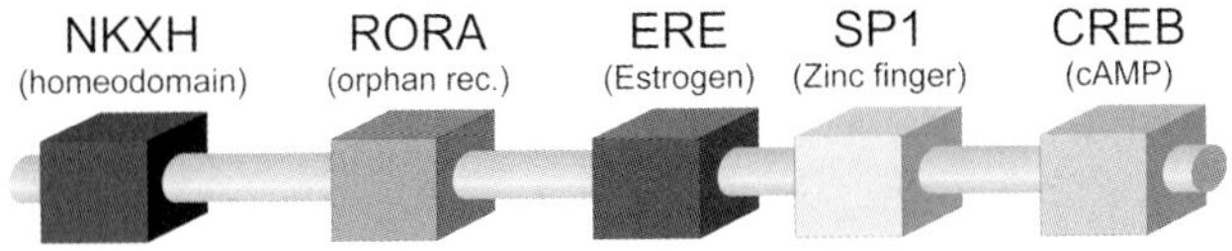

Figure 3 GFAP promoter model conserved in human mouse and rat promoter. The boxes indicate the individual TFBSs found and the bar indicates the genomic DNA.

for further development, as also became evident from follow-up publications of the same authors [30, 94].

Generally, this group of methods achieves much higher specificity than the first-generation programs following general models. However, the price for this increase in specificity is usually restriction of the promoter models to a small subset (class) of promoters.

The model of the glial fibrillary acidic protein (GFAP) promoter shown in Figure 3 was derived from a comparison of the human, mouse and rat GFAP promoters. This model contains five different TFBSs and was derived from the set of three sequences using GEMS Launcher (Genomatix). This model recognizes a single sequence, the GFAP promoter, when searched against more than 36 000 human promoter sequences and thus is absolutely gene specific. Interestingly, if the search is carried out with relaxed stringency (allowing for less-perfect matches) only a second sequence comes up, the DGAT2 gene, which is the diacylglycerol *O*-acetyltransferase homolog 2 (homolog to mouse). From the literature it becomes immediately evident that both genes are brain-expressed (GFAP is brain/astrocyte specific) and both are genes associated with insulin signaling. Thus the promoter model-based search found biologically linked genes.

4.4.4 **Early Attempts at Promoter Prediction**

There are various programs that might be called first-generation programs for promoter prediction, some of which were absolutely instrumental in paving the way towards the newer developments, but are no longer of practical use. For that reason they will only be summarized here and not discussed in detail. The first exception to this rule will be Promoter Scan, as this was really the first program ever published for promoter prediction in mammalian sequences and served as a role model for a number of other developments.

Several of the general promoter prediction programs followed the basic design of Prestridge who used the EPD by Bucher's group [78] to train his software for promoter recognition. His program Promoter Scan was the first published method to tackle this problem [70]. He utilized primate nonpromoter sequences from GenBank as a negative training set and included the proximal promoter region in the prediction. The program uses individual profiles for the TFBSs indicative of their relative frequency in promoters to accumulate scores for DNA sequences analyzed. Promoter Scan employs the SIGNAL SCAN IUPAC library of TFBSs [70], introducing a good deal of biological knowledge into the method, although modular organization of the proximal region is necessarily ignored. Results of the first version were combined with the Bucher NWM for the TATA box, which served as a representation of the core promoter module [12].

Other methods following a similar design will not be discussed in detail, but should be mentioned. These include PromFD by Chen and coworkers [18], and the programs TSSG/TSSW from Solovyev's group, which are basically gene prediction methods that include promoter prediction [82]. Other programs in that category are XLandscape [60] and PromFind [44]. Michael Zhang published a new method to detect TATA-box containing core promoters by discrimination analysis.

5 Annotation of Large Genomic Sequences

Many of the methods discussed above were developed before the databases started to be filled with sequence contigs exceeding 100 000 nucleotides in length. The complete human genome draft now contains more than 3 billion nucleotides and many more genomic sequences of similar size are entering the databases. This changes the paradigm for sequence annotation. While complete annotation remains an important goal, specific annotation becomes mandatory when even individual sequences exceed the capabilities of researchers for manual inspection. Annotation of genomic sequences has to be fully automatic in order to keep pace with the rate of generation of new sequences. Simultaneously, annotations are embedded into a large natural context rather than residing within relatively short isolated stretches of DNA. This has several quite important consequences.

5.1 Balance between Sensitivity and Specificity

We will confine the discussion here to regulatory regions, but the problems are general. A very sensitive approach will minimize the amount of false-negative predictions and thus is oriented towards a complete annotation. However, this inevitably requires accepting large numbers of false-positive hits, which easily outnumber the true-positive predictions by an order of magnitude.

In order to avoid this problem methods can be designed to yield the utmost specificity (e.g. specific promoter modeling as discussed above). Here, the catch is inevitably a high number of false-negative results, which also may obscure 70–90% of the true-positive regions. The newer developments of general, but still specific, promoter finding (especially Refs. [3, 77]) may provide a way out of the dilemma. Once a rough annotation has been achieved, other methods can come in to locate promoters reliably in more restricted search spaces such as the FirstExon Finder [24] and Eponym [26]. There was a recent survey of promoter finding in genomic sequences emphasizing that suitability of methods for analysis of large genomic sequences cannot be inferred from limited tests with short samples, which could be referred to analysis in a "sheltered environment" [4].

Gene (or gene group)-specific methods were shown to produce more that 50% true-positive matches in their total output (e.g. Ref. [35]), but recognize just a small fraction of all promoters, which is inevitable for a function-specific model. A single specific model like the phylogenetically conserved GFAP promoter model (Figure 3) matched only once in the human genome, indicating that it is absolutely specific for the GFAP gene.

Definition of the required number of specific models based on current technologies was not a feasible task until recently. However, new developments have already been initiated to overcome the current obstacles and Genomatix is actually working on a genome-wide library of evolutionarily conserved organizational promoter models.

It is quite evident that functional promoter analysis in laboratories is capable of dealing specifically with several hundred or even thousand predicted regions, whereas predicting several hundred thousand or even millions of regions remains out of reach. However, recent improvements of laboratory high-throughput technologies such as location of the TSS by the so-called oligo-capping method [87] have provided an unprecedented amount of verified TSSs (which by definition are located within the promoters). Nevertheless, enhancements of the specificity of promoter recognition *in silico* will also be required as the oligo-capping method has an inherent error rate of 20–30%. Both developments will meet sometime in the future to close the gap in our knowledge about the location of promoters in the genome. More elaborative approaches will be required both in the laboratory as well as in bioinformatics in order to also understand the functionality hidden within these regulatory sequences. It is clear from the past and present developments that bioinformatics will probably cover significantly more than half of that path.

5.2 Genes – Transcripts – Promoters

Originally, the notion was that one gene would represent one function. We learned in the early days of molecular biology during the 1980s that this is not quite true and that one gene may very well have several functions. However, it did not become clear how this is realized until the large-scale

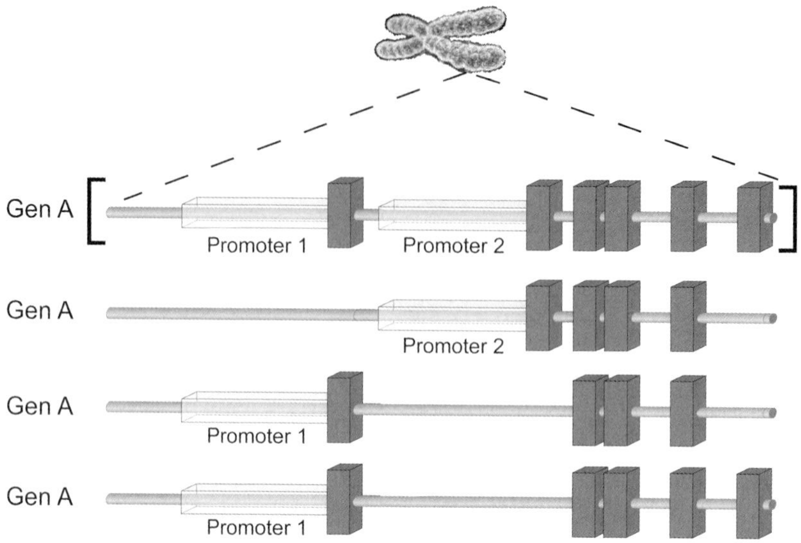

Figure 4 Genomic organization of genes, promoters and transcripts. The transparent boxes indicate promoters, the grey boxes indicate exons and the grey bars indicate the genomic sequence. The brackets delineate the locus of the gene.

mapping and sequencing effort provided us with a better insight into genomic organization. This knowledge has changed our perception of a gene. A gene is no longer an entity, but rather a container with individual transcripts representing the entities. Figure 4 illustrates this new notion schematically. The area in brackets indicates the genomic locus of the gene. This region can be larger than a million base pairs in some cases, providing the space for the complex inner organization. The line with the brackets indicates the genomic structure, such that both promoters and all exons are in a linear arrangement with no clue about the functional links between the individual elements. The lines below refer to individual transcripts, with two transcripts exhibiting alternative splicing originating from promoter 1, while another transcript originates from promoter 2. The important consequence is that this gene may behave like two independent genes with respect to regulation, and the transcript originating from promoter 2 can be completely independent in terms of regulation and function from the other transcripts. They may even encode quite different proteins.

From this it is immediately evident why the paradox of humans having only a moderate amount of genes in excess to *Drosophila* or *Caenorhabditis elegans* is not a real paradox. The inner complexity of transcript and regulatory combinations more than compensates for the apparent lack in total gene numbers. If we count transcripts rather than genes, mammalians do have close to or even above 100 000 – a number earlier pondered for genes as required for the observed complexity. It just turns out that within regulatory sequences as well as within the whole genome, complex hierarchical organization prevails over simple numbers of elements. This is not surprising as the hierarchical principle allows a much more economic utilization of genomic sequences.

5.3 Sources for Finding Alternative Transcripts and Promoters

Of course, once we realized that alternative transcripts as well as alternative promoters are important in general, the question arises how to cope with this extra level of complexity. There are several consequences that need to be taken into account. First, many expressed sequence tags (ESTs) sequences so far simply dismissed as "genomic contamination" may in fact indicate alternative transcripts, as what is an exon in one transcript can be an intron in another. The same is true for a predicted or experimentally verified promoter. If such a promoter was located inside a well-known gene, it was readily dismissed as a false positive, because we already "knew" that the promoter was further upstream. We have seen many cases in which the "false" promoter has found its own transcript in the meantime and was promoted from false to alternative. However, this has blurred the line between "true" and "false" considerably. What is apparently true for one condition (e.g. in one tissue) may be "false" for another condition (e.g. in another tissue). This dilemma is far from being finally solved, but as a practical approach we have adopted a policy of "multiple-evidence" support. The idea is very simple – both theoretical as well as laboratory-based approaches may yield false results. However, if two or more *independent* methods suggest the same conclusion, it is much more likely to be true than that both methods made exactly the same mistake. For example, if oligo-capping indicates a TSS, which happens to be located right inside a predicted promoter, we take this as evidence for a real promoter. Both methods are totally independent of each other and the chance of a result converging by chance is absolutely minimal. Based on this concept, ElDorado (Genomatix) has accumulated more than 150 000 primary transcripts as well as promoters for five mammalian species so far and we are quite confident that we have not yet seen the end of the story.

5.4 Comparative Genomics of Promoters

We have alluded to the "multiple-evidence" approach already in the previous section. However, there is very powerful line of evidence that has not yet been mentioned – the evolutionary conservation of gene regulation. This is one of the most direct lines of evidence towards the functional conservation of promoters as functional regions or elements are far better conserved that the sequence in general. We took advantage of this fact and developed a complete strategy affording the identification and subsequent mapping and analysis of orthologous promoters (Genomatix, patent pending). On top of identification of promoters of orthologous genes, this also includes finding the individual promoters within each species that correspond to each other, which we termed orthologous promoter sets. This is very important for subsequent analysis as functional elements, because functional element conservation is only detectable within orthologous promoter sets. Of course, this approach becomes stronger and stronger as the number of available genomes rises. As of 2004, this enabled us to detect or confirm more than 10 000 promoters in the human genome, making comparative genomics of promoters a major source of promoter annotation (as taken from the ElDorado statistics; Genomatix).

6 Genome-wide Analysis of Transcription Control

If the focus is broadened from individual genes or small gene groups towards looking at the whole genome it is no longer sufficient to just take promoters into consideration. On a genome-wide scale the hierarchy of gene regulation comes into the picture in full force. First, expression of genes on the mRNA level by transcription requires the locus of the gene to be accessible. Regulation of gene expression at the DNA level effected not by TFs, but by other factors elsewhere in the genome, is generally termed epigenetic regulation. This includes regulation by alternation of the chromatin structure, where DNA and histone modifications (e.g. DNA methylation or histone acetylation) play a role and the S/MAR elements discussed above become important. Whether the chromatin structure is open or closed determines whether a promoter becomes available for transcription or not. Thus, a gene with the perfect setup within its promoter(s) can be silent even if all the required TFs are present, provided the chromatin is closed, thus blocking access of these factors to the promoter. Let us assume that the chromatin is in an open, i.e. accessible, state. Even this does no guarantee active transcription of the embedded genes. Local DNA methylation can interfere, an active silencer can specifically block individual genes or one crucial factor may be missing or sequestered (e.g. the nuclear factor NFκB can be blocked by its inhibitor IκB, rendering it nonfunctional the despite presence of the protein). Active transcription is only observed when all conditions are right: the chromatin is open, no repressor is active, and all crucial factors can actually access their respective binding sites on the promoter and enhancer, if one is required. There is also a very old mechanism that seems to gain importance in the regulation of gene expression again – antisense transcription [98]. This means that the same region is transcribed in both directions, resulting in complementary RNAs that can form dimers and thus cancel each other out, as RNA dimers are prone to be destructed immediately. This very complex situation is a formidable safeguard against spurious expression of genes, which could be disastrous for a cell.

6.1 Context-specific Transcripts and Pathways

The many conditions that have to be met to enable the expression of a gene are also behind the differential expression of individual transcripts often coupled to particular pathways. Transcripts can be cell/tissue specific, pathway specific (or better associated as complete specificity is rare) or tied to a particular developmental stage of an organism. This emphasizes the important fact that biological function is tied to the transcript/protein, not to the gene, which may well encode various functions in various transcripts. There is also an important consequence for the analysis of regulatory networks behind signaling or metabolic pathways. It is not sufficient to identify which genes are involved in that pathway, but of utmost importance to identify the promoters associated with that particular transcript/pathway. This is also the reason that the very same pathway containing the same genes can still be differently regulated in different tissues, if different transcripts/promoters are involved in the different tissues. The upside of this complicated situation is that regulatory analysis based on the correct promoters is as close to the real biological situation as we currently can get with *in silico* methods. As it does not make any sense to simplify biology to fit our generic models it is well worth the effort to identify the conditionally important transcripts and promoters as this assures biological importance of the results.

6.2 Consequences for Microarray Analysis

Another field to which the bioinformatics of regulatory DNA regions can be expected to contribute significantly is the analysis of results from high-throughput experiments in expression analysis (e.g. all forms of expression arrays). Due to the discontinuous nature of regulatory regions there is no way of deducing common regulatory features from the expression data directly which are usually based on coding regions. However, the general availability of the corresponding genomic regulatory regions for many (and very soon all) of the genes analyzed in an expression array experiment enables attempts to elucidate the genomic structures underlying common expression patterns of genes. Expression arrays (described in detail in Chapters 24–28) directly deliver information, *which* genes are expressed *where* under the conditions tested. However, they cannot provide any clue to *why* this happens or how the same genes would behave under yet untested conditions. Identification of functional features by comparative sequence analysis (e.g. promoter modules) can reveal different functional subgroups of promoters despite common regulation under specific conditions. Consequently, the detection of known functional modules can suggest expression patterns under yet untested conditions [95]. Moreover, the organizational structures of promoters can also be used to identify additional potential target genes either within the same organism in other genomes or via comparative genomics. Given the exponential number of possibilities for combinations of conditions, bioinformatics of regulatory sequences will also become instrumental for the rational design of expression arrays as well as for selection of experimental conditions.

While this basic conduct of analysis of microarray data remains unchanged, our growing knowledge of alternative transcripts and alternative promoters has far-reaching consequences on strategies employed to analyze transcript levels on a large scale – the microarrays of DNA chips. The most obvious consequences of course are for the analysis of microarray data based on current chip designs that can be purchased from several vendors. As this is the most clear-cut consequence, let us focus on this point first. If there is a single transcript from a single promoter for a given gene, there is no problem, as none of the above complications applies. However, according to current knowledge probably more than 80% of all genes have alternative transcripts and maybe more than half also alternative promoters [99]. Both numbers are rough estimates from what we already know and can be expected to rise even further. This illustrates nicely that the carefree situation of single transcripts and promoters is most likely the exception, not the rule, for genes represented on micorarrays. It has already been recognized that this may cause problems with the traditional way of probe selection, rendering part of the probes on a microarray uninformative [39]. The problem with alternative splicing has been recognized already and studies in that direction have been carried out [49]. There are also efforts under way by microarray manufacturers to take alternative splicing into account. Fortunately, it became possible to check which probes can be reliably used and which probes might cause problems thanks to the high-quality genomic sequences available and our increasing knowledge about alternative transcripts. It should be noted that the set of useful probes depends to some extent on the experimental conditions, not the array used. Some probes might be very informative, whereas the alternative transcripts also recognized by the same probes are not expressed. Use of such probes might cause problems under conditions in which such alternative transcripts are coexpressed.

However, the case of alternative promoters is much less well recognized, but is of equal importance as in many cases transcripts appear to be the same, but originate from different promoters. For example, the CYP19A (also known as aromatase) gene that has at least seven promoters (probably even 10), all of which appear to encode the same transcript. The reason for that paradox is that all promoters are linked to alternative noncoding first exons of almost identical length all of which splice invariably to the identical coding region comprised of nine additional exons. Thus, basically all probes recognize any of the transcripts indifferently. However, events important in breast cancer include a switch of promoter usage not detectable that way [19]. Only transcript-specific probes will help here and they can only be designed based on knowledge about the alternative promoters.

As already mentioned, it is possible to reduce the amount of potentially ambiguous probes by utilizing the existing knowledge of alternative transcript structures [49]. Based on the huge promoter collection in ElDorado, Genomatix is currently evaluating genome-wide probe sets that are specific for alternative promoter usage in order to afford the design of microarrays that will directly indicate promoter selection. This will be of great use for subsequent promoter analyzes as it will take the guesswork out of the selection of promoters. This will also be the only way to tackle the problems of closely related transcripts such as in the case of the CYP19A gene discussed above. It is safe to assume that transcript- and promoter-specific microarrays will become the standard in the near future, bringing the results obtained with such arrays a lot closer to the underlying molecular mechanisms that present-day arrays allow.

7 Conclusions

The experimental dissection of functional mechanisms of transcription control has gained an enormous momentum over recent years. The ever-increasing number of publications on this topic bears witness to this development, which found one early hallmark manifestation in the introduction of a new section in the *Journal of Molecular and Cellular Biology* entirely devoted to analysis of transcription control, which just spearheaded widespread publication of similar articles in most other leading journals. The complex interleaved networks of transcription control certainly represent one of the cornerstones on which to build our understanding of how life functions, in terms of embryonic development, tissue differentiation, and maintenance of the shape and fitness of adult organisms throughout life (see also Chapter 21). This is also the reason why both the experimental analysis and the bioinformatics of transcription control will move more and more into the focus of medical/pharmaceutical research. A considerable number of diseases are directly or indirectly connected to alterations in cellular transcription programs (e.g. most forms of cancer). We recently demonstrated how promoter analysis can be used to elucidate some underlying molecular networks in insulin signaling with relevance to the maturity onset of diabetes of the young (MODY [25]). Furthermore, many drugs influence transcription control via signaling pathways (triggering TFs) [47], which could also be connected to certain side-effects of drugs [73]. The various genome-sequencing projects will provide us with a complete catalog of the components of a number of mammalian species probably within a few years. This will complement the blueprint of the material basis of a human already derived from the human genome sequence. However, only the analysis of the regulatory part of the genome and the corresponding expression patterns and the complex metabolic networks will provide deeper insight into how the complex machinery called life actually works. Definition and detection of regulatory regions by bioinformatics will contribute to this part of the task, and will become instrumental in guiding experimental approaches as well.

As a final note it should be emphasized that transcriptional regulation necessarily involves thousands of proteins, which is why proteomics analyses will also make important contributions to our understanding of regulatory events (see Chapter 28). However, despite its much longer history, protein research has not yet reached the level where it can be readily merged with the DNA-based analysis of transcription control. Nevertheless, we are quite confident that in the very near future protein research will be as integrated into the analysis of genome regulation as are nucleotide sequence- based methods today. Biology simply cannot be divided into DNA, RNA and protein "fields" as all of this is required to define and support the wonderful concerted action called life.

References

1 Arnone, M. I. and E. H. Davidson. 1997. The hardwiring of development: organization and function of genomic regulatory systems. Development **124**: 1851–64.

2 Audic, S. and J. M. Claverie. 1997. Detection of eukaryotic promoters using Markov transition matrices. Comput. Chem. **21**: 223–7.

3 Bajic, V. B., S. H. Seah, A. Chong, G. Zhang, J. L. Koh and V. Brusic. 2002. Dragon PromoterFinder: recognition of vertebrate RNA polymerase II promoters. Bioinformatics **18**: 198–9.

4 Bajic, V. B., S. L. Tan, Y. Suzuki and S. Sugano. 2004. Promoter prediction analysis on the whole human genome. Nat. Biotechnol. **22**: 1467–73.

5 Bell, P. J., V. J. Higgins, I. W. Dawes and P. H. Kissinger. 1997. Tandemly repeated 147 bp elements cause structural and functional variation in divergent MAL promoters of *Saccharomyces cerevisiae*. Yeast **13**: 1135–44.

6 Bergers, G., P. Graninger, S. Braselmann, C. Wrighton and M. Busslinger. 1995. Transcriptional activation of the *fra-1* gene by AP-1 is mediated by regulatory sequences in the first intron. Mol. Cell. Biol. **15**: 3748–58.

7 Biberman, Y. and O. Meyuhas. 1997. Substitution of just five nucleotides at and around the transcription start site of rat beta-actin promoter is sufficient to render the resulting transcript a subject for translational control. FEBS Lett. **405**: 333–6.

8 BLANCHETTE, M. AND M. TOMPA. 2003. FootPrinter: a program designed for phylogenetic footprinting. Nucleic Acids Res. **31**: 3840–2.

9 BODE, J., S. GOETZE, H. HENG, S. A. KRAWETZ AND C. BENHAM. 2003. From DNA structure to gene expression: mediators of nuclear compartmentalization and dynamics. Chromosome Res. **11**: 435–45.

10 BOHJANEN, P. R., Y. LIU AND M. A. GARCIA-BLANCO. 1997. TAR RNA decoys inhibit tat-activated HIV-1 transcription after preinitiation complex formation. Nucleic Acids Res. **25**: 4481–6.

11 BOUTILLIER, A. L., D. MONNIER, D. LORANG, J. R. LUNDBLAD, J. L. ROBERTS AND J. P. LOEFFLER. 1995. Corticotropin-releasing hormone stimulates proopiomelanocortin transcription by cFos-dependent and -independent pathways: characterization of an AP1 site in exon 1. Mol. Endocrinol. **9**: 745–55.

12 BUCHER, P. 1990. Weight matrix descriptions of four eukaryotic RNA polymerase II promoter elements derived from 502 unrelated promoter sequences. J. Mol. Biol. **212**: 563–78.

13 BUSSEMAKER, H. J., H. LI AND E. D. SIGGIA. 2001. Regulatory element detection using correlation with expression. Nat. Genet. **27**: 167–71.

14 CAM, H., E. BALCIUNAITE, A. BLAIS, A. SPEKTOR, R. C. SCARPULLA, R. YOUNG, Y. KLUGER AND B. D. DYNLACHT. 2004. A common set of gene regulatory networks links metabolism and growth inhibition. Mol. Cells **16**: 399–411.

15 CARTHARIUS, K., K. FRECH, K. GROTE, *et al.* 2005. MatInspector and beyond: promoter analysis based on transcription factor binding sites. Bioinformatics **21**: 2933–42.

16 CAVENER, D. R. 1987. Comparison of the consensus sequence flanking translational start sites in *Drosophila* and vertebrates. Nucleic Acids Res. **15**: 1353–61.

17 CHEN, Q. K., G. Z. HERTZ AND G. D. STORMO. 1995. MATRIX SEARCH 1.0: a computer program that scans DNA sequences for transcriptional elements using a database of weight matrices. Comput. Appl. Biosci. **11**: 563–6.

18 CHEN, Q. K., G. Z. HERTZ AND G. D. STORMO. 1997. PromFD 1.0: a computer program that predicts eukaryotic Pol II promoters using strings and IMD matrices. Comput. Appl. Biosci. **13**: 29–35.

19 CHEN, S., T. ITOH, K. WU, D. ZHOU AND C. YANG. 2002. Transcriptional regulation of aromatase expression in human breast tissue. J. Steroid. Biochem. Mol. Biol. **83**: 93–9.

20 CHETOUANI, F., P. MONESTIE, P. THEBAULT, C. GASPIN AND B. MICHOT. 1997. ESSA: an integrated and interactive computer tool for analyzing RNA secondary structure. Nucleic Acids Res. **25**: 3514–22.

21 COLGAN, J. AND J. L. MANLEY. 1995. Cooperation between core promoter elements influences transcriptional activity *in vivo*. Proc. Natl Acad. Sci. USA **92**: 1955–9.

22 CONAWAY, J. W. AND R. C. CONAWAY. 1991. Initiation of eukaryotic messenger RNA synthesis. J Biol Chem **266**: 17721–4.

23 CRAMER, P., C. G. PESCE, F. E. BARALLE AND A. R. KORNBLIHTT. 1997. Functional association between promoter structure and transcript alternative splicing. Proc. Natl Acad. Sci. USA **94**: 11456–60.

24 DAVULURI, R. V., I. GROSSE AND M. Q. ZHANG. 2001. Computational identification of promoters and first exons in the human genome. Nat. Genet. **29**: 412–7.

25 DOEHR, S., A. KLINGENHOFF, H. MAIER, M. HRABE DE ANGELIS, T. WERNER AND R. SCHNEIDER. 2005. Linking disease-associated genes to regulatory networks via promoter organization. Nucleic Acids Res. **33**: 864–72.

26 DOWN, T. A. AND T. J. HUBBARD. 2002. Computational detection and location of transcription start sites in mammalian genomic DNA. Genome Res. **12**: 458–61.

27 DVIR, A., J. W. CONAWAY AND R. C. CONAWAY. 2001. Mechanism of transcription initiation and promoter escape by RNA polymerase II. Curr. Opin. Genet. Dev. **11**: 209–14.

28 FICKETT, J. W. 1996. Coordinate positioning of MEF2 and myogenin binding sites. Gene **172**: GC19–32.

29 FICKETT, J. W. AND A. G. HATZIGEORGIOU. 1997. Eukaryotic promoter recognition. Genome Res. **7**: 861–78.

30 FICKETT, J. W. AND W. W. WASSERMAN. 2000. Discovery and modeling of transcriptional regulatory regions. Curr. Opin. Biotechnol. **11**: 19–24.

31 FRECH, K., J. DANESCU-MAYER AND T. WERNER. 1997. A novel method to develop highly specific models for regulatory units detects a new LTR in GenBank which contains a functional promoter. J. Mol. Biol. **270**: 674–87.

32 FRECH, K., P. DIETZE AND T. WERNER. 1997. ConsInspector 3.0: new library and enhanced functionality. Comput. Appl. Biosci. **13**: 109–10.

33 FRECH, K., G. HERRMANN AND T. WERNER. 1993. Computer-assisted prediction, classification, and delimitation of protein binding sites in nucleic acids. Nucleic Acids Res. **21**: 1655–64.

34 FRECH, K., K. QUANDT AND T. WERNER. 1997. Finding protein-binding sites in DNA sequences: the next generation. Trends. Biochem. Sci. **22**: 103–4.

35 FRECH, K., K. QUANDT AND T. WERNER. 1998. Muscle actin genes: a first step towards computational classification of tissue specific promoters. In Silico Biol. **1**: 29–38.

36 FRECH, K., K. QUANDT AND T. WERNER. 1997. Software for the analysis of DNA sequence elements of transcription. Comput. Appl. Biosci. **13**: 89–97.

37 FRISCH, M., K. FRECH, A. KLINGENHOFF, K. CARTHARIUS, I. LIEBICH AND T. WERNER. 2002. *In silico* prediction of scaffold/matrix attachment regions in large genomic sequences. Genome Res. **12**: 349–54.

38 GALM, O. AND M. ESTELLER. 2004. Beyond genetics – the emerging role of epigenetic changes in hematopoietic malignancies. Int. J. Hematol. **80**: 120–7.

39 GAUTIER, L., M. MOLLER, L. FRIIS-HANSEN AND S. KNUDSEN. 2004. Alternative mapping of probes to genes for Affymetrix chips. BMC Bioinformatics **5**: 111.

40 GIANGRANDE, P. H., W. ZHU, R. E. REMPEL, N. LAAKSO AND J. R. NEVINS. 2004. Combinatorial gene control involving E2F and E Box family members. EMBO J. **23**: 1336–47.

41 GILINGER, G. AND J. C. ALWINE. 1993. Transcriptional activation by simian virus 40 large T antigen: requirements for simple promoter structures containing either TATA or initiator elements with variable upstream factor binding sites. J. Virol. **67**: 6682–8.

42 HENG, H. H., S. GOETZE, C. J. YE, et al. 2004. Chromatin loops are selectively anchored using scaffold/matrix-attachment regions. J. Cell Sci. **117**: 999–1008.

43 HOFACKER, I. L., B. PRIWITZER AND P. F. STADLER. 2004. Prediction of locally stable RNA secondary structures for genome-wide surveys. Bioinformatics **20**: 186–90.

44 HUTCHINSON, G. B. 1996. The prediction of vertebrate promoter regions using differential hexamer frequency analysis. Comput. Appl. Biosci. **12**: 391–8.

45 IMPEY, S., S. R. MCCORKLE, H. CHA-MOLSTAD, et al. 2004. Defining the CREB regulon: a genome-wide analysis of transcription factor regulatory regions. Cell **119**: 1041–54.

46 IOSHIKHES, I., A. BOLSHOY, K. DERENSHTEYN, M. BORODOVSKY AND E. N. TRIFONOV. 1996. Nucleosome DNA sequence pattern revealed by multiple alignment of experimentally mapped sequences. J. Mol. Biol. **262**: 129–39.

47 JAMORA, C., R. DASGUPTA, P. KOCIENIEWSKI AND E. FUCHS. 2003. Links between signal transduction, transcription and adhesion in epithelial bud development. Nature **422**: 317–22.

48 JIANG, J. G. AND R. ZARNEGAR. 1997. A novel transcriptional regulatory region within the core promoter of the hepatocyte growth factor gene is responsible for its inducibility by cytokines via the C/EBP family of transcription factors. Mol. Cell. Biol. **17**: 5758–70.

49 Johnson, J. M., J. Castle, P. Garrett-Engele, et al. 2003. Genome-wide survey of human alternative pre-mRNA splicing with exon junction microarrays. Science **302**: 2141–4.

50 Keich, U. and P. A. Pevzner. 2002. Finding motifs in the twilight zone. Bioinformatics **18**: 1374–81.

51 Kel-Margoulis, O. V., A. G. Romashchenko, N. A. Kolchanov, E. Wingender and A. E. Kel. 2000. COMPEL: a database on composite regulatory elements providing combinatorial transcriptional regulation. Nucleic Acids Res. **28**: 311–5.

52 Kel, A., O. Kel-Margoulis, V. Babenko and E. Wingender. 1999. Recognition of NFATp/AP-1 composite elements within genes induced upon the activation of immune cells. J. Mol. Biol. **288**: 353–76.

53 Kel, A. E., Y. V. Kondrakhin, A. Kolpakov Ph, O. V. Kel, A. G. Romashenko, E. Wingender, L. Milanesi and N. A. Kolchanov. 1995. Computer tool FUNSITE for analysis of eukaryotic regulatory genomic sequences. Proc. ISMB **3**: 197–205.

54 Kim, J., S. Klooster and D. J. Shapiro. 1995. Intrinsically bent DNA in a eukaryotic transcription factor recognition sequence potentiates transcription activation. J. Biol. Chem. **270**: 1282–8.

55 Kleinjan, D. A. and V. van Heyningen. 2005. Long-range control of gene expression: emerging mechanisms and disruption in disease. Am. J. Hum. Genet. **76**: 8–32.

56 Klingenhoff, A., K. Frech, K. Quandt and T. Werner. 1999. Functional promoter modules can be detected by formal models independent of overall nucleotide sequence similarity. Bioinformatics **15**: 180–6.

57 Kodadek, T. 1998. Mechanistic parallels between DNA replication, recombination and transcription. Trends Biochem. Sci. **23**: 79–83.

58 Lavorgna, G., A. Guffanti, G. Borsani, A. Ballabio and E. Boncinelli. 1999. TargetFinder: searching annotated sequence databases for target genes of transcription factors. Bioinformatics **15**: 172–3.

59 Lee, G. E., J. H. Kim and I. K. Chung. 1998. Topoisomerase II-mediated DNA cleavage on the cruciform structure formed within the 5′ upstream region of the human beta-globin gene. Mol. Cells **8**: 424–30.

60 Levy, S., L. Compagnoni, E. W. Myers and G. D. Stormo. 1998. Xlandscape: the graphical display of word frequencies in sequences. Bioinformatics **14**: 74–80.

61 Levy, S. and S. Hannenhalli. 2002. Identification of transcription factor binding sites in the human genome sequence. Mamm. Genome **13**: 510–4.

62 Li, H. and Y. Capetanaki. 1994. An E box in the desmin promoter cooperates with the E box and MEF-2 sites of a distal enhancer to direct muscle-specific transcription. EMBO J. **13**: 3580–9.

63 Lian, J. B., J. L. Stein, G. S. Stein, M. Montecino, A. J. van Wijnen, A. Javed and S. Gutierrez. 2001. Contributions of nuclear architecture and chromatin to vitamin D-dependent transcriptional control of the rat osteocalcin gene. Steroids **66**: 159–70.

64 Loots, G. G. and I. Ovcharenko. 2004. rVISTA 2.0: evolutionary analysis of transcription factor binding sites. Nucleic Acids Res. **32**: W217–21.

65 Mahony, S., D. Hendrix, A. Golden, T. J. Smith and D. S. Rokhsar. 2005. Transcription factor binding site identification using the self-organizing map. Bioinformatics.

66 Novina, C. D. and A. L. Roy. 1996. Core promoters and transcriptional control. Trends Genet. **12**: 351–5.

67 Ohler, U. and H. Niemann. 2001. Identification and analysis of eukaryotic promoters: recent computational approaches. Trends Genet. **17**: 56–60.

68 Panne, D., T. Maniatis and S. C. Harrison. 2004. Crystal structure of ATF-2/c-Jun and IRF-3 bound to the interferon-beta enhancer. EMBO J. **23**: 4384–93.

69 Pilpel, Y., P. Sudarsanam and G. M. Church. 2001. Identifying regulatory

networks by combinatorial analysis of promoter elements. Nat. Genet. **29**: 153–9.

70 Prestridge, D. S. 1996. SIGNAL SCAN 4.0: additional databases and sequence formats. Comput. Appl. Biosci. **12**: 157–60.

71 Pritsker, M., Y. C. Liu, M. A. Beer and S. Tavazoie. 2004. Whole-genome discovery of transcription factor binding sites by network-level conservation. Genome Res. **14**: 99–108.

72 Quandt, K., K. Frech, H. Karas, E. Wingender and T. Werner. 1995. MatInd and MatInspector: new fast and versatile tools for detection of consensus matches in nucleotide sequence data. Nucleic Acids Res. **23**: 4878–84.

73 Rhee, K. and E. A. Thompson. 1996. Glucocorticoid regulation of a transcription factor that binds an initiator-like element in the murine thymidine kinase (Tk-1) promoter. Mol. Endocrinol. **10**: 1536–48.

74 Rothenberg, E. V. and S. B. Ward. 1996. A dynamic assembly of diverse transcription factors integrates activation and cell-type information for interleukin 2 gene regulation. Proc. Natl Acad. Sci. USA **93**: 9358–65.

75 Sap, J., A. Munoz, J. Schmitt, H. Stunnenberg and B. Vennstrom. 1989. Repression of transcription mediated at a thyroid hormone response element by the v-*erb-A* oncogene product. Nature **340**: 242–4.

76 Scherf, M., A. Klingenhoff, K. Frech, et al. 2001. First pass annotation of promoters on human chromosome 22. Genome Res. **11**: 333–40.

77 Scherf, M., A. Klingenhoff and T. Werner. 2000. Highly specific localization of promoter regions in large genomic sequences by PromoterInspector: a novel context analysis approach. J. Mol. Biol. **297**: 599–606.

78 Schmid, C. D., V. Praz, M. Delorenzi, R. Perier and P. Bucher. 2004. The Eukaryotic Promoter Database EPD: the impact of *in silico* primer extension. Nucleic Acids Res. **32**: D82–5.

79 Singh, G. B., J. A. Kramer and S. A. Krawetz. 1997. Mathematical model to predict regions of chromatin attachment to the nuclear matrix. Nucleic Acids Res. **25**: 1419–25.

80 Sinha, S., M. Blanchette and M. Tompa. 2004. PhyME: a probabilistic algorithm for finding motifs in sets of orthologous sequences. BMC Bioinformatics **5**: 170.

81 Sloan, L. S. and A. Schepartz. 1998. Sequence determinants of the intrinsic bend in the cyclic AMP response element. Biochemistry **37**: 7113–8.

82 Solovyev, V. and A. Salamov. 1997. The Gene-Finder computer tools for analysis of human and model organisms genome sequences. Proc. ISMB **5**: 294–302.

83 Staden, R. 1984. Computer methods to locate signals in nucleic acid sequences. Nucleic Acids Res. **12**: 505–19.

84 Stein, G. S., A. J. van Wijnen, J. Stein, J. B. Lian and M. Montecino. 1995. Contributions of nuclear architecture to transcriptional control. Int. Rev. Cytol. **162A**: 251–78.

85 Stormo, G. D. 2000. DNA binding sites: representation and discovery. Bioinformatics **16**: 16–23.

86 Stormo, G. D. and G. W. Hartzell, 3rd. 1989. Identifying protein-binding sites from unaligned DNA fragments. Proc. Natl Acad. Sci. USA **86**: 1183–7.

87 Suzuki, Y., R. Yamashita, S. Sugano and K. Nakai. 2004. DBTSS, DataBase of Transcriptional Start Sites: progress report 2004. Nucleic Acids Res. **32**: D78–81.

88 Tompa, M., N. Li, T. L. Bailey, et al. 2005. Assessing computational tools for the discovery of transcription factor binding sites. Nat. Biotechnol. **23**: 137–44.

89 Tronche, F., F. Ringeisen, M. Blumenfeld, M. Yaniv and M. Pontoglio. 1997. Analysis of the distribution of binding sites for a tissue-specific transcription factor in the vertebrate genome. J. Mol. Biol. **266**: 231–45.

90 Usheva, A. and T. Shenk. 1996. YY1 transcriptional initiator: protein interactions and association with a DNA site containing unpaired strands. Proc. Natl Acad. Sci. USA **93**: 13571–6.

91 van Drunen, C. M., R. G. Sewalt, R. W. Oosterling, P. J. Weisbeek, S. C.

SMEEKENS AND R. VAN DRIEL. 1999. A bipartite sequence element associated with matrix/scaffold attachment regions. Nucleic Acids Res. **27**: 2924–30.

92 WANG, W., T. CHI, Y. XUE, S. ZHOU, A. KUO AND G. R. CRABTREE. 1998. Architectural DNA binding by a high-mobility-group/kinesin-like subunit in mammalian SWI/SNF-related complexes. Proc. Natl Acad. Sci. USA **95**: 492–8.

93 WASSERMAN, W. W. AND J. W. FICKETT. 1998. Identification of regulatory regions which confer muscle-specific gene expression. J. Mol. Biol. **278**: 167–81.

94 WASSERMAN, W. W., M. PALUMBO, W. THOMPSON, J. W. FICKETT AND C. E. LAWRENCE. 2000. Human–mouse genome comparisons to locate regulatory sites. Nat. Genet. **26**: 225–8.

95 WERNER, T. 2001. Cluster analysis and promoter modelling as bioinformatics tools for the identification of target genes from expression array data. Pharmacogenomics **2**: 25–36.

96 WERNER, T. 1999. Models for prediction and recognition of eukaryotic promoters. Mamm. Genome **10**: 168–75.

97 YAMAUCHI, M., Y. OGATA, R. H. KIM, J. J. LI, L. P. FREEDMAN AND J. SODEK. 1996. AP-1 regulation of the rat bone sialoprotein gene transcription is mediated through a TPA response element within a glucocorticoid response unit in the gene promoter. Matrix Biol. **15**: 119–30.

98 YELIN, R., D. DAHARY, R. SOREK, et al. 2003. Widespread occurrence of antisense transcription in the human genome. Nat. Biotechnol. **21**: 379–86.

99 ZAVOLAN, M., S. KONDO, C. SCHONBACH, J. ADACHI, D. A. HUME, Y. HAYASHIZAKI AND T. GAASTERLAND. 2003. Impact of alternative initiation, splicing, and termination on the diversity of the mRNA transcripts encoded by the mouse transcriptome. Genome Res. **13**: 1290–300.

100 ZAWEL, L. AND D. REINBERG. 1995. Common themes in assembly and function of eukaryotic transcription complexes. Annu. Rev. Biochem. **64**: 533–61.

7
Finding Repeats in Genome Sequences

Brian J. Haas and Steven L. Salzberg

1 Introduction

An essential component of genome sequence analysis is the identification of repetitive sequences (repeats). A repeat is a substring that occurs multiple times within a sequence or collection of sequences. Repeats are commonly found in the genomes of both prokaryotes and eukaryotes, although generally to a lesser extent in the compact genomes of prokaryotes. In some cases, the number of repeats and their contribution to overall genome size and content is staggering, e.g. ;ore than half of the human genome is composed of repetitive sequences [34]. In addition, the large genomes of higher plants including maize and wheat are composed mostly of repetitive sequences [3, 7, 19]. In some bacteria, repetitive plasmid sequences are so similar to one another that it is extremely difficult even to determine how many plasmids are present, as in the case of the Lyme disease spirochete [20].

Although repeated sequences represent a diverse group of features, they tend to fall into one of two broad categories: tandem repeats or dispersed repeats. Tandem repeats are those that are found directly adjacent to one another, contiguously arrayed. These are often termed "satellite" DNA. Simple sequence repeats (SSRs or microsatellites) are tandem repeats where the repeat unit is very short, typically 1–6 nucleotides. SSRs tend to have a uniform distribution within genomes, and are sometimes found within protein coding and untranslated regions of genes. Trinucleotide repeats within genes are of special interest since they have been linked to several human genetic disorders, including Fragile-X mental retardation, Huntington's disease and myotonic dystrophy [37, 50, 51]. The term "satellite repeat" typically refers to repeat units greater than 100 bp, which are found as contiguous stretches that can span up to tens of thousands or even millions of base pairs of chromosomal DNA. Such satellite repeats include the 170- to 180-bp repeat units found at centromeres of higher eukaryotes [12, 23] and the long contiguous rDNA cassettes that comprise nucleolus organizer regions [47]. The term

Bioinformatics - From Genomes to Therapies Vol. 1. Edited by Thomas Lengauer

ISBN: 978-3-527-31278-8

minisatellites is used to refer to tandem repeats with unit lengths intermediate to SSRs and satellites.

Dispersed repeats predominantly consist of transposable elements – mobile sequences that can cut and paste or copy themselves to other locations in a genome (for details, see Refs. [11, 13]). Complete autonomous transposons encode one or more proteins that are required for their mobility, and can exceed 10 kb in length. Nonautonomous transposable elements also parasitize genomes; these numerous elements lack the machinery for their own transposition and rely on the proteins encoded by other complete elements to mediate their transposition. Transposons tend to be most abundant in regions of heterochromatin, typically in pericentromeric regions mostly devoid of expressed genes. These elements are sometimes found within introns of genes or interrupting an exon, in which case the gene is likely rendered nonfunctional. Dispersed repeats account for a majority of the repeat content in large eukaryotic genomes. Little is known about the purpose of transposons; they are often regarded as "selfish" elements that provide no benefit to the host organism.

Molecular events resulting in gene duplication, including unequal crossing over during meiotic recombination, recombinational repair or, at the extreme, whole genome duplications, also generate repeated sequences. On the smallest level, slippage of the DNA replication machinery can result in short repeats gaining additional copies. Depending on the event responsible for the genomic rearrangement and the resulting configuration of the genetic material, the duplicated segments may appear in tandem or at remote locations typical of dispersed repeat families. Repeat regions are involved in multiple human diseases, the most well-known being Down's syndrome, which involves an extra copy of chromosome 21. Both Huntington's disease and Fragile X syndrome result from an expansion of trinucleotide repeats. At another level, repeats are used for the new science of microbial forensics; as these regions are among the most highly variable in many species, they provide unique DNA-based signatures that distinguish bacteria from one another, including very closely related strains of organisms such as the anthrax bacterium, *Bacillus anthracis* [49].

Rigorous studies of genome sequence repeats involve identifying similar sequence pairs, grouping the related elements to examine their number and distribution within the genome, differentiating repeats of known function from those of unknown function (and genes from nongenes), and unraveling the details of the length, number of copies and orientation of repeat elements. Each step of the analysis is complicated by the nature of the underlying repetitive sequences, including the degree of divergence between related elements, the background of genomic architectural rearrangements which disrupt or conceal the original repeats and the resulting mosaic nature of repeat elements such that related repeats may share only a subsequence in common. In particular, the boundaries of repeats are notoriously difficult to resolve, complicated by issues described above coupled with the difficulty in obtaining pairwise alignments which terminate precisely at repeat boundaries. An often-cited statement made by Bao and Eddy [4] nicely summarizes the state of automated repeat finding: "the problem of automated repeat sequence family classification is inherently messy and ill-defined and does not appear to be amenable to clean algorithmic attack". This remains true today, although new algorithms, tools and ideas regarding repeat analysis continue to shed light on the problem.

Repetitive sequences impose formidable challenges to sequence analysis in the postgenomic era. They create havoc for genome assembly; regions rich in repeats are difficult if not impossible to assemble correctly using currently available tools and algorithms, and often lead to misassembly of regions flanked by repeats or excessive fragmentation of what would otherwise be a more cohesive genome sequence (see also Chapter 2). Subsequent to sequence assembly, genome annotation is also confounded by repeats. In particular, the transposon sequences found between genes and within introns can be easily mistaken for exons of protein-coding genes by gene-finding programs. This "junk" DNA requires prior recognition and exclusion to facilitate more accurate identification of the coveted host genes localized to the remaining sequence (see also Chapter 5).

This chapter focuses on the algorithms and tools commonly used for identifying repeats in genome sequences. The impact of repeats on genome assembly and methods used by assemblers to circumvent associated problems are described. Additional topics include methods for clustering elements to organize repeat families, resolving repeat boundaries, efforts to untangle the mosaic nature of related repeats and the annotation of repeat sequences.

2 Algorithms and Tools for Mining Repeats

Sequence alignment is at the very core of repeat identification. In contrast to aligning sequences from different genomes to identify regions of homology, sequence alignment is applied to a single genome, as a single sequence or collection of sequences, to identify significant intra- and inter-sequence similarities. The more general application of repeat analysis is summarized as first finding all pairwise alignments, then clustering related elements. Finding all pairwise alignments is relegated to standard sequence alignment tools and algorithms, a topic described in Chapter 3 and so minimally covered here. Clustering of repeat elements into repeat families is a major challenge, and recent efforts to deconvolute pairwise alignments into more meaningful repeat sets are described. Finding tandem repeats is a related, but distinct challenge; a vast amount of literature exists on this topic, describing algorithms and tools which are specially designed for this aspect of repeat structure. Because of this, tandem repeats are the focus of a separate section in this chapter.

2.1 Finding Intra- and Inter-sequence Repeats as Pairwise Alignments

Pairwise sequence alignment algorithms are well suited to the problem of repeat identification; due to the enormous complexity of the problem in large genomes, heuristics are required to improve efficiency. The Smith–Waterman alignment algorithm [54] finds the single best-scoring local alignment between two sequences. If the sequences being compared are two distinct entries from the collection of sequences corresponding to the single genome under study, this best local alignment would suffice as a repeat. This approach cannot be used, however, when the genome is a single contiguous sequence, as is often the case for complete bacterial, archaeal and viral genomes. Comparing a sequence to itself to find the best local alignment would yield only the obvious perfect alignment along the diagonal corresponding to the alignment of the sequence matching itself from beginning to end. A modification to the Smith–Waterman algorithm, as described by Waterman and Eggert [58], affords the identification of all nonintersecting high scoring alignments, rather than just the single best local alignment between two sequences (see also Chapter 3). This modification unravels the internal repetitive structure of a sequence when aligned to itself. Huang and Miller [24] describe the sim algorithm, which yields all high scoring nonintersecting alignments using linear space, and the lalign utility of Bill Pearson's fasta2 toolsuite (http://ftp.virginia.edu/pub/fasta/) is a popular tool that implements this algorithm. The accompanying plalign utility generates an illustration of the repetitive structure as a postscript file.

Although these algorithms are well suited to repeat finding, they are simply not fast enough to tackle large genomes and so we turn to heuristics. The "seed and extend" heuristic is perhaps the most common strategy to quickly ascertain significant pairwise alignments between sequences. Early uses of this strategy include the FASTA algorithm [43], followed by the hugely popular BLAST algorithm [1,2], among other database search and alignment tools including MUMmer [14,15,32], PatternHunter [35,36] and BLAT [29], and a less well known but similarly useful tool for studying repeats called ICAass [41], the repeat mining utility of the Miropeats software [42]. Here, matches to exact words of predefined length provide the seeds for alignments, which are extended in both directions to extract the maximal scoring alignment containing the seed (see also Chapter 3). A major limitation of these methods is the requirement of a predefined seed length. A seed length that is too short requires numerous extensions, few of which lead to significant alignments. A seed length that is too long involves fewer extensions, but many significant alignments may lack such a seed and are missed. The programs MUMmer and Reputer [31,33] take a more sophisticated approach, employing a suffix tree data structure to find exact word matches. The suffix tree is not limited to finding seeds of a constant length; all exact matches are found regardless of length and this is done very fast, in linear time and space.

The focus of repeat-finding software can vary; the focus may be to find all matching substrings within a sequence or among a collection of sequences, or the focus may be the postprocessing of pairwise alignments to cluster related elements into families, resolve repeat boundaries or to illustrate the mosaic nature of repeated sequences. Progress in repeat analysis, as in other areas, builds upon previous contributions to the field. As such, we present an overview of each contribution in roughly chronological order.

2.2 Miropeats (alias Printrepeats)

Miropeats [42] is perhaps one of the earliest and most popular repeat-finding analysis tools to find widespread use in genome sequence analysis (see, e.g. Refs [52]). The repeat-finding engine of Miropeats is the program ICAass of the ICAtools suite [41]. ICAass finds maximal gap-less aligned segment pairs (MSPs) within a single sequence and/or among a collection of sequences using a "seed and extend" strategy similar to that used in BLASTN [1b]. All overlapping 8-mers are loaded into a hash table, using two-bits per base encoding and so allowing 4 bases per byte. All sequences to be examined are indexed and the 8-mers are then used to seed potentially longer alignments. Those ungapped alignments meeting the minimum score threshold are reported. While the ICAass program includes additional components, only the MSP identification steps are utilized by Miropeats. Miropeats is a Unix C-shell script that calls ICAass to identify MSPs as repeats, and then writes a postscript file which illustrates the positions and associations among the repeats. Arcs are drawn between the matching end-points of each repeat pair and arcs are drawn in such a way to help ascertain their relative orientation and overlap. Although ICAass is used, in theory, any program capable of generating meaningful pairwise alignments could be employed, including BLAST (i.e. WU-BLAST with the -span option selected), BLAT or PatternHunter, although Miropeats would require some minor customization to accept this. The strength of Miropeats as a repeat analysis tool lies in its illustration capabilities, particularly with respect to those repeats found in close proximity along the nucleotide sequence, e.g. the structures of transposable elements typically include some form of terminal repeat, either direct or inverted, at the elements' boundaries. The illustration of repeats within transposon-rich regions helps to elucidate their terminal repeat structures as well as their relative abundance along the genomic contig; an example is provided in Figure 1, where the long terminal repeats (LTRs) of a gypsy-family retrotransposon are nicely illustrated by the Miropeats software). (A "contig" is a contiguous stretch of DNA without gaps. Whole-genome shotgun (WGS) sequencing projects generally produce nearly complete genomes that consist of a set of contigs separated by gaps.) As with other repeat-finding applications, the use of Miropeats extends beyond repeat analysis and includes additional areas of sequence comparison, such as to position a small set of bacterial artificial chromosomes (BACs) in a section of a genome BAC tiling path by defining the overlaps among their ends (see also Chapter 2). Due to the relatively slow ICAass repeat-finding step and because of the static illustration of the repeat structures provided by Miropeats, the software is limited in practice to analyzing sequences whose length is no more than a few hundred thousand base pairs, although application to longer sequences is not restricted.

2.3 REPuter

Of the methods available for finding repeated strings in genomic sequences, those based on suffix trees are most efficient and practical for large-scale genome analyses. A suffix tree is a data structure specifically designed to capture a text string and all of its substrings, which makes it well suited for capturing DNA and protein sequences. The tree itself is set of nodes and edges, where each edge is labeled with a string. A suffix of a string S (which might be an entire genome, for example) is simply a substring that starts within S and extends to the end of S. A suffix tree represents all suffixes of S implicitly; each suffix is a path from the root node to a leaf node of S. Internal nodes of S represent other (nonsuffix) substrings; in fact, a suffix tree contains all substrings of S. Suffix trees represented a major advance over previous sequence analysis techniques because of two key properties: (i) the size of the tree is a linear function of the size of the sequence, and (ii) the tree can be constructed and searched in linear time. This contrasts with alternative sequence alignment methods, which are quadratic in time, space or both. Details regarding construction and search algorithms for suffix trees are described in Ref. [22].

Stefan Kurtz's REPuter software [31,33], the first production quality repeat-finding software to employ suffix trees, can mine complete eukaryotic genomes (megabase pairs) for all maximal repeats in a matter of seconds on a personal computer. REPuter has existed in two versions. The earlier version, first described in 1999, was limited to finding identical maximal repeats. An enhanced tool suite was released under the package name REPuter in 2001 with the search engine named REPfind, capable of extending the problem of

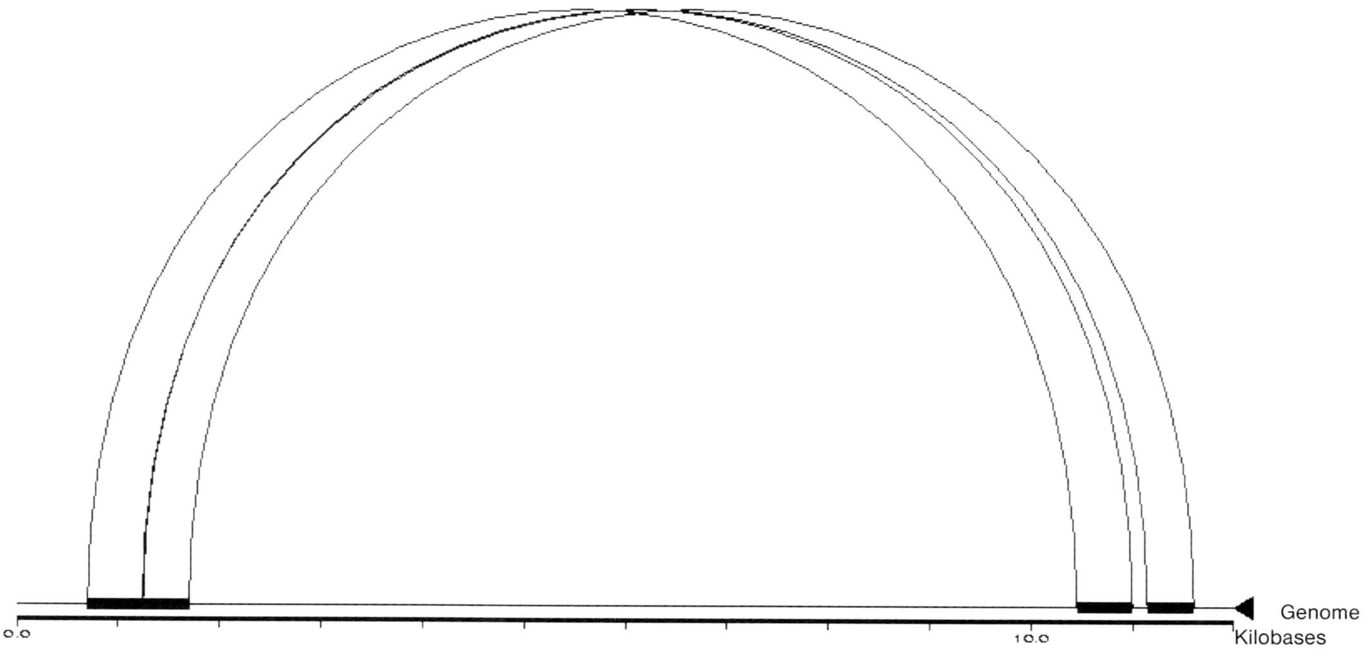

Figure 1 LTRs of gypsy-like retrotransposon family (Athila) element of *Arabidopsis* (locus At1g40077) found and illustrated using Miropeats. Arcs are drawn connecting pairwise matches found in an approximately 12-kb stretch of genome sequence, corresponding here to the LTRs of an *Arabidopsis* gypsy-family retrotransposon.

repeat finding from identical repeats only, to approximate repeats, allowing for mismatches and indels. Each version introduced key concepts and contributions to repeat analysis, so each is described here in order of their availability. To distinguish between the two versions, the earlier version is referred to as REPuter and the later version as REPfind.

As stated earlier, REPuter is limited to finding exactly identical maximal repeats; the repeats are maximal in that extending the alignment between two paired sequence regions would introduce a mismatch and violate the requirement of identical sequence pairs. REPuter can find the following four classifications of repeats, each exemplified using the 4-mer "`gcta`" and the forward sequence orientation (top strand only):

Forward:	5′-`gcta`-3′ with 5′-`gcta`-3′
Palindromic (reverse complemented or inverted):	5′-`gcta`-3′ with 5′-`tagc`-3′
Complemented:	5′-`gcta`-3′ with 5′-`cgat`-3′
Reversed:	5′-`gcta`-3′ with 5′-`atcg`-3′

Although cataloguing all identical sequence substrings is a useful component of repeat analysis, few repeats, unless very recently duplicated, will be free of mismatches or indels. As a result, these repeated identical substrings are often parts of larger repeat units that are nonidentical, although detectibly similar in sequence. REPfind takes this into account and is able to find degenerate repeats, allowing for mismatches and insertions/deletions (indels) as part of the larger repeats.

REPfind exhaustively finds all degenerate repeats in a genome sequence given a user specified minimum length and maximum number of errors. Errors are measured by one of two methods: hamming distance or edit distance. Hamming distance corresponds to the number of mismatches in a gap-free sequence alignment. Edit distance includes the number of differences in an alignment possibly containing indels. The identification of approximate repeats relies on the basis that every degenerate repeat contains a substring of identical sequence.

To find approximate repeats, REPfind locates all exact word matches followed by an extension process to determine if the word match is part of a longer degenerate repeat. Approximate repeats of two types are found: maximal mismatch repeats using the MMR algorithm and maximal difference repeats using the MDR algorithm, as described below. Both types of repeats rely on the existence of an exact word match; the exact word matches are found as described earlier by the original REPuter software.

The MMR algorithm finds a gap-less maximal mismatch repeat by looking for the longest alignment that contains the seed and has no more than k mismatches (the degenerate repeat in this context called a maximal k-mismatch repeat; here k is a specified parameter. This is done by identifying the first $k + 1$

mismatched nucleotides to the left of the seed [ordered from left to right (l_1, $l_2, \ldots, l_{(k+1)}$)], followed by identifying the first $k + 1$ mismatches to the right of the seed (ordered $r_1, r_2, \ldots, r_{(k+1)}$), with the mismatches l_1 and $r_{(k+1)}$ bounding a sequence region of k mismatches from the left or right of the seed boundary, respectively. For all values of i, from 1 to $k + 1$, the substring with coordinates $l_i + 1$ to $r_i - 1$ contains exactly k mismatches. The k-mismatch substring with the greatest length is reported.

The MDR algorithm extends seeds taking into account insertions and deletions. The idea is similar to the MMR algorithm in that k-differences are explored to each side of the seed, and the combination of coordinates within this range which maximize repeat length and satisfy the k-mismatch criteria are chosen. The primary difference is that, instead of searching for nucleotide differences along a single dimension as with MMR, a search is performed in two dimensions allowing for insertions and deletions. A dynamic programming matrix banded at $\pm\pm$ $(k + 1)$, extending from both ends of the seed, is used to find all alignment termini yielding each of 1 to k maximum number of mismatches. Each pair of alignment termini are examined and the pair of left and right termini providing the longest repeat length and a maximum of k-differences is reported.

It is often the case that a single maximal k-difference repeat will contain multiple seeds. To avoid outputting distinct maximal k-difference repeats which contain seeds of neighboring k-difference repeats, the alignment extensions to the left of a target seed are restricted to the right of any previously occurring seed. This guarantees that each maximal k-difference repeat will derive from the extension of its left-most containing seed.

By default, REPfind reports only exact matches, as done by the earlier REPuter program. Options are available to pursue either k-mismatch repeats using the MMR algorithm or k-difference repeats using the MDR algorithm. Unless there is a keen interest in obtaining gap-less repeats only, it is sensible to mine maximum difference repeats exclusively using the MDR algorithm, given that it will report maximum matches with or without gaps, whichever provides the maximal k-difference repeat. Rather than setting the k-value directly, the user can specify the parameter values of minimum repeat length and a maximum error rate, from which the value of k is computed internally.

An improvement over the earlier REPuter software is the inclusion of statistical significance for each of the repeats found in the form of an E-value (see Chapter 3 for an explanation of the concept of an E-value). In the case that multiple solutions exist for the maximal k-mismatch or k-difference repeat, the single repeat yielding the sequence with the smallest E-value is reported. By selecting the option "-allmax", each solution is reported in the case of ties among candidates meeting the maximum length k-difference criteria.

REPuter is available to researchers in several forms: a set of command-line driven utilities for local installations, a comfortable web interface for more interactive and targeted analyses and most recently as a web service enabling distributed computing environments with repeat analysis capabilities.

We should note that the REPuter package, although still widely used and available to researchers, has more recently been subsumed by the Vmatch large-scale sequence analysis software (Kurtz, unpublished; http://www.vmatch.de). Improvements include the use of suffix arrays in place of suffix trees, which reduces memory requirements and processing time. Also, the alphabet of sequences to be aligned is no longer restricted to nucleotide characters, allowing one to examine protein sequences as well.

2.4 RepeatFinder

Given fast and efficient methods to detect pairwise similarities within or among sequences, some repeat analysis software is devoted to the postprocessing of pairwise alignment data to collect and organize the repetitive sequences identified. An early example of this is the Repeat Pattern Toolkit (Agarwal and States [1a]) applied to the clustering of WU-BLAST ungapped alignments derived from 3.6 Mbp of the *Caenorhabditis elegans* genome, placing the alignments into a graph, and finding the minimum spanning tree for connected components to represent the relationships between repeats. A more modern approach involves the postprocessing of repeats found using suffix trees.

Natalia Volfovsky's RepeatFinder [56] uses a catalog of exactly repeated strings to further refine the definitions of individual repeated elements followed by the construction of repeat classes. In contrast to our canonical definition of a repeat as a pair of sequences which share similarity from beginning to end, RepeatFinder describes merged repeats where a merged repeat is found elsewhere in the genome at least once, and may be found in partial copies. The exactly repeated strings are found using the original REPuter software; a newer version of RepeatFinder uses REPfind. These exact matches compose the initial repeat set and these are redefined as repeat elements using a merging procedure. Since repeated sequences are expected to contain mismatches and indels, few complete repeats will be reported as exact matches. The merging procedure serves to consolidate regions defined as repeats that are found in close proximity or overlapping along each genomic sequence. By doing so, indels and mismatches fragmenting single repeats into disparate word matches are merged into larger degenerate repeats, and the dispersive and the fragmented nature of repeat regions is accounted for (i.e. portions of a larger repeat may be found as separate fragments elsewhere in the genome). The merging procedure to redefine repeat regions is restricted to either merging overlapping repeats, or merging neighboring repeats, with minimum overlap or gap size as user-specified parameters, respectively.

Each merged repeat retains a list of all the originally identified repeats (considered subrepeats) contained by it. Clustering of related merged repeats is done by grouping those merged repeats containing subrepeats in common into the same class. In order to further collapse clusters of similar elements, an "all-vs-all" BLASTN search is performed and separate clusters containing elements with sequence similarity (below a specified *E*-value threshold) are grouped into a single cluster.

Although the software is useful for rapidly extracting repetitive sequences from the genome and grouping related elements, the boundaries of the repeats remain ill-defined and all members of each cluster are not guaranteed to be similar to each other given the transitive relationships established via the clustering algorithm employed. More sophisticated clustering methods employed by RECON [4] address these issues more satisfactorily.

2.5 RECON

Bao and Eddy's development of RECON [4] for repeat analysis was viewed as a pioneering effort, as it represented the first tool to attempt to delineate boundaries of repeat elements in a biologically meaningful way. The algorithm of RECON is broken down into the following major tasks: obtaining pairwise alignments among the input sequences, defining elements based on the pairwise alignments and, finally, grouping elements into families. In contrast to RepeatFinder which obtains the pairwise alignment data using REPuter, RECON uses BLASTN of the WU-BLAST package [21]. The process of defining repeat elements based on pairwise alignment data is illustrated in Figure 2.

Repeat elements are initially defined by collapsing the overlapping pairwise alignments along the genomic sequence (Step II in Figure 2). Multiple alignment information is used to infer the boundaries of the element, and also to recognize and partition those elements found to be composed of multiple distinct repeat units. Given the set of overlapping alignments that initially define a repeat element in the genomic sequence, a preponderance of alignment ends found clustered to a short region of genomic sequence signifies a boundary of an element. Some candidate boundaries may be misleading because they derive from related but distinct repeat elements, those which share subrepeats in common, but are otherwise different. Misleading alignments between pairs of elements are identified by their proportionally large amount of unaligned sequence when compared to the entire element lengths (not shown). These are then discarded from subsequent element boundary refinement methods.

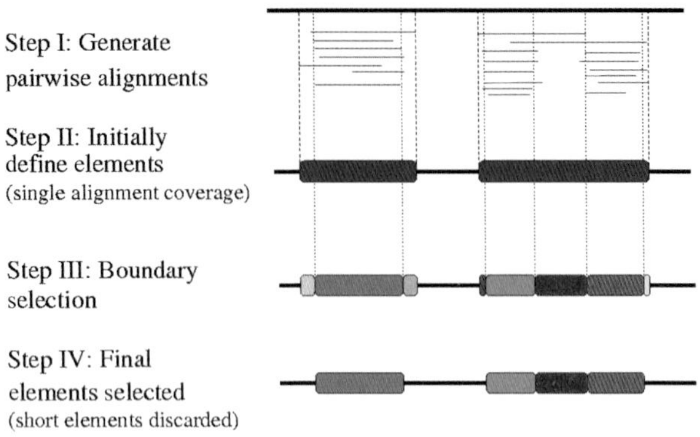

Figure 2 RECON's algorithm for defining repeat elements. In the first stage, WU-BLAST is used to generate pairwise alignments. These pairwise alignments are collapsed along the genomic sequence to define regions of alignment coverage. Clusters of alignment boundaries within short windows are used to redefine element boundaries and initial elements are repartitioned at these boundaries into separate elements. Short elements likely resulting from falsely extended alignments (from the first step) are removed to yield the refined final element set.

After eliminating the misleading alignments, the remainders are examined for the purpose of boundary refinement. Aggregations of alignment endpoints are identified by sliding a short window of predefined length (default 30 nucleotides) along the repeat element, clustering all neighboring alignment ends found separated by no more than the window size. The ratio of alignments with clustered ends to the total number of alignments spanning the corresponding region is used as an indicator of the significance of an aggregation point. A ratio above a specified threshold (default of 2.0) infers a boundary condition and the boundary is defined as the mean coordinate value for the clustered ends. Upon finding a significant aggregation point, the original element is considered composite. The composite element and its underlying supporting alignments are split at the boundary, and the split alignments are reassigned to their corresponding split element (Step III of Figure 2). Elements without significant aggregation points remain as originally defined. Split elements or the split supporting alignments found shorter than a minimum length cutoff are presumed artifacts due to the short random extensions that occur in pairwise alignments and these are discarded (Step IV of Figure 2). The remaining elements provide the set of repeats with defined boundaries.

Following the identification of the individual repeat elements as described above, the elements are classified into families. Special effort is taken to group related but distinct families separately. First, candidates for family membership are chosen by examining alignments between element pairs. For the purpose of clustering the elements, a graph is constructed in which elements are represented by nodes and relationships between nodes represented by edges. Edges are classified into two types: primary edges are used to link elements of the same family, those elements found to align to more than a specified threshold of length coverage of either element (default 90%); secondary edges link elements of different but related families that contain significant alignments but below the threshold of alignment coverage required for family membership. Before clustering family members based on the primary edges, all edges require reevaluation because some edges may have been falsely classified as primary edges. Partial elements are easily misclassified with primary edges during edge assignment since they pass the alignment coverage test with complete elements to which they are compared. It is by virtue of the secondary edges that false primary edges are identified and remedied. False primary edges are found via triangles of inequality: for example, elements *A* and *B* are deemed from the same family (primary edges), and elements *A* and *C* are deemed from the same family (primary edges), but elements *B* and *C* are deemed from separate families (secondary edge). In this case, element *A* is presumed partial given that it aligns with high coverage separately to the two elements *B* and *C*, which themselves lack significant coverage of alignments between them. To prevent element *A* from grouping the two related but distinct families together, all but the single primary edge extending from *A*, corresponding to its most similar element, are removed. Following the conversion of false primary edges to secondary edges, all the secondary edges are removed and families are generated by transitive closure of the remaining primary edges.

The algorithm of RECON addresses the problem of delineating the boundaries of individual repeat elements as they are found in the genomic sequence, but it does not describe the mosaic nature of the related repeat elements nor the consensus boundaries and length of a prototypical element among a family of elements. A rough consensus sequence for large RECON-defined repeat families can be derived from alignments of the longest repeat elements within each large family. This is a useful approximation and works well for some repeat families, but is not rigorous enough to yield a consensus for each repeat family in a biologically meaningful way.

2.6 PILER

As discussed in the Introduction, repeats are diverse features, with wide variety in size, location and biological function. Major classes of repeats, including dispersed or tandem repeats, yield specific patterns in the context of whole-genome self-alignments. Bob Edgar's PILER [16] includes a suite of tools each of which focuses on specific patterns evident in sets of alignments to reliably identify elements of the corresponding repeat class. Examples of patterns

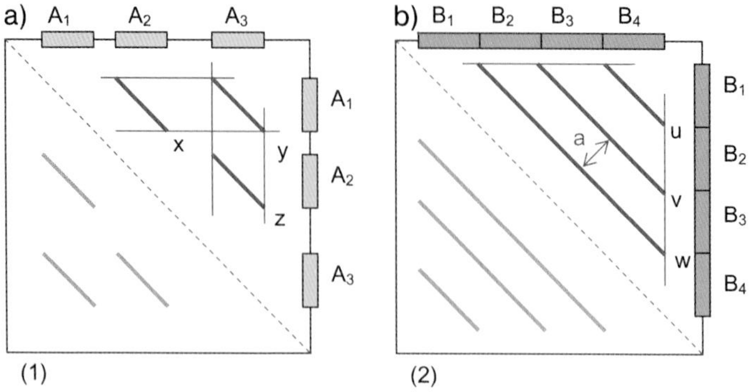

Figure 3 Patterns of sequence alignments targeted by PILER. Dot plots are shown for a comparison of a genome sequence against itself, with the dotted line as an indicator of the main diagonal of the plot. (a) Patterns of alignments generated by alignments of members of a family of dispersed repeats A_1, A_2 and A_3. (b) The "pyramid" pattern generated by alignments from a stretch of tandem repeats B_1 through B_4, indicating a repeat length of a. Figures were derived from Ref. [16], and reproduced here with permission from author Bob Edgar and Oxford University Press.

sought by PILER are illustrated in Figure 3. The individual tools of PILER and corresponding repeat classes are: PILER-DF for detecting individual intact elements of a dispersed repeat family, PILER-PS to find pseudo-satellites, PILER-TA to find tandem arrays, and PILER-TR to find repeat elements that have terminal repeats (a common characteristic of intact transposable elements).

Similarly to RepeatFinder, RECON, and other repeat finding and clustering tools, a set of intra-genome alignments is required. Rather than rely on REPUTER or BLAST to generate alignments, PILER includes an efficient alignment program called PALS (Pairwise Alignment of Long Sequences), which is specially designed with optimizations for detecting repetitive sequences; optimizations targeted towards searching a sequence against itself, limiting searches to banded regions and unrestricted reporting of numerous colocalized alignments, among others. After generating all alignments, overlapping hits along the genome sequence are linked together into a "pile" of contiguously overlapping alignments. These piles of hits are further subjected to specific analyses provided by the PILER-* utilites.

PILER-DF is designed to detect "Dispersed Families" of repeats with characteristics of transposable elements. The specific signature of a dispersed family as revealed by pairwise alignments is illustrated in Figure 3(a). The isolated elements are found as globally alignable regions, all with alignments of similar lengths. The goal of PILER-DF is to find aligned pairs that have similar characteristics. This is done by analyzing aligned sequence pairs such that each aligned region is found in a different pile (dispersed). Given a pairwise alignment between X and Y, such that X and Y are in different piles, an edge is drawn between X and Y if each aligned region spans most of its corresponding pile [(length of X)/(length of pile containing X), (length of Y)/(length of pile containing Y)]. After examining all pairwise alignments in this matter, all connected components are found in the resulting graph. The connected components are interpreted as dispersed families of complete repetitive elements.

PILER-PS searches for "Pseudo-Satellites; – repeats with features of satellite sequences in that they are found clustered locally in the genome. The algorithm here is identical to PILER-DF with the exception that the pairwise alignments result from a banded search, requiring that the alignments be in close proximity to one another.

PILER-TA finds "Tandem Arrays". The repeat-finding tools described so far are mostly limited or specially tuned to find dispersed repeats. PILER-TA is an exception in that it purposely mines these features from the genome. Please note that the general topic of finding tandem repeats is the major focus of the next section of this chapter and so it will be mentioned only briefly here as it relates to PILER. Sequences arrayed in tandem leave a specific signature in the pairwise alignments termed "pyramids" (see Figure 3b). The first observation is that pairwise alignments in a given pyramid are restricted to the same pile since all of them overlap. A banded search is used to find pairs of alignments within a pile that have the following characteristics: the shorter alignment pair is at least half the length of the longer alignment pair, and the distances between the alignments' respective start and end coordinates are each within a predefined percentage of the shorter alignment length. All pairs of alignments meeting such criteria are connected by an edge and, at completion, all connected components are gathered. Each connected component is interpreted as a tandem array. Simple heuristics are employed to define boundaries between the individual repeat elements. Diagonal distances that are in good agreement define the element length and sequences of this length from hit end-points provide representative elements of the array.

PILER-TR finds families of elements with "Terminal Repeats". This search is geared towards finding transposable elements with terminal repeats, such as the long LTR retrotransposons. The signature of these features is a set of repeats, about 50–2000 bp, separated by anywhere from 50 to 15 000 bp (all default parameters). A banded search is used to find candidate terminal repeats. To avoid reporting tandem repeats and pseudo-satellites that would also be found via a banded search, these are found and masked as a prerequisite to this search. After finding candidate terminal repeats, a second search is carried out to find different elements with matching terminal repeats, in which case a nonbanded regular search is performed. All candidates with matching terminal repeats are clustered and reported as families of terminal repeat elements.

2.7 RepeatScout

Alkes Price and Pavel Pevzner's RepeatScout [46] takes a distinct approach to repeat family identification, which circumvents some of the difficulties associated with the more traditional approach involving the postprocessing of pairwise alignments. Generating pairwise alignments as the first step of repeat sequence identification can take a long time, utilize many CPU cycles and generate copious output that can consume an enormous amount of disk space. With large genomes, this can be intractable and further effort is required to partition data sets into more manageable inputs, all of which can adversely affect the results obtained. In contrast, RepeatScout employs a repeat family search stage heuristic similar to that used in database-searching algorithms like BLASTN [1b]. Where BLASTN requires that two potentially homologous sequences share at least one exact word match in common, RepeatScout requires that all initially targeted members of a repeat family share at least one exact word in common. In basic terms, RepeatScout uses an exact word match (seed) to identify potential members of a repeat family, and then maximally extends alignments of all targeted regions to the left and right of the seed to compute a consensus sequence representation of the repeat family with repeat boundaries optimized. The specifics of this approach are described below.

The first phase of the RepeatScout algorithm involves scanning the genome for frequently occurring words. A collection of genomic sequences are scanned and the positions of all words of user specified length (i.e. 13-mers) are catalogued. Closely spaced repeat word occurrences are ignored to avoid tandem repeats [tandem repeat finding is relegated to Tandem Repeat Finder (TRF) [9] and is not an objective of RepeatScout]. After the scanning is complete, frequently occurring word matches are fed to the final phase of RepeatScout – the repeat family identification and consensus sequence construction stage.

Starting with the most frequently occurring word, RepeatScout attempts to extend all such word occurrences to the left and to the right, terminating the extension at what are considered to be the most appropriate boundaries of the repeated element, and simultaneously generating a consensus sequence for this repeat family. The extension phase is perhaps the most distinctive and critical feature of the RepeatScout algorithm, and it is the extension algorithm that rigorously defines the repeat boundaries. The consensus sequence generated by the word extension phase is optimally aligned to all members of that repeat family and cannot be further extended without reducing the total alignment score. This is accomplished by the following objective function, which computes the score of the consensus sequence as the sum of the scores of each individual repeat element aligned to the consensus:

$$A(Q; S_1, \ldots, S_n) = \left\lfloor \sum_k \max\{a(Q, S_k), 0\} \right\rfloor - c * \text{length}(Q)$$

where $a(Q, S_k)$ corresponds to the score of an alignment between the consensus sequence Q and a genome sequence substring (S_k) of equal length that extends from both ends of the seed. The constant c imposes a minimum threshold on the number of individual repeat elements (S_n) that must align with the consensus sequence to provide a suitable representation of a repeat family.

The choice of alignment function $a(Q, S_k)$ determines how the consensus sequence boundaries are positioned. With a Smith–Waterman local alignment function [54], short partial repeats would not be penalized and spurious alignment extensions to the more complete elements could drive the consensus boundary position beyond more appropriate repeat boundaries. Towards the other extreme, a fit-alignment algorithm, which fits one sequence into another [57] could be used to force all underlying complete and partial elements to match the consensus, but this can have the affect of yielding consensus boundaries that underrepresent the true boundaries. As a more suitable compromise between these two scenarios, the authors introduce a *fit-preferred* alignment function that yields a consensus sequence shared by some but not all of the underlying complete and partial copies. The fit-preferred alignment function is described below:

$$\begin{aligned}
f(i,0) &= \max(-\gamma i, -p)\,, \\
f(0,j) &= 0\,, \\
f(i,j) &= \max \begin{cases} f(i-1, j-1) + \mu_{ij} \\ f(i, j-1) - \gamma \\ f(i-1, j) - \gamma \\ -p \end{cases}, \\
a(Q,S) &= \max_{i,j} \begin{cases} f(i,j) & \text{if} \quad i = |Q| \\ f(i,j) - p & \text{if} \quad i < |Q| \end{cases}
\end{aligned}$$

where the match/mismatch score is provided by μ_{ij}, the gap penalty score by $\gamma\gamma$ and the fixed incomplete-fit penalty provided by p. Here, $f(i, j)$ is the score of a best alignment between the $1, \ldots, i$ characters in the consensus sequence Q and $1, \ldots, j$ characters in the repeat element S. The fit-preferred alignment score $a(Q, S)$ is simply $f(i, j)$ if the best alignment includes the entire consensus sequence. If not, the incomplete-fit penalty is subtracted from the best alignment score, penalizing the alignment for not including the entire consensus sequence.

The fit-preferred alignment algorithm is used by RepeatScout to generate alignments separately to the left and to the right of the word match. The fixed incomplete-fit penalty (p) is subtracted from the score of any optimal alignment to the consensus sequence that fails to extend all the way to the left boundary of the consensus sequence; an analagous penalty applies to the right boundary. If the alignment is incomplete on both boundaries, the penalty is subtracted twice. The result is that partial copies of the element are penalized in the presence of longer, more complete elements, and the consensus sequences that are generated are more suitable representations of the underlying repeat copies targeted by the exact word match. False-positive candidate elements targeted by the initial word matching strategy do not pose problems for RepeatScout; these will acquire negative alignment scores and are eliminated from contributing to the consensus by virtue of the main objective function.

The most rigorous approach to generating the consensus sequence would involve n-dimensional dynamic programming (where n is the number of sequences), but this would not be practical or even possible for more than a few sequences given that this task would be NP-hard. Instead, a heuristic approach is taken to generate the consensus whereby the word match is extended to the left and right one nucleotide at a time. A single nucleotide extension is attempted using each of the four nucleotides (G, A, T and C) and the single nucleotide extension providing the optimal alignment score is chosen. The consensus sequence is constructed greedily in this way until a maximal score is obtained and a predetermined number of subsequent iterations fails to improve upon this maximal score. The consensus sequence providing the maximal score is chosen to represent the underlying set of repeats and the termini of the consensus sequence delimit the repeat boundaries.

The RepeatScout algorithm, as described, is applied to each frequently occurring word match, beginning with those most frequent. As a single repeat family is likely to contain many exact word matches, effort must be taken to prevent re-identifying the same repeat family based on other yet-to-be processed frequently occurring words. In an attempt to prevent this effect, the counts of words found within approximate occurrences of the consensus sequence are readjusted within the set of frequently occurring words, decreasing the chance but not absolutely preventing the possibility of finding the same (or a portion of a) repeat family identified previously. A future release of RepeatScout may improve upon this functionality for identifying approximate occurrences of the consensus sequence, in order to more completely preclude repeat family rediscovery based on subsequent word matches.

The task of finding occurrences of repeat family members in the genome is relegated to searching the genome with the database of consensus sequences using RepeatMasker, BLAST or another sequence search and alignment utility. These homology searches may find repeat elements that have diverged considerably from those used by RepeatScout to generate the consensus (i.e. they lacked an exact word match required to be included in the consensus sequence construction stage of RepeatScout). This procedure provides a powerful mechanism to rigorously identify and annotate the individual elements of a larger repeat family.

3 Tandem Repeats

Tandem repeats form a special class of repetitive sequences, composed of a contiguous stretch of two or more copies of a repeat pattern. The length of the repeat pattern is called the period. This class of repeats, termed satellites, is of great biological relevance, found to correspond to specialized structures within eukaryotic genomes such as the short pentamer to heptamer repeats that form telomeres, and the longer repeats that form centromeres (e.g. 180-bp period repeats in *Arabidopsis*). Microsatellites (SSRs), found both within and between genes, are of great interest to those studying biodiversity and population genetics, and for DNA fingerprinting studies. Expansions in trinucleotide repeats have been correlated with various disease states, including Huntington's disease and Friedreich's ataxia, among others.

The problem of finding tandem repeats has received much attention from computer scientists and biologists alike, due to both the tractability of the problem from an algorithmic perspective and because of the importance of tandem repeats in their diverse biological roles. The challenge of finding tandem repeats involves identification of the repeated pattern and the number of times the pattern is repeated. Over the past decade, many algorithms have been proposed for the identification of tandem repeats, some of which seem to be academic exercises in algorithm development and few are found implemented in publicly available software for the general application of tandem repeat finding in the postgenomic era. One exception is Gary Benson's TRF [9], which has seen widespread use in genome sequence analysis and remains the most popular tandem repeat analysis software today. Alternative tools for finding tandem repeats have recently become available and extend the repertoire of essential software available to genome researchers. Here, we survey a few of these tools and describe the algorithms employed for finding tandem repeats.

3.1 TRF

Gary Benson's TRF is a powerful software tool capable of finding exact and approximate (containing mismatches or indels) tandem repeats in genomic sequences [9]. The algorithm employed in TRF is broken into two stages; first, the detection component which identifies candidate tandem repeats using a set of statistical criteria, followed by the analysis component calculates the consensus repeat pattern and period size using sequence alignment methods.

TRF initially detects tandem repeats based on the premise that similar sequences found contiguously arrayed are likely to share exact substrings, and the distance between paired substrings will be approximately the same and correspond to the period of the tandem repeat. In a search for these exact substrings that may target a tandem repeat, the genomic sequence is scanned from left to right for exact word matches, with some fixed word length w.

All further analyses of the candidate tandem repeat during this detection phase rely on statistical analyses whereby the alignment between candidate tandem repeats are modeled as a sequence of independent and identically distributed (iid) Bernoulli trials, equivalent to a sequence of coin tosses such that heads correspond to matching pairs of nucleotides, and tails correspond to mismatches or indels. The probability pM of a match and the probability pI of an indel are user-defined parameters that provide an upper limit to the allowed divergence between candidate tandem repeats.

The probabilistic model of the iid Bernoulli sequence is used with several statistical criteria to evaluate candidate tandem repeats: the *sum of heads distribution* to dictate the number of matches required among candidate repeats; the *random walk distribution* to model the indels between tandem repeats that might cause variability in the apparent period length; the *apparent size distribution* to distinguish between tandem repeats and dispersed repeats by analyzing the distribution of matches along the proposed period length; and the *waiting time distribution* to choose match search criteria that are most suitable for different period lengths. Each of the above distributions depend on the period length, word length used for scanning matches, and the user-defined cutoffs of pM and pI.

During the scan of the genomic sequence from left to right, word matches are accumulated. The position of each word is kept in a history list and the distance between word matches is kept in a distance list. Once a word match is found, the distance separating the words is presumed a candidate period length for a tandem repeat. The candidate tandem repeat would require additional matches along the remainder of its period length. These additional matches are found by querying the distance list, searching for word matches separated by the same period length, with the leading word match positioned between the triggering word matches. The statistical test using the sum of heads distribution determines the minimum number of matches required for a candidate tandem repeat. Here, the normal distribution is used to determine the minimum number x of matches, such that 95% of the time, at least x nucleotides (heads) are counted as part of exact word matches (head runs of length w) along the period length. To account for indels between word pairs of approximate tandem repeats, the period length is not fixed at a constant during this phase, but allowed to vary slightly, consistent with the random walk distribution. The allowed variation is restricted to that expected within 95% of random occurrences based on the indel probability pI, under the hypothesis of a random walk along one dimension with maximal displacement equal to d^*pI.

Candidate tandem repeats which pass the sum of heads test are further analyzed to differentiate tandem repeats from local repeats that are not arrayed in tandem. Tandem repeats are distinguished from nontandem direct repeats, i.e. repeats found in close proximity but not directly adjacent, by the distribution of matches along the period length of the candidate tandem repeat. Nontandem direct repeats will tend to have matches concentrated on the right side of the period length (because the algorithm processes them from left to right), whereas the tandem repeats should have leading word matches distributed throughout. The apparent repeat length is calculated as the maximal distance between the first and last run of matches found using the exact word match scan. This apparent repeat length is likely to be smaller than the actual repeat length, but provides a useful approximation for this analysis. A minimum apparent repeat length threshold for a tandem repeat is determined by simulation. An apparent size distribution is generated from random Bernoulli sequences using the pM value to model an alignment between two genuine tandem repeats with period length d, and the distances between the first and last runs of exact word matches are collected. A minimum apparent repeat length is chosen such that 95% of the time, the apparent repeat length determined for random Bernoulli sequences with pM exceeds this cutoff length. Candidate tandem repeats passing the minimum apparent repeat length threshold are further subjected to the analysis component.

The waiting time distribution is used to pick word lengths used during the initial genome scan. Random Bernoulli sequences are used to determine the minimum number of aligned residues (coin tosses) to find an exact word match (run of heads) of length w, 95% of the time, given a probability of a match (pM or probability of heads). As with other sequence alignment software, such as BLAST, the choice of word length is very important and affects the sensitivity and running time of the analysis. Short word lengths accumulate large history lists and many false-positive matches, unlikely to be indicative of tandem repeats. Alternatively, large word lengths accumulate few false positives, but are unlikely to detect short approximate tandem repeats. Therefore, the word length needs to be chosen in accordance with the repeat length under consideration. The waiting time distribution is used to pick a set of word lengths to apply to different ranges of pattern sizes, given pM. Word lengths of 3–7 bp are chosen to detect tandem repeats with periods of up to 500 bp and at least 75–80% identity.

Those candidate tandem repeats passing the above statistical criteria are further examined under the analysis component of TRF. The analysis component involves aligning the interval of the candidate tandem repeat to the surrounding genomic sequence using the technique of wrap-around dynamic programming (WDP) [18], nicely described in Ref. [10] and Appendix A of Ref. [8]. The technique of WDP provides a method whereby a single copy of the tandem repeat can be aligned with all copies in a larger stretch of genomic sequence, such that the alignment is allowed to wrap around the tandem repeat from end to beginning again, to continue alignments to the subsequent copies of the repeat. The candidate tandem repeat used in WDP may not be the optimal sequence, as the consensus among the repeat copies may contain nucleotide differences or indels when compared to the initial candidate used to generate the alignments. A consensus pattern is generated from the alignment, and this consensus is realigned to calculate the period length and number of copies of the tandem repeat. TRF is limited to finding tandem repeats with unit lengths up to 500 bp.

3.2 STRING (Search for Tandem Repeats IN Genomes)

The algorithm underlying STRING relies almost exclusively on sequence alignment methods to identify tandem repeats [40]. As with TRF, the algorithm consists of two stages; first, the identification of candidate tandem repeats, followed by a more detailed analysis stage to resolve the tandem repeat structures. The identification of candidate tandem repeats involves what are referred to as autoalignments, which involves aligning a sequence to itself. A variation of the Waterman–Eggert algorithm [58] is implemented to identify all nonintersecting local alignments, with a modification to avoid reporting the trivial alignment of the complete sequence to itself along its entirety. Features of some autoalignments are found characteristic of tandem repeats: aligned sequence pairs with overlapping or neighboring coordinates indicate a tandem repeat with a period equal to the distance between coordinates of aligned residues. Such autoalignments do not rigorously define the tandem repeat, but rather highlight the regions of genomic sequences which are strong candidates for containing tandem repeats, to be analyzed in a subsequent tandem repeat finding search stage. Candidate regions are selected by grouping all autoalignments with overlapping coordinates and including those nonoverlapping autoalignments that are found in close proximity that could be extensions of tandem repeats not captured during the autoalignment stage.

Each candidate region is further subjected to the tandem repeat search stage, as follows. A set of words are chosen such that each distinct word is a potential isolated element of larger tandem repeat. A variation of WDP is used to align each word against the larger candidate region. This involves performing the Waterman–Eggert-style alignment using a cyclically addressed word, capturing all high scoring non-intersecting local alignments. Each word and alignment, as a unit, is referred to as a Single-Expansion Interpretative Pattern (SIP). All SIPs found within a candidate region are compared to each other in a pairwise fashion in order to eliminate redundancy and resolve conflicts between overlapping SIPs (some SIPs may be found as insignificant versions of larger SIPs). Remaining SIPs are reported as tandem repeat tracts with the triggering word as the consensus for the tandemly repeated element. STRING is limited to finding tandem repeats of unit length smaller than 100 bp.

3.3 MREPS

A distinguishing feature of the newer MREPS program is its ability to find tandem repeats of any length, from microsatellites to large tandem segmental genome duplications [30]. At the heart of MREPS is a very efficient combinatorial algorithm based on advanced string processing techniques, which finds approximate tandem repeats, also called k-mismatch repeats, running in linear time $O(nk\log(k) + S)$ for a sequence of length n containing S repeats with at most k mismatches per tandem repeat copy. MREPS finds all k-mismatch repeats for values of k up to a user-specified maximal resolution parameter, enabling the program it to find highly divergent repeats. Additional processing time is spent refining the results of this search to report biologically meaningful repeats, coping with artifacts resulting from the algorithmic definition of k-mismatch repeats and consolidating redundant repeats found at different k-mismatch runs, as described below.

The mathematical definition of the k-mismatch repeat requires that the repeat be maximal. For this purpose, mismatches are sometimes added to the repeat boundaries, extending the repeat length to enforce the maximal k-mismatch repeat definition. MREPS attempts to identify these unwanted extensions and trim them from the repeat termini, retaining the more meaningful and longer core of the repeat.

Another postprocessing step consolidates redundancy among repeats and computes their optimal period value. The same region of genomic sequence can be reported as having tandem repeat sequences with different periods. For example, a tandem repeat with period of 2 may also be reported with periods that are multiples of 2. Each period may be associated with a different degree of degeneracy, based on the k-mismatch limit for reporting the repeat. The optimal period for the repeat is chosen as that which minimizes the number of mismatches between tandem copies with period p.

This is done for every period from 1 to p and the period with minimum error rate is reported. After computing the optimal period, repeats with the same period and overlapping by at least two periods are merged together to form a single repeat. By doing this, repeats originally found as k-mismatch repeats are redefined in a more satisfying way.

The tandem repeats found as a result of this process are filtered to retain only those that are considered to be statistically significant. In this case, the statistically significant repeats are those that are unlikely to be found within random sequences. Empirical thresholds for minimum length and maximum error rates were determined for various resolution parameters using shuffled genomic sequences, and these thresholds are applied to the collection of repeats to remove insignificant entries.

4 Repeats and Genome Assembly Algorithms

Genome assembly is perhaps the computationally most demanding task in genomics, requiring days or weeks of computation time for the largest genomes, even on the latest vintage computers. The assembly problem itself is simple enough to state: given a collection of input sequences, compute how these sequences overlap one another and use these overlaps to reconstruct the original chromosomes. A large mammalian genome assembly generated from a WGS sequencing project might include over 20 million input sequences of approximately 800 bp in length. Most of the sequences are generated in pairs, by sequencing both ends of a larger DNA fragment; these fragments are grouped into "libraries" with a characteristic fragment size. The assembly algorithm must keep track of those sizes in order to place the sequence pairs (or "mates") approximately the right distance apart in the final assembly. A thorough account of the computational assembly of genomes is given in Chapter 2.

Repetitive sequences make genome sequence assembly hard; without repeats, almost any algorithm can correctly assembly a genome. This follows from the fact that without repeats, any overlapping sequence shared by two or more individual sequence "reads" clearly implies that the reads came from the same chromosomal location and can be assembled together. As repeats are so central to the assembly problem, much effort has been dedicated within large-scale assembly systems to the repeat identification problem.

Assembly systems are only looking for repeats that will confuse them, which are a subclass of all repeats. First of all, assembly algorithms must compare large numbers of reads (as many as 30 million) looking for overlaps. The fundamental goal of assembly from a WGS sequencing project is to unify overlapping reads if they originated from the same place on the same chromosome. Consequently, the reads should be identical up to the limits of sequencing error. Thus, assemblers look for shared sequences that are nearly identical – a typical threshold is to require that two reads must overlap by at least 40 bp and the overlapping region must be at least 98% identical. This leads naturally to the observation that any pair of repetitive sequences that is less than 98% identical will not cause any serious problems for assembly. Such divergent repeats can be sorted out and placed into the correct locations in the genome. Of course, this is a somewhat simplistic view; e.g. it is often the case that short regions in the middle of long repeats are identical – and therefore confusing – even if the entire repeat is not.

Second, any repeat that is contained entirely within a sequencing read does not cause a problem, because the unique sequence flanking the repeat will allow the read to be placed correctly in the genome assembly. Current sequencing technology generates reads of 800 bp or longer; therefore a repeat region that spans less than 800 bp rarely presents a problem. Note that the phrase "repeat region" here refers both to single-copy repeats and to tandem repeats. If a repeat occurs in 20-bp units, but those units occur in tandem arrays spanning 100 copies, then the repeat region spans 2000 bp and is definitely a problem for assembly, even though the repeat unit itself is quite short.

Thus, it should be clear that assembly algorithms must identify sequence reads that are comprised entirely of repetitive sequence, and they must handle these repeat reads differently. For the sake of discussion, we will describe how they are handled in the Celera Assembler [38], although many of these strategies are similar to those employed by Arachne [5, 25] and other current assemblers. (Note that the Celera Assembler is now open source and includes many enhancements not described in the original paper; the code is available at http://sourceforge.net/projects/wgs-assembler.)

4.1 Repeat Management in the Celera Assembler and other Assemblers

There are two main tasks in repeat processing for assembly: (i) one must identify repeats and (ii) one must attempt to place them in the assembled genome. We will discuss these two issues in order.

4.2 Repeat Identification by k-mer Counts

The first major computation in most assembly algorithms is the overlap step, in which all reads must be compared to all other reads. In order to compute this essentially quadratic operation efficiently, most assemblers employ a hashing strategy: they create a hash table and record in it all k-mers of a certain length. Each k-mer entry stores the read identifier and the position within that read where the k-mer occurred. Typical values of k are 22 (used in the TIGR Assembler [55] and Celera Assembler) and 24 (used in the Arachne assembler); this is long enough that in a random DNA sequence, the vast majority of k-mers will not occur at all.

The k-mer hash array provides a simple and natural vehicle for identifying repeats. Recall that we want to identify sequence reads that are entirely repetitive; i.e. satisfying the condition that no unique sequence can be found within the read. After scanning all reads for all k-mers, it is a trivial matter to note the average depth of coverage by computing the mean number of entries for each k-mer in the table. Based on this value, one can determine a threshold above which a k-mer can safely be assumed to be repetitive, e.g. 3 times the mean. The Celera Assembler then scans the reads a second time and for each read looks at the counts of each k-mer in the read. If all k-mers in a read have a count above the threshold for repeats, then the read itself is labeled as a repeat. Arachne takes a slightly different approach, eliminating all k-mers that are overrepresented so that they will not be used in the overlap calculation.

4.3 Repeat Identification by Depth of Coverage (Arrival Rates)

A second method for identifying repeats occurs later during the assembly process, after at least one round of contig creation. Once the assembler has a set of contigs built, it can ask whether entire contigs are repetitive. The simplest method here is based on coverage: for a genome covered at, for example, 8 times coverage, any contig with significantly deeper coverage is highly likely to be repetitive. As the average coverage is easy to compute, it is also easy to detect any contigs whose coverage is 2 or 3 times normal. However, such a simple approach fails to account for the fact that, statistically, a longer contig is likely to have coverage closer to the mean than a short contig. For example, in an 8 times assembly, a long contig with 15 times coverage is far more likely to represent a repeat than is a short contig with the same coverage.

The Celera Assembler models the expected coverage of a contig using an "arrival rate" statistic. The idea is the following: assuming that the WGS reads are generated by a uniform random process that samples every location in the genome equally, then the reads should "arrive" at a contig (i.e. they should align to it) at a rate that can be modeled as a Poisson process. This arrival rate statistic is computed as follows [38]. Suppose that the genome size is G, the sequencing project generated F reads and we are examining a contig containing k of those reads. Consider the positions where all k reads begin in the contig; these are the arrival locations. If the contig occurs just once in the genome, then we should have sampled it at the same rate as any other interval on the genome; in this case, if we look at the region of length r between the first and last arrival locations, then the probability of seeing $k-1$ arrivals in that interval is $[(rF/G)^k/k!]e^{(-rF/G)}$. If the contig occurs twice in the genome (and therefore we expect twice the arrival rate) then the probability of seeing $k-1$ arrivals is $[(2rF/G)^k/k!]e^{(-2rF/G)}$. The "$A$ statistic" in the Celera Assembler is computed as the log ratio of these two probabilities; i.e. $A = (\log e)rF/G - (\log 2)k$. In simple terms, this statistic computes whether a contig is more likely to occur once than twice in the genome. By adjusting this single parameter, the assembler can be more or less cautious about what it considers a repeat.

4.4 Repeat Identification by Conflicting Links

A third way of identifying repeats is to notice that a contig has two or more adjacent contigs according to mate-pair information. If a repeat occurs at multiple distinct locations (i.e. not in tandem), then the paired sequences that align to the repeat will have mates (links) that point to different loci. This can be recognized relatively easily during the scaffolding stage of assembly.

4.5 Repeat Placement: Rocks and Stones

Once a repeat has been identified, the assembler must decide what to do with it. A number of assemblers, including Celera Assembler, first assemble the obviously nonrepetitive sequences and then try to place the repeats. One strategy for doing this in Celera Assembler is called the "rocks and stones" approach. The idea is to first build scaffolds from the unique sequences, linking contigs together only if at least two mate-pairs (i.e. the pair of sequences that comes from opposite ends of a single DNA fragment) agree on the linkage. Rocks and stones are repetitive contigs (based on the A statistic) that are "thrown" into these gaps if mate-pair links indicate they belong there; rocks must have at least two links and stones only need one. The assembler then attempts to join the flanking contigs together by finding a tiling of reads across the contigs and the newly placed stones. Note that one weakness of this approach is that it is sometimes difficult to determine whether one or multiple copies of a stone belong in a gap and it is possible that the wrong number of copies of a tandem repeat will end up in the assembly.

4.6 Repeat Placement: Surrogates

One final issue surrounding repeats in an assembly is the precise mapping of the reads to the final consensus sequence. Even if the consensus sequence, i.e. the genome sequence that ends up being deposited in public archives, is correct, it may be impossible to determine exactly which sequences belong to particular regions of the genome. For regions that are 100% identical and that are longer than a single read, multiple reads can be mapped to multiple distinct genomic locations and it may simply be impossible to tell where each read goes, since all the repetitive reads can go in each copy of the repeat. The Celera Assembler algorithm handles this problem through the use of "surrogate" contigs: the reads are assembled into a contig which is labeled as a surrogate, meaning that it apparently occurs more than once in the genome. This surrogate can then be placed in multiple locations in the genome based on mate pairs and on reads that span the repeat boundary. However, internal reads cannot be mapped to the genome, so the final consensus assembly points only to the surrogate, but not to the multi-alignment of all the reads. An alternative strategy, used in Arachne, is to place the reads multiply, allowing them to occur at two or more locations in the genome. Neither solution is entirely satisfactory, but in both cases the genome assembly can be constructed correctly.

4.7 Repeat Resolution in Euler

A different approach to repeat identification is taken by the Euler assembler [45]. In this unusual assembly algorithm, the normal overlap computation is handled quite differently from the hash table approach of Celera Assembler and Arachne. Instead, an overlap graph is created in which nodes represent overlap and edges represent k-mers. Contigs can be created by finding an Eulerian path through such a graph, called a de Bruijn graph, a problem that can be solved in linear time. (A Eulerian path through a graph is a path that uses each edge in the graph exactly once, here meaning that each overlap is realized exactly once – from left to right – as it should be in assembling genomes.) To reconstruct the sequence of the contig, the algorithm follows the Euler path and "reads off" the k-mers found on each edge in the path.

A fuller description of the Euler assembler is beyond the scope of this review, but it is worth mentioning how it handles repeats. In the de Bruijn graph created for the purpose of assembly, repeats appear as edges that share the same label. These edges can be superposed, yielding a new data structure (the A-Bruijn graph, see below) in which some nodes have many edges entering or exiting them. Then repetitive sequences correspond to edges whose boundaries are at such nodes. A critical aspect of the Euler algorithm is its error correction: in any large genome project, the small number of sequencing errors in individual reads can easily be confused with slight variations between repeat copies. By reducing the error rate, the algorithm can more easily tell if two near-identical sequences represent two different repeat copies. Euler takes advantage of the fact that errors tend to be random and, therefore, that short k-mers with a very low count probably represent such errors. This enables the algorithm to identify and correct a large majority of sequencing errors, and this in turn enables it to separate repeats that are very close to identical.

Despite all the efforts in Celera Assembler, Arachne, Euler and other assemblers, some classes of repeats continue to confound large-scale genome assembly algorithms. Although this is not widely known or discussed, the genomes available today in public archives likely contain numerous assembly errors. The most common errors are collapses of tandem repeats into too few copies; in addition, gross rearrangements around repeats have also been discovered (and in some cases corrected). The continuing problems highlight the fact that more work needs to be done to continue to improve the quality of "finished" genomes.

5 Untangling the Mosaic Nature of Repeats (The A-Bruijn Graph)

Pavel Pevzner and coworkers introduce a graph data structure, the A-Bruijn graph, to describe repeated sequences ascertained from pairwise alignments and to reveal the complex mosaic structure commonly encountered among related sequences [44]. This A-Bruijn graph has found use in several applications, including genome fragment assembly, multiple sequence alignment and *de novo* repeat finding [44, 48]. The graph glues similar sequence regions together into edges of the graph. When applied to a single genome, the glued-together edges correspond to repeated sequences, and when applied to related sequence sets from different genomes, homologous sequence regions are found glued together. The graph itself provides a compact view of the related regions among a collection of sequences, in addition to the sequence regions found to be unique to a single sequence or subset of sequences. Software packages implementing the A-Bruijn graph include the ABA multiple sequence aligner [48], and the *de novo* repeat-finding program RepeatGluer [44], which is the focus here.

The A-Bruijn graph is constructed from a set of pairwise alignments. All genome alignments generated by an alignment program (those tested include BLAST, PatternHunter and BLAT, among others) are decomposed into the A-Bruijn graph by "gluing" paired genome regions together as edges in a graph bounded by nodes, with similar sequences forming a single edge and forking at a node into separate edges where sequences diverge. After constructing the graph, edges with a multiplicity greater than one correspond to repeated regions of sequences. By removing all nonrepetitive sequences from the graph by discarding all edges with multiplicity of one, the A-Bruijn graph is broken into sets of connected components termed tangles. The tangles represent

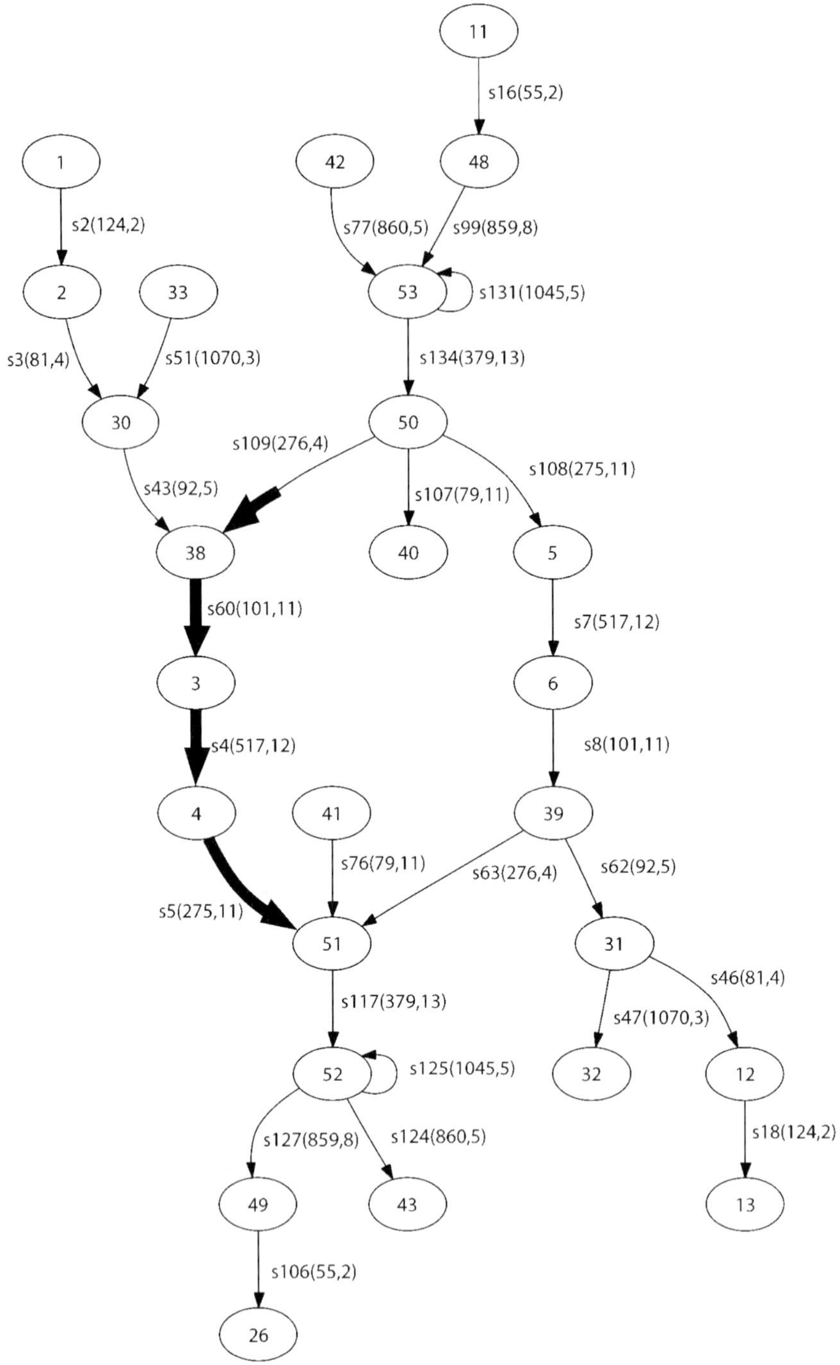
11
s16(55,2)
1
42
48
s2(124,2)
s77(860,5)
s99(859,8)
2
33
53
s131(1045,5)
s3(81,4)
s51(1070,3)
s134(379,13)
30
50
s109(276,4)
s108(275,11)
s43(92,5)
s107(79,11)
38
40
5
s60(101,11)
s7(517,12)
3
6
s4(517,12)
s8(101,11)
4
41
39
s76(79,11)
s63(276,4)
s62(92,5)
s5(275,11)
51
31
s46(81,4)
s117(379,13)
s47(1070,3)
52
s125(1045,5)
32
12
s127(859,8)
s124(860,5)
s18(124,2)
49
43
13
s106(55,2)
26

the repeat elements, specifically the structure of subrepeats that form larger mosaic repeats. Each complete repeat element found in the genome can be reconstructed by traversing edges of valid paths in the corresponding tangle, and the relationships among different mosaic repeats are elucidated by virtue of their subrepeats found in common.

The identity and structure of each representative repeat element, termed by Pevzner and coworkers as the "elementary repeat", is provided by the A-Bruijn graph as a maximal simple path with multiplicity greater than one. RepeatGluer generates the repeat graph in a text file format compatible with graph viewing software (i.e. dotty program of the Graphviz package [17]). Additionally, the genomic sequences corresponding to each subrepeat are extracted and a consensus for each is provided in FASTA format. An example of a repeat graph is shown in Figure 4, highlighting the largest repeat tangle found in the repeat-rich genome of *Deinococcus radiodurans*.

6 Repeat Annotation in Genomes

Given the great diversity and functional significance of repetitive sequences, their annotation in completed genomes is an important task. Annotating repeats is similar to annotating other features in the genome in that a set of coordinates is required to delimit the feature location, along with a description of the biological significance of that feature, if known. The location of repetitive sequences can be obtained in two different ways or in their combination. Repeats are mined directly from the genome sequence based on algorithms that locate repeated sequences based exclusively on the genome sequence composition; these tools and algorithms were the focus of previous

Figure 4 Largest tangle in the repeat graph for *Deinococcus radiodurans* as constructed by RepeatGluer. Repeat consensus sequences yielded for edges in this tangle mostly correspond to parts of transposons (often called "insertion sequences" in bacterial genomics). The coding sequence for the transposase of one transposon (annotated gene identifier DRB0020, gi10957398) is shown threaded through the graph with wide edges. This graph was illustrated using graphviz. Unique identifiers are assigned to each node and edge. Edges correspond to subrepeats and nodes bound the subrepeats that are found in common between larger repeats. The length of the subrepeat and its multiplicity are specified to the right of the edge identifier. Only edges with multiplicity above 1 are shown. As repeats on the forward strand can be merged into single edges with repeats on the reverse strand, the graph is constructed using both strands as if they were independent sequences, which yields symmetry in the resulting graph; the transposase coding sequence described above is highlighted on only one side of the symmetric graph.

discussions. Alternatively, repeats can be identified based on homology to entries in a preexisting library of known repeat sequences.

The advantage of the former method is that no prior knowledge is required. Any newly sequence genome can be applied to the earlier described *de novo* repeat finding methods to rapidly identify the repeats. These are considered *de novo* methods simply because repeats are found based on an analysis of the genome sequence alone, without prior knowledge of the location or sequence composition of the repeats. The disadvantage is that other than knowing that a sequence is a repeat, we do not know what the repeat sequence represents biologically (i.e. gene, transposon, satellite, segmental duplication).

The latter method, based on repeat libraries, is currently the most widely used and most trusted method for annotating repeat sequences. Entries of repeat libraries are typically well annotated with some indication of function when functional information is known, and homology found to a known repeat reveals both its location and identity. Both software and repeat libraries are available as resources for genome annotation. Repbase began as a collection of human representative repeat sequences and fragments, and grew into a large collection of repeats from a variety of (mostly) model organisms, now known as Repbase Update [27]. Repbase Update includes separate repeat libraries for primates, rodents, zebrafish, *C. elegans*, *Drosophila* and *Arabidopsis*, with prototype sequences that correspond to consensus of large families and subfamilies of repeats. A "Simple" library, which can be applied to any genome, provides entries that help identify low complexity microsatellite sequences.

The program originally used to search genome sequences against Repbase libraries is CENSOR [28]. Due to the growing data volume in the libraries and the need for faster searching programs, CENSOR was eventually replaced by the more efficient alignment program RepeatMasker [53]. RepeatMasker identifies regions of homology to Repbase entries using Phil Green's cross_match algorithm (http://www.genome.washington.edu/UWGC/analysistools/Swat.cfm) and then replaces these homologous regions in the genomic sequence with "N" characters, effectively masking them in the genome sequence. A further 30-fold speed increase is obtained by using WU-BLAST in place of cross_match, as implemented by MaskerAid [6] as an enhancement to RepeatMasker. By masking the sequence, these repeat regions are precluded from subsequent sequence analyses in a larger annotation pipeline; hidden from gene finding programs, and transcript and protein homology searches, focusing subsequent analyses on the remaining unique sequence. For perspective, almost half of the human genome is masked due to the repeat content and this step is incredibly important when trying to find components of genes that remain hidden in the unmasked nonrepetitive regions.

The disadvantage with the repeat library-based scanning method is that comprehensive repeat libraries are available only for those organisms that have been well studied, and have attained the status of model organisms. Repbase is a tremendous resource for repeat annotation when corresponding repeat libraries are available, but it is of limited utility for many organisms whose genomes are currently being sequenced because of the limits of homology detection at the level of the nucleotide sequence coupled with the rate of divergence within repeats across evolutionary boundaries. Independent efforts are sometimes necessary to generate comprehensive repeat libraries to supplement that offered by Repbase [39].

A use of *de novo* repeat-finding programs is to automate the generation of repeat libraries that can be used with RepeatMasker. RepeatScout, in particular, yields consensus sequences for repeat families that are to be subsequently searched against the genome using RepeatMasker to identify locations of individual members of the family. Care must be taken with this approach if the repeats are to be masked from the genomic sequence and hidden from subsequent analyses. The Repbase libraries include many transposable elements and other repeat features which are excluded from the host gene set, and so masking homologous features from the genome should not interfere with subsequent efforts to find genes. The output from *de novo* repeat finding tools contains transposable elements but may also include repetitive features such as members of large gene families, and by blindly masking these "repeat" features from the genome, important features will be inadvertently disguised. None of the *de novo* repeat-finding programs directly address the problem of deducing the biological significance of the repeats that are found computationally. This is a difficult problem, currently left to the biologists and bioinformaticians, and examined on a case-by-case basis.

Searching for occurrences of repeats using representative repeat sequences or consensus sequences is limited by the information provided by that single sequence. By searching with profile representations of repeat families, the sensitivity of a search can be improved; a study by Juretic and coworkers demonstrates improved sensitivity in the detection of transposable elements in Rice by using hidden Markov model (HMM) profiles created for known transposable element families [26]. Although this methodology or similar profile methods are popular for finding members of protein families or occurrences of protein domains, HMM profiles for repeats are not currently widely employed, but do show great promise for repeat detection and analysis, and should be considered along with the existing alternatives.

Acknowledgments

We would like to thank the following individuals for reviewing an earlier version of the manuscript: Stefan Kurtz, Pavel Pevzner, Bob Edgar, Zhirong Bao, and Alkes Price. Your comments were invaluable. S.L.S. was supported in part by the US National Institutes of Health under grants R01-LM06845 and R01-LM007938.

Repeat Finding Tools and Resources

Miropeats and ICAtools	http://www.littlest.co.uk/software/bioinf/index
mreps	http://mreps.loria.fr
PILER	http://www.drive5.com/piler
RECON	http://www.genetics.wustl.edu/eddy/recon
Repbase Update	http://www.girinst.org/Repbase_Update
RepeatFinder	http://www.tigr.org/software
RepeatGluer	http://nbcr.sdsc.edu/euler/intro_tmp
RepeatMasker	http://www.repeatmasker.org
RepeatScout	http://www-cse.ucsd.edu/groups/bioinformatics/software
REPuter	http://bibiserv.techfak.uni-bielefeld.de/reputer
STRING	http://www.caspur.it/~castri/STRING/index.htm.old
Tandem Repeat Finder	http://tandem.bu.edu/trf/trf

References

1 **a.** AGARWAL, P., D. J. STATES. 1994. The Repeat Pattern Toolkit (RPT): analyzing the structure and evlution of the *C. elegans* genome. Proc. Int. Conf. Intell. Syst. Mol. Biol. **2**: 1–9.
b. ALTSCHUL, S. F., W. GISH, W. MILLER, E. W. MYERS AND D. J. LIPMAN. 1990. Basic local alignment search tool. J. Mol. Biol. **215**: 403–10.

2 ALTSCHUL, S. F., T. L. MADDEN, A. A. SCHAFFER, J. ZHANG, Z. ZHANG, W. MILLER AND D. J. LIPMAN. 1997. Gapped BLAST and PSI-BLAST: a new generation of protein database search programs. Nucleic Acids Res. **25**: 3389–402.

3 ARUMUGANATHAN, K. AND E. D. EARLE. 1991. Nuclear DNA content of some important plant species. Plant Mol. Biol. Rep. **9**: 208–218.

4 BAO, Z. AND S. R. EDDY. 2002. Automated *de novo* identification of repeat sequence families in sequenced genomes. Genome Res. **12**: 1269–76.

5 BATZOGLOU, S., D. B. JAFFE, K. STANLEY, et al. 2002. ARACHNE: a whole-genome shotgun assembler. Genome Res. **12**: 177–89.

6 BEDELL, J. A., I. KORF AND W. GISH. 2000. MaskerAid: a performance enhancement to RepeatMasker. Bioinformatics **16**: 1040–1.

7 BENNETZEN, J. L., P. SANMIGUEL, M. CHEN, A. TIKHONOV, M. FRANCKI AND Z. AVRAMOVA. 1998. Grass genomes. Proc. Natl Acad. Sci. USA **95**: 1975–8.

8 BENSON, G. 1997. Sequence alignment with tandem duplication. J. Comput. Biol. **4**: 351–67.

9 BENSON, G. 1999. Tandem repeats finder: a program to analyze DNA sequences. Nucleic Acids Res. **27**: 573–80.

10 BENSON, G. AND M. S. WATERMAN. 1994. A method for fast database search for all k-nucleotide repeats. Nucleic Acids Res. **22**: 4828–36.

11 CAPY, P., C. BAZIN, D. HIGUET, AND T. LANGIN. 1998. *Dynamics and Evolution of Transposable Elements*. Landes Bioscience, Austin, TX.

12 CHOO, K. H., B. VISSEL, A. NAGY, E. EARLE AND P. KALITSIS. 1991. A survey of the genomic distribution of alpha satellite DNA on all the human chromosomes, and derivation of a new consensus sequence. Nucleic Acids Res. **19**: 1179–82.

13 CRAIG, N. L. 2002. *Mobile DNA II*. ASM Press, Washington, DC.

14 DELCHER, A. L., S. KASIF, R. D. FLEISCHMANN, J. PETERSON, O. WHITE AND S. L. SALZBERG. 1999. Alignment of whole genomes. Nucleic Acids Res. **27**: 2369–76.

15 DELCHER, A. L., A. PHILLIPPY, J. CARLTON AND S. L. SALZBERG. 2002. Fast algorithms for large-scale genome alignment and comparison. Nucleic Acids Res. **30**: 2478–83.

16 EDGAR, R. C. AND E. W. MYERS. 2005. PILER: identification and classification of genomic repeats. Bioinformatics **21** (Suppl. 1): i152–8.

17 ELLSON, J., E. GANSER, Y. KOREN, et al. 2005. Graphviz – Graph Visualization Software. http://www.graphviz.org/.

18 FISCHETTI, V., G. LANDAU, J. SCHMIDT AND P. SELLERS. 1992. Identifying periodic occurrences of a template with applications to protein structure (presented at the 3rd Annual Symposium on Combinatorial Pattern Matching). Lecture Notes Comput. Sci. **644**: 111–120.

19 FLAVELL, R. B., M. D. BENNETT, J. B. SMITH AND D. B. SMITH. 1974. Genome size and the proportion of repeated nucleotide sequence DNA in plants. Biochem. Genet. **12**: 257–69.

20 FRASER, C. M., S. CASJENS, W. M. HUANG, et al. 1997. Genomic sequence of a Lyme disease spirochaete, *Borrelia burgdorferi*. Nature **390**: 580–6.

21 GISH, W. AND D. J. STATES. 1993. Identification of protein coding regions by database similarity search. Nat. Genet. **3**: 266–72.

22 GUSFIELD, D. 1997. *Algorithms on Strings, Trees, and Sequences: Computer Science and Computational Biology*. Cambridge University Press, Cambridge;

23 HESLOP-HARRISON, J. S., M. MURATA, Y. OGURA, T. SCHWARZACHER AND F. MOTOYOSHI. 1999. Polymorphisms and genomic organization of repetitive DNA from centromeric regions of *Arabidopsis* chromosomes. Plant Cell **11**: 31–42.

24 HUANG, X. AND W. MILLER. 1991. A time-efficient, linear-space local similarity algorithm. Adv. Appl. Math. **12**: 337–357.

25 JAFFE, D. B., J. BUTLER, S. GNERRE, E. MAUCELI, K. LINDBLAD-TOH, J. P. MESIROV, M. C. ZODY AND E. S. LANDER. 2003. Whole-genome sequence assembly for Mammalian genomes: arachne 2. Genome Res. **13**: 91–6.

26 JURETIC, N., T. E. BUREAU AND R. M. BRUSKIEWICH. 2004. Transposable element annotation of the rice genome. Bioinformatics **20**: 155–60.

27 JURKA, J. 2000. Repbase update: a database and an electronic journal of repetitive elements. Trends Genet. **16**: 418–20.

28 JURKA, J., P. KLONOWSKI, V. DAGMAN AND P. PELTON. 1996. CENSOR – a program for identification and elimination of repetitive elements from DNA sequences. Comput. Chem. **20**: 119–21.

29 KENT, W. J. 2002. BLAT – the BLAST-like alignment tool. Genome Res. **12**: 656–64.

30 KOLPAKOV, R., G. BANA AND G. KUCHEROV. 2003. mreps: efficient and

flexible detection of tandem repeats in DNA. Nucleic Acids Res. **31**: 3672–8.

31 Kurtz, S., J. V. Choudhuri, E. Ohlebusch, C. Schleiermacher, J. Stoye and R. Giegerich. 2001. REPuter: the manifold applications of repeat analysis on a genomic scale. Nucleic Acids Res. **29**: 4633–42.

32 Kurtz, S., A. Phillippy, A. L. Delcher, M. Smoot, M. Shumway, C. Antonescu and S. L. Salzberg. 2004. Versatile and open software for comparing large genomes. Genome Biol. **5**: R12.

33 Kurtz, S. and C. Schleiermacher. 1999. REPuter: fast computation of maximal repeats in complete genomes. Bioinformatics **15**: 426–7.

34 Lander, E. S., L. M. Linton, B. Birren, et al. 2001. Initial sequencing and analysis of the human genome. Nature **409**: 860–921.

35 Li, M., B. Ma, D. Kisman and J. Tromp. 2004. Patternhunter II: highly sensitive and fast homology search. J. Bioinform. Comput. Biol. **2**: 417–39.

36 Ma, B., J. Tromp and M. Li. 2002. PatternHunter: faster and more sensitive homology search. Bioinformatics **18**: 440–5.

37 Margolis, R. L., M. G. McInnis, A. Rosenblatt and C. A. Ross. 1999. Trinucleotide repeat expansion and neuropsychiatric disease. Arch. Gen. Psychiatry **56**: 1019–31.

38 Myers, E. W., G. G. Sutton, A. L. Delcher, et al. 2000. A whole-genome assembly of *Drosophila*. Science **287**: 2196–204.

39 Ouyang, S. and C. R. Buell. 2004. The TIGR Plant Repeat Databases: a collective resource for the identification of repetitive sequences in plants. Nucleic Acids Res. **32**: D360–3.

40 Parisi, V., V. De Fonzo and F. Aluffi-Pentini. 2003. STRING: finding tandem repeats in DNA sequences. Bioinformatics **19**: 1733–8.

41 Parsons, J. D. 1995. Improved tools for DNA comparison and clustering. Comput. Appl. Biosci. **11**: 603–13.

42 Parsons, J. D. 1995. Miropeats: graphical DNA sequence comparisons. Comput. Appl. Biosci. **11**: 615–9.

43 Pearson, W. R. and D. J. Lipman. 1988. Improved tools for biological sequence comparison. Proc. Natl Acad. Sci. USA **85**: 2444–8.

44 Pevzner, P. A., H. Tang and G. Tesler. 2004. *De novo* repeat classification and fragment assembly. Genome Res. **14**: 1786–96.

45 Pevzner, P. A., H. Tang and M. S. Waterman. 2001. An Eulerian path approach to DNA fragment assembly. Proc. Natl Acad. Sci. USA **98**: 9748–53.

46 Price, A. L. and P. A. Pevzner. 2005. *De novo* identification of repeat families in large genomes. Bioinformatics.

47 Prokopowich, C. D., T. R. Gregory and T. J. Crease. 2003. The correlation between rDNA copy number and genome size in eukaryotes. Genome **46**: 48–50.

48 Raphael, B., D. Zhi, H. Tang and P. Pevzner. 2004. A novel method for multiple alignment of sequences with repeated and shuffled elements. Genome Res. **14**: 2336–46.

49 Read, T. D., S. L. Salzberg, M. Pop, et al. 2002. Comparative genome sequencing for discovery of novel polymorphisms in *Bacillus anthracis*. Science **296**: 2028–33.

50 Richards, R. I. and G. R. Sutherland. 1997. Dynamic mutation: possible mechanisms and significance in human disease. Trends Biochem. Sci. **22**: 432–6.

51 Richards, R. I. and G. R. Sutherland. 1996. Repeat offenders: simple repeat sequences and complex genetic problems. Hum. Mutat. **8**: 1–7.

52 Romero, D., J. Martinez-Salazar, E. Ortiz, C. Rodriguez and E. Valencia-Morales. 1999. Repeated sequences in bacterial chromosomes and plasmids: a glimpse from sequenced genomes. Res. Microbiol **150**: 735–43.

53 Smit, A. F., R. Hubley and P. Green. 1996–2004. RepeatMsker Open-3.0.

54 Smith, T. F. and M. S. Waterman. 1981. Identification of common molecular subsequences. J. Mol. Biol. **147**: 195–7.

55 SUTTON, G., O. WHITE, M. ADAMS AND A. R. KERLAVAGE. 1995. TIGR Assembler: a new tool for assembling large shotgun sequencing projects. Genome Sci. Technol. **1**: 9–19.

56 VOLFOVSKY, N., B. J. HAAS AND S. L. SALZBERG. 2001. A clustering method for repeat analysis in DNA sequences. Genome Biol. **2**: RESEARCH0027.

57 WATERMAN, M. S. 2000. *Introduction to Computational Biology: Maps, Sequences And Genomes: Interdisciplinary Statistics*. Chapman & Hall/CRC Press, Boca Raton, FL.

58 WATERMAN, M. S. AND M. EGGERT. 1987. A new algorithm for best subsequence alignments with application to tRNA–rRNA comparisons. J. Mol. Biol. 197: 723–8.

8
Analyzing Genome Rearrangements

Guillaume Bourque

1 Introduction

The study of comparative maps and the rearrangements they evidence was pioneered in the late 1910s at the Morgan *Drosophila* lab [45, 74]. In the context of phylogenetics, the analysis of genome rearrangements was first introduced by Dobzhansky and Sturtevant in a study of inversions in *Drosophila pseudoobscura* [22]. What followed was a succession of developments in the fields of comparative mapping and comparative genomics. In particular, breakthroughs in mapping and sequencing afforded genome-wide analyses of gene order in various sets of genomes [3,12,20,27,51,53,54,63]. Recently, the considerable investments in large sequencing projects have made accessible detailed sequences and maps for many eukaryotic genomes [26,32,37,79,84]. One of the stated purpose of these endeavors is to further our understanding of these species through comparative analyses [50]. The availability of these large genomes leads to great opportunities, but also challenges, in the study of genome rearrangements.

The exploration of large-scale events shaping whole-genome architecture provides a complementary perspective on the evolution of these organisms as compared to more traditional molecular studies focused on the analysis of individual genes. In fact, rearrangement studies allow detailed reconstructions of evolutionary scenarios, including ancestral reconstructions of entire eukaryotic genomes [13, 14]. Furthermore, such analyses can lead to the identification of regions of genomic instability (high rates of rearrangements, breakpoint reuse, etc.) that challenge and help refine our understanding of the dynamics of chromosome evolution [46, 57]. A related problem, also associated with genomic instability, is the study of cancer. Rearrangements in a tumor genome can be analyzed very much as if the tumor was a new organism that had recently diverged from the normal human genome. The interest is that although cancer progression is frequently associated with genome rearrangements, the forces behind these alterations are still poorly understood.

Bioinformatics - From Genomes to Therapies Vol. 1. Edited by Thomas Lengauer

ISBN: 978-3-527-31278-8

This chapter is organized as follows. Section 2 presents some of the basic concepts required for the analysis of genome rearrangements such as how the genomes are modeled and what types of rearrangements are considered. Section 3 presents three criteria that can be used to compute the distance between a pair of genomes: the breakpoint distance, the rearrangement distance and the conservation distance. Section 4 shows how the same criteria can be use to infer phylogenies when multiple genomes are considered using three different approaches: distance-based, maximum parsimony and maximum likelihood. Section 5 presents a few recent applications of analysis of genome rearrangements in large genomes and also recent work studying genome rearrangements in cancer. Finally, Section 6 concludes with some remarks on important challenges and promising new developments for the comparative analyses of gene order.

2 Basic Concepts

2.1 Genome Representation

Initially, the focus of genome rearrangement studies was on the comparative analyses of small genomes such as mitochondria [12,53,54,63], chloroplasts [20,51,54], viruses [27] and small region of larger genomes [3]. In this context, the relative order of homologous genes in different organisms was used to infer phylogenetic relationships and even rearrangement scenarios. An example showing differences between the order of homologous genes in two mitochondria is given in Figure 1.

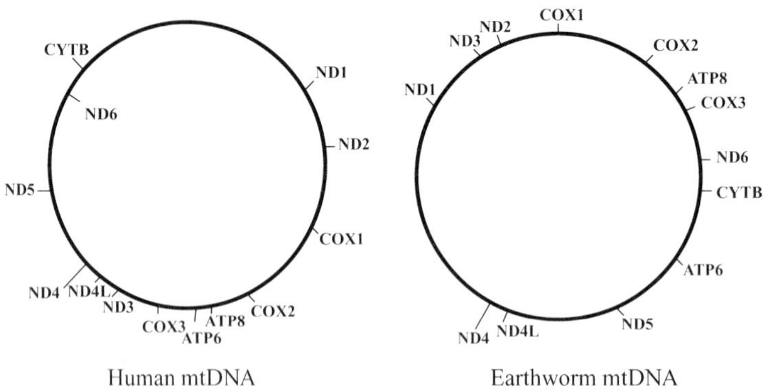

Figure 1 Coding genes on the human and on the earthworm mitochondrial DNA (mtDNA). The list of genes is the same, but their order differs. For instance, ND1 and ND2 are adjacent in Human but they are seperated by ND3 in the earthworm. tRNA genes have been left out of this figure to simplify the example. GenBank accession numbers: NC_001807 and NC_001673.

For this purpose, the relative gene order of different genomes can be encapsulated into a set of *signed permutations*. One of the genomes is identified as the reference genome and is associated to the identity permutation where each integer corresponds to one of its genes. The permutation associated to each other genome can directly be obtained from the order of appearance of the homologous genes. Furthermore, a sign corresponding to the relative orientation (strand) of the gene, as compared to the reference genome, is given to each integer of the new permutation. To continue with the example shown in Figure 1, if the mtDNA is selected as the reference genome, and we label the genes starting with COX1 in Human as "1" and going clockwise until ND2 is assigned "13", we obtain the permutations shown in Table 1. Of all the genes in the two mtDNAs, only ND6 in Human was on the reverse DNA strand, which is why it is represented by "−10" in Earthworm as its relative orientation is reversed.

Table 1 Signed permutations associated with the two mitochondria genomes shown in Figure 1

Human	1	2	3	4	5	6	7	8	9	10	11	12	13
Earthworm	1	2	3	5	−10	11	4	9	7	8	12	6	13

This genome representation can be adapted and generalized for data sets with other distinctive features such as multiple chromosomes, unsigned gene orders, unequal gene content and different source of homology markers. We briefly present these variants.

2.1.1 Circular, Linear and Multichromosomal Genomes

A genome can consist of a single chromosome or a collection of chromosomes and is called *unichromosomal* or *multichromosomal* accordingly. There are two types of chromosomes: circular and linear. The mitochondria shown in Figure 1 are circular genomes. Linear chromosomes have two separated endpoints. Unless otherwise stated, multichromosomal genomes will be assumed to have linear chromosomes. The different types of genomes can also be represented by permutations, but additional markers are required for multichromosomal genomes to mark the boundaries of the chromosomes.

It is important to specify the type of chromosomes we are considering because they will lead to different equivalent representations. For instance, consider the three genomes:

$$\begin{aligned} G_1 &= \quad 1 \quad 2 \quad 3 \\ G_2 &= \quad 2 \quad 3 \quad 1 \\ G_3 &= -3 \ -2 \ -1 \end{aligned}$$

As circular chromosomes, all three representation are equivalent; however, as linear chromosomes, only G_1 and G_3 correspond to the same representation (there is usually no distinction between the two end-points of a linear chromosome and so a complete flip leads to an equivalent representation).

2.1.2 Unsigned Genomes

If the orientation of genes in a set of genomes is unknown, the relative gene order can still be encapsulated into a set of unsigned permutations. Unfortunately, many genome-rearrangement problems, such as calculating the reversal distance, are significantly harder for unsigned permutations [17] compared to signed permutations (see Section 3.2). For this reason, but also because the relative orientation is usually obtainable, we will assume that we are dealing with signed genomes in the rest of this chapter.

2.1.3 Unequal Gene Content

In representing genomes with permutations, we have assumed that a set of n genes was found with a unique homologous counterpart in all genomes. In fact, in many cases this assumption will be violated: a genome may have gained additional genes through rearrangements events such as insertions or duplications and it may have lost genes following deletions. To encapsulate the relative gene orders of genomes with unequal gene content we need to generalize the representation to account for this variable alphabet. Although models that are not restricted to equal gene content are more complete and realistic (see Ref. [24] for a review), they are also more challenging algorithmically and have been limited to few applications [23, 66, 69]. We will focus on genomes with equal gene content in the rest of this chapter except for the presentation of some rearrangement events affecting gene content in Section 2.2.

2.1.4 Homology Markers

So far, it was implicit that the signed permutations representing the genomes were constructed based on the relative position of homologous genes. Actually, however, any type of marker with an homologous counterpart in all genomes can be used to construct similar permutations. This is important because, especially in large eukaryotic genomes (e.g. human, mouse) where genes only cover a small fraction of the genome, the ability to use markers extracted from raw DNA sequence allows to study rearrangement events that occur anywhere in the genome. This will be covered in more detail in Section 5.1.

2.2 Types of Genome Rearrangements

During evolution, an assortment of events can modify the genome. These events are known as mutations and they can occur especially during DNA replication. Mutations are divided into two major categories: point mutations and chromosomal mutations. Point mutations are at the single-base level. Although point mutations can have a significant impact on the genome (e.g. a base change could be responsible to the insertion of an early stop codon that completely annihilates a gene), they will not be considered further here as they mostly affect individual genes. In contrast, chromosomal mutations affect directly the architecture of genomes by modifying the gene order or the gene content. There are various types of chromosomal mutations, but the most common are reversals (or inversions), translocations, fusions, fissions, transpositions, inverted transpositions, insertions, deletions and duplications. See Figure 2 for a cartoon example of how reversals and deletions could occur during DNA replication. Only translocations, fusions and fissions are specific to multichromosomal genomes. The first six types of chromosomal mutations rearrange the genes, but they do not modify the set of genes present in a given genome. Insertions, deletions and duplications, on the other hand, modify the gene content of a genome by adding, or removing, some genes or by generating multiple copies of the same gene. The effect of these different chromosomal mutations are exemplified in Table 2.

Table 2 Examples of chromosomal mutations that impact either gene order or gene content

Mutation type	Before		After	Impact
Reversal	1 2 [3 4 5] 6	⇒	1 2 −5 −4 −3 6	gene order
Translocation	1 2 [3 4 5] 6 7 [8]	⇒	1 2 8 6 7 3 4 5	gene order
Fusion	[1 2 3 4] [5 6]	⇒	1 2 3 4 5 6	gene order
Fission	[1 2 3 4] [5 6]	⇒	1 2 3 4 5 6	gene order
Transposition	1 [2 3] 4 5 [ ] 6	⇒	1 4 5 2 3 6	gene order
Inverted transposition	1 [2 3] 4 5 [ ] 6	⇒	1 4 5 −3 −2 6	gene order
Insertion	1 2 3 4 [ ] 5 6	⇒	1 2 3 4 7 5 6	gene content
Deletion	1 [2 3] 4 5 6	⇒	1 4 5 6	gene content
Duplication	1 2 [3 4] 5 6	⇒	1 2 3 4 3′ 4′ 5 6	gene content

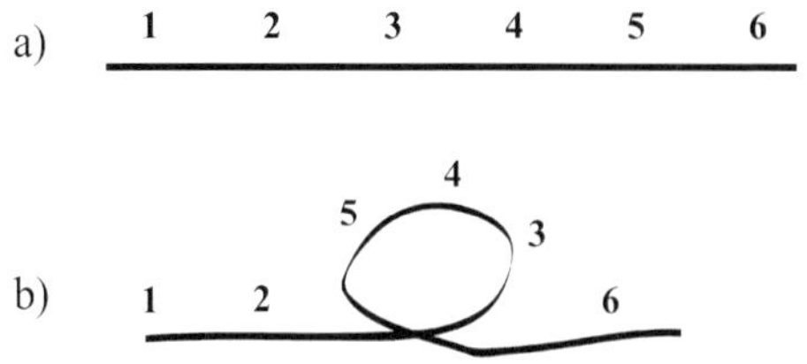

Figure 2 (a) DNA fragment with each integer corresponding to a gene. (b) The same DNA fragment but twisted. During replication, if the twisted loop is copied, it leads to a *reversal* and the fragment becomes 1 2 −5 −4 −3 6. Note that the sign or the strand of the genes is modified at the same time as the order. If the twisted loop is ignored, it results in a *deletion* and the fragment is transformed into 1 2 6.

3 Distance between Two Genomes

In this section, we review different criteria that can be use to measure the distance between two genomes based on comparative gene order. The first criterion, the *breakpoint distance*, counts the number of disruptions of the relative gene order between a pair of genomes. The second criterion, the *rearrangement distance*, relies on the *a priori* definition of a set of permissible operations (e.g. only reversals) and then minimizes the number of such operations required to convert one gene order into the next. The final criterion described is the *conservation distance*, which, similar to the breakpoint distance, circumvents the requirement of a rearrangement model. Under this criterion, the disruption of the relative gene order is measured by the number of conserved or common intervals.

3.1 Breakpoint Distance

The breakpoint distance [48,85] compares two permutations by directly counting the number of gene order disruptions between two genomes. Formally, given two signed permutations of size n, π and γ, the first step to compute the breakpoint distance is to extend both permutations so that they start with 0 and end with $n+1$: $\pi = 0, \pi_1, \pi_2 \dots \pi_n, n+1$ and $\gamma = 0, \gamma_1, \gamma_2 \dots \gamma_n, n+1$. Then, the *breakpoint distance*, $d_{\text{break}}(\pi, \gamma)$, is defined as the number of pairs (γ_i, γ_{i+1}), $0 \le i \le n$, such that neither the pair (γ_i, γ_{i+1}) nor $(-\gamma_{i+1}, -\gamma_i)$ appears in π. For instance, using the example from Table 1 and setting $\pi =$ human and $\gamma =$ earthworm, we get $d_{\text{break}}(\pi, \gamma) = 9$. The nine breakpoints are displayed in γ using arrows:

$$\begin{array}{ccccccccccccccccccccccccccccc}
0 & & 1 & & 2 & & 3 & & 5 & & -10 & & 11 & & 4 & & 9 & & 7 & & 8 & & 12 & & 6 & & 13 & & 14 \\
 & & & & & & & \uparrow & & \uparrow & & \uparrow & & \uparrow & & \uparrow & & \uparrow & & & & \uparrow & & \uparrow & & \uparrow & & &
\end{array}$$

Two important strengths of this criterion measuring the degree of similarity are that (i) it is easily computable in linear time and (ii) it does not require any assumptions about the underlying rearrangement mechanisms.

3.2 Rearrangement Distance

Given two permutations π and γ and a set of permissible rearrangements, the rearrangement distance, $d_{\text{rear}}(\pi, \gamma)$, is defined as the minimum number of operations required to convert one permutation into the other. For example, given that reversals are the only allowed operations, what is the minimum number of events required to convert the permutation associated with the earthworm mtDNA into the one associated with the human mtDNA shown in Table 1? The problem is quite challenging. In this particular case, the answer is seven and Table 3 shows one such scenario. We will use d_{rev} for the special case of d_{rear} when reversals are the only permissible operations.

The interest in looking for the minimum number of steps is that, under the assumption that such events are rare (and that our rearrangement model is correct), we hope to recover the sequence of rearrangements that really occurred. The caveat is that it is well known that the most parsimonious scenario underestimates the actual number of operations when this number is above a threshold of θn, where n is the size of the permutation and θ is in the range from 1/3 to 2/3 [15,35,83].

Table 3 Example of a most parsimonious rearrangement scenario with seven reversals between earthworm and human mtDNA

Earthworm	1	2	3	5	−10	11	4	9	7	8	12	6	13
$\rho(6,11)$	1	2	3	5	−10	11	4	9	7	8	12	6	13
$\rho(9,12)$	1	2	3	5	−10	−12	−8	−7	−9	−4	−11	6	13
$\rho(7,10)$	1	2	3	5	−10	−12	−8	−7	−6	11	4	9	13
$\rho(6,12)$	1	2	3	5	−10	−12	−11	6	7	8	4	9	13
$\rho(4,6)$	1	2	3	5	−10	−9	−4	−8	−7	−6	11	12	13
$\rho(4,7)$	1	2	3	9	10	−5	−4	−8	−7	−6	11	12	13
$\rho(6,10)$	1	2	3	4	5	−10	−9	−8	−7	−6	11	12	13
Human	1	2	3	4	5	6	7	8	9	10	11	12	13

Based on different sets of permissible rearrangements, various methods have been proposed to efficiently compute the rearrangement distance and sort a pair of genomes. Of all the choices of permissible operations, the reversal-only model is probably the most extensively studied. The work was pioneered by Sankoff and Kececioglu [68], but was followed by the development of increasingly efficient polynomial-time algorithms [1,8,9,28, 34]. Other studied sets of permissible operations include transpositions [2,81], inversions, translocations, fusions and fissions [29, 52, 75], and more recently block interchange (a more general type of transposition) [19, 41, 86].

In the remainder of this section, we review a methodology that was developed to compute the distance between a pair of genomes using reversals only (for unichromosomal genomes) or reversals, translocations, fusions and fissions (for multichromosomal genomes). We will refer to it as the Hannenhalli–Pevzner (HP) theory. This methodology was developed in Bafna and Pevzner [4] and in Hannenhalli and Pevzner [30], it was summarized in Pevzner [58], it was improved in Tesler [75], and, finally, it was implemented in a program called GRIMM [76].

3.2.1 **HP Theory**

We first describe the methodology for unichromosomal genomes where reversals are the only permissible operations. Assume we have a permutation γ that we wish to sort with respect to the identity permutation π. The first step is to convert γ, a signed permutation, into γ', an unsigned permutation, by mimicking every directed element i by two undirected elements i^{t} and i^{h} representing the tail and the head of i. Since γ is a permutation of size n, γ' will be a permutation of size $2n$. We now extend the permutation γ' by adding $\gamma'_0 = 0$ and $\gamma'_{2n+1} = n+1$. The next step is to construct the breakpoint graph associated with γ. The *breakpoint graph* of γ, $G(\gamma)$, is an edge-colored graph with $2n+2$ vertices. Black edges are added between vertices γ'_{2i} and γ'_{2i+1} for $0 \le i \le n$. Grey edges are added between i^{h} and $(i+1)^{\mathrm{t}}$ for $0 < i < n$, between 0 and 1^t, and between n^h and $n+1$. Black edges correspond to the actual state of the permutation while grey edges correspond to the sorted permutation we seek. See Figure 3 for an example.

Bafna and Pevzner [4], and later Hannenhalli and Pevzner [30], showed that $G(\gamma)$ contains all the necessary information for efficiently sorting the permutation γ. The first step is to look at the maximal cycle decomposition of the breakpoint graph. Finding the maximal cycle decomposition of a graph in general can be a very difficult problem; however, fortunately, because of

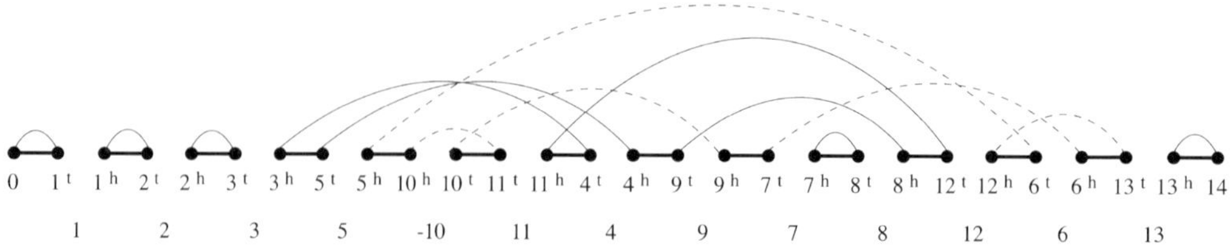

Figure 3 Breakpoint graph associated with the two permutations from Table 1. Black edges are shown using think lines. All other lines (both solid and dashed) correspond to grey edges. Dashed lines are used to show the only nontrivial oriented cycle.

the way the breakpoint graph was constructed for a signed permutation, each vertex has degree two and so the problem is trivial. Suppose $c(\gamma)$ is the maximum number of edge-disjoint alternating cycles in $G(\gamma)$. The cycles are alternating because, in the breakpoint graph of a signed permutation, each pair of consecutive edges always has different colors.The important lower bound:

$$d_{\text{rev}}(\pi, \gamma) = d(\gamma) \geq n + 1 - c(\gamma).$$

was first presented by Kececioglu and Sankoff [35].

A few additional concepts on the breakpoint graph are required to present the result of Hannenhalli and Pevzner [28]. A grey edge in $G(\gamma)$ is said to be *oriented* if it spans an odd number of vertices (when the vertices of $G(\gamma)$ are arranged in the canonical order $\gamma'_0, \ldots, \gamma'_{2n+1}$). A cycle is said to be *oriented* if it contains at least one oriented grey edge. Cycles which are not oriented are said to be *unoriented* unless they are of size 2, in which case they are said to be *trivial*. The term "oriented" comes from the fact that if we traverse an oriented cycle we will traverse at least one black edge from left to right and one black edge from right to left. In the breakpoint graph shown in Figure 3, there are only two nontrivial cycles: one where the grey edges are displayed using solid lines and one where the grey edges are displayed using dashed lines. The cycle with solid lines is unoriented since it does not contain an oriented edge but the cycle with dashed lines is oriented because it contains an oriented edge [e.g. $(10^{\text{h}}, 11^{\text{t}})$].

For each grey edge in $G(\gamma)$ we will now create a vertex v_e in the *overlap graph*, $O(G(\gamma))$. Whenever two grey edges e and e′ overlap or cross in the canonical representation of $G(\gamma)$, we will connect the corresponding vertices v_{e} and $v_{\text{e}'}$. A *component* will mean a connected component in $O(G(\gamma))$. A component will be oriented if it contains a vertex v_e for which the corresponding grey edge e is oriented. As for cycles, a component which consists of a single vertex (grey edge) will be said to be trivial. In Figure 3, there are five trivial components and one larger oriented component since at least one of its grey edge is oriented. The difficulty in sorting permutations comes from unoriented components.

Unoriented components can be classified into two categories: hurdles and protected nonhurdles. A *protected nonhurdle* is an unoriented component that separates other unoriented components in $G(\gamma)$ when vertices in $G(\gamma)$ are placed in canonical order. A *hurdle* is any unoriented component that is not a protected nonhurdle. A hurdle is a *superhurdle* if deleting it would transform a protected nonhurdle into a hurdle, otherwise it is said to be a *simple hurdle*. Finally, γ is said to be a *fortress* if there exists an odd number of hurdles and all are superhurdles in $O(G(\gamma))$ [71].

The main result fromHannenhalli and Pevzner [28] is that:

$$d_{\text{rev}}(\pi, \gamma) = d(\gamma) = n + 1 - c(\gamma) + h(\gamma) + f(\gamma),$$

where $h(\gamma)$ is the number of hurdles in γ, and $f(\gamma)$ is 1 if γ is a fortress and 0 otherwise. However, the machinery to recover an optimal sequence of sorting reversals was also presented. The fact that the distance between the human and earthworm is 7 can directly be extracted from this formula and from the breakpoint graph shown in Figure 3 since there are 13 genes, seven cycles and no hurdles or fortress.

Finally, Hannenhalli and Pevzner [29] derived a related equation to compute the rearrangement distance between two multichromosomal genomes when permissible operations are: reversals, translocations, fusions and fissions. We refer the reader to Pevzner [58] and Tesler [75] for the details of the calculation, but we will briefly present how the formula can be obtained.

The main idea to compute the rearrangement distance between two multichromosomal genomes Π and Γ is to concatenate their chromosomes into two permutations π and γ. The purpose of these concatenated genomes is that every rearrangement in a multichromosomal genome Γ can be mimicked by a reversal in a permutation γ. In an *optimal* concatenate, sorting γ with respect to π actually corresponds to sorting Γ with respect to Π. Tesler [75] also showed that when such an optimal concatenate does not exist, a *near-optimal* concatenate exists such that sorting this concatenate mimics sorting the multichromosomal genomes and uses a single extra reversal which corresponds to a reordering of the chromosomes.

3.3 Conservation Distance

Recently, two criteria were proposed to measure the level of similarity between sets of genomes: common intervals [31,78] and conserved intervals [7]. In a way, both of these criteria represent a generalization of the breakpoint distance but consider intervals instead of only adjacencies. There are two important properties that common/conserved distances share with the breakpoint distance:

(i) It can be directly defined on a set of more than two genomes and allows the identification of shared features in a family of organisms.

(ii) It does not rely on an a priori model of rearrangements.

3.3.1 Common Intervals

Given two signed permutations, π and γ, a *common interval* is a set of two or more integers that is an interval in both permutations [31,78]. Using the example from Table 1, we get that there are 14 common intervals, eight of which are shown in the earthworm using boxes:

$$1 \quad 2 \quad 3 \quad 5 \quad -10 \quad 11 \quad 4 \quad 9 \quad 7 \quad 8 \quad 12 \quad 6 \quad 13$$

The additional common intervals not displayed are: $[2,3]$, $[2,3,5,\ldots 6]$, $[2,3,5,\ldots 13]$, $[3,5,\ldots 6]$, $[3,5,\ldots 13]$ and $[5,10,\ldots 13]$.

Suppose $C(\pi,\gamma)$ and $C_i(\pi,\gamma)$ are the number of common intervals and the number of common intervals of size i in π and γ, respectively. We note that the maximum number of common intervals for two permutations of size n is achieved for identical permutations and is simply:

$$\sum_{i=2}^{n} C_i(\pi,\pi) = \quad (n-1)+(n-2)+\ldots+1 = \frac{n(n-1)}{2}.$$

Of course, the more common intervals between two permutations, the higher the conservation. In the example above, there is only 14 common intervals while the maximum achievable is 78.

3.3.2 Conserved Intervals

Given two permutations, π and γ, a *conserved interval* is an interval $[a,b]$ such that a precedes b or $-b$ precedes $-a$, in both π and γ, and the set of elements, without signs, between a and b is the same in both π and γ [7]. Continuing with the example from Table 1, there are only five conserved intervals between the human and earthworm mtDNA:

$$1 \quad 2 \quad 3 \quad 5 \quad -10 \quad 11 \quad 4 \quad 9 \quad 7 \quad 8 \quad 12 \quad 6 \quad 13$$

Although, initially the definition of conserved intervals may seem unnatural, it is tightly connected to the HP theory (it corresponds to *subpermutations* in Ref. [29]) and it was shown that it can be used to efficiently sort permutations by reversals [5].

4 Genome Rearrangement Phylogenies

An important challenge in the comparative analysis of gene order is the construction of phylogenies based on genome rearrangements that describe the genetic relationships between the organisms. Phylogenies are represented by unrooted binary trees such that the leaf nodes of the trees correspond to contemporary genomes and the internal nodes correspond to their extinct ancestors (see Figure 5 for an example). Phylogenetic tree reconstruction is difficult largely because the number of unrooted trees grows at a rate that is more than exponential with the number of leaf nodes.

We review three main classes of approaches that can be used for phylogenetic tree reconstruction based on relative gene order: distance-based methods, maximum parsimony methods and maximum likelihood methods. These main classes of approaches are very similar in spirit to the ones developed for phylogenetic tree reconstruction based on sequence evolution with point mutations instead of chromosomal mutations (see Chapter 4). Links for some of the programs available to analyze genome rearrangements described in this section are provided in Table 4.

Table 4 Links for some of the software tools available to analyze genome rearrangements

BPAnalysis	http://www.cs.washington.edu/homes/blanchem/software
GRAPPA	http://www.cs.unm.edu/~moret/GRAPPA
GOTREE	http://www.mcb.mcgill.ca/~bryant/GoTree
GRIMM	http://www-cse.ucsd.edu/groups/bioinformatics/GRIMM
MGR	http://www-cse.ucsd.edu/groups/bioinformatics/MGR
BADGER	http://badger.duq.edu

4.1 Distance-based Methods

These approaches construct trees strictly based on the pairwise distances between the leaf nodes of the tree. The first step computes the pairwise distance matrix for the genomes of interest using one of the criterion described in Section 3 or from other criterion such as EDE, the "empirically derived estimator", that attempts to correct the bias in the parsimony assumption for large distances [83]. Distance-based methods differ in the second step in how they make use of the distance matrix to reconstruct the trees. Currently, the most common family of distance-based methods is probably "neighbor-joining" which was first proposed by Saitou and Nei [62].

Methods in this class are typically very efficient; in many cases phylogenies can be inferred in polynomial time. When applied to gene order data, one of the limitations of distance-based approaches is that they do not label the internal nodes and they do not associate a rearrangement scenario to the phylogeny. For challenging data sets, this may lead to infeasible or less accurate solutions [82]. This limitation is addressed both by maximum parsimony and by maximum-likelihood methods.

4.2 Maximum Parsimony Methods

Methods seeking the most parsimonious scenario attempt to recover the tree, and its internal nodes, that minimizes the number of events on its branches. It corresponds to the Steiner Tree Problem [33] on various metrics. The first methods of this type were developed for sequence data [25] but they were later adapted for gene order data [27,64]. Formally, given a set of m genomes, the problem is to find an unrooted tree T, where the m genomes are leaf nodes, and assign internal ancestral nodes such that $D(T)$ is minimized:

$$D(T) = \sum_{(\pi,\gamma)\in T} d(\pi,\gamma),$$

where $d(\pi,\gamma)$ can be any of the distances described in Section 3. The special case of three genomes ($m = 3$) is called the median problem. Although the tree topology for this problem is trivial, the assignment of the optimal internal node can still be challenging.

If a rearrangement distance is used, a detailed rearrangement scenario could also be associated to the tree that will describe every intermediate step of the evolution of these genomes. Again, under the assumption that rearrangements events are rare [15,61], its reasonable to seek the most parsimonious scenario to recover the actual tree.

Although many of the pairwise distances can be computed in polynomial time (e.g. the breakpoint distance d_{break} and the reversal distance d_{rev}, see Section 3.2), it was shown that both the median problem for d_{break} and the median problem for d_{rev} are NP-hard [16,18,55]. Nevertheless, there are a few efficient heuristics to tackle both the median problem [67,72] and the full phylogeny problem [10,15,44] under different sets of assumptions. We briefly present some of these methods.

Sankoff and Blanchette [67] studied the median problem for the breakpoint distance; they described a clever reduction of this problem to the Traveling Salesman Problem for which reasonably efficient algorithms are available. Using this result, Blanchette and coworkers [10] developed BPAnalysis, a method to recover the most parsimonious scenario for m genomes also under the breakpoint distance. That method first looked for the optimal assignment of internal nodes for a given topology by solving a series of median problem (this is also known as the small parsimony problem). The next step was to scan the space of all possible tree topologies to find the best tree (large parsimony problem). One of the drawbacks of this approach is that, as we have seen, the tree space quickly becomes prohibitive. This limitation was partially addressed by Moret and coworkers [44] who develop GRAPPA which improves on BPAnalysis by computing tight bounds and efficiently pruning the tree space. Another program to reconstruct phylogenies based on the breakpoint distance is GOTREE (see Table 4). A special feature of this last tool is that it is not restricted to genomes with equal gene content.

Siepel and Moret [72] studied a different problem: the median problem for the reversal distance. They derived a branch-and-bound algorithm to prune the search space using simple geometric properties of the problem and the linear-time machinery to compute the reversal distance [1]. Concurrently, Bourque and Pevzner [15] developed a method called MGR for both the median and the full phylogeny problem that made use of properties of additive or nearly additive trees. This algorithm, combined with GRIMM [76], is applicable to unichromosomal genomes for the reversal distance and to multichromosomal genomes for a rearrangement distance that allows reversals, translocations, fusions and fissions. The main idea of the algorithm is to look for rearrangements in the starting genomes that reduce the total distance to the other genomes and iteratively "reverse history". The key is to use good criterion to chose the order in which the rearrangements are selected.

The first method that used the conservation distance as the criterion to be minimized in the phylogenetic reconstruction problem was presented by Bergeron and coworkers [6]. Even though the problem was restricted to finding an assignment of internal nodes on a fixed phylogeny (small parsimony problem), this is a promising and active area of research.

4.3 Maximum Likelihood Methods

If we make assumptions about the mechanisms of evolution and the rates at which these changes occur, we can seek the tree which is the most likely to have generated the data observed. Such methods are called maximum likelihood methods. They tend to be computationally intensive but they have the advantage of providing a global picture of the solution space in contrast to maximum parsimony which provides a unique solution for instance.

In the context of the comparative analysis of gene order, a maximum likelihood approach turns out to be quite challenging because of our incomplete understanding of the frequency of rearrangement events but mostly because of the significantly large number of potential states at internal nodes and of phylogenetic trees [70]. Nevertheless, Dicks [21] developed one such method for gene order data, but the method presented was restricted to small instances of the problem. Other promising approaches involve the construction of a Bayesian framework and the use of Markov chain Monte Carlo to sample parameter space for two unichromosomal genomes [42, 87] or m unichromosomal genomes [38, 39]. Specifically, Larget and coworkers [39] developed the program BADGER and used it to quantify the uncertainty among the relationships of metazoan phyla on the basis of mitochondrial gene orders. So far, although these frameworks are propitious, their range of applications has been limited. It will be interesting to see if these approaches can be further applied and adapted to larger and also multichromosomal genomes.

5 Recent Applications

We have already seen some applications in which genome rearrangements acted as complementary phylogenetic characters to study evolutionary relationships in a group of organisms such as mitochondria, chloroplasts, viruses or small regions of larger genomes [3, 12, 20, 27, 51, 53, 54, 63]. We will now show how the same concepts and methodologies can be applied to compare entire eukaryotic genomes. Apart from the topology of the phylogeny, interesting questions arise from studying rates of rearrangements, types of rearrangements and predictions at ancestral nodes. We will also present some preliminary work studying genome rearrangements in cancer.

5.1 Rearrangements in Large Genomes

Genome rearrangements studies have traditionally been based on the relative order of homologous genes; however, as hinted at in Section 2.1, they can also be based on the relative order of a common set of homology synteny blocks (HSBs). These blocks can be defined either directly from sequence similarity [36, 56] or from the clustering of homologous genes [88]. In this context, rearrangement studies for large genomes will be reconfigured into a two-step process:

(i) Identification of HSBs shared by the set of genomes under study.

(ii) Genome rearrangement analysis of the HSBs.

In Step (i), both for sequence-based and gene-based HSBs, thresholds need to be set to allow the HSBs to extend over minor local inconsistencies that could stem from different sources: sequencing and assembly errors, small rearrangement events not enclosed in the rearrangement model of Step (ii) (e.g. transposons), inaccurate prediction of orthologous genes (e.g. in the presence of many paralogous copies), etc. For the identification of HSBs, there are advantages to using both sequence and gene data.

The most important benefit of using raw sequence data is probably to circumvent the limitation of analyzing strictly coding regions (these regions only cover a small portion of the eukaryotic genomes). Other benefits include that it avoids annotation problems, it is less sensitive to gene families and, finally, it preserves additional information on *micro-rearrangements* (rearrangements within HSBs) that can then be used as additional independent phylogenetic characters [13]. Advantages of using gene-based HSBs are that it focuses

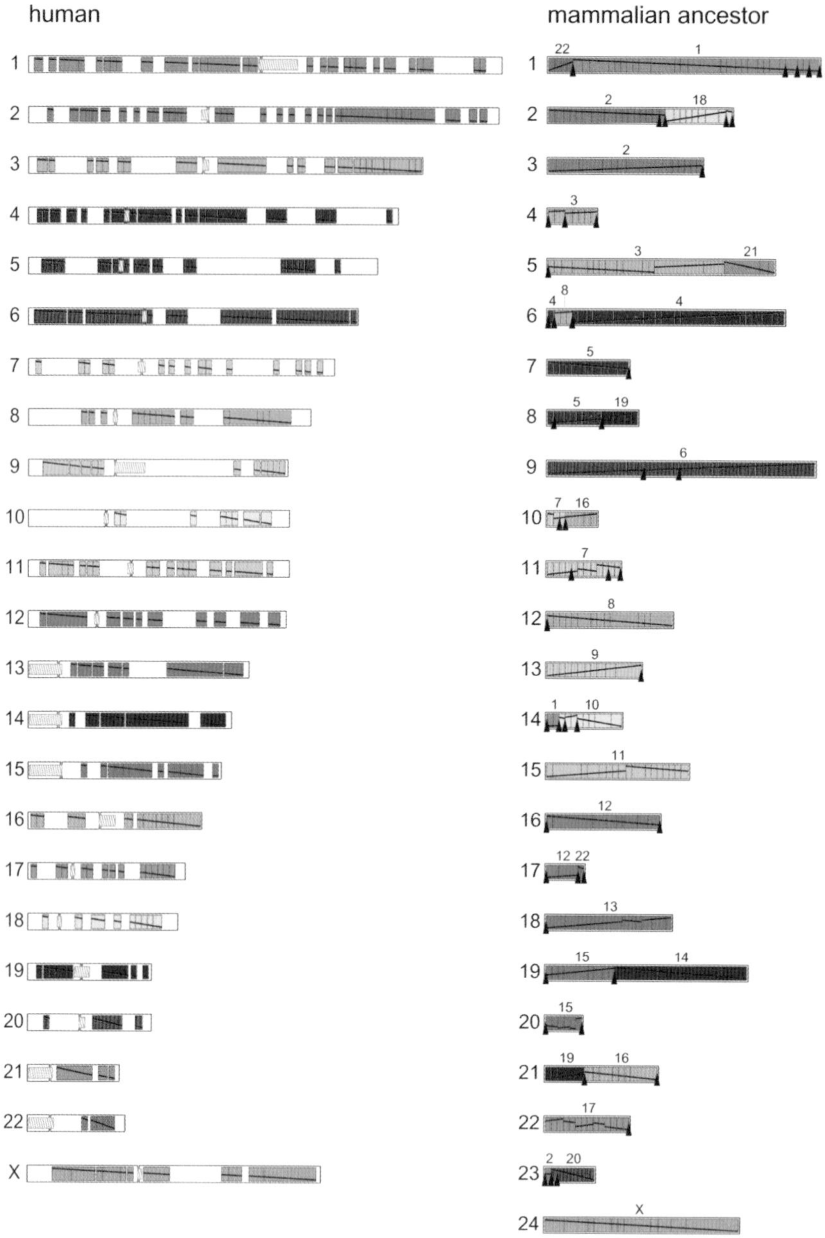
human
mammalian ancestor
1
2
3
4
5
6
7
8
9
10
11
12
13
14
15
16
17
18
19
20
21
22
X
23
24

the investigations on critical regions of the genome, the thresholds are length independent and it avoids some of the noise created by repeat regions.

After Step (i), the comparison of the respective arrangements of the HSBs in the different genomes can be performed using the models, algorithms and programs described in Sections 3 and 4. This two-step analysis was used to compare the human with the mouse genome [56] and suggested a larger number inversions than previously expected [48]. It also helped motivate a model for chromosome evolution in which some breakpoints are reused nonrandomly [40, 57].

When many genomes are compared, rearrangement analysis provides information not only on phylogenetic relationships, but also on rates of rearrangements and on putative genomic architecture of ancestral genomes [13, 14, 46, 47]. For instance, the availability of the rat genome [26] allowed a comparative study with the human and the mouse [14] that confirmed an observation made using lower-resolution studies that rodent genomes have had an accelerated rate of inter-chromosomal rearrangements (e.g. translocations, fusions and fissions). The same study also conjectured on the genomic architecture of the putative murid rodent ancestor. The addition of the chicken genome [32] acting as an outgroup allowed us to look further back in time and predict the potential architecture of the mammalian ancestor [13]. This analysis also suggested:

- Variable rates of inter-chromosomal rearrangements across lineages.
- High ratio of intra-chromosomal versus inter-chromosomal rearrangements in the chicken lineage.
- Low rate of rearrangements in chicken, in the early mammalian ancestor or in both.

More recently, a comprehensive analysis of eight mammalian genomes, three sequenced genomes (human–mouse–rat) and five with high-resolution

Figure 4 Inferred genomic architecture of the mammalian ancestor (adapted from Ref. [46]). Each human chromosome is assigned a unique color and is divided into HSBs. These HSBs correspond to stretches of DNA for which sufficient similarity has been retain to unambiguously allow the identification of the homologous regions in all other species. The size of each block is approximately proportional to the actual size of the block in human. In human, blocks are arranged on each chromosome from left (p-arm) to right (q-arm) and physical gaps between blocks are shown to give an indication of coverage. Numbers above the rec onstructed ancestral chromosomes indicate the human chromosome homolog. Diagonal lines within each block indicate their relative order and orientation. Black arrows under the ancestral chromosome indicate that the two adjacent HSBs separated by the arrow were not found in every one of the most parsimonious solutions explored; these are considered *weak adjacencies*.

radiation-hybrid maps (cat–dog–cow–pig–horse), afforded a detailed analysis of the dynamics of mammalian chromosome evolution [46]. This study also produced a refined model of the genomic architecture of the mammalian ancestor, see Figure 4.

Applications focusing on specific areas of the genomes allow for the identification of very detailed scenarios. For instance, in the results of the study by Murphy and coworkers [46], it is possible to focus exclusively on the HSBs found on human chromosome 17; there are 14 such blocks. Chromosome 17 is interesting because, similarly to the X chromosome, it has seldom exchanged genetic material with other chromosomes during mammalian evolution. Specifically, the 14 blocks are found in one contiguous segment on a single chromosome in mouse, rat, cat and pig. They are found in two contiguous pieces on two chromosomes in cow and in three contiguous pieces on three chromosomes in dog (horse is left out of this analysis because of insufficient data). See Figure 5 for a parsimonious rearrangement scenario describing the mammalian history of this chromosome. This example once again seems to point towards uneven rates of rearrangements with no rearrangement between the cetartiodactyl ancestor and pig, but five rearrangements in cow during the same period of evolution. According to this reconstruction, the pig chromosome 12 (the homolog of human chromosome 17) is ancestral in the sense that no large-scale rearrangement has occurred on it since the divergence of these species.

5.2 Genomes Rearrrangements and Cancer

The previous section described examples of the use of genome rearrangements to study the evolution of a group of organisms. Now, because a rapid increase of chromosomal mutations is frequently observed in cancer cells, it is possible to study the cancer genome very much like as it was a new organism that had recently diverged from the normal human genome. The interest is that although cancer progression is frequently associated with genome rearrangements, the mechanisms behind these rearrangements are still poorly understood. There are many challenges in studying rearrangements in cancer cells: the heterogeneity of the cells, the complexity of the rearrangements (which include translocations, but also frequent duplications), but mostly the fact that detailed sequence is only sparsely available. So far, the cost of sequencing has been a prohibitive factor preventing large cancer genome sequencing projects, but new emerging sequencing techniques such as End Sequence Profiling [80] and Ditags [49] might help alleviate this problem. Such techniques justify the development of algorithms and tools, related to the analysis of genome rearrangement, to extract detailed tumor architecture from such data sets [59,60].

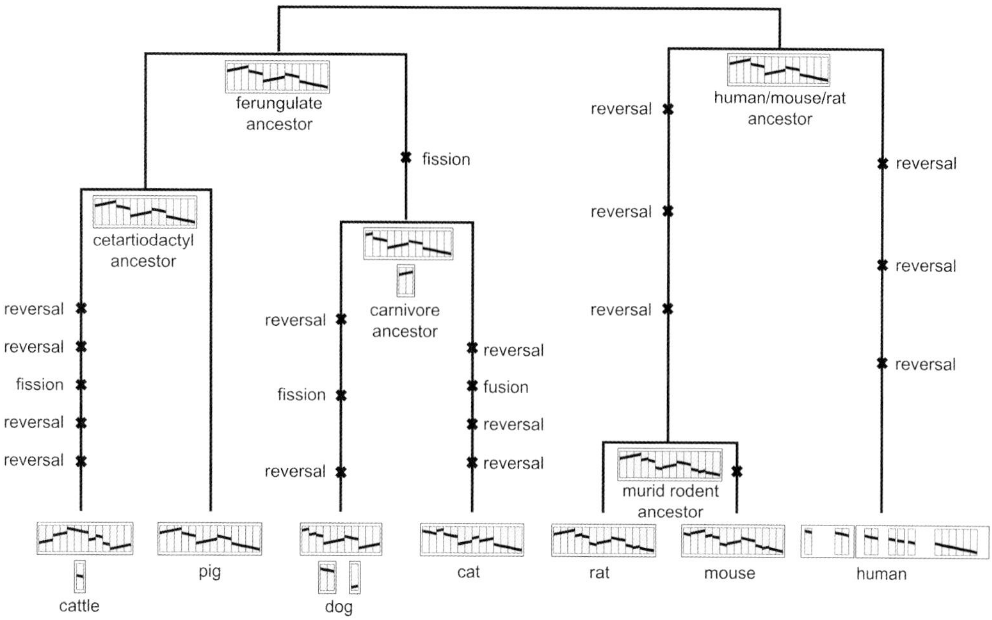

Figure 5 Mammalian history of human chromosome 17. The arrangements of 14 blocks (stretches of DNA) from human chromosome 17 with syntenic counterparts in seven other mammalian genomes (mouse, rat, cat, dog, pig and cattle) are shown at the bottom of the tree. Blocks are drawn proportionally to their size in human. A diagonal line traverses the blocks to show their order and relative orientation. In human, blocks are arranged from left (p-arm) to right (q-arm) and physical gaps between blocks are shown to give an indication of coverage. In other species, the same blocks are drawn also from left to right but in some cases these blocks are found on multiple chromosomes [cattle (2), dog (3) and carnivore ancestor (2)]. Crosses on the edges of the tree are labeled and indicate putative rearrangement events even though their exact timing is unknown. Data adapted from Ref. [46].

A complementary approach for the study of rearrangements in cancer involves looking at breakpoint regions. Many such regions have already been characterized in a large population of cancer patients [43]. Studying their distribution with respect to either chromosomal location [65] or evolutionary breakpoints [46] (identified from multispecies comparisons) is likely to provide invaluable information on the forces acting on these aberrant genomes.

6 Conclusion

6.1 Challenges

Comparative analyses of gene order would greatly benefit from established benchmarking data sets. These instances could be use to compare and refine current approaches for the study of genome rearrangements. Of course, the challenge is that for data sets generated from real genomes, the actual rearrangement history for these organisms is unknown. Thus, recovered scenarios can only be evaluated with respect to some limited aspects of their solution such as topology of the recovered tree [12, 20]. This is a suitable criterion to evaluate the merits of an approach because the topology can also be inferred from alternative, more traditional, approaches such as the comparison of individual genes. Even then, ambiguities will remain since for many interesting sets of species, some aspects of the topology are debatable (e.g. especially when deep branches are involved) and the information extracted from genome rearrangements might be different from that provided from sequence analysis but not necessarily erroneous.

Other criteria that can be used for the evaluation of solutions are some of the coarse features of the recovered ancestors such as the ancestral chromosomal associations. These are associations between modern chromosomes (e.g. human chromsomome) that are inferred to have been present in the ancestors [73]. Unfortunately, once again, definitive evaluation is difficult for two reasons: (i) the expected associations rely on low-density comparative maps and are likely to be incomplete, and (ii) multiple alternative ancestors are typically recovered in rearrangement studies making more than a single prediction [13, 14]. Now that high-quality sequences are increasingly becoming available for many genomes, one would actually expect to see the knowledge on such associations to be expanded and refined, especially after carrying out combinatorial analyses that take into account more than just co-occurrences.

A logical alternative to real benchmarking data sets with unknown rearrangement history is provided by simulated data sets. Unfortunately, there are drawbacks inherent to this approach as well. In particular, simulated data sets will always bias the evaluation towards approaches that have an underlying rearrangement model that is most compatible with the model that was used to generate the data. Such data sets can be a great asset in evaluating alternative methods that have the same assumptions, but they are of limited value in identifying whether a particular method will be successful on real data.

Another desired development would be a more systematic study and comparison of different distance criterion. Specifically, with the development of new measures [7,31,78], a detailed analysis of the strengths and weaknesses of the different approaches is needed to assess the context in which they are most applicable. For instance, model-free measures such as the breakpoint distance and the conservation distance are probably the most appropriate when the underlying rearrangements follow uncharacterized rules.

Finally, a key challenge associated with this type of analysis involves studying the causes and consequences of genome rearrangements. Although these events are well characterize in both evolution and cancer, the extent of the biological repercussions is still unclear. For instance, large rearrangement events can have a significant impact at the population level by creating subpopulations for which recombination in the affected region will be impossible but the question of whether such events also play a role in speciation for instance is still debated. On a different topic, is there a faster phenotypic evolution associated with a faster rearrangement rate? Given the amount of comparative data recently made available [26,32,37,79,84], the hope is that some answers might be within reach.

In order to start exploring these questions, looking at sets of highly diverse genomes spanning long evolutionary distances is not the most appropriate. Inherent to such data sets will always be ambiguities such as the accuracy of the rearrangement model, the quality of the solution obtained, the order of rearrangements found on edges of the phylogeny, the presence of alternative ancestors and the presence of alternative rearrangement scenarios. A more practical framework in which to ask questions about the impact of genome rearrangements would probably involve looking at more closely related species where the inferred rearrangement scenario is less disputable.

6.2 Promising New Approaches

The rearrangement model will always have a critical impact on the reconstructed scenario. In many of the applications presented [13, 14, 46], the rearrangement model includes reversals, translocations, fusions and fissions, but these events are considered equally likely (i.e. the weight of each of the events is the same when the distance is computed). In reality, short reversals are probably more common than fusions for instance. Consider the carnivore ancestor shown Figure 5; in the displayed solution, there is a fission between the ferungulate ancestor and the carnivore ancestor followed by a fusion in the cat lineage. An alternative solution exists with the same total number of rearrangements, but in which this fission plus fusion is replaced by a single fission on the dog lineage and a reversal in the cat lineage. Such a scenario is probably more realistic than the one displayed but it is masked by our assumption of equally likely events. Perhaps approaches with weighted events, such as in Blanchette and coworkers [11], or approaches that make use of a maximum likelihood framework, such as Larget and coworkers [39], could help alleviate some of these ambiguities.

Although strictly incorporating transpositions into the rearrangement model remains computationally challenging, there is renewed interest in allowing block interchanges, an operation which includes all types of transposition [19, 41,86]. The inclusion of this process actually allows a dramatic simplification of the HP theory (see Section 3.2) [86] and is likely to enable new applications.

Nevertheless, because rearrangement models are always debatable, model-free approaches that make use of breakpoint or conservation distance, such as Bergeron and coworkers [6], are also attractive and interesting. Hopefully these approaches will be extended and applied to a larger variety of problems.

Finally, another promising area of research is the analysis of breakpoint regions. These regions typically contain an unusual mosaic of content [36,77] and they are also likely to harbor information on the mechanisms behind the rearrangements that created them. In the context of cancer, these are also the regions that have the potential to host the destructive fusion genes. Comparing cancer breakpoints with evolutionary breakpoints [46] might provide some information on the forces shaping the genomic architecture of modern organisms.

Acknowledgments

The author is supported by funding from the Biomedical Research Council (BMRC) of the Agency for Science, Technology and Research (A*STAR) in Singapore.

References

1 Bader, D., B. Moret and M. Yan. 2001. A linear-time algorithm for computing inversion distance between signed permutations with an experimental study. In Proc. 7th Int. Workshop on Algorithms and Data Structures, Providence, RI, USA: 365–76.

2 Bafna, V. and P. A. Pevzner. 1998. Sorting by transpositions. SIAM J. on Discrete Mathematics **11**: 224–40.

3 Bafna, V. and P. Pevzner. 1995. Sorting by reversals: genome rearrangements in plant organelles and evolutionary history of X chromosome. Mol. Biol. Evol. **12**: 239–46.

4 Bafna, V. and P. Pevzner. 1996. Genome rearrangements and sorting by reversal. SIAM J. Comput. **25**: 272–89.

5 Bergeron, A., J. Mixtacki and J. Stoye. 2004. Reversal distance without hurdles and fortresses. In Proc. Annu. Symp. on Combinatorial Pattern Matching, Istanbul, Turkey: 388–99.

6 Bergeron, A., M. Blanchette, A. Chateau and C. Chauve. 2004. Reconstructing ancestral gene orders using conserved intervals. in Proc. WABI, Bergen, Norway: 14–25.

7 Bergeron, A. and J. Stoye. 2003. On the similarity of sets of permutations and its applications to genome comparison. In Proc. COCOON, Big Sky, MT, USA: 68–79.

8 Bergeron, A. 2001. A very elementary presentation of the Hannenhalli–Pevzner theory. In Proc. Annu. Symp. on Combinatorial Pattern Matching, Jerusalem, Israel: 106–17.

9 Berman, P. and S. Hannenhalli. 1996. Fast sorting by reversal. In Proc. Annu. Symp. on Combinatorial Pattern Matching, Laguna Beach, CA, USA: 168–85.

10 Blanchette, M., G. Bourque and D. Sankoff. 1997. Breakpoint phylogenies. In Proc. Genome Informatics Workshop, Tokyo, Japan: 25–34.

11 Blanchette, M., T. Kunisawa and D. Sankoff. 1996. Parametric genome rearrangement. Gene **172**: GC11–7.

12 BLANCHETTE, M., T. KUNISAWA AND D. SANKOFF. 1999. Gene order breakpoint evidence in animal mitochondrial phylogeny. J. Mol. Evol. **49**: 193–203.

13 BOURQUE, G., E. ZDOBNOV, P. BORK, P. PEVZNER AND G. TESLER. 2005. Comparative architectures of mammalian and chicken genomes reveal highly variable rates of genomic rearrangements across different lineages. Genome Res. **15**: 98–110.

14 BOURQUE, G., P. PEVZNER AND G. TESLER. 2004. Reconstructing the genomic architecture of ancestral mammals: lessons from human, mouse, and rat genomes. Genome Res. **14**: 507–16.

15 BOURQUE, G. AND P. PEVZNER. 2002. Genome-scale evolution: reconstructing gene orders in the ancestral species. Genome Res. **12**: 26–36.

16 BRYANT, D. 1998. The complexity of breakpoint median problem. *Technical Report CRM-2579*. Centre de recherches mathématiques, Université de Montréal.

17 CAPRARA, A. 1997. Sorting by reversals is difficult. Proc. RECOMB **1**: 75–83.

18 CAPRARA, A. 1999. Formulations and complexity of multiple sorting by reversals. Proc. RECOMB **3**: 84–93.

19 CHRISTIE, D. A. 1996. Sorting permutations by block-interchanges. Inf. Process. Lett. **60**: 165–9.

20 COSNER, M., R. JANSEN, B. MORET, L. RAUBESON, L. WANG, T. WARNOW AND S. WYMAN. 2002. A new fast heuristic for computing the breakpoint phylogeny and experimental phylogenetic analyses of real and synthetic data. Proc. ISMB **8**: 104–15.

21 DICKS, J. 2002. CHROMTREE: maximum likelihood estimation of chromosomal phylogenies. In SANKOFF D. AND J. H. NADEAU (eds.), *Comparitive Genomics (DCAF-2000)*. Kluwer, Dordrecht: 333–42. 2000.

22 DOBZHANSKY, T. AND A. STURTEVANT. 1938. Inversions in the chromosomes of *Drosophila pseudoobscura*. Genetics **23**: 28–64.

23 EARNEST-DEYOUNG, J., E. LERAT AND B. MORET. 2004. Reversing gene erosion – reconstructing ancestral bacterial genomes from gene-content and order data. In Proc. WABI, Bergen, Norway: 1–13.

24 EL'MABROUK, N. 2005. Genome rearrangement with gene families. In GASCUEL O. (ed.), *Mathematics of Evolution and Phylogeny*. Oxford University Press, Oxford: 291–313.

25 FITCH, W. 1977. On the problem of discovering the most parsimonious tree. Amer. Natur. **111**: 223–57.

26 GIBBS, R., G. WEINSTOCK, M. METZKER, ET AL. 2004. Genome sequence of the Brown Norway rat yields insights into mammalian evolution. Nature **428**: 493–521.

27 HANNENHALLI, S., C. CHAPPEY, E. KOONIN AND P. PEVZNER. 1995. Genome sequence comparison and scenarios for gene rearrangements: a test case. Genomics **30**: 299–311.

28 HANNENHALLI, S. AND P. PEVZNER. 1995. Transforming cabbage into turnip (polynomial algorithm for sorting signed permutations by reversals). Proc. 27th Annu. ACM–SIAM Symp. on the Theory of Computing, Las Vegas, NV, USA: 178–89.

29 HANNENHALLI, S. AND P. PEVZNER. 1995. Transforming men into mice: polynomial algorithm for genomic distance problem. Proc. 36th IEEE Symp. on Foundations of Computer Science, Los Alamitos, CA, USA: 581–92.

30 HANNENHALLI, S. AND P. PEVZNER. 1999. Transforming cabbage into turnip: polynomial algorithm for sorting signed permutations by reversals. J. ACM **46**: 1–27.

31 HEBER, S. AND J. STOYE. 2001. Finding all common intervals of *k* permutations. In Proc. Annu. Symp. on Combinatorial Pattern Matching, Jerusalem, Israel: 207–18.

32 HILLIER, L., W. MILLER, E. BIRNEY, ET AL. 2004. Sequence and comparative analysis of the chicken genome provide unique perspectives on vertebrate evolution. Nature **432**: 695–716.

33 JARNIK, V. 1934. Sur les graphes minima, contenant n points donnés. Cas. Pest. Mat. **63**: 223–35.

34 KAPLAN, H., R. SHAMIR AND R. TARJAN. 1997. Faster and simpler algorithm for sorting signed permutations by reversals. Proc. 8th Annu. ACM–SIAM Symp. on Discrete Algorithms, New Orleans, LA, USA: 344–51.

35 KECECIOGLU, J. AND D. SANKOFF. 1994. Efficient Bounds for oriented chromosome inversion distance. In Proc. Annu. Symp. on Combinatorial Pattern Matching, Asilomar, CA, USA: 307–25.

36 KENT, W., R. BAERTSCH, A. HINRICHS, W. MILLER AND D. HAUSSLER. 2003. Evolution's cauldron: duplication, deletion, and rearrangement in the mouse and human genomes. Proc. Natl Acad. Sci. USA **100**: 11484–9.

37 LANDER, E. S., L. M. LINTON, B. BIRREN, ET AL. 2001. Initial sequencing and analysis of the human genome. Nature **409**: 860–921.

38 LARGET, B., D. SIMON AND J. KADANE. 2002. Bayesian phylogenetic inference from animal mitochondrial genome arrangements (with discussion). J. R. Stat. Soc. **B**: 681–93.

39 LARGET, B., D. SIMON, J. KADANE AND D. SWEET. 2005. A Bayesian analysis of metazoan mitochondrial genome arrangements. Mol. Biol. Evol. **22**: 486–95.

40 LARKIN, D., A. E. VAN DER WIND, M. REBEIZ ET AL. 2003. A cattle–human comparative map built with cattle BAC-ends and human genome sequence. Genome Res. **13**: 1966–72.

41 LIN, Y. C., C. L. LU, H. Y. CHANG AND C. Y. TANG. 2005. An efficient algorithm for sorting by block-interchanges and its application to the evolution of *Vibrio* species. J. Comput. Biol. **12**: 102–12.

42 MIKLOS, I. 2003. MCMC genome rearrangement. Bioinformatics **19**: ii130–7.

43 MITELMAN, F., B. JOHANSSON AND F. MERTENS. 2005. Mitelman Database of Chromosome Aberrations in Cancer. http://cgap.nci.nih.gov/Chromosomes/Mitelman.

44 MORET, B., S. WYMAN, D. BADER, T. WARNOW AND M. YAN. 2001. A new implementation and detailed study of breakpoint analysis. Pac. Symp. Biocomput. **6**: 583–94. 2001.

45 MORGAN, T. AND C. BRIDGES. 1916. Sex-linked inheritance in *Drosophila*. Carnegie Inst. Washington Publ. **237**: 1–88.

46 MURPHY, W., D. LARKIN, A. E. VAN DER WIND, G. BOURQUE, ET AL. 2005. Dynamics of mammalian chromosome evolution inferred from multispecies comparative maps. Science **309**: 613–17.

47 MURPHY, W., G. BOURQUE, G. TESLER, P. PEVZNER AND S. O'BRIEN. 2003. Reconstructing the genomic architecture of mammalian ancestors using multispecies comparative maps. Hum. Genomics **1**: 30–40.

48 NADEAU, J. AND B. TAYLOR. 1984. Lengths of chromosomal segments conserved since divergence of man and mouse. Proc. Natl Acad. Sci. USA **81**: 814–8.

49 NG, P., C. WEI, W. SUNG ET AL. 2005. Gene identification signature (GIS) analysis for transcriptome characterization and genome annotation. Nat. Methods **2**: 105–11.

50 O'BRIEN, S., M. MENOTTI-RAYMOND, W. MURPHY, ET AL. 1999. The promise of comparative genomics in mammals. Science **286**: 458–81.

51 OLMSTEAD, R. AND J. PALMER. 1994. Chloroplast DNA systematics: a review of methods and data analysis. Am. J. Bot. **81**: 1205–24.

52 OZERY-FLATO, M. AND R. SHAMIR. 2003. Two notes on genome rearrangement. J. Bioinf. Comput. Biol. **1**: 71–94.

53 PALMER, J. AND L. HERBON. 1988. Plant mitochondrial DNA evolves rapidly in structure, but slowly in sequence. J. Mol. Evol. **27**: 87–97.

54 PALMER, J. 1992. Chloroplast and mitochondrial genome evolution in land plants. In HERRMANN, R. (ed.), *Cell Organelles*. Springer, Berlin: 99–133.

55 PE'ER, I. AND R. SHAMIR. 1988. The median problem for breakpoints are NP-complete. Electronic Colloqium on Computational Complexity. Technical Report **TR98-071**.

56 PEVZNER, P. AND G. TESLER. 2003. Genome rearrangements in mammalian evolution: lessons from human and mouse genomes. Genome Res. **13**: 37–45.

57 PEVZNER, P. AND G. TESLER. 2003. Human and mouse genomic sequences reveal extensive breakpoint reuse in mammalian evolution. Proc. Natl Acad. Sci. USA **100**: 7672–7.

58 PEVZNER, P. 2000. *Computational Molecular Biology: An Algorithmic Approach.* MIT Press, Cambridge, MA.

59 RAPHAEL, B., S. VOLIK, C. COLLINS AND P. PEVZNER. 2003. Reconstructing tumor genome architectures. Bioinformatics **19**: ii162–71.

60 RAPHAEL, B. AND P. PEVZNER. 2004. Reconstructing tumor amplisomes. Bioinformatics **20 (Suppl 1)**: i265–73.

61 ROKAS, A. AND P. HOLLAND. 2000. Rare genomic changes as a tool for phylogenetics. Trends Ecol. Evol. **15**: 454–9.

62 SAITOU, N. AND M. NEI. 1987. The neighbor-joining method: A new method for reconstructing phylogenetic trees. Mol. Biol. Evol. **4**: 406–25.

63 SANKOFF, D., G. LEDUC, N. ANTOINE, B. PAQUIN, B. LANG AND R. CEDERGREN. 1992. Gene order comparisons for phylogenetic inference: evolution of the mitochondrial genome. Proc. Natl Acad. Sci. USA **89**: 6575–9.

64 SANKOFF, D., G. SUNDARAM AND J. KECECIOGLU. 1996. Steiner points in the space of genome rearrangements. Int. J. Found. of Comp. Sci. **7**: 1–9.

65 SANKOFF, D., M. DENEAULT, P. TURBIS AND C. ALLEN. 2002. Chromosomal distributions of breakpoints in cancer, infertility, and evolution. Theor. Popul. Biol. **61**: 497–501.

66 SANKOFF, D. AND J. H. NADEAU (EDS). 2000. *Comparative Genomics: Gene Order Dynamics, Comparative Maps and Multigene Families.* Kluwer, Dordrecht.

67 SANKOFF, D. AND M. BLANCHETTE. 1997. The median problem for breakpoints in comparative genomics. In Proc. COCOON, Shanghai, China: 251–63.

68 SANKOFF, D. 1992. Edit distance for genome comparison based on non-local operations. In: Proc. Annu. Symp. on Combinatorial Pattern, Tucson, AZ, USA: 121–35.

69 SANKOFF, D. 1999. Genome rearrangement with gene families. Bioinformatics **15**: 909–17.

70 SAVVA, G., J. DICKS AND I. ROBERTS. 2003. Current approaches to whole genome phylogenetic analysis. Brief. Bioinformat. **4**: 63–74.

71 SETUBAL, J. AND J. MEIDANIS. 1997. *Introduction to Computational Molecular Biology.* PWS Publishing, Boston, MA.

72 SIEPEL, A. AND B. MORET. 2001. Finding an optimal inversion median: experimental results. In Proc. WABI, Aarhus, Denmark: 189–203.

73 STANYON, R., G. STONE, M. GARCIA AND L. FROENICKE. 2003. Reciprocal chromosome painting shows that squirrels, unlike murid rodents, have a highly conserved genome organization. Genomics **82**: 245–9.

74 STURTEVANT, A. H. 1921. Genetic studies on *Drosophila simulans.* II. Sex-linked group of genes. Genetics **6**: 43–64.

75 TESLER, G. 2002. Efficient algorithms for multichromosomal genome rearrangements. J. Comput. Syst. Sci. **65**: 587–609.

76 TESLER, G. 2002. GRIMM: genome rearrangements web server. Bioinformatics **18**: 492–3.

77 TRINH, P., A. MCLYSAGHT AND D. SANKOFF. 2004. Genomic features in the breakpoint regions between syntenic blocks. Bioinformatics **20 (Suppl. 1)**: I318–25.

78 UNO, T. AND M. YAGIURA. 2000. Fast algorithms to enumerate all common intervals of two permutations. Algorithmica **26**: 290–309.

79 VENTER, J., M. ADAMS, E. MYERS, ET AL. 2001. The sequence of the human genome. Science **291**: 1304–51.

80 VOLIK, S., S. ZHAO, K. CHIN, J. BREBNER, ET AL. 2003. End-sequence profiling: sequence-based analysis of

aberrant genomes. Proc. Natl Acad. Sci. USA **100**: 7696–701.

81 Walter, M., Z. Dias and J. Meidanis. 2000. A new approach for approximating the transposition distance. In Proc. Seventh International Symposium on String Processing Information Retrieval, La Coruna, Spain: 199–208.

82 Wang, L., R. Jansen, B. Moret, L. Raubeson and T. Warnow. 2002. Fast phylogenetic methods for the analysis of genome rearrangement data: an empirical study. Pac. Symp. Biocomput. **7**: 524–35.

83 Wang, L.-S. and T. Warnow. 2001. Estimating true evolutionary distances between genomes. Proc. 33rd Symp. on Theory of Computing, Heraklion, Crete, Greece: 637–46.

84 Waterston, R., K. Lindblad-Toh, E. Birney, et al. 2002. Initial sequencing and comparative analysis of the mouse genome. Nature **420**: 520–62.

85 Watterson, G., W. Ewens, T. Hall and A. Morgan. 1982. The chromosome inversion problem. J. Theor. Biol. **99**: 1–7.

86 Yancopoulos, S., O. Attie and R. Friedberg. 2005. Efficient sorting of genomic permutations by translocation, inversion and block interchange. Bioinformatics **21**: 3340–6.

87 York, T., R. Durrett and R. Nielsen. 2002. Bayesian estimation of the number of inversions in the history of two chromosomes. J. Comput. Biol. **9**: 805–18.

88 Zdobnov, E., C. von Mering, I. Letunic, et al. 2002. Comparative genome and proteome analysis of *Anopheles gambiae* and *Drosophila melanogaster*. Science **298**: 149-59.

Part 4 Molecular Structure Prediction

9
Predicting Simplified Features of Protein Structure

Dariusz Przybylski and Burkhard Rost

1 Introduction

1.1 Protein Structures are Determined Much Slower than Sequences

At the end of 2005 there were about 30 000 experimentally determined protein three-dimensional (3-D) structures in public databases [17]. At the same time there were almost 40 million genes known [16] and approximately 1.5 million verified [11] protein sequences. This gap between structure and sequence continues to grow – despite successful efforts at large-scale structure determination ("structural genomics" [118, 150]), the rate of new structures (thousands per year) continues to increase much slower than the rate of new sequences (many millions per year). Moreover, experimental structure determination has been largely or entirely unsuccessful for important classes such as cell membrane proteins.

1.2 Reliable and Comprehensive Computations of 3-D Structures are not yet Possible

In principle, we could compute 3-D structures from sequences using basic physical principles [9]. However, the complexity of the problem exceeds by far today's computational resources. Speeding up molecular dynamics by a factor of 1000 appears an objective within reach to Schroedinger Inc. While this would undoubtedly yield important insights into the problem, it may still not bring reliable predictions of 3-D structures from sequence. Even given infinite CPU resources, another serious obstacle is raised by the

Bioinformatics - From Genomes to Therapies Vol. 1. Edited by Thomas Lengauer

ISBN: 978-3-527-31278-8

minute energy differences between native and unfolded structures (around 1 kcal mol^{-1}). This minute difference along with the uncertainty in estimating constants needed for calculations based on first principles makes it very difficult to find an approximate approach that is both simple and sufficiently accurate. Although we cannot model from sequence, comparative modeling yields rather accurate predictions based on sequence homology to proteins of known structure [101]. Such modeling is based on the fact that proteins with similar sequences usually have similar structures. Assume we know the structure for K and that we want to predict the structure for U that is sequence-similar to K. Comparative modeling simply predicts U to have the same structure as K and models the structure of U based on the known backbone of K. However, for the majority of protein sequences no sufficiently detailed structural information is available or computable.

1.3 Predictions of Simplified Aspects of 3-D Structure are often very Successful

In the absence of experimental or predicted 3-D structures, many researchers concentrate on trying to simplify the problem and predict particular structural features. One of the first well-defined problems was the prediction of protein secondary structure. Progress in this field has been steady and current secondary structure predictions are useful for many biological applications. Techniques that were developed in the context of secondary structure predictions were successfully applied to the prediction of many other aspects of protein structure such as solvent accessibility, inter-residue contact maps, disordered regions, domain organization and specialized for distinctive cases such as transmembrane regions of proteins.

2 Secondary Structure Prediction

2.1 Assignment of Secondary from 3-D Structure

2.1.1 Regular Secondary Structure Formation is Mostly a Local Process

Three-dimensional structures exhibit extensive local conformational regularities known as regular secondary structure. These local structures (most importantly helices and sheets) can be described as ordered arrangements of a polypeptide chain without reference to amino acid type or actual 3-D conformations. They are stabilized primarily by hydrogen bonds formed between the atoms present in the polypeptide backbone, but interactions with solvent and other protein atoms also play an important role. It is believed that the formation of secondary structure is an important step toward folding. Identifying the rules for packing the elements of secondary structure against each other would afford the derivation of a very limited number of possible stable conformations. Unfortunately, the formation of secondary structure is not entirely a local process. Thus, a perfect prediction of secondary structure without knowledge of nonlocal information is unlikely. Note that secondary structure can be written in a string of assignments for each residue, i.e. it is essentially a 1-D feature of protein 3-D structure. (Unfortunately, some authors are lured into misusing the term 2-D structure, possibly in response to a misunderstanding of the word "secondary".)

2.1.2 Secondary Structures can be Somehow Flexible

Regular secondary structure is a striking, macroscopically visible aspect of 3-D structure. However, secondary structures are not rigid. Calculations and experiments indicate that structural shifting occurs, especially in surface regions. The adoption of a particular structure may depend on many environmental factors. This is illustrated by the fact that sometimes the secondary structure states differ among various crystals of the same protein as well as various nuclear magnetic resonance (NMR) models by as much as 5–15%. This variability constrains the upper limit of what we can expect from prediction methods – arguably levels of about 90% (percentage of residues predicted correctly in either of the three states helix, strand, other). While many residues can be confidently classified into one of the secondary structure types, there are also those for which classification is ambiguous. This problem is especially evident at terminal locations of secondary structure elements; it is just another aspect of the observation that protein structures are dynamic objects. Historically, assignments were carried out through visual inspection by experimentalists. That approach introduced a human-based inconsistency. In 1983, this inconsistency was first addressed by an objective, automatic assignment method [Dictionary of Secondary Structure of Proteins (DSSP), see below]. Many such methods followed; they all apply criteria consistent for all proteins but they often differ between each other.

2.1.3 Automatic Assignments of Secondary Structure

The first assignments of protein secondary structure were carried out by Pauling and others [126] even before experimental 3-D structures of proteins became available. They were based on intra-backbone hydrogen bonds. One of the first and most popular automatic methods, DSSP [76], used a similar approach. The DSSP method calculates the interaction energy between backbone atoms based on an electrostatic model [76]. It assigns a hydrogen bond if the interaction energy is below a chosen threshold (–0.5 kcal mol^{-1}). The structure assignments are defined such that visually appealing and unbroken structures are formed from groups of hydrogen bonds. Another popular automatic assignment method, the STRuctural IDEntification method (STRIDE [51]) uses ϕ–ψ torsion angles and empirically derived hydrogen bond energy. The parameters used by this method are optimized to reproduce visual assignments provided by experimentalists determining 3-D structures and so in effect the method averages out human bias. The method DEFINE [143] assigns secondary structure using C_α coordinates. The assignment is carried out through comparison of observed C_α distances with those derived from ideal secondary structures. If the distances are within set discrepancy limits, then the secondary structure is assigned. The method P-Curve [173] makes assignments based on geometrical analysis of protein curvature. It uses differential geometry-based representations of standard structural motifs and through a set of geometrical transformations tries to match these motifs with those found in known 3-D structures. P-Curve assignments differ significantly from those based on hydrogen bonds and/or ϕ–ψ torsion angles. DSSP, P-Curve and Define assignment methods agree for only about two-thirds of all residues [30]. There are various reasons for disagreements; the most important one may simply be that secondary structure is dynamic, i.e. that there simply is no such thing as a secondary structure "state". This problem is reflected in the DSSPcont method that introduces continuous secondary structure assignments [8]. The continuum results from calculations of weighted averages of DSSP assignments that are based on various hydrogen bond energy thresholds. As a result, each protein residue is assigned with likelihoods of all secondary structure states. Residues that have a higher probability for a single "state" appear to also be more rigid according to NMR measurements of motions on timescales important for protein function [8]. Other, more application-oriented approaches to defining local structures are possible. For example, one may try to define a new secondary structure alphabet with a goal of improving fold recognition algorithms [78]. The numerical values of prediction accuracy presented in this chapter are based on the most widely used DSSP assignment. Evaluations based on STRIDE tend to yield higher values and no state-of-the-art prediction method has been evaluated on P-Curve.

2.1.4 Reduction to Three Secondary Structure States

DSSP distinguishes eight different "states": three types of helical structures [α-helix ("H", four-residue period), π-helix ("I", five-residue period) and 3_{10}-helix ("G", 3-residue period)], extended β-sheet ("E"), β-bridge ("B"), turn ("T"), bend ("S") and other nonregular states (blank). Of those, α-helix and β-strand (Figure 1) comprise more than 50% of all protein residues. Some prediction methods attempt to predict all eight states. However, a widely used strategy is to map the eight "states" into three major "classes": helical, extended and other (often imprecisely referred to as "nonregular", "coil"

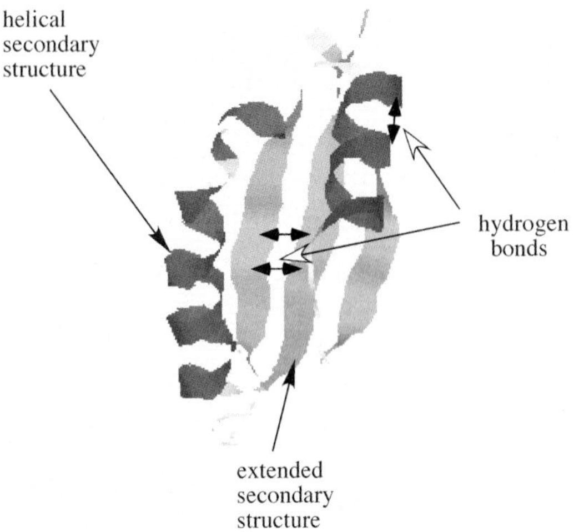

Figure 1 Ribbon diagram of protein secondary structure. Secondary structures are local arrangements of the protein backbone without reference to the amino acid type or the 3-D conformation. They are stabilized by hydrogen bonds between atoms of the main chain (backbone). Very roughly, secondary structures can be classified into three classes: helical (H), extended (E) (strand) and loopy (other) L. The figure contains a schematic representation of the E2 DNA-binding domain [21] (Protein Data Bank [17] code 1a7g).

or "turn"). Different maps are possible, but the most popular one (which incidentally is most difficult to predict [35]) is the following: [GHI] = helical ("h"), [EB] = extended ("e") and [TS] = nonregular (coil) ("l"). The alternative translation that results in seemingly higher prediction accuracies, i.e. [H] = helical, [E] = extended and [GITS] = nonregular, is sometimes used.

2.2 Measuring Performance

2.2.1 Performance has Many Aspects Relating to Many Different Measures

Depending on the application there are various views as to what constitutes a high-quality prediction. On the one hand, it is important to correctly predict the secondary structure "state" for each residue (per-residue accuracy); on the other hand, it may be more relevant to predict the coarse-grained presence of, for example, a helix than all residues in the helix (segment-based accuracy). Accordingly, many measures have been used to assess prediction quality: simple percentages of per-residue accuracy (Eq. 1), Matthew's correlation coefficients, percentage of confusion between strand and helix states [38] (Eq. 2); simple segment-based measures such as the number of correctly predicted segments, the average ratio of predicted to observed segment lengths, the difference between the distribution of predicted and observed segment lengths [156]; or the more elaborated and widely used segment overlap score *SOV* [160, 187] (Eq. 3). These are only some of the measures that have been applied. In this chapter, we focus on two measures for per-residue accuracy, i.e. percentages Q_K (Eq. 1) and the *BAD* score (Eq. 2), and one measure for per-segment accuracy, i.e. *SOV*.

2.2.2 Per-residue Percentage Accuracy: Q_K

Perhaps the most intuitive and simplest measure for performance is the average percentage of correctly predicted states. For a protein composed of L residues and for K possible secondary structure states the per-residue prediction accuracy Q_K is defined as:

$$Q_K = 100 \times \sum_{i=1}^{K} C_i / L \tag{1}$$

where C_i is the number of residues correctly predicted in secondary structure state i. For a three-state alphabet this translates into a Q_3 measure. The average accuracy can be computed as an average per protein or an average per residue in which case the number of all residues is used for L.

2.2.3 Per-residue Confusion between Regular Elements: *BAD*

Not all secondary structure prediction mistakes are equal. For instance, when using secondary structure predictions to model 3-D structure, confusing helix and extended (strand) is more detrimental than confusing regular with non-regular states. The percentage of such "bad" predictions constitutes the *BAD* score. If L is a total number of amino acid residues in a protein and Bh (Be) is the number of helical (strand) residues predicted in strand (helix) state, then the *BAD* score is expressed as:

$$BAD = 100 * \frac{Bh + Be}{L} \,. \tag{2}$$

Two predictions with equal Q_3 and/or *SOV* scores can have very different *BAD* scores.

2.2.4 Per-segment Prediction Accuracy: *SOV*

Regular secondary structure elements are built of continuous stretches of residues belonging to the same state, e.g. most helices are about 10 residues long. It can be argued that mis-predicting two residues at either end of a helix is not an important mistake (note: 2 + 2 out of 10 means 60% accuracy). In contrast, only predicting 60% of the helices in a protein is a severe problem. Such realities are reflected in segment-based measures. The most widely used is the segment overlap (*SOV*) measure [160,187]:

$$SOV = 100 \times \frac{1}{N} \sum_{i}^{K} \sum_{S(i)} \frac{minov(s_{obs}, s_{pred}) + \delta(s_{obs}, s_{pred})}{\mathrm{maxov}(s_{obs} s_{pred})} \times len(s_{obs}) \tag{3}$$

where K is the number of different secondary structure types; the second summation is over all overlapping secondary structure segments of observed s_{obs} and predicted s_{pred} secondary structure of the same type; *minov* is the number of positions at which segments overlap; *maxov* is the number of overlapping positions plus the number of remaining residues from each segment of the given pair; $len(s_{obs})$ is the length of a reference secondary structure segment (observed experimentally); N is the total number of overlapping segments pairs of the same type; and $\delta(s_{obs}, s_{pred})$ is the accepted variation between segments that assures ratio of 1.0 when the variations between s_{obs} and s_{pred} are minor. One can easily envision two different secondary structure predictions that have the same Q_3 and different *SOV* scores. For example, if instead of a observed long helix of length n one prediction consists of a shorter helix of length m and the second prediction comprises two short helices of combined length equal to m (other residues predicted as coil), then the Q_3 scores of both predictions are going to be the same while the *SOV* scores are going to be different.

2.3 Comparing Different Methods

2.3.1 Generic Problems

In this section we describe problems with the evaluation of prediction methods that are entirely generic, i.e. valid for all prediction methods. Although many ideas and concepts have been introduced to predict secondary structure and have then been used for other purposes, many of the mistakes in comparing methods have also been unraveled first and most clearly for the example of secondary structure predictions. Secondary structure prediction methods may be the only example of publications with claims to performance accuracy that survived more than a decade. (To put this into perspective: our section focuses on methods for which performance has, on average, been unusually well estimated; nevertheless, the only other field that we review for which *any* estimate survived 5 years was the prediction of solvent accessibility and the vast majority of publications in that field heavily overestimated performance!)

2.3.2 Numbers can often not be Compared between Two Different Publications

Prediction methods are often published with estimates of performance that are supported by cross-validation experiments. However, the terms "cross-validation" or the related term "jackknife" are by no means sufficiently well-defined to translate into "estimate ok". In fact, most publications make some serious mistakes as is demonstrated by the simple fact that very few estimates of performance have survived. One problem is the overlap between "training" and "testing" sets. It is trivial to reach very high performance by training on proteins that are very similar to those in the testing set. There are various strategies that deal with the similarity problem [67,188]. Another issue is that of using the performance of the test set to choose some parameters by, for example, reporting full cross-validation results for N different parameters and then concluding that the best of those N is the performance of the final method. Instead, performance estimates should always be based on a data set that was not used in ANY step of the development. However, even if we had two publications that both used cross-validation "correctly", we still cannot necessarily compare the numbers published by both directly. First, both have to have used the same standard of truth (here, the same assignment method, e.g. DSSP, and the same conversion of the eight DSSP states into three prediction classes). Second, they both have to have been based on identical test sets. Often, the test sets used by developers are not representative and differ from each other. Proteins vary in their structural complexity and such variation is correlated with prediction difficulty. We could argue that test sets should be frozen (and this has indeed been done in many cases). Such a set should be sufficiently large to allow proper evaluation of statistical differences among methods. Although a *sine qua non*, this freezing strategy does not suffice – data sets in biology change constantly, almost always more recent sets are more reliable and representative. Therefore, we also need evaluations based on sets that are as recent as possible. One way of merging these two demands is by carrying out two tests: one on a frozen set used by others and the other on a more recent set. As an aside, it is not necessary to use n-fold cross-validation experiments with the largest possible n. The exact value of n is not important as long as the test set is not misused for adjusting a method's parameters and it is representative of the entire structure space.

2.3.3 Appropriate Comparisons of Methods Require Large, "Blind" Data Sets

One of the solutions to the problem of comparing methods is to use a sufficiently large test set composed of proteins that were neither used nor are similar to any protein that was used for development of any method. This idea was first realized in the field of structure prediction through the Critical Assessment of Structure Prediction (CASP) experiments in which various prediction methods are tested over the course of a few months on sequences of proteins the 3-D structure of which is unknown at the time of the prediction ("blind" prediction). Those experiments evaluate fully automatic methods as well as human experts (see also Chapter 11 for a more detailed description of CASP and CAFASP). Due to a variety of reasons, CASP cannot be based on sufficiently large, representative data sets. Servers that automatically evaluate methods whenever new data is available address this shortcoming. Such servers base their comparisons on thousands instead of tens of test cases (as does CASP). Two such servers exist: EVA and LiveBench. EVA

[44] continuously evaluates automatic prediction methods (servers) providing results based on a large, statistically significant and, subsequently, more representative data sets. One of its principles is to facilitate comparisons on identical sets and to render comparisons on different sets very difficult. Another principle is to never distinguish in the rank between two methods if the difference in their performance is not statistically significant. Both principles are in stark contrast to what most CASP assessors did.

2.4 History

2.4.1 First Generation: Single-residue Statistics

First attempts to correlate amino acid residue frequency with secondary structure type can be traced to correlating the content of certain amino acids (e.g. proline) with the content of α-helix [176]. This was done even before the first crystallographic structures were available [81, 127]. Attempts to correlate the content of all amino acids with the content of α-helix and β-strand opened the field of secondary structure prediction [19, 20]. The early methods were usually based on single-residue statistics obtained from very limited data sets of known protein structures. As such they were not very accurate (Figure 2) and in addition their accuracy was overestimated at the time.

2.4.2 Second Generation: Segment Statistics

As the number of experimentally determined protein structures grew it became possible to estimate propensities for secondary structure based on consecutive segments of residues. Various numbers of adjacent residues (typically 11–21) were considered in assigning secondary structure to a central residue of a segment. Many different algorithms were applied, but they did not achieve per-residue prediction accuracies higher than slightly above 60% (Figure 2). Reports of higher accuracies were due to small data sets and did not hold for long. The main approaches used were (i) statistical information, (ii) physicochemical properties, (iii) sequence patterns, (iv) artificial neural networks, (v) graph theory, (vi) expert rules, (vii) nearest-neighbor algorithms and (viii) hybrid approaches of various algorithms.

2.4.3 Third Generation: Evolutionary Information

Proteins with similar sequences adopt similar structures [27, 166]. In fact, proteins can change more than 70% of their residues without altering the basic fold [1, 15, 125, 189]. However, the vast majority of possible sequences supposedly do not adopt globular structures at all. Rather, the exact substitution pattern of which residues can be changed and how is indicative of particular structural details. Consequently, the evolutionary information

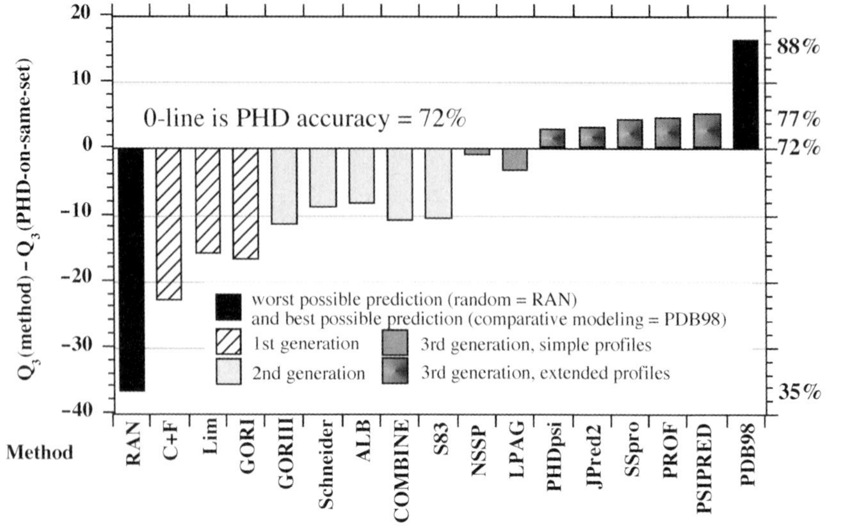

Figure 2 Three-state per-residue accuracy of various prediction methods. Included are only those methods for which we could run independent tests. Unfortunately, for most old methods this was not possible. However, for each method we had independent results from PHD (third generation, 1993) [151,154,159] available. We normalized the differences between data set by simply compiling levels of accuracy with respect to PHD. For comparison, we added the expected accuracy of a random prediction (RAN), and the best currently possible prediction accuracy achieved through comparative modeling of close homolog (PDB98). The methods were: C+F (Chou and Fasman; first generation, 1974) [28,29]; Lim (first, 1974) [93]; GORI (first, 1978) [53]; Schneider (second, 1989) [169]; ALB (second, 1983) [140]; GORIII (second, 1987) [57]; COMBINE (second, 1996) [52]; S83 (second, 1983) [77]; LPAG (third, 1993) [92]; NSSP (third, 1994) [175]; PHDpsi (third, 2001) [137]; JPred2 (third, 2000) [34]; SSpro (third, 1999) [12]; PROF (third, 2001) [149]; PSIPRED (third, 1999) [73].

contained in sequence alignments can aid structure prediction. In particular this approach improves prediction of β-strands. For the first and second generation of prediction methods β-strand prediction was particularly bad (often only slightly better than random). The pioneering method that used alignment information was proposed in the 1970s [41]. The first approaches were based on visual gathering of information from sequence alignments. In one of the first automatic algorithms making use of alignment information [107,189] the final secondary structure prediction was an average over all predictions compiled for each sequence in the alignment. The first method that succeeded in significantly improving performance by automatically using alignment information was PHD [151, 154, 157] (Figure 3). This method used a residue profile extracted from a multiple sequence alignment as an input to the artificial neural network. Many other methods used artificial neural networks [73, 123, 133], but various other algorithms were also applied successfully [38,

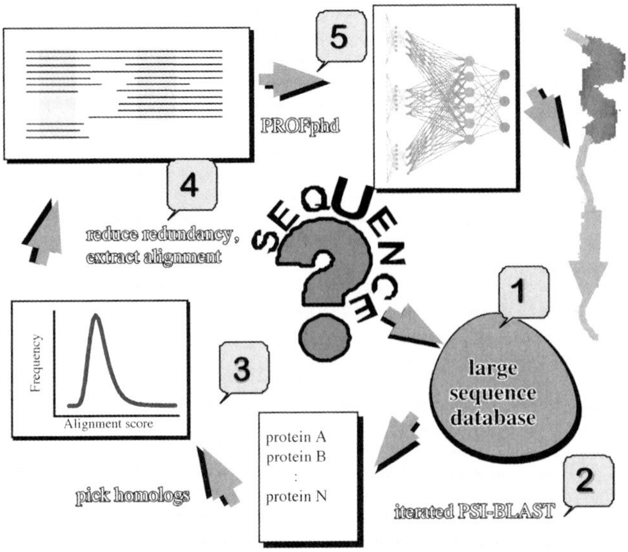

Figure 3 Using evolutionary information to predict secondary structure. Starting from a sequence of unknown structure (SEQUENCE) the following steps are required to feed evolutionary information into the PROFsec neural networks (upper right): (1 and 2) a database search for homologs through iterated PSI-BLAST [6,7] (protocol from Ref. [137]), (3) a decision for which proteins will be considered as homologs, (4) a reduction of redundancy (purge too many too similar proteins), and (5) a final refinement and extraction of the resulting multiple alignment. Numbers 1–5 illustrate where users of the PredictProtein server [151,161] can interact to improve prediction accuracy without changes made to the actual prediction method PROFsec.

39, 51, 91, 109, 146, 163] including support vector machines (SVMs) [68,181], hidden Markov models (HMMs) [79], nearest-neighbor algorithms [163].

2.4.4 Recent Improvements of Third-generation Methods

PHD tore down what once was a magical wall of 70% accuracy. The mark has been put much higher since. The first significant improvement was achieved by training neural networks on more diverse sequence alignments [73]. The alignments were generated by a new alignment method – PSI-BLAST [7]. It has been shown that a major improvement can be achieved by using previous types of neural networks with PSI-BLAST alignments [34]. Interestingly, it was also shown that a significant part of the improvement was simply due to the growth of sequence databases that resulted in more diverse profiles [137]. In general, the more divergent the alignment the better the prediction can be obtained. The input quality is also dependent on alignment quality. This is especially important for divergent homologous proteins where alignment methods tend to make many mistakes. Yet another simple source of improvement is related to the growth of the database of protein structures [17]. Apart from improvements in alignments, there is a lot of research pursuing development of more sophisticated and accurate algorithms. Those include new network architectures or learning techniques [3, 12, 78, 132, 133], SVMs [181] and many others.

2.4.5 Meta-predictors Improve Somehow

Different methods often make different mistakes. As long as those errors are not purely systematic, combining any number of methods can lead to improvements in prediction accuracy [62]. For example, the PHD method utilized this observation by combining differently trained neural networks. Various implementations of the similar concept were used in many other methods [24, 34, 128]. Alternatively, or in addition, different methods can be combined [5, 35, 36, 60, 83, 158, 170]. Overall, combinations of independent methods tend to top the single best method. However, it probably is not beneficial to use all of the available prediction methods in the meta-methods. For example, averaging over all methods evaluated by EVA evaluation server [44, 46] decreased accuracy over the best individual methods (Rost, unpublished). It is not fully straightforward how to decide whether to include a given method or not [5]. Concepts weighing the individual method based on its accuracy and "entropy" [128] appear to be successful only for large numbers of methods. More rigorous studies for the optimal combination may provide a better picture. An interesting approach resulted from attempts to improve meta-methods by developing new methods that are algorithmically different from the methods already used [85, 171]. Recently, an observation has been made indicating that optimizing meta-servers to achieve highest per-residue prediction accuracy is not always beneficial when using the final predictions in various applications [108]. Another issue that has first been introduced for secondary structure prediction is the measurement for the reliability of a prediction. To make an extreme point: a method that has 50% accuracy, but that always correctly identifies in which of the cases it is right and it which it errs (before knowing the answer), is more useful than a method with 75% accuracy and no notion about which 25% of the residues are wrong. State-of-the-art methods reliably estimate the reliability of a prediction. This is not the case for any of the existing meta-methods.

2.5 State-of-the-art Performance

2.5.1 Average Predictions Have Good Quality

Today's best methods reach average levels of almost 78% in Q_3 (Eq. 1) [44, 86]. They are able to accurately predict most segments (*SOV* scores around 76%).

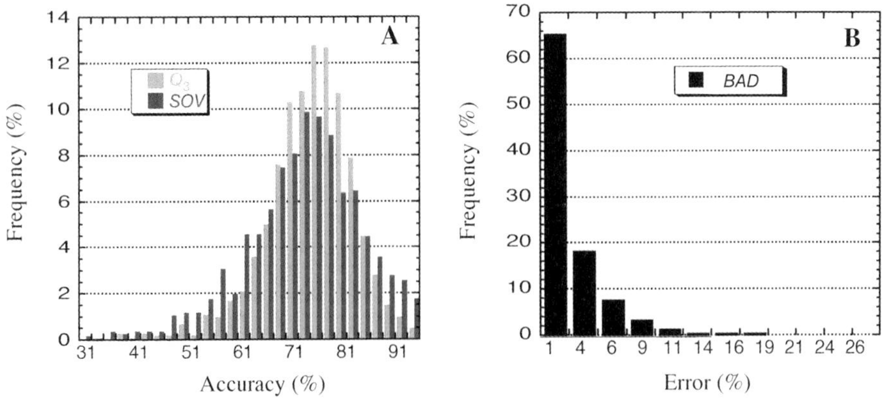

Figure 4 Expected variation of prediction accuracy for PROFsec. (A) Three-state per-residue (Q_3) and segment overlap (*SOV*) accuracies. (B) Percentage of *BAD* predictions, i.e. residues either predicted in helix and observed in strand or predicted in strand and observed in helix.

In addition the confusion between helices and strands is low (*BAD* score of less than 3%).

2.5.2 Prediction Accuracy Varies among Proteins

The standard deviation of three-state-per-residue accuracy computed on the per-protein basis is about 13% [44, 86] (Figure 4). Thus, some of the proteins are predicted very well (above 90%), while others are predicted very badly (even below 40% accuracy levels). The relatively large deviations are also found in prediction quality measured by other measures. The standard deviation of the *SOV* score is about 15% and that of the *BAD* score is about 5%. In particular, proteins having no sequence homologs (no alignment input) are poorly predicted. This is an important issue for the applicability of secondary structure predictions since badly predicted secondary structure is not very valuable.

2.5.3 Reliability of Prediction Correlates with Accuracy

For the user interested in a particular protein *U*, the fact that the prediction accuracy varies from protein to protein implies a rather unfortunate message: the accuracy for *U* could be lower than 40% or it could be higher than 90% (Figure 4). Is there any way to provide an estimate at which end of the distribution the accuracy for *U* is likely to be? Indeed, many methods provide numerical estimates of the expected quality of their predictions through so called reliability indices. Those indices correlate with accuracy. In other words, residues with higher reliability index are predicted with higher ac-

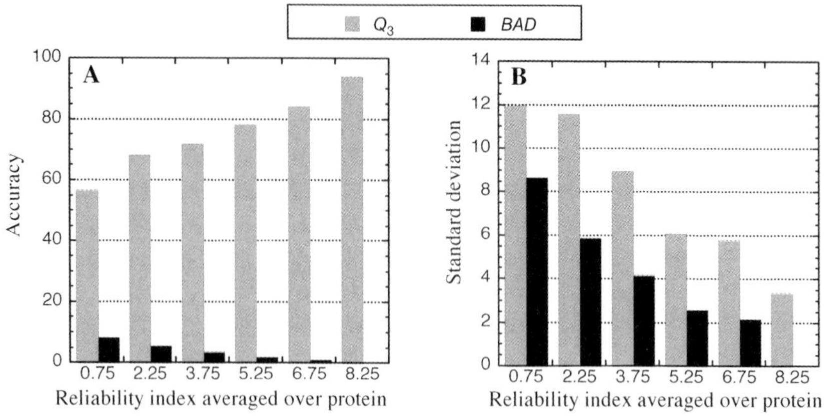

Figure 5 Prediction quality correlates with reliability indices. (A) Average three-state per-residue accuracy and *BAD* score at different reliability index thresholds (averaged over entire protein) as predicted by PROFsec [149]. (B) Corresponding values of standard deviation.

curacy [151,154,157]. Thus, the reliability index offers an excellent tool to focus on some key regions predicted at high levels of expected accuracy. Furthermore, the reliability index averaged over an entire protein correlates with the overall prediction accuracy for this protein (Figure 5).

2.5.4 Understandable Why Certain Proteins Predicted Poorly?

It is not easy to anticipate performance of a secondary structure prediction method based on overall structural features of proteins. However, prediction accuracy is correlated with alignment quality. Poor alignments (i.e. noninformative and/or falsely aligned residues) result in inaccurate predictions. Another interesting observation is that frequently the *BAD* predictions, i.e. the confusion between helix and strand are observed in regions that are stabilized by long-range interactions. Furthermore, helices and strands that are confused despite a high reliability index often have functional properties or are correlated to disease states (Rost, unpublished data). Regions predicted with equal propensity in two different states often correlate with "structural switches".

2.6 Applications

2.6.1 Better Database Searches

Initially, three groups independently applied secondary structure predictions for fold recognition, i.e. the detection of structural similarities between proteins of unrelated sequences [50,152,162]. A few years later, almost every other fold recognition/threading method has adopted this concept [10, 37, 40, 63, 72, 74, 80, 87, 122, 124]. Two recent methods extended the concept by not only refining the database search, but by actually refining the quality of the alignment through an iterative procedure [65,71]. A related strategy has been employed to improve predictions and alignments for membrane proteins [117]. It has also been indicated that prediction mistakes tend to correlate among structurally related proteins [138], and that alignments based on purely predicted secondary structure have comparable quality with those based on matching predicted and observed states. Thus predicted secondary structure may prove useful in searching sequence databases.

2.6.2 One-dimensional Predictions Assist in the Prediction of Higher-dimensional Structure

Secondary structure predictions are now accurate enough to be used as input for methods that target the prediction of higher order aspects of protein structure automatically. A few successful applications include the following. Contact map predictions [13] have recently improved the level of accuracy significantly; an important contribution was the inclusion of secondary structure predictions [141]. They also help in the prediction of folding rates [69,142]. Secondary structure predictions have also become a popular first step toward predicting 3-D structure. Ortiz and coworkers [121] successfully use secondary structure predictions as one component of their 3-D structure prediction method. Eyrich and coworkers [47,48] minimized the energy of arranging predicted rigid secondary structure segments. Lomize and coworkers [103] also started from secondary structure segments. Chen and coworkers [25] suggested using secondary structure predictions to reduce the complexity of molecular dynamics simulations. Levitt and coworkers [164,165] combined secondary structure-based simplified presentations with a particular lattice simulation attempting to enumerate all possible folds.

2.6.3 Predicted Secondary Structure Helps Annotating Function

Secondary structure predictions are also useful to annotate/predict protein function. For example, secondary structure predictions have been used successfully in completely automatic predictions of subcellular localization [116]. A more typical use of secondary structure prediction is in aiding experts in finding similarities among proteins with insignificant sequence similarity. In this way functional annotation is sometimes transferred from one protein to another [184].

2.6.4 Secondary Structure-based Classifications in the Context of Genome Analysis

Proteins can be classified into families based on predicted and observed secondary structure [56, 139]. However, such procedures have been limited to a very coarse-grained grouping only sometimes useful for inferring function. Nevertheless, predictions of membrane helices and coiled-coil regions are crucial for genome analysis. More than one fifth of all eukaryotic proteins appear to have regions longer than 60 residues apparently lacking any regular secondary structure [102]. Most of these regions were not of low complexity, i.e. not composition biased. Surprisingly, these regions appeared evolutionarily as conserved as all other regions in the respective proteins. This application of secondary structure prediction may aid in classifying proteins, and in separating domains, possibly even in identifying particular functional motifs.

2.6.5 Regions Likely to Undergo Structural Change Predicted Successfully

Prions and prion-like proteins appear to aggregate through the transition of a regular secondary structure: what is "usually" a helical region switches to a strand that becomes the root of aggregation in the case of disease mutants. The reliability of the PHD secondary structure predictions combined with experimental evidence gave the first hint where this expected transition might occur [136]. Interestingly, it is still difficult to actually observe the strand in structures of even the mutant prion, while state-of-the-art prediction methods always predict the region with an observed helix to be in a strand. This example casts some light on the importance of transitions and the usefulness of predictions to capture such transitions. Young and coworkers [84] have pushed this observation further by unraveling an impressive correlation between local secondary structure predictions and global conditions. The authors monitor regions for which secondary structure prediction methods give equally strong preferences for two different states. Such regions are processed combining simple statistics and expert rules. The final method has been tested on 16 proteins known to undergo structural rearrangements and on a number of other proteins (one of those was a prion). The authors report no false positives and identify most known structural switches. Subsequently, the group applied the method to the myosin family identifying putative switching regions that were not known before, but appeared reasonable candidates [84]. This method is remarkable in two ways: (i) it is a very general method using predictions of protein structure to predict some aspects of function and (ii) it illustrates that predictions may be useful even when structures are known (as in the case of the myosin family). While the method is tailored to catch more subtle changes than occur in prions, there is some evidence that amyloid aggregation is also captured to some extent.

2.7 Things to Remember when using Predictions

2.7.1 Special Classes of Proteins

Prediction methods are usually derived from knowledge contained in proteins from subsets of current databases. Consequently, they should not be applied to classes of proteins not included in these subsets, e.g. methods for predicting helices in globular proteins are likely to fail when applied to predict transmembrane helices. In general, results should be taken with caution for proteins with unusual features, such as proline-rich regions, unusually many cysteine bonds or for domain interfaces.

2.7.2 Better Alignments Yield Better Predictions

Multiple alignment-based predictions are substantially more accurate than single sequence-based predictions [14, 39, 151]. How many sequences are needed in the alignment for an improvement; and how sensitive are prediction methods to errors in the alignment? The more sequences contained in the alignment diverge, the better (two distantly related sequences often improve secondary structure predictions by several percentage points). Regions with few aligned sequences yield less reliable predictions. The sensitivity to alignment errors depends on the methods, e.g. secondary structure prediction is less sensitive to alignment errors than solvent accessibility prediction.

2.8 Resources

2.8.1 Internet Services are Widely Available

Programs for the prediction of secondary structure available as Internet services have mushroomed since the first prediction service PredictProtein went on line in 1992 [159, 161]. The META-PredictProtein server [45] enables users to access a number of the best prediction methods through one single interface. Unfortunately, not all methods available have been sufficiently tested and some are not very accurate. This problem is addressed by the EVA server that evaluates prediction servers continuously and automatically [44, 86].

2.8.2 Interactive Services

The PHD/PROF prediction methods are automatically available via the Internet service PredictProtein [45]. Users have the choice between the fully automatic procedure taking the query sequence through the entire cycle or expert intervention into the generation of the alignment. Indeed, without spending much time users typically can improve prediction accuracy easily by choosing "good" alignments. A few of the state-of-the-art methods are also available to run locally. Note, however, that one crucial step is the generation of appropriate alignments; usually this is not "done for you" when you run the prediction method locally!

2.8.3 Servers

The following servers are publicly available (most links given by the EVA server): PROFsec [149], PHDsec [159], PHDpsi [137], PSIPRED [73], SSPRO [133], PORTER [132], SABLE [3], SAM-T02 [79], Jpred [34], APSSP, JUFO [110], PROF [123], YASPIN [94].

3 Transmembrane Regions

3.1 Transmembrane Proteins are an Extremely Important Class of Proteins

Approximately 15–30% of all proteins are estimated to contain transmembrane regions [97, 111]. Those proteins are responsible for the communication between the cell and its surroundings, and are of great importance to biomedicine. The cell membrane environment, composed of a lipid bilayer, is very different from one found in most cellular compartments. The transmembrane segments of proteins tend to be hydrophobic which enables them to remain within a membrane by avoiding the solvent present at both boundaries. The special features of transmembrane protein sequences serve as the basis for identifying them by computational methods. As in case of globular proteins, the transmembrane segments form regular secondary structures and can be assigned to two broad classes: those composed entirely of helices and those composed of strands (despite ardent searches and putative evidence, we still do not have any proof for the existence of a mixture of the two). By far the majority of all membrane proteins appear to be of the helical type [18]. An important characteristic of transmembrane proteins is the orientation of membrane segments with respect to the N-terminus of a protein, often referred to as the topology. Usually, the successful prediction of transmembrane segments requires proper identification of transmembrane regions in sequence, actual prediction of the secondary structure and deciphering the topology. It is very difficult to experimentally determine 3-D structures for transmembrane proteins. Despite considerable advances over the last decade, we still have experimental structures or theoretical models for supposedly less than 10% of, for example, all human membrane proteins (Punta, Liu and Rost, unpublished). Useful predictions of structural and functional aspects are therefore highly needed.

3.2 Prediction Methods

Although all known transmembrane regions constitute of regular secondary structures, most secondary structure prediction methods developed for non-membrane proteins mostly fail to correctly predict membrane regions. Furthermore, very few methods have been developed for proteins with β-strands in the membrane. The first and most basic methods for helical membrane regions focused on identification of transmembrane segments based simply on residue hydrophobicity [90]. It was observed that positively charged residues are more abundant on the inside of the membrane (the "positive-in" rule). A simple Kyte–Doolittle hydrophobicity plot [90] can thus provide much information on the presence of such transmembrane segments. This led to the development of the method that predicted positions of helices and the topology of helical membrane proteins [179]. Next, neural networks were applied to better identify transmembrane helices, and differentiate between membrane and nonmembrane proteins [153]. Among other approaches were HMM methods attempting to match the sequence to the predefined "grammar" of transmembrane proteins [88] (see Chapter 3 for basics on HMMs) and many others [33,66]. Recently, groups have begun to venture into the development of methods that predict membrane regions with β-strands [18, 42, 59, 70].

3.3 Performance

Estimates for the accuracy in predicting membrane regions are extremely problematic because there are so few high-resolution structures available. Consequently, all methods in the past were evaluated by also using low-resolution information from biochemical experiments that provide some evidence for the location of transmembrane regions. Unfortunately, such experiments can be more inaccurate than prediction methods [26]. This was one of the reasons why the performance of prediction methods had been significantly overestimated by the end of the last millennium [26, 113]. It now appears that the best prediction methods correctly predict all membrane helices for about 50–70% of all proteins, very few methods avoid the confusion between very hydrophobic signal peptides and membrane proteins, and the best methods falsely identify membrane helices in about 10% of all nonmembrane proteins [26, 113]. However, results can be far worse, e.g. most hydrophobicity-based methods misclassify over 50% (!) of all globular proteins as "containing membrane helices" [26]. Overestimates in publications are also a very serious problem – even over the last few years, methods have been published in prominent journals with estimated levels of above 95% accuracy that failed to reach significantly above 50% and misclassified over 30% of the globular proteins [26]. Note also that there are a few top methods available at the moment; all of these have their own strengths and weaknesses, i.e. there is no single one "best method". Predictions of β-barrel membrane regions currently appear to be more accurate than those for helical membrane regions; however, this may likely turn out to be an overestimate caused by the fact that we have too limited experimental information.

3.4 Servers

There are many more methods than the following available; however, the methods listed here have sustained many evaluations. Helical membrane proteins: PHDhtm [153], SOSUI [66], TopPred [179], TMHMM [88], DAS [33]; β-barrel membrane regions: ProfTMB [18].

4 Solvent Accessibility

4.1 Solvent Accessibility Somehow Distinguishes Structurally Important from Functionally Important

In 3-D structures of globular proteins some of residues are buried deep inside, whereas others are located on the surface and thus are more exposed to the surrounding solvent. Residues that are more exposed to solvent are also more accessible to other biological agents and, consequently, are much more likely to be involved in functional interactions which require spatial accessibility such as enzymatic activity, DNA binding, signal transduction, etc. However, buried residues are much more likely to play important roles in stabilizing structures of proteins. Thus, a good distinction between exposed and buried residues can be very useful to distinguish residues that are important for function (conserved and exposed) from those that are important for structure (conserved and buried).

4.2 Measuring Solvent Accessibility

Solvent accessibility is usually measured in terms of the surface area accessible to water molecules. The values can range from 0 Å for entirely buried residues to around 300 Å for the largest residues on the surfaces of proteins. A measure that is not dependent on the size of the amino acid residue is the relative solvent accessibility expressed as a percentage of the residue surface that is exposed to solvent. It appears that among homologous proteins the relative solvent accessibility is less conserved than secondary structures [155]. In addition, the solvent accessibility of protein residues is strongly influenced by nonlocal interactions, where residues located far away along a protein sequence can be in spatial proximity resulting in mutual screening from solvent. Thus, predicting solvent accessibility appears to be more difficult than prediction of secondary structure. It addition, it was shown that among the evolutionarily related proteins of similar structure buried residues (less than 10% accessible surface area) tend to be much more highly conserved than highly exposed residues (more than 60%) [155]. Thus, for methods that use evolutionary information derived from alignments of related proteins it should be easier to closely predict accessibility for buried residues than for the exposed ones. A simplified approach is to try to distinguish between residues below a certain solvent accessibility threshold ("buried") and those above it ("exposed"). There is no biophysical reason to choose one threshold over another, and different researchers often choose different thresholds (7, 9, 16 and 25% are used). On average, about half of all protein residues have more than 25% of their surfaces exposed.

4.3 Best Methods Combine Evolutionary Information with Machine Learning

Some of the methods that predict secondary structure also have the capability of predicting solvent accessibility, since essentially the same basic concepts apply to building a solvent accessibility predictor. For example, PHDacc [155] and PROFacc [149] methods, which are part of the PredictProtein [159, 161] server, use the same sequence profile input as do their respective secondary structure prediction counterparts (PHDsec and PROFsec). They use a neural network that assigns relative solvent accessibility into one of the 10 states corresponding to squares of relative solvent accessibility (state 10 corresponds to a range 81–100% of solvent accessibility). This 10-state scheme can be converted to a two-state scheme or to a prediction in terms of actual value of the exposed surface. Another well known method is Jpred [36]. It is also a server that predicts both secondary structure and solvent accessibility. The method uses alignments generated by HMMs and PSI-BLAST as input to a neural network. The output of predictions from two different networks is combined to give a final relative solvent accessibility. Many other variations and similar approaches have been attempted which include various types of neural networks [2, 4, 34, 131], SVMs [82], Bayesian networks [177], information-theoretic approaches [115] and simple baseline approaches [144]. Most recently the relation between secondary structure and accessibility was explored to develop methods that combine both predictions explicitly to improve each one [2, 149].

4.4 Performance

Unlike the prediction of secondary structure that is continuously assessed and monitored on identical data sets, methods for the prediction of solvent accessibility are not. Given that different groups use widely different data sets and different conventions to convert actual values of solvent accessibility into prediction states, it is impossible to compare and reasonably summarize levels of performance. However, two-state predictions (either buried or exposed) are predicted at levels above 75% accuracy. Whatever values you read, note that advanced methods are significantly more accurate than simple methods based on simple features such as hydrophobicity, polarity or simple statistics.

4.5 Servers

PROFacc [149], PHDacc [155], SABLE [2], Jpred [34], ACCpro [131].

5 Inter-residue Contacts

5.1 Two-dimensional Predictions may be a Step Toward 3-D Structures

Directly predicting 3-D structure still fails. Predictions of 1-D aspects of protein structure, such as secondary structure and solvent accessibility, provide very valuable information. However, 1-D predictions are far too simplified. There is a path seemingly in between these two extremes (1-D/3-D), i.e. the prediction of inter-residue distances. In fact, 3-D structures can be reconstructed more or less completely from 2-D distance maps. The catch is that distance maps are as hard to guess as 3-D coordinates. As a consequence, existing methods try to solve the simplified problem of predicting contact maps, where two residues are considered to be in contact if they are located within a certain spatial cut-off distance (this results in a binary classification of residue pairs, i.e. contact/noncontact pairs).

5.2 Measuring Performance

There is no widely accepted threshold for the maximal distance between two residues that are considered as "in contact". While the smallest physically possible distance could be agreed upon, the limit beyond which the interaction between two residues can be considered negligible is more difficult to define. However, the distance of 8 Å between C_β atoms is the most widely used threshold for the evaluation of the performance of these prediction methods. The output of contact prediction programs is generally a list of residue pairs, ranked according to some internal confidence score. Usually, only contacts between pairs that exceed a minimal sequence separation are evaluated. Although many different thresholds have been used, minimal separations of six and 24 sequence positions are most common for prediction of medium- and long-range contacts, respectively. These parameters are important as the task becomes more difficult with increasing separation (this tendency levels off for separations over 20).

5.3 Prediction Methods

One line of methods was based upon the observation that evolutionary pressure on maintaining protein structure would sometimes require correlation in the mutations of amino acid residues that are in spatial proximity to each other. In principle, such patterns of correlation could be discerned in the multiple alignments of protein sequences. Some of the early contact prediction methods have indeed used only correlated mutations computed from multiple sequence alignments [58, 119]. The currently best methods make also use of other protein features, such as evolutionary profiles of the nearest neighbors of the residue pair being predicted, sequence separation, secondary structure and solvent accessibility predictions. Further improvement of predictions was achieved through machine learning techniques such as: neural networks that use [49, 61, 120] or do not use [130, 141] correlated mutations, HMMs [22, 129, 172], SVMs [188] and genetic programming [104].

5.4 Performance and Applications

As the prediction of nonlocal contacts is difficult, progress in the field had been slow until recently when two promising new methods entered the CASP6 competition in 2004. When $L/2$ predictions are considered, the accuracy of state-of-the-art methods is around 30% for sequence separation of at least six and around 20% for sequence separation of at least 24. Although predicted contact maps are not very accurate, they are nevertheless better on average than the contact maps obtained from the best *de novo* predictions of 3-D structures [46]. As a result, the automatically predicted contact maps were successfully used in prediction of 3-D protein structures [119, 121, 174].

5.5 Servers

PROFcon [141], CORNET [120], CMAPpro [129], GPCPRED [104], Hamilton's server [61].

6 Flexible and Intrinsically Disordered Regions

6.1 Local Mobility, Rigidity and Disorder all are Features that Relate to Function

In crystal structures of proteins, the uncertainty of atomic positions can be represented by B-factors (Debye–Waller factors) [32]. B-factors represent the combined effects of thermal variation and static disorder. In general, the higher the B-factor of a residue, the higher is its flexibility. Further, it has been demonstrated that many proteins and protein regions lack a unique 3-D structure [180]. Those regions are often characterized as an ensemble of rapidly changing alternative structures with differing backbone torsion angles. Estimates indicate that a substantial fraction of all proteins (as much as 25%) may contain disordered regions or be entirely disordered [43, 102, 148, 182]. Many important functional interactions, such as cell-cycle regulation, signal transduction, gene expression and chaperon action, are associated with proteins containing very flexible and disordered regions. Determination of these regions also plays an important role in structural genomics, since such regions can be a source of problems in protein expression, purification and crystallization.

6.2 Measuring Flexibility and Disorder

Protein flexibility can be derived from normalized B-factors [23]. Characterization of disordered regions can be provided by many experimental techniques, but in particular by NMR spectroscopy. Regions of protein X-ray structures without atomic coordinates are often considered as intrinsically disordered regions. Successful predictions should be able to simply indicate intrinsically disordered regions, or in case of protein flexibility to assign accurate normalized B-factors to protein residues.

6.3 Prediction Methods

Methods predicting regions of low compositional complexity in protein sequences (SEG [185] and CAST [135]) can be considered as the first methods predicting disordered regions in proteins. However, the correlation between low-complexity regions and disorder is far from perfect. The low-complexity regions are highly repetitive in their amino acid composition but many of them have well defined 3-D structures [167]. There are methods that attempt to predict if entire proteins are in "natively unfolded" configurations based on hydrophobicity and charge information derived from sequences [178]. The disordered regions can be predicted based on disorder propensity assigned to each amino acid [95]. Other methods use machine learning algorithms such as neural networks [75, 95, 147] or SVMs [182]. The NORSp method [99] predicts extended nonregular secondary structure segments that often correlate with disorder. Predictions of B-factors were also carried out by methods using artificial neural networks [168] or support vector regression [186]. The prediction accuracy of those methods was not experimentally verified on the large scale yet.

6.4 Servers

PROFbval [168], PONDR [147], DISOPRED [75], DISOPRED2 [182], GlobPlot [95], NORSp [99], FoldIndex [134], DisEMBL [95].

7 Protein Domains

7.1 Independent Folding Units

The visual inspection of 3-D structures of large proteins often reveals compact structural subunits referred to as protein domains. Such domains are assumed to often constitute units that fold independently. Studies indicate that some of those proteins can be viewed as combinatorial arrangements of protein domains that are genetically mobile. Often, the structural domains are associated with particular biological functions. It is postulated that domains are independent folding units of large proteins. Knowledge of the domain organization of proteins of unknown 3-D structures can help experimental and computational attempts to elucidate their structure and function. Recent analyses of sequence-structure families suggest that over two-thirds of all proteins have more than one domain and that most domains span over about 100 residues [96].

7.2 Prediction Methods

The prediction of the domain organization is a challenging problem if we do not know the 3-D structure (and automatic assignment methods disagree much more than secondary structure assignment methods even if we know the structure). Many sequence-based methods predict domains that are significantly shorter than actual structural domains [98]. The first automatic prediction methods, such as ProDom [31], attempted to determine domains based on "boundaries" in multiple alignments of protein sequences. This approach often results in fragmentation of actual structural domains since sequence similarity conservation often does not extend over entire domains.

In a similar approach, domain constraints can also be obtained from sequence alignment databases such as BLOCKS [64]. Attempts to explicitly elongate sequence alignments were also made [54]. Other automatic prediction methods apply concepts from protein structure prediction [55] or try do derive domains from predicted contact maps [145]. There are methods that use statistics of domain size distributions [183] or a statistical approach toward combining various sources of information [89]. Some of the methods use artificial neural networks [112,114]. Others explore alternative ways of using sequence alignment information [105] or alignments of predicted secondary structure elements [106]. The most accurate methods (e.g. CHOP [96]) simply use sequence homology to proteins with known domain assignments. The downside of such methods is the low coverage, i.e. that they often do not find domains. None of these more recent methods has yet been experimentally verified on large scale.

7.3 Servers

CHOP (homology based) [96], CHOPnet [100], ProDom (homology based) [31], DOMAINATION [54], SnapDRAGON [55], DomSSEA [106].

Acknowledgments

We are grateful to Kaz Wrzeszczynski, Marco Punta and Avner Schlessinger (all from Columbia University) for valuable input. Thanks to Volker Eyrich (Schroedinger Inc.) and Ingrid Koh (Columbia) for their help in setting up the EVA server. The work of D. P. and B. R. was supported by the grant RO1-LM07329-01 from the National Library of Medicine. Last, not least, thanks to all those who deposit their experimental data in public databases and to those who maintain these databases.

References

1 Abagyan, R. A. and S. Batalov. 1997. Do aligned sequences share the same fold? J. Mol. Biol. **273**: 355–68.

2 Adamczak, R., A. Porollo and J. Meller. 2004. Accurate prediction of solvent accessibility using neural networks-based regression. Proteins **56**: 753–67.

3 Adamczak, R., A. Porollo and J. Meller. 2005. Combining prediction of secondary structure and solvent accessibility in proteins. Proteins **59**: 467–75.

4 Ahmad, S. and M. M. Gromiha. 2002. NETASA: neural network based prediction of solvent accessibility. Bioinformatics **18**: 819–24.

5 Albrecht, M., S. C. Tosatto, T. Lengauer and G. Valle. 2003. Simple consensus procedures are effective and sufficient in secondary structure prediction. Protein Eng. **16**: 459–62.

6 ALTSCHUL, S. F. AND W. GISH. 1996. Local alignment statistics. Methods Enzymol. **266**: 460–80.

7 ALTSCHUL, S. F., T. L. MADDEN, A. A. SCHAEFFER, J. ZHANG, Z. ZHANG, W. MILLER AND D. J. LIPMAN. 1997. Gapped Blast and PSI-Blast: a new generation of protein database search programs. Nucleic Acids Res. **25**: 3389–402.

8 ANDERSEN, C. A. F., A. G. PALMER, S. BRUNAK AND B. ROST. 2002. Continuum secondary structure captures protein flexibility. Structure **10**: 175–84.

9 ANFINSEN, C. B. 1973. Principles that govern the folding of protein chains. Science **181**: 223–30.

10 AYERS, D. J., P. R. GOOLEY, A. WIDMER-COOPER AND A. E. TORDA. 1999. Enhanced protein fold recognition using secondary structure information from NMR. Protein Sci. **8**: 1127–33.

11 BAIROCH, A. AND R. APWEILER. 2000. The SWISS-PROT protein sequence database and its supplement TrEMBL in 2000. Nucleic Acids Res. **28**: 45–8.

12 BALDI, P., S. BRUNAK, P. FRASCONI, G. SODA AND G. POLLASTRI. 1999. Exploiting the past and the future in protein secondary structure prediction. Bioinformatics **15**: 937–46.

13 BALDI, P., G. POLLASTRI, C. A. ANDERSEN AND S. BRUNAK. 2000. Matching protein beta-sheet partners by feedforward and recurrent neural networks. Proc. ISMB **8**: 25–36.

14 BARTON, G. J. 1995. Protein secondary structure prediction. Curr. Opin. Struct. Biol. **5**: 372–76.

15 BENNER, S. A. AND D. GERLOFF. 1991. Patterns of divergence in homologous proteins as indicators of secondary and tertiary structure: a prediction of the structure of the catalytic domain of protein kinases. Adv. Enzyme Regul. **31**: 121–81.

16 BENSON, D. A., I. KARSCH-MIZRACHI, D. J. LIPMAN, J. OSTELL AND D. L. WHEELER. 2003. GenBank. Nucleic Acids Res. **31**: 23–27.

17 BERMAN, H. M., J. WESTBROOK, Z. FENG, G. GILLLILAND, T. N. BHAT, H. WEISSIG, I. N. SHINDYALOV AND P. E. BOURNE. 2000. The Protein Data Bank. Nucleic Acids Res. **28**: 235–42.

18 BIGELOW, H. R., D. S. PETREY, J. LIU, D. PRZYBYLSKI AND B. ROST. 2004. Predicting transmembrane beta-barrels in proteomes. Nucleic Acids Res. **32**: 2566–77.

19 BLOUT, E. R. 1962. The dependence of the conformation of polypetides and proteins upon amino acid composition. In STAHMAN, M. (ed.), *Polyamino Acids, Polypeptides, and Proteins*. University of Wisconsin Press, Madison, WI: 275–79.

20 BLOUT, E. R., C. DE LOZÉ, S. M. BLOOM AND G. D. FASMAN. 1960. Dependence of the conformation of synthetic polypeptides on amino acid composition. J. Am. Chem. Soc. **82**: 3787–9.

21 BUSSIERE, D. E., X. KONG, D. A. EGAN, K. WALTER, T. F. HOLZMAN, F. LINDH, T. ROBINS AND V. L. GIRANDA. 1998. Structure of the E2 DNA-binding domain from human papillomavirus serotype 31 at 2.4 Å. Acta Crystallogr. D **54**: 1367–76.

22 BYSTROFF, C. AND Y. SHAO. 2002. Fully automated ab initio protein structure prediction using I-SITES, HMMSTR and ROSETTA. Bioinformatics **18**: S54–61.

23 CARUGO, O. AND P. ARGOS. 1997. Correlation between side chain mobility and conformation in protein structures. Protein Eng. **10**: 777–87.

24 CHANDONIA, J. M. AND M. KARPLUS. 1999. New methods for accurate prediction of protein secondary structure. Proteins **35**: 293–306.

25 CHEN, C. C., J. P. SINGH AND R. B. ALTMAN. 1999. Using imperfect secondary structure predictions to improve molecular structure computations. Bioinformatics **15**: 53–65.

26 CHEN, C. P., A. KERNYTSKY AND B. ROST. 2002. Transmembrane helix predictions revisited. Protein Sci. **11**: 2774–91.

27 CHOTHIA, C. AND A. M. LESK. 1986. The relation between the divergence of sequence and structure in proteins. EMBO J. **5**: 823–26.

28 Chou, P. Y. and G. D. Fasman. 1974. Prediction of protein conformation. Biochemistry **13**: 211–5.

29 Chou, P. Y. and G. D. Fasman. 1978. Prediction of the secondary structure of proteins from their amino acid sequence. Adv. Enzymol. **47**: 45–148.

30 Colloc'h, N., C. Etchebest, E. Thoreau, B. Henrissat and J. P. Mornon. 1993. Comparison of three algorithms for the assignment of secondary structure in proteins: the advantages of a consensus assignment. Protein Eng. **6**: 377–82.

31 Corpet, F., F. Servant, J. Gouzy and D. Kahn. 2000. ProDom and ProDom-CG: tools for protein domain analysis and whole genome comparisons. Nucleic Acids Res. **28**: 267–9.

32 Creighton, T. 1992. *Proteins: Structures and Molecular Properties*. Freeman, San Francisco, CA.

33 Cserzo, M., F. Eisenhaber, B. Eisenhaber and I. Simon. 2004. TM or not TM: transmembrane protein prediction with low false positive rate using DAS-TMfilter. Bioinformatics **20**: 136–7.

34 Cuff, J. A. and G. J. Barton. 2000. Application of multiple sequence alignment profiles to improve protein secondary structure prediction. Proteins **40**: 502–11.

35 Cuff, J. A. and G. J. Barton. 1999. Evaluation and improvement of multiple sequence methods for protein secondary structure prediction. Proteins **34**: 508–19.

36 Cuff, J. A., M. E. Clamp, A. S. Siddiqui, M. Finlay and G. J. Barton. 1998. JPred: a consensus secondary structure prediction server. Bioinformatics **14**: 892–3.

37 de la Cruz, X. and J. M. Thornton. 1999. Factors limiting the performance of prediction-based fold recognition methods. Protein Sci. **8**: 750–9.

38 Defay, T. and F. E. Cohen. 1995. Evaluation of current techniques for *ab initio* protein structure prediction. Proteins **23**: 431–45.

39 Di Francesco, V., J. Garnier and P. J. Munson. 1996. Improving protein secondary structure prediction with aligned homologous sequences. Protein Sci. **5**: 106–13.

40 Di Francesco, V., P. J. Munson and J. Garnier. 1999. FORESST: fold recognition from secondary structure predictions of proteins. Bioinformatics **15**: 131–40.

41 Dickerson, R. E., R. Timkovich and R. J. Almassy. 1976. The cytochrome fold and the evolution of bacterial energy metabolism. J. Mol. Biol. **100**: 473–91.

42 Diederichs, K., J. Freigang, S. Umhau, K. Zeth and J. Breed. 1998. Prediction by a neural network of outer membrane beta-strand protein topology. Protein Sci. **7**: 2413–20.

43 Dunker, A. K., C. J. Brown, J. D. Lawson, L. M. Iakoucheva and Z. Obradovic. 2002. Intrinsic disorder and protein function. Biochemistry **41**: 6573–82.

44 Eyrich, V. A., M. A. Martí-Renom, D. Przybylski, A. Fiser, F. Pazos, A. Valencia, A. Sali and B. Rost. 2001. EVA: continuous automatic evaluation of protein structure prediction servers. Bioinformatics **17**: 1242–3.

45 Eyrich, V. A. and B. Rost. 2003. META-PP: single interface to crucial prediction servers. Nucleic Acids Res. **31**: 3308–10.

46 Eyrich, V. A., D. Przybylski, I. Y. Koh, O. Grana, F. Pazos, A. Valencia and B. Rost. 2003. CAFASP3 in the spotlight of EVA. Proteins **53 (Suppl. 6)**: 548–60.

47 Eyrich, V. A., D. M. Standley, A. K. Felts and R. A. Friesner. 1999. Protein tertiary structure prediction using a branch and bound algorithm. Proteins **35**: 41–57.

48 Eyrich, V. A., D. M. Standley and R. A. Friesner. 1999. Prediction of protein tertiary structure to low resolution: performance for a large and structurally diverse test set. J. Mol. Biol. **288**: 725–42.

49 Fariselli, P., O. Olmea, A. Valencia and R. Casadio. 2001. Prediction of contact maps with neural networks and correlated mutations. Protein Eng. **14**: 835–43.

50 FISCHER, D. AND D. EISENBERG. 1996. Fold recognition using sequence-derived properties. Protein Sci. **5**: 947–55.

51 FRISHMAN, D. AND P. ARGOS. 1995. Knowledge-based protein secondary structure assignment. Proteins **23**: 566–79.

52 GARNIER, J., J.-F. GIBRAT AND B. ROBSON. 1996. GOR method for predicting protein secondary structure from amino acid sequence. Methods Enzymol. **266**: 540–53.

53 GARNIER, J., D. J. OSGUTHORPE AND B. ROBSON. 1978. Analysis of the accuracy and Implications of simple methods for predicting the secondary structure of globular proteins. J. Mol. Biol. **120**: 97–120.

54 GEORGE, R. A. AND J. HERINGA. 2002. Protein domain identification and improved sequence similarity searching using PSI-BLAST. Proteins **48**: 672–81.

55 GEORGE, R. A. AND J. HERINGA. 2002. SnapDRAGON: a method to delineate protein structural domains from sequence data. J. Mol. Biol. **316**: 839–51.

56 GERSTEIN, M. AND M. LEVITT. 1997. A structural census of the current population of protein sequences. Proc. Natl Acad. Sci. USA **94**: 11911–6.

57 GIBRAT, J.-F., J. GARNIER AND B. ROBSON. 1987. Further developments of protein secondary structure prediction using information theory. New parameters and consideration of residue pairs. J. Mol. Biol. **198**: 425–43.

58 GOBEL, U., C. SANDER, R. SCHNEIDER AND A. VALENCIA. 1994. Correlated mutations and residue contacts in proteins. Proteins **18**: 309–17.

59 GROMIHA, M. M., R. MAJUMDAR AND P. K. PONNUSWAMY. 1997. Identification of membrane spanning beta strands in bacterial porins. Protein Eng. **10**: 497–500.

60 GUERMEUR, Y., C. GEOURJON, P. GALLINARI AND G. DELEAGE. 1999. Improved performance in protein secondary structure prediction by inhomogeneous score combination. Bioinformatics **15**: 413–21.

61 HAMILTON, N., K. BURRAGE, M. A. RAGAN AND T. HUBER. 2004. Protein contact prediction using patterns of correlation. Proteins **56**: 679–84.

62 HANSEN, L. K. AND P. SALAMON. 1990. Neural network ensembles. IEEE Trans. Pattern Anal. Machine Intell. **12**: 993–1001.

63 HARGBO, J. AND A. ELOFSSON. 1999. Hidden Markov models that use predicted secondary structures for fold recognition. Proteins **36**: 68–76.

64 HENIKOFF, J. G. AND S. HENIKOFF. 1996. Blocks database and its applications. Methods Enzymol. **266**: 88–105.

65 HERINGA, J. 1999. Two strategies for sequence comparison: profile-preprocessed and secondary structure-induced multiple alignment. Comput. Chem. **23**: 341–64.

66 HIROKAWA, T., S. BOON-CHIENG AND S. MITAKU. 1998. SOSUI: classification and secondary structure prediction system for membrane proteins. Bioinformatics **14**: 378–79.

67 HOBOHM, U., M. SCHARF, R. SCHNEIDER AND C. SANDER. 1992. Selection of representative protein data sets. Protein Sci. **1**: 409–17.

68 HUA, S. AND Z. SUN. 2001. A novel method of protein secondary structure prediction with high segment overlap measure support vector machine approach. J. Mol. Biol. **308**: 397–407.

69 IVANKOV, D. N. AND A. V. FINKELSTEIN. 2004. Prediction of protein folding rates from the amino acid sequence-predicted secondary structure. Proc. Natl Acad. Sci. USA **101**: 8942–4.

70 JACOBONI, I., P. L. MARTELLI, P. FARISELLI, V. DE PINTO AND R. CASADIO. 2001. Prediction of the transmembrane regions of beta-barrel membrane proteins with a neural network-based predictor. Protein Sci. **10**: 779–87.

71 JENNINGS, A. J., C. M. EDGE AND M. J. STERNBERG. 2001. An approach to improving multiple alignments of protein sequences using predicted secondary structure. Protein Eng. **14**: 227–31.

72 JONES, D. T. 1999. GenTHREADER: an efficient and reliable protein fold recognition method for genomic sequences. J. Mol. Biol. **287**: 797–815.

73 JONES, D. T. 1999. Protein secondary structure prediction based on position-specific scoring matrices. J. Mol. Biol. **292**: 195–202.

74 JONES, D. T., M. TRESS, K. BRYSON AND C. HADLEY. 1999. Successful recognition of protein folds using threading methods biased by sequence similarity and predicted secondary structure. Proteins **37**: 104–11.

75 JONES, D. T. AND J. J. WARD. 2003. Prediction of disordered regions in proteins from position specific score matrices. Proteins **53 (Suppl. 6)**: 573–8.

76 KABSCH, W. AND C. SANDER. 1983. Dictionary of protein secondary structure: pattern recognition of hydrogen bonded and geometrical features. Biopolymers **22**: 2577–637.

77 KABSCH, W. AND C. SANDER. 1983. How good are predictions of protein secondary structure? FEBS Lett. **155**: 179–82.

78 KARCHIN, R., M. CLINE, Y. MANDEL-GUTFREUND AND K. KARPLUS. 2003. Hidden Markov models that use predicted local structure for fold recognition: alphabets of backbone geometry. Proteins **51**: 504–14.

79 KARPLUS, K., R. KARCHIN, J. DRAPER, J. CASPER, Y. MANDEL-GUTFREUND, M. DIEKHANS AND R. HUGHEY. 2003. Combining local-structure, fold-recognition, and new fold methods for protein structure prediction. Proteins **53**: 491–6.

80 KELLEY, L. A., R. M. MACCALLUM AND M. J. STERNBERG. 2000. Enhanced genome annotation using structural profiles in the program 3D-PSSM. J. Mol. Biol. **299**: 499–520.

81 KENDREW, J. C., R. E. DICKERSON, B. E. STRANDBERG, R. J. HART, D. R. DAVIES AND D. C. PHILLIPS. 1960. Structure of myoglobin: a three-dimensional Fourier synthesis at 2 Å resolution. Nature **185**: 422–7.

82 KIM, H. AND H. PARK. 2004. Prediction of protein relative solvent accessibility with support vector machines and long-range interaction 3D local descriptor. Proteins **54**: 557–62.

83 KING, R. D., M. OUALI, A. T. STRONG, A. ALY, A. ELMAGHRABY, M. KANTARDZIC AND D. PAGE. 2000. Is it better to combine predictions? Protein Eng. **13**: 15–9.

84 KIRSHENBAUM, K., M. YOUNG AND S. HIGHSMITH. 1999. Predicting allosteric switches in myosins. Protein Sci. **8**: 1806–15.

85 KLOCZKOWSKI, A., K. L. TING, R. L. JERNIGAN AND J. GARNIER. 2002. Combining the GOR V algorithm with evolutionary information for protein secondary structure prediction from amino acid sequence. Proteins **49**: 154–66.

86 KOH, I. Y., V. A. EYRICH, M. A. MARTI-RENOM, et al. 2003. EVA: evaluation of protein structure prediction servers. Nucleic Acids Res. **31**: 3311–5.

87 KORETKE, K. K., R. B. RUSSELL, R. R. COPLEY AND A. N. LUPAS. 1999. Fold recognition using sequence and secondary structure information. Proteins **37**: 141–8.

88 KROGH, A., B. LARSSON, G. VON HEIJNE AND E. L. SONNHAMMER. 2001. Predicting transmembrane protein topology with a hidden Markov model: application to complete genomes. J. Mol. Biol. **305**: 567–80.

89 KULIKOWSKI, C. A., I. MUCHNIK, H. J. YUN, A. A. DAYANIK, D. ZHANG, Y. SONG AND G. T. MONTELIONE. 2001. Protein structural domain parsing by consensus reasoning over multiple knowledge sources and methods. Medinfo **10**: 965–9.

90 KYTE, J. AND R. F. DOOLITTLE. 1982. A simple method for displaying the hydrophathic character of a protein. J. Mol. Biol. **157**: 105–32.

91 LEVIN, J. M. 1997. Exploring the limits of nearest neighbour secondary structure prediction. Protein Eng. **10**: 771–6.

92 LEVIN, J. M., S. PASCARELLA, P. ARGOS AND J. GARNIER. 1993. Quantification of secondary structure prediction improvement using multiple alignment. Protein Eng. **6**: 849–54.

93 LIM, V. I. 1974. Structural principles of the globular organization of protein

chains. a stereochemical theory of globular protein secondary structure. J. Mol. Biol. **88**: 857–72.

94 LIN, K., V. A. SIMOSSIS, W. R. TAYLOR AND J. HERINGA. 2005. A simple and fast secondary structure prediction method using hidden neural networks. Bioinformatics **21**: 152–9.

95 LINDING, R., R. B. RUSSELL, V. NEDUVA AND T. J. GIBSON. 2003. GlobPlot: Exploring protein sequences for globularity and disorder. Nucleic Acids Res. **31**: 3701–8.

96 LIU, J. AND B. ROST. 2004. CHOP: parsing proteins into structural domains. Nucleic Acids Res. **32**: W569–71.

97 LIU, J. AND B. ROST. 2001. Comparing function and structure between entire proteomes. Protein Sci. **10**: 1970–9.

98 LIU, J. AND B. ROST. 2003. Domains, motifs and clusters in the protein universe. Curr. Opin. Chem. Biol. **7**: 5–11.

99 LIU, J. AND B. ROST. 2003. NORSp: predictions of long regions without regular secondary structure. Nucleic Acids Res. **31**: 3833–5.

100 LIU, J. AND B. ROST. 2004. Sequence-based prediction of protein domains. Nucleic Acids Res. **32**: 3522–30.

101 LIU, J. AND B. ROST. 2002. Target space for structural genomics revisited. Bioinformatics **18**: 922–33.

102 LIU, J., H. TAN AND B. ROST. 2002. Loopy proteins appear conserved in evolution. J. Mol. Biol. **322**: 53–64.

103 LOMIZE, A. L., I. D. POGOZHEVA AND H. I. MOSBERG. 1999. Prediction of protein structure: the problem of fold multiplicity. Proteins **Suppl. 3**: 199–203.

104 MACCALLUM, R. M. 2004. Striped sheets and protein contact prediction. Bioinformatics **20 (Suppl. 1)**: i224–31.

105 MARCOTTE, E. M., M. PELLEGRINI, M. J. THOMPSON, T. O. YEATES AND D. EISENBERG. 1999. A combined algorithm for genome-wide prediction of protein function. Nature **402**: 83–6.

106 MARSDEN, R. L., L. J. MCGUFFIN AND D. T. JONES. 2002. Rapid protein domain assignment from amino acid sequence using predicted secondary structure. Protein Sci. **11**: 2814–24.

107 MAXFIELD, F. R. AND H. A. SCHERAGA. 1979. Improvements in the prediction of protein topography by reduction of statistical errors. Biochemistry **18**: 697–704.

108 MCGUFFIN, L. J. AND D. T. JONES. 2003. Benchmarking secondary structure prediction for fold recognition. Proteins **52**: 166–75.

109 MEHTA, P. K., J. HERINGA AND P. ARGOS. 1995. A simple and fast approach to prediction of protein secondary structure from multiply aligned sequences with accuracy above 70%. Protein Sci. **4**: 2517–25.

110 MEILER, J., M. MUELLER, A. ZEIDLER AND F. SCHMAESCHKE. 2001. Generation and evaluation of dimension-reduced amino acid parameter representation by artificial neural networks. J. Mol. Model. **7**: 360–9.

111 MELEN, K., A. KROGH AND G. VON HEIJNE. 2003. Reliability measures for membrane protein topology prediction algorithms. J. Mol. Biol. **327**: 735–44.

112 MIYAZAKI, S., Y. KURODA AND S. YOKOYAMA. 2002. Characterization and prediction of linker sequences of multi-domain proteins by a neural network. J Struct. Funct. Genomics **2**: 37–51.

113 MÖLLER, S., D. R. CRONING AND R. APWEILER. 2001. Evaluation of methods for the prediction of membrane spanning regions. Bioinformatics **17**: 646–53.

114 MURVAI, J., K. VLAHOVICEK, C. SZEPESVARI AND S. PONGOR. 2001. Prediction of protein functional domains from sequences using artificial neural networks. Genome Res. **11**: 1410–7.

115 NADERI-MANESH, H., M. SADEGHI, S. ARAB AND A. A. MOOSAVI MOVAHEDI. 2001. Prediction of protein surface accessibility with information theory. Proteins **42**: 452–9.

116 NAIR, R. AND B. ROST. 2003. Better prediction of sub-cellular localization by combining evolutionary and structural information. Proteins **53**: 917–30.

117 NG, P. C., J. G. HENIKOFF AND S. HENIKOFF. 2000. PHAT: a transmembrane-specific substitution matrix. Predicted hydrophobic and transmembrane. Bioinformatics **16**: 760–6.

118 Norvell, J. C. and A. Z. Machalek. 2000. Structural genomics programs at the US National Institute of General Medical Sciences. Nat. Struct. Biol. **7 (Suppl.)**: 931.

119 Olmea, O., B. Rost and A. Valencia. 1999. Effective use of sequence correlation and conservation in fold recognition. J. Mol. Biol. **293**: 1221–39.

120 Olmea, O. and A. Valencia. 1997. Improving contact predictions by the combination of correlated mutations and other sources of sequence information. Fold. Des. **2**: S25–32.

121 Ortiz, A. R., A. Kolinski, P. Rotkiewicz, B. Ilkowski and J. Skolnick. 1999. Ab initio folding of proteins using restraints derived from evolutionary information. Proteins **Suppl. 3**: 177–85.

122 Ota, M., T. Kawabata, A. R. Kinjo and K. Nishikawa. 1999. Cooperative approach for the protein fold recognition. Proteins **37**: 126–32.

123 Ouali, M. and R. D. King. 2000. Cascaded multiple classifiers for secondary structure prediction. Protein Sci. **9**: 1162–76.

124 Panchenko, A., A. Marchler-Bauer and S. H. Bryant. 1999. Threading with explicit models for evolutionary conservation of structure and sequence. Proteins **Suppl. 3**: 133–40.

125 Park, J., K. Karplus, C. Barrett, R. Hughey, D. Haussler, T. Hubbard and C. Chothia. 1998. Sequence comparisons using multiple sequences detect three times as many remote homologues as pairwise methods. J. Mol. Biol. **284**: 1201–10.

126 Pauling, L. and R. B. Corey. 1951. Configurations of polypeptide chains with favored orientations around single bonds: two new pleated sheets. Proc. Natl Acad. Sci. USA **37**: 729–40.

127 Perutz, M. F., M. G. Rossmann, A. F. Cullis, G. Muirhead, G. Will and A. T. North. 1960. Structure of haemoglobin: a three-dimensional Fourier synthesis at 5.5 Å resolution, obtained by X-ray analysis. Nature **185**: 416–22.

128 Petersen, T. N., C. Lundegaard, M. Nielsen, H. Bohr, J. Bohr, S. Brunak, G. P. Gippert and O. Lund. 2000. Prediction of protein secondary structure at 80% accuracy. Proteins **41**: 17–20.

129 Pollastri, G. and P. Baldi. 2002. Prediction of contact maps by GIOHMMs and recurrent neural networks using lateral propagation from all four cardinal corners. Bioinformatics **18 (Suppl. 1)**: S62–70.

130 Pollastri, G., P. Baldi, P. Fariselli and R. Casadio. 2001. Improved prediction of the number of residue contacts in proteins by recurrent neural networks. Bioinformatics **17**: S234–42.

131 Pollastri, G., P. Baldi, P. Fariselli and R. Casadio. 2002. Prediction of coordination number and relative solvent accessibility in proteins. Proteins **47**: 142–53.

132 Pollastri, G. and A. McLysaght. 2005. Porter: a new, accurate server for protein secondary structure prediction. Bioinformatics **21**: 1719–20.

133 Pollastri, G., D. Przybylski, B. Rost and P. Baldi. 2002. Improving the prediction of protein secondary structure in three and eight classes using recurrent neural networks and profiles. Proteins **47**: 228–35.

134 Prilusky, J., C. E. Felder, T. Zeev-Ben-Mordehai, E. Rydberg, O. Man, J. S. Beckmann, I. Silman and J. L. Sussman. 2005. FoldIndex©: a simple tool to predict whether a given protein sequence is intrinsically unfolded. Bioinformatics **21**: 3435–8.

135 Promponas, V. J., A. J. Enright, S. Tsoka, D. P. Kreil, C. Leroy, S. Hamodrakas, C. Sander and C. A. Ouzounis. 2000. CAST: an iterative algorithm for the complexity analysis of sequence tracts. Complexity analysis of sequence tracts. Bioinformatics **16**: 915–22.

136 Prusiner, S. B., M. R. Scott, S. J. DeArmond and F. E. Cohen. 1998. Prion protein biology. Cell **93**: 337–48.

137 Przybylski, D. and B. Rost. 2002. Alignments grow, secondary structure prediction improves. Proteins **46**: 195–205.

138 PRZYBYLSKI, D. AND B. ROST. 2004. Improving fold recognition without folds. J. Mol. Biol. **341**: 255–69.

139 PRZYTYCKA, T., R. AURORA AND G. D. ROSE. 1999. A protein taxonomy based on secondary structure. Nat. Struct. Biol. **6**: 672–82.

140 PTITSYN, O. B. AND A. V. FINKELSTEIN. 1983. Theory of protein secondary structure and algorithm of its prediction. Biopolymers **22**: 15–25.

141 PUNTA, M. AND B. ROST. 2005. PROFcon: novel prediction of long-range contacts. Bioinformatics **21**: 2960–8.

142 PUNTA, M. AND B. ROST. 2005. Protein folding rates estimated from contact predictions. J. Mol. Biol. **348**: 507–12.

143 RICHARDS, F. M. AND C. E. KUNDROT. 1988. Identification of structural motifs from protein coordinate data: secondary structure and first-level supersecondary structure. Proteins **3**: 71–84.

144 RICHARDSON, C. J. AND D. J. BARLOW. 1999. The bottom line for prediction of residue solvent accessibility. Protein Eng. **12**: 1051–4.

145 RIGDEN, D. J. 2002. Use of covariance analysis for the prediction of structural domain boundaries from multiple protein sequence alignments. Protein Eng. **15**: 65–77.

146 RIIS, S. K. AND A. KROGH. 1996. Improving prediction of protein secondary structure using structured neural networks and multiple sequence alignments. J. Comput. Biol. **3**: 163–83.

147 ROMERO, P., Z. OBRADOVIC, C. R. KISSINGER, J. E. VILLAFRANCA AND A. K. DUNKER. 1997. Identifying disordered regions in proteins from amino acid sequence. In Proc. IEEE Int. Conf. on Neural Networks, Houston, TX. Volume 1: 90–5.

148 ROMERO, P., Z. OBRADOVIC, C. R. KISSINGER, J. E. VILLAFRANCA, E. GARNER, S. GUILLIOT AND A. K. DUNKER. 1998. Thousands of proteins likely to have long disordered regions. Pac. Symp. Biocomput.: 437–48.

149 ROST, B. 2005. How to use protein 1D structure predicted by PROFphd. In WALKER, J. E. (ed.), *The Proteomics Protocols Handbook*. Humana, Totowa NJ: 875–901.

150 ROST, B. 1998. Marrying structure and genomics. Structure **6**: 259–63.

151 ROST, B. 1996. PHD: predicting one-dimensional protein structure by profile based neural networks. Methods Enzymol. **266**: 525–39.

152 ROST, B. 1995. TOBITS: threading one-dimensional predictions into three-dimensional structures. Proc. Int. Conf. Intell. Syst. Mol. Biol. **3**: 314–21.

153 ROST, B., R. CASADIO AND P. FARISELLI. 1996. Refining neural network predictions for helical transmembrane proteins by dynamic programming. Proc. Int. Conf. Intell. Syst. Mol. Biol. **4**: 192–200.

154 ROST, B. AND C. SANDER. 1994. Combining evolutionary information and neural networks to predict protein secondary structure. Proteins **19**: 55–72.

155 ROST, B. AND C. SANDER. 1994. Conservation and prediction of solvent accessibility in protein families. Proteins **20**: 216–26.

156 ROST, B. AND C. SANDER. 1993. Improved prediction of protein secondary structure by use of sequence profiles and neural networks. Proc. Natl Acad. Sci. USA **90**: 7558–62.

157 ROST, B. AND C. SANDER. 1993. Prediction of protein secondary structure at better than 70% accuracy. J. Mol. Biol. **232**: 584–99.

158 ROST, B. AND C. SANDER. 2000. Third generation prediction of secondary structures. Methods Mol. Biol. **143**: 71–95.

159 ROST, B., C. SANDER AND R. SCHNEIDER. 1994. PHD – an automatic server for protein secondary structure prediction. CABIOS **10**: 53–60.

160 ROST, B., C. SANDER AND R. SCHNEIDER. 1994. Redefining the goals of protein secondary structure prediction. J. Mol. Biol. **235**: 13–26.

161 ROST, B. AND J. LIU. 2003. The PredictProtein server. Nucl. Acids Res. **31**: 3300–4.

162 RUSSELL, R. B., R. R. COPLEY AND G. J. BARTON. 1996. Protein fold recognition

by mapping predicted secondary structures. J. Mol. Biol. **259**: 349–65.

163 SALAMOV, A. A. AND V. V. SOLOVYEV. 1995. Prediction of protein secondary structure by combining nearest-neighbor algorithms and multiple sequence alignment. J. Mol. Biol. **247**: 11–5.

164 SAMUDRALA, R., E. S. HUANG, P. KOEHL AND M. LEVITT. 2000. Constructing side chains on near-native main chains for *ab initio* protein structure prediction. Protein Eng. **13**: 453–57.

165 SAMUDRALA, R., Y. XIA, E. HUANG AND M. LEVITT. 1999. *Ab initio* protein structure prediction using a combined hierarchical approach. Proteins **Suppl. 3**: 194–98.

166 SANDER, C. AND R. SCHNEIDER. 1991. Database of homology-derived structures and the structural meaning of sequence alignment. Proteins **9**: 56–68.

167 SAQI, M. 1995. An analysis of structural instances of low complexity sequence segments. Protein Eng. **8**: 1069–73.

168 SCHLESSINGER, A. AND B. ROST. 2005. Protein flexibility and rigidity predicted from sequence. Proteins **61**: 115–26.

169 SCHNEIDER, R. 1989. Sekundärstrukturvorhersage von Proteinen unter Berücksichtigung von Tertiärstrukturaspekten. *Diploma Thesis*. Department of Biology, University of Heidelberg.

170 SELBIG, J., T. MEVISSEN AND T. LENGAUER. 1999. Decision tree-based formation of consensus protein secondary structure prediction. Bioinformatics **15**: 1039–46.

171 SEN, T. Z., R. L. JERNIGAN, J. GARNIER AND A. KLOCZKOWSKI. 2005. GOR V server for protein secondary structure prediction. Bioinformatics **21**: 2787–8.

172 SHAO, Y. AND C. BYSTROFF. 2003. Predicting interresidue contacts using templates and pathways. Proteins **53**: 497–502.

173 SKLENAR, H., C. ETCHEBEST AND R. LAVERY. 1989. Describing protein structure: a general algorithm yielding complete helicoidal parameters and a unique overall axis. Proteins **6**: 46–60.

174 SKOLNICK, J., Y. ZHANG, A. K. ARAKAKI, A. KOLINSKI, M. BONIECKI, A. SZILAGYI AND D. KIHARA. 2003. TOUCHSTONE: a unified approach to protein structure prediction. Proteins **53 (Suppl. 6)**: 469–79.

175 SOLOVYEV, V. V. AND A. A. SALAMOV. 1994. Predicting α-helix and β-strand segments of globular proteins. Comput. Appl. Biol. Sci. **10**: 661–9.

176 SZENT-GYÖRGYI, A. G. AND C. COHEN. 1957. Role of proline in polypeptide chain configuration of proteins. Science **126**: 697.

177 THOMPSON, M. J. AND R. A. GOLDSTEIN. 1996. Predicting solvent accessibility: higher accuracy using Bayesian statistics and optimized residue substitution classes. Proteins **25**: 38–47.

178 UVERSKY, V. N., J. R. GILLESPIE AND A. L. FINK. 2000. Why are "natively unfolded" proteins unstructured under physiologic conditions? Proteins **41**: 415–27.

179 VON HEIJNE, G. 1992. Membrane protein structure prediction. J. Mol. Biol. **225**: 487–94.

180 VUCETIC, S., Z. OBRADOVIC, V. VACIC, *et al.* 2005. DisProt: a database of protein disorder. Bioinformatics **21**: 137–40.

181 WARD, J. J., L. J. MCGUFFIN, B. F. BUXTON AND D. T. JONES. 2003. Secondary structure prediction with support vector machines. Bioinformatics **19**: 1650–5.

182 WARD, J. J., J. S. SODHI, L. J. MCGUFFIN, B. F. BUXTON AND D. T. JONES. 2004. Prediction and functional analysis of native disorder in proteins from the three kingdoms of life. J. Mol. Biol. **337**: 635–45.

183 WHEELAN, S. J., A. MARCHLER-BAUER AND S. H. BRYANT. 2000. Domain size distributions can predict domain boundaries. Bioinformatics **16**: 613–8.

184 WHISSTOCK, J. C. AND A. M. LESK. 2003. Prediction of protein function from protein sequence and structure. Q. Rev. Biophys. **36**: 307–40.

185 WOOTTON, J. C. AND S. FEDERHEN. 1996. Analysis of compositionally biased regions in sequence databases. Methods Enzymol. **266**: 554–71.

186 YUAN, Z., T. L. BAILEY AND R. D. TEASDALE. 2005. Prediction of protein B-factor profiles. Proteins **58**: 905–12.

187 ZEMLA, A., C. VENCLOVAS, K. FIDELIS AND B. ROST. 1999. A modified definition of *SOV*, a segment-based measure for protein secondary structure prediction assessment. Proteins **34**: 220–3.

188 ZHAO, Y. AND G. KARYPIS. 2003. Clustering in life sciences. Methods Mol. Biol. **224**: 183–218.

189 ZVELEBIL, M. J., G. J. BARTON, W. R. TAYLOR AND M. J. E. STERNBERG. 1987. Prediction of protein secondary structure and active sites using alignment of homologous sequences. J. Mol. Biol. **195**: 957–61.

10 Homology Modeling in Biology and Medicine

Roland L. Dunbrack, Jr.

1 Introduction

1.1 The Concept of Homology Modeling

To understand basic biological processes such as cell division, cellular communication, metabolism and development, knowledge of the three-dimensional (3-D) structure of the active components is crucial. Proteins form the key players in all of these processes, and the study of their diverse and elegant designs is a mainstay of modern biology. The Protein Databank (PDB) of experimentally determined protein structures [14] now contains nearly 40 000 entries, which can be grouped into about 1500 superfamilies [5]. The fact that proteins that share very little or no sequence similarity can have quite similar structures has led to the hypothesis that there are in fact only a few thousand different superfamilies [46, 84, 233] which have been adapted by a process of duplication, mutation and natural selection to perform all the biological functions that proteins accomplish.

Since it was first recognized that proteins can share similar structures [156], computational methods have been developed to build models of proteins of unknown structure based on related proteins of known structure [24]. Most such modeling efforts, referred to as homology modeling or comparative modeling, follow a basic protocol laid out by Greer [72, 73]: (i) identify a template structure related to the target sequence of unknown structure, and align the target sequence to the template sequence and structure; (ii) for core secondary structures and all well-conserved parts of the alignment, borrow the backbone coordinates of the template according to the sequence alignment of the target and template; (iii) for segments of the target sequence for which coordinates cannot be borrowed from the template because of insertions and deletions in the alignment (usually in loop regions of the protein) or because of missing coordinates in the template, build these segments using some construction method based on our knowledge of the determinants of protein structure; (iv) build side chains determined by the target sequence on to the

Bioinformatics - From Genomes to Therapies Vol. 1. Edited by Thomas Lengauer

ISBN: 978-3-527-31278-8

backbone model built from the template structure and loop construction; (v) refinement of the model from the template backbone and toward the target structure.

The alignment step may involve a number of different strategies, including manual adjustment, even after the template structure or structures have been identified. Steps (iii) and (iv), backbone and side chain modeling, may be coupled, since certain backbone conformations may be unable to accommodate the required side-chains in any low-energy conformation. The refinement step involves moving beyond the aligned part of the backbone fixed in the template position and instead allowing it to adjust to the new sequence. For instance, two helices packed against each other may move apart to accommodate larger side-chains.

An alternative strategy has been developed by Blundell and colleagues, based on averaging a number of template structures, if these exist, rather than using a single structure [18,207,208]. More complex procedures based on reconstructing structures (rather than perturbing a starting structure) by satisfying spatial restraints using distance geometry [78] or molecular dynamics and energy minimization [118, 173, 174, 180] have also been developed.

Many methods have been proposed to perform each of the steps in the homology modeling process. There are also a number of research groups that have developed complete packages that take as input a sequence alignment or even just a sequence and develop a complete model. In this chapter, we describe some of the basic ideas that drive loop and side-chain modeling individually as well as the complete modeling process. This chapter is a revised version of one that was published in 2001 [48]. In this revision, we emphasize those methods for which *usable* programs are currently publicly available. We also discuss more extensively the concept of modeling from the biological unit, including complexes of proteins with other proteins, DNA and ligands. The identification and alignment steps are covered in Chapters 3 and 11.

1.2 How do Homologous Protein Arise?

By definition, homologous proteins arise by evolution from a common ancestor. However, there are several different mechanisms for this and these are illustrated in Figure 1. The first is random mutation of individual nucleotides that change protein sequence, including missense mutations (changing the identity of a single amino acid) as well as insertions and deletions of a number of nucleotides that result in insertion and deletion of amino acids. As a single species diverges into two species, a gene in the parent species will continue to exist in the divergent species and over time will gather mutations that change the protein sequence. In this case, the genes in the different organisms will

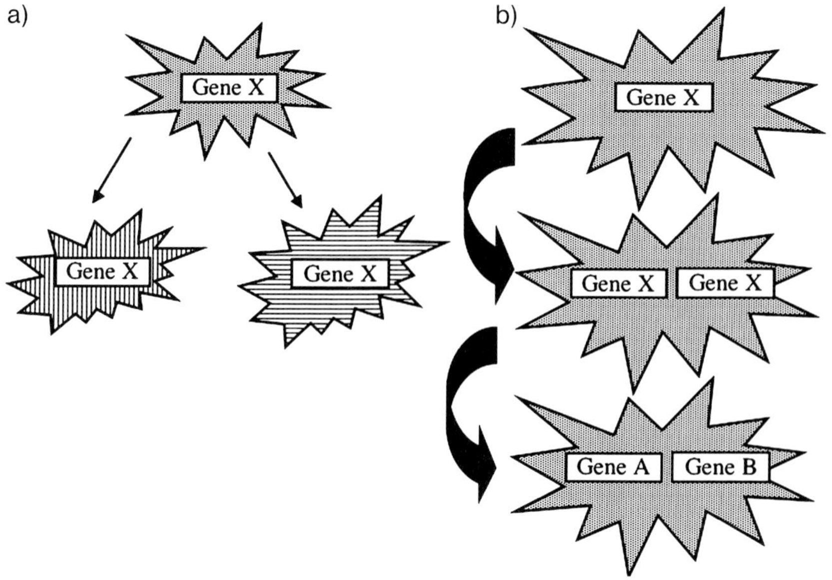

Figure 1 Orthologs versus paralogs. Schematic of the evolutionary process that gives rise to homologous proteins. (a) A single gene X in one species is retained as the species diverges into two separate species. The genes in these two species are *orthologous*. (b) A single gene X in one species is duplicated. As each gene gathers mutations, it may begin to perform new functions, or the two genes may specialize in carrying out two or more functions of the ancestral gene, thus improving the fitness of the organism. These genes in one species are paralogous. If the species diverges, each daughter species may maintain the duplicated genes, and therefore each species contains an ortholog and a paralog to each gene in the other species.

usually maintain the same function. These genes are referred to as orthologs of one another. A second mechanism is duplication of a gene or of a gene segment within a single organism or germ line cell. As time goes by, the two copies of the gene may begin to gather mutations. If the template gene performed more than one function, e.g. similar catalytic activity on two different substrates, one of the duplicated genes may gain specificity for one of the reactions, while the other gene gains specific activity for the other. If this divergence of specificity in the two proteins is advantageous, the duplication will become fixed in the population. These two genes are paralogs of one another. If the species with the pair of paralogs diverges into two species, each species will contain the two paralogs. Each gene in each species will now have an ortholog and a paralog in the other species.

1.3 The Purposes of Homology Modeling

Homology modeling of proteins has been of great value in interpreting the relationships of sequence, structure and function. In particular, orthologous proteins usually show a pattern of conserved residues that can be interpreted in terms of 3-D models of the proteins. Conserved residues often form a contiguous active site or interaction surface of the protein, even if they are distant from each other in the sequence. With a structural model, a multiple alignment of orthologous proteins can be interpreted in terms of the constraints of natural selection and the requirements for protein folding, stability, dynamics and function.

For paralogous proteins, 3-D models can be used to interpret the similarities and differences in the sequences in terms of the related structure, but different functions of the proteins concerned [121]. In many cases, there are significant insertions and deletions and amino acid changes in the active or binding site between paralogs. However, by grouping a set of related proteins into individual families, orthologous within each group, the evolutionary process that changed the function of the ancestral sequences can be observed. Indeed, homology models can serve to help us identify which protein belongs to which functional group by the conservation of important residues in the active or binding site [62]. A number of recent papers have been published that use comparative modeling to predict or establish protein function [95,106,142,222, 225]; see also Chapter 33.

Another important use of homology modeling is to interpret point mutations in protein sequences that arise either by natural processes or by experimental manipulation. The human genome project has produced significant amounts of data concerning polymorphisms and other mutations potentially related to differences in susceptibility, prognosis and treatment of human disease. There are now many such examples, including the Factor V/Leiden R506Q mutation [247] that causes increased occurrence of thrombosis, mutations in cystathionine β-synthase that cause increased levels of homocysteine in the blood, a risk factor for heart disease [101], and BRCA1 for which many sequence differences are known, some of which may lead to breast cancer [34]. At the same time, there are many polymorphisms in important genes that have no discernible effect on those who carry them. At least for some of these, there may be some effect that has yet to be measured in a large enough population of patients and therefore the risk of cancer, heart disease or other illness to these patients is unknown. This is yet another important application of homology modeling, since a good model may indicate readily which mutations pose a likely risk and which do not [92].

Homology models may also be used in computer-aided drug design, especially when a good template structure is available for the target sequence. For enzymes that maintain the same catalytic activity, the active site may be sufficiently conserved such that a model of the protein provides a reasonable target for computer programs which can suggest the most likely compounds that will bind to the active site (see also Chapter 16). This has been used successfully in the early development of HIV protease inhibitors [223, 224] and in the development of anti-malarial compounds that target the cysteine protease of *Plasmodium falciparum* [166].

1.4 The Effect of the Genome Projects

The many genome projects now completed or underway have greatly affected the practice of homology modeling of protein structures. First, the many new sequences have provided a large number of targets for modeling. Second, the large amount of sequence data makes it easier to establish remote sequence relationships between proteins of unknown structure and those of known structure on which a model can be built. The most commonly used methods for establishing sequence relationships such as PSI-BLAST [3] are dependent on aligning many related sequences to compile a pattern or profile of sequence variation and conservation for a sequence family. This profile can be used to search among the sequences in the PDB for a relative of the target sequence (see Chapter 11). The more numerous and more varied sequences there are in the family, the more remote are the homologous relationships that can be determined and the more likely it is that a homologous template for a target sequence can be found. Third, it is likely that the accuracy of sequence alignments between the sequence of unknown structure that we are interested in and the protein sequence of a template are also greatly improved with profiles established from many family members of the target sequence [184]. Fourth, the completion of a number of microbial genomes has prodded a similar effort among structural biologists to determine the structures of representatives of all common protein sequence families, or all proteins in a prototypical genome, such as *Mycobacterium tuberculosis* [15, 126, 163, 210, 241]. Protein structures determined by X-ray crystallography or NMR spectroscopy are being solved at a much faster pace than was possible even 10 years ago. The great increase in the number of solved protein structures has a great impact on the field of homology modeling, since it becomes ever more likely that there will be a template structure in the PDB for any target sequence of interest [221] (see also Chapter 13).

Given the current sequence and structure databases, it is of interest to determine what fraction of sequences might be modeled and the range of sequence identities between target sequences and sequences of known structure. In Figure 2, we show histograms of sequence identities of the sequences in several genomes and their nearest relatives of known structure in the PDB. These relationships were determined with PSI-BLAST as described in the legend. PSI-BLAST is fairly sensitive in determining distant homology relationships [85, 184, 232], although more sensitive techniques exist (see Chapter 11). The results indicate that on average 30–40% of genomic protein sequences are

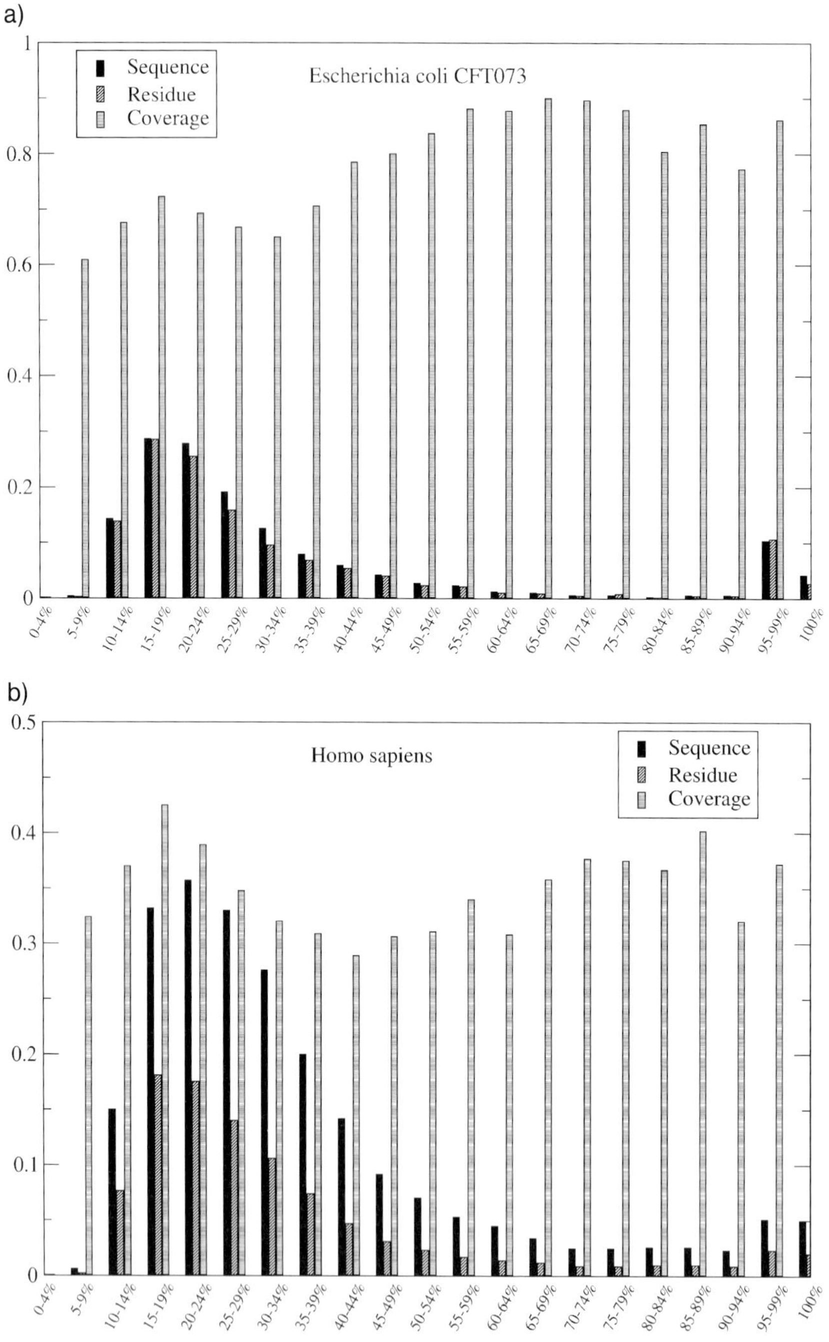
a)
Escherichia coli CFT073
Sequence
Residue
Coverage
1
0.8
0.6
0.4
0.2
0
0-4%
5-9%
10-14%
15-19%
20-24%
25-29%
30-34%
35-39%
40-44%
45-49%
50-54%
55-59%
60-64%
65-69%
70-74%
75-79%
80-84%
85-89%
90-94%
95-99%
100%
b)
Homo sapiens
Sequence
Residue
Coverage
0.5
0.4
0.3
0.2
0.1
0
0-4%
5-9%
10-14%
15-19%
20-24%
25-29%
30-34%
35-39%
40-44%
45-49%
50-54%
55-59%
60-64%
65-69%
70-74%
75-79%
80-84%
85-89%
90-94%
95-99%
100%

easily identified as related to proteins of known structure, which presents a large number of potential targets for homology modeling. However, it should also be pointed out that the average sequence identity between target sequences and template structures in the PDB is less than 25%.

The low sequence identity between target and template sequences in Figure 2 presents a major challenge for homology modeling practitioners, since a major determinant in the accuracy of homology modeling is the sequence identity between the target sequence and the sequence of the template structure. At levels below 30% sequence identity, related protein structures diverge significantly and there may be many insertions and deletions in the sequence [31]. At 20% sequence identity, the average RMSD of core backbone atoms is 2.4 Å [31, 169]. However, as demonstrated in Figure 2, it is likely that we will most often face a situation where the target and template sequences are remotely related. Most widely used homology modeling methods have been predicated on much higher sequence identities between template and target, usually well above 30% [43, 155, 181]. What methods should be used at sequence identities in the 10–30% range is of crucial importance in this post-genomic era.

2 Input Data

To produce a protein model that will be useful and informative requires more than placing a new sequence onto an existing structure. A large amount of sequence data and other kinds of experimental data can often be gathered on the target sequence and on its homolog of known structure to be used for model building. This information can be used to build a better model *and* as the data to be interpreted in light of the model. The goal is to forge an

Figure 2 Distribution of sequence identities between protein in four genomes and their closest homologs in the PDB for those sequences in genomes with homologs in the PDB. PSI-BLAST was used to search the nonredundant protein sequence database with a representative set of PDB sequences as queries. The program was run for four iterations, with a maximum E-value of 0.0001 used to determine sequences which are included in the position-specific similarity matrix. After four iterations, each matrix was used to search each of the four genomes. Coiled-coil and low-sequence complexity sequences were removed from each genome and the nonredundant sequence database. All hits in the genomes with E-values less than 0.001 were saved and the histograms were built from the PSI-BLAST-derived sequence identities.

integrated model of the protein sequence, structure, and function, not merely to build a structure. In Table 1, we list the kinds of information that might be available for a target protein and how these data might be processed.

Table 1 Input information for homology model building

Target sequence

- Target orthologous relatives (from PSI-BLAST)
- Target paralogous relatives (from PSI-BLAST)
- Multiple sequence alignment of orthologs and paralogs (either BLAST multiple alignment or (preferably) other multiple alignment program)
- Sequence profile of ortho/paralogs

Template sequences and structures

- Homolog(s) of known structure [template(s)] determined by database search methods (BLAST, PSI-BLAST, intermediate sequence search methods, HMMs, fold recognition methods)
- Template orthologous sequences
- Template paralogous sequences
- Multiple sequence alignment of template orthologs and paralogs
- Biological units of available templates from RCSB and EBI/PQS

Alignment of target sequence to template sequence and structure

- Pairwise alignment
- Profile alignment
- Multiple sequence alignment of target and template sequence relatives
- Profile–profile alignment
- Fold recognition alignment
- Visual examination of proposed alignments and manual adjustment
- Assessment of confidence in alignment by residue (some regions will be more conserved than others)

Structure alignment of multiple templates, if available

- Align by structure (fssp, VAST, CE, etc.)
- Compare sequence alignments from structure to sequence alignments from multiple sequence alignments (see above)

Experimental information

- Mutation data (site directed, random, naturally occurring)
- Functional data, e.g. DNA binding, ligands, metals, catalysis, etc.
- Oligomer data, e.g. analytical ultracentrifugation, native gel electrophoresis

Since proteins act through their interactions with other molecules, it is important to gather information on known or putative ligands or binding partners of the target. Indeed, the target of the modeling may not be a single protein but a protein complex. As the number of structures of multi-protein complexes increases, there are more and more templates for this kind of modeling. Many proteins act as homo-multimers and so it is important to know whether the goal of modeling is a dimer or tetramer or other multimer of the target. While this may not be known for the modeling target itself, it may be known experimentally for homologs of the target through various experiments, including analytical ultracentrifugation, native gel electrophoresis and of course X-ray crystallography (see below in this section). Information on protein–protein interactions of the target, DNA binding, and other ligands such as ions and organic substrates or cofactors is also important and may be included in the modeling.

With the large amount of sequence information available, it is almost always possible to produce a multiple alignment of sequences related to the target protein. The first step in modeling therefore is to use a database search program such as PSI-BLAST [3] against a nonredundant protein sequence database such as NCBI's *nr* database [13] or the curated UniProt database [7]. With some care, a list of relatives to the target sequence can be gathered and aligned. PSI-BLAST provides reasonable multiple alignments, but it may be desirable to take the sequences identified by the database search and realign them with a multiple sequence alignment program such as ClustalW [211] and Muscle [55]. PSI-BLAST tends to create multiple sequence alignments with many gaps, because insertions relative to the query may be placed at slightly different positions (see also Chapter 3).

It may be that a database search consisting of several rounds of PSI-BLAST will provide one or more sequences of known 3-D structure. If this is not the case then more sensitive methods based on fold recognition or hidden Markov models (HMMs) [6, 8, 23, 53, 54, 93] of protein superfamilies may identify a suitable template structure (see Chapter 11). Once a template structure is identified, a sequence database search will provide a list of relatives of the template, analogous to searches for relatives of the target. At this stage it is useful to divide the sequences related to the target into orthologs of either the target or the template (or both). The sequence variation within the set of proteins that are orthologous to the target provides information as to what parts of the sequence are most conserved and therefore likely to be most important in the model. Similarly variation in the set of proteins that are orthologous to the template provide a view of the template protein family that can be used to identify features in common or distinct in the template and target families. These features can be used to evaluate and adjust a joint multiple alignment of both families.

If there are multiple structures in the PDB that are homologous to the target sequence, then it is necessary to evaluate them to determine which PDB entry will provide the best template structure and whether it will be useful to use more than one structure in the modeling process. In the case of a single sequence that occurs in multiple PDB entries, it is usually a matter of selecting the entry with the highest resolution or the most appropriate ligands (DNA, enzyme inhibitors, metal ions). In other cases, there may be more than one homolog related to the target sequence, and the task is to select the one more closely related to the target or to combine information from more than one template structure to build the model. To do this, a structure alignment of the potential templates can be performed with one of a number of available computer programs (Dali [82], CE [194], etc.). From alignments of the target to the available templates, the location of insertions and deletions can be observed, and often it will be clear that one template is better than others. This may not be uniform, however, such that some regions of the target may have no insertions or deletions with respect to one template, but other regions are more easily aligned with the other template. In this case, a hybrid structure may be constructed [207].

As noted above, it may be desirable to build a particular multimer of the target sequence. It is therefore important to gather information on the biological units for the available template structures. The biological unit is defined as the likely oligomeric state of a protein in its relevant biological context. By contrast, the asymmetric unit is the object for which there is independent experimental information in the crystallographic experiment. The asymmetric unit may be a monomer or dimer or higher multimer of the protein or proteins in the crystal. Quite often the biological unit is present within the crystal and may or may not coincide with the asymmetric unit. In some cases it may be made of parts or all of more than one asymmetric unit. In other cases, the asymmetric unit is composed of more than one biological unit.

The possibilities are illustrated in Figure 3, where the asymmetric units from three different crystal structures of hemoglobin are shown. Hemoglobin is a tetramer consisting of two α- and two β-chains. In the first structure, the asymmetric unit consists of an entire tetramer and therefore coincides with the biological unit. The second structure contains only an α–β dimer and therefore the biological unit is constructed with the space group symmetry operators to form a tetramer. In the third case, the biological unit consists of two tetramers and therefore contains two copies of the biological unit.

The probable biological units are obtainable from both the PDB and the European Bioinformatics Institute (EBI) from their Protein Quaternary Server (PQS) [80]. Often these two sources do not agree on the biological unit for a particular PDB entry and they should be interpreted as hypothetical

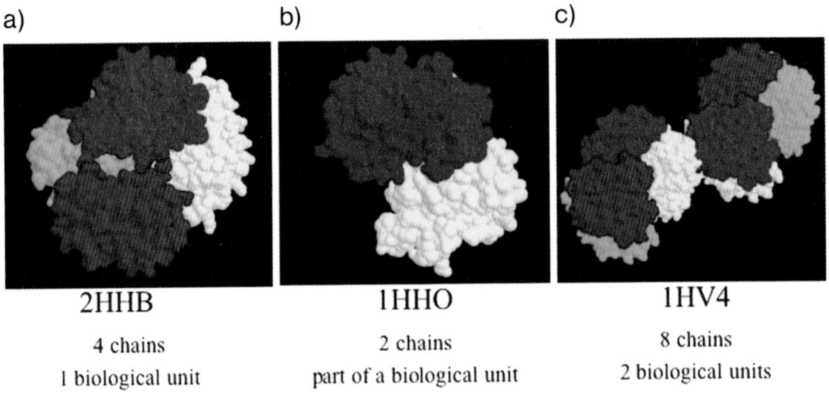

Figure 3 Asymmetric units for hemoglobin from three different structures. The biological unit consists of four chains (two α- and two β-chains). Three scenarios are shown: (a) the asymmetric unit consists of exactly one biological unit, (b) the asymmetric unit is smaller than a biological unit (in this case, it is one half of a biological unit) and (c) the asymmetric unit is larger than the biological unit (in this case, it is two biological units).

oligomers. By comparing the asymmetric units with those from the PDB and the EBI, we found that for over 50% of structures, the asymmetric unit does not correspond to the biological unit for PDB or PQS or both. The PDB and PQS agree 80% of the time on the biological unit (see Section 4.3). It is therefore important to choose a template that has the correct multimer status in its biological unit and to use this biological unit in the modeling process, rather than the asymmetric unit.

Finally, any other experimental data available on the target or template proteins may be very helpful in producing and interpreting a structural model. These can include inhibitor studies, DNA binding and sequence motifs, proteolysis sites, metal binding, mutagenesis data, etc. A number of databases are available on the web that summarize information on particular genes or that collect information on mutations and polymorphisms linked to disease, including: the Cancer Genome Anatomy Project [201], the Online Mendelian Inheritance in Man (OMIM) [76] and the Human Gene Mutation Database [102, 200].

3 Methods

3.1 Modeling at Different Levels of Complexity

Once an alignment is obtained between the target and a protein of known structure (as described in Section 2, and in Chapters 3 and 10), it is possible to build a series of models of increasing sophistication.

(i) *Simple model*: keep backbone and conserved side chains by renaming and renumbering coordinates in the template structure with the new sequence using the alignment of target and template; rebuild other side chains using a side chain modeling program (e.g. SCWRL [22, 30, 47]); do not model insertions or deletions (i.e. do not build new loops and do not close up gaps).

(ii) *Stepwise model*: borrow core backbone from template structure, minus coil regions with insertions or deletions in the sequence alignment; rebuild core side chains; rebuild coil regions with loop prediction method in conjunction with side-chain prediction method. Core backbone and side chains may or may not be held fixed during loop prediction. The entire model may be refined using energy minimization, Monte Carlo or molecular dynamics techniques.

(iii) *Jigsaw model*: borrow backbone from a common core of several structurally aligned templates, using loop regions from different templates according to the alignments, usually keeping those loops for which there is no gap in the alignment with the target sequence. Some loops may need to be modeled.

(iv) *Global model*: build entire protein from spatial restraints drawn from known structure(s) and sequence alignment (e.g. MODELLER [174,180]).

It is not always the case that more sophisticated models are better than simpler, less-complete ones. If elements of secondary structures are allowed to move away from their positions in the template and large changes are made to accommodate insertions and deletions, it may be the case that the model is further away from the target structure (if it were known) than the template structure was to begin with. This is the "added value" problem discussed by John Moult at the Critical Assessment of Protein Structure Prediction (CASP) meetings [138,140,141]. We would like methods that move the template structure closer to the target structure, such that they "add value" to a simple model or unrefined stepwise model based on an unaltered template structure, with side-chains replaced. Extensive energy minimization or molecular dynamics simulations often bring a model further away from the correct structure than toward it [59,98].

The simple model is sometimes justified when there are no insertions and deletions between the template and target or when these sequence length changes are far from the active site or binding site of the protein to be modeled. This often occurs in orthologous enzymes that are under strong selective pressure to maintain the geometry of the active site. Even in nonorthologous enzymes, sometimes we are most interested in an accurate prediction of the active-site geometry and not in regions of the protein distant from the active site.

A stepwise model is probably the most common method used in homology modeling, since it is conceptually simpler than the more complex models and since each piece can be constructed and examined in turn. Some programs therefore proceed by taking the sequence–structure alignment, removing all regions where there are insertions and deletions, and reconstructing loops and side-chains against the fixed template of the remaining atoms. Some methods may also allow all parts of the template structure to adjust to the changes in sequence and insertions and deletions. This usually takes the form of a Monte Carlo or molecular dynamics simulation [118]. A global model, as described above, rebuilds a structure according to constraints derived from the known template structure or structures. This is in contrast to stepwise models that proceed essentially by replacing parts of the template structure and perhaps perturbing the structure.

Many computer programs for homology modeling are developed to solve a single problem, such as loop or side-chain building, and may not be set up to allow all atoms of the protein to adjust or to model many components simultaneously. In many cases these methods have been tested by using simplified modeling situations. Such examples include experiments with removing and rebuilding loops onto single protein structures, and stripping and rebuilding all side chains. In the next sections we review some of the work in these two areas.

3.2 Side-chain Modeling

3.2.1 Input Information

Side-chain modeling is a crucial step in predicting protein structure by homology, since side-chain identities and conformations determine the specificity differences in enzyme active sites and protein binding sites. The problem has been described as "solved" [117], although new methods [120, 133, 157, 193, 234] or improvements on older ones [30] continue to be published. Some side-chain prediction methods stand on their own and are meant to be used with a fixed backbone conformation and sequence to be modeled given as input. Other methods have been developed in the context of general homology modeling methods, including the prediction of insertion-deletion regions. Even when using general modeling procedures, such as MODELLER, it may be worthwhile subsequently to apply a side-chain modeling step with other programs optimized for this purpose [220]. This is especially the case when side-chain conformations may be of great importance to interpretation of the model. It is also often the case that insertion-deletion regions are far away from the site of interest and loop modeling may be dispensed with. Indeed, significant alterations of the backbone of the template, if they are not closer to the target to be modeled (if it were known) than the template, may in fact result in poorer side-chain modeling than if no loop modeling were performed. As described above, the choice of template may depend not only on sequence identity but also on the absence of insertions and deletions near the site of interest. If this is successful, side-chain modeling rises in importance in relation to loop prediction.

Side-chain prediction methods described in detail in the literature have a long history although only a small number of programs are currently publicly available (see Table 2). Nearly all assume a fixed backbone, which may be from a homologous protein of the structure to be modeled, or may be the actual X-ray backbone coordinates of the protein to be modeled. Many methods have in fact only been tested by replacing side-chains onto backbones taken from the actual 3-D coordinates of the proteins being modeled ("self-backbone predictions"). Nevertheless, these methods can be used for homology modeling by first substituting the target sequence onto the template backbone and then modeling the side chains. When a protein is modeled from a known structure, information on the conformation of some side chains may be taken from the template [22, 30, 204]. This is most frequently the case when the template and target residue are identical, in which case the template residue's Cartesian coordinates may be used. These may be kept fixed as the other side chains are placed and optimized or they may be used only as a starting conformation and optimized with all other side chains. Only a small number of methods use information about nonidentical side chains borrowed from the template. For instance, Phe $\leftrightarrow$ Tyr substitutions only require the building or removal of a hydroxyl group while Asn $\leftrightarrow$ Asp substitutions require changing one of δ-atoms from NH_2 to O or vice versa. Summers and Karplus [203, 204] used a more detailed substitution scheme, by which for instance the χ_1 angle of very different side-chain types (e.g. Lys $\leftrightarrow$ Phe) might be used in building side chains. In the long run, this is probably not advantageous, since the conformational preferences of nonsimilar side-chain types may be quite different from each other [50].

Table 2 Publicly available side-chain prediction programs

Program	Availability	Website
SMD	download	http://condor.urbb.jussieu.fr/Smd.php
Confmat, Decorate	web	http://lorentz.immstr.pasteur.fr/website/projects
CARA/GeneMine	download	http://www.bioinformatics.ucla.edu/genemine
RAMP	download	http://www.ram.org/computing/ramp
SCAP	download	http://honiglab.cpmc.columbia.edu/programs/sidechain
SCWRL	download	http://dunbrack.fccc.edu/scwrl
Maxsprout/Torso	web	http://www.ebi.ac.uk/maxsprout
SCATD	download	http://www.bioinformatics.uwaterloo.ca/%7Ej3xu
PLOP	download	http://francisco.compbio.ucsf.edu/~jacobson/plop_manual/plop_overview

3.2.2 **Rotamers and Rotamer Libraries**

Nearly all side-chain prediction methods depend on the concept of side-chain *rotamers* (reviewed in Ref. [49]). From conformational analysis of organic molecules, it was predicted long ago [182,183] that protein side chains should attain a limited number of conformations because of steric and dihedral strain within each side chain, and between the side chain and the backbone. Dihedral strain occurs because of Pauli exclusion between bonding molecular orbitals in eclipsed positions [94]. For sp^3–sp^3 hybridized bonds, the energy minima for the dihedral are at the staggered positions that minimize dihedral strain at approximately 60°, 180°, and –60°. For sp^3–sp^2 bonds, the minima are usually narrowly distributed around +90° or –90° for aromatics and widely distributed around 0° or 180° for carboxylates and amides (e.g. Asn/Asp χ_2 and Glu/Gln χ_3).

As crystal structures of proteins have been solved in increasing numbers, a variety of rotamer libraries have been compiled with increasing amounts of detail and greater statistical soundness, i.e. with more structures at higher resolution [12,17,50–52,88,125,132,161,187,212]. The earliest rotamer libraries were based on a small number of structures [12, 17, 88, 161]. Even the widely used Ponder and Richards library was based on only 19 structures, including only 16 methionines [161]. The most recent libraries are based on over 850 structures with resolution of 1.7 Å or better and mutual sequence identity less than 50% between any two chains used.

Most rotamer libraries are backbone-conformation-independent. In these libraries, the dihedral angles for side chains are averaged over all side chains of a given type and rotamer class, regardless of the local backbone conformation or secondary structure. The most recent of these is by Lovell and coworkers [125], who derived a more accurate backbone-independent rotamer library by eliminating side chains of low stereochemical quality, including those with high B-factors, steric conflicts in the presence of predicted hydrogen atom locations, and other factors. The statistical analysis does not rely on a parametric distribution function such as the normal model, and hence can model factors like skew in an unbiased way.

Several libraries have been proposed that are dependent on the conformation of the local backbone [50–52,132,187]. McGregor and coworkers [132] and Schrauber and coworkers [187] compiled rotamer probabilities and dihedral angle averages in different secondary structures We have used Bayesian statistical methods to compile a backbone-dependent rotamer library with rotamer probabilities and average angles and standard deviations at all values of the backbone dihedral angles ϕ and ψ in 10° increments [49–52]. The current version of this library is based on 850 chains with resolution better than 1.7 Å and less than 50% mutual sequence identity.

Finally, there is an alternative form of a rotamer library that includes large numbers of conformations of each side-chain type in the form of Cartesian coordinates. These libraries therefore include variation in bond lengths and bond angles, as well as dihedral angles. They are generally used for fine sampling of side-chain positions in the context of side-chain prediction. For instance, Xiang and Honig [234] produced a library consisting of 7560 conformations for use in their side-chain prediction method from a set of 297 high-resolution structures. The variation in bond angles and dihedrals away from average values is particularly useful for larger side chains for which a small change in an angle near the base of the side chain may cause large motions of atoms at the far end of the side chain. Other groups have also used large rotamer libraries to introduce flexibility about mean dihedral angles of rotamers as well as variation in bond lengths and bond angles [157,193].

3.2.3 Side-chain Prediction Methods

Side-chain prediction methods can be classified in terms of how they treat side-chain dihedral angles (rotamer library, grid or continuous dihedral angle distribution), bond lengths and bond angles (fixed, variable, sampled from Cartesian conformers), potential energy function used to evaluate proposed conformations, and search strategy.

The potential energy functions in side-chain prediction methods have varied tremendously from simple steric exclusion terms to full molecular mechanics potentials. In most cases, the potential energy function is a standard Lennard–Jones potential:

$$E(r) = 4\varepsilon\left[\left(\frac{\sigma}{r}\right)^{12} - \left(\frac{\sigma}{r}\right)^{6}\right]. \tag{1}$$

In this equation, r is the distance between two nonbonded atoms, and ε and σ are parameters that determine the shape of the potential. This potential has a minimum at the distance $r = 2^{1/6}\sigma$ and a well depth of ε. Different values of σ and ε may be chosen for different pairs of atom types. Some potential energy functions for side chains may also include a hydrogen bond term. Depending on the potential parameters, these potentials may not accurately model the relative energies of rotamers for each side-chain type that are determined from local interactions within each side chain and between the side chain and the local backbone. For instance, in molecular mechanics potentials, interactions between atoms connected by three covalent bonds (atoms i and $i+3$ in a chain) are not usually treated by van der Waals terms, but rather in torsion terms of the form [127]:

$$E(\tau) = \sum_{m} K_m \cos(m\tau + a_m) \tag{2}$$

where the sum over m may include 1-, 2-, 3-, 4- and 6-fold cosine terms. The K_m and a_m are constants specific for each dihedral angle and each term in the sum. These torsion terms are included in some side-chain prediction methods, but ignored in others [96].

Electrostatic interactions in the form of a Coulomb potential have been included in methods that rely on full molecular mechanics potentials, usually with a distance-dependent dielectric, $\varepsilon(r) = r$:

$$E = \frac{q_i q_j}{\varepsilon(r) r} \tag{3}$$

Solvent interactions are also usually ignored, since these can be difficult or expensive to model properly (for exceptions, see [120, 185, 228]).

A number of side-chain methods use an energy term based on the probability of rotamers as a function of backbone conformation. These probabilities are given in the backbone-dependent rotamer library, and the energy function is usually of the form:

$$E_i = -K \ln \left(\frac{p_i(\phi, \psi, R)}{p_{\max}(\phi, \psi, R)} \right) \tag{4}$$

where the energy of rotamer i is expressed as a function of the probability of this rotamer given the backbone dihedrals ϕ and ψ and the residue type R, and the probability of the most common rotamer for the same backbone dihedrals and residue type. The constant K is empirical and can be optimized given the other terms in the energy function.

Side-chain conformation prediction incurs the risk of combinatorial explosion, since there are on the order of n_{rot}^N possible conformations, where n_{rot} is the average number of rotamers per side chain and N is the number of side chains. However, in fact, the space of conformations is much smaller than that, since side chains can only interact with a small number of neighbors, and in most cases clusters of interacting side chains can be isolated and each cluster can be solved separately [22, 212]. Also, many rotamers have prohibitively large interactions with the backbone and are at the outset unlikely to be part of the final predicted conformation. These can be eliminated from the search early on.

Many standard search methods have been used in side-chain conformation prediction, including Monte Carlo simulation [83, 109, 116, 120, 137, 167], simulated annealing [86], self-consistent mean field calculations [96, 133, 134], the dead-end elimination (DEE) method [40–42, 70, 107, 123, 159], neural networks [99] and graph theory [30, 111, 236].

Self-consistent mean field calculations represent each side chain as a set of conformations, each with its own probability. Each rotamer of each side chain has a certain probability, $p(r_i)$. The total energy is a weighted sum of the interactions with the backbone and interactions of side chains with each other:

$$E_{\text{tot}} = \sum_{i=1}^{N} \sum_{r_i=1}^{n_{\text{rot}}(i)} p(r_i) E_{bb}(r_i) + \sum_{i=1}^{N-1} \sum_{r_i=1}^{n_{\text{rot}}(i)} \sum_{j=i+1}^{N} \sum_{r_j=1}^{n_{\text{rot}}(j)} p(r_i) p(r_j) E_{\text{sc}}(r_i, r_j) \quad (5)$$

In this equation, $p(r_i)$ is the density or probability of rotamer r_i of residue i, $E_{bb}(r_i)$ is the energy of interaction of this rotamer with the backbone, and $E_{\text{sc}}(r_i,r_j)$ is the interaction energy (van der Waals, electrostatic) of rotamer r_i of residue i with rotamer r_j of residue j. Some initial probabilities are chosen for the ps in Eq. (5) and the energies calculated. New probabilities $p\prime(r_i)$ can then be calculated with a Boltzmann distribution based on the energies of each side chain and the probabilities of the previous step:

$$\begin{aligned} E(r_i) &= E_{bb}(r_i) + \sum_{j=1, j\neq i}^{N} \sum_{r_j=1}^{n_{\text{rot}}(j)} p(r_j) E_{\text{sc}}(r_i, r_j) \\ p'(r_i) &= \frac{\exp(-E(r_i)/kT)}{\sum_{r_i=1}^{n_{\text{rot}}(i)} \exp(-E(r_i)/kT)} \end{aligned} \quad (6)$$

Alternating steps of new energies and new probabilities can be calculated from the expressions in Eq. (6) until the changes in probabilities and energies in each step become smaller than some tolerance.

The DEE algorithm is a method for pruning the number of rotamers used in a combinatorial search by removing rotamers that cannot be part of the global minimum energy conformation [41, 42, 70, 107, 108, 123]. This method can be used for any search problem that can be expressed as a sum of single-residue terms and pairwise interactions. Goldstein's improvement on the original DEE can be expressed as follows [70]. If the total energy for all side chains is expressed as the sum of singlet and pairwise energies:

$$E = \sum_{i=1}^{N} E_{bb}(r_i) + \sum_{i=1}^{N-1} \sum_{j>i}^{N} E_{\text{sc}}(r_i, r_j) \quad (7)$$

then a rotamer r_i can be eliminated from the search if there is another rotamer s_i for the same side chain that satisfied the following equation:

$$E_{bb}(r_i) - E_{bb}(s_i) + \sum_{j=1, j\neq i}^{N} \min_{r_j} \{E_{\text{sc}}(r_i, r_j) - E_{\text{sc}}(s_i, r_j)\} > 0 \quad (8)$$

In words, rotamer r_i of residue i can be eliminated from the search if another rotamer of residue i, s_i, always has a lower interaction energy with all other side chains regardless of which rotamer is chosen for the other side chains. More powerful versions have been developed that eliminate certain pairs of rotamers from the search [42,70,123]. DEE-based methods have also proved very useful in protein design, where there is variation of residue type as well as conformation at each position of the protein [37,71,218].

The current SCWRL algorithm [30] uses graph theory to solve the combinatorial problem. In this method, each side chain in the protein is considered a node in an undirected graph. An edge exists between two nodes i and j if at least one rotamer of residue i and one rotamer of residue j interact with each other, i.e. have a nonzero interaction energy. This produces a number of separate graphs that are not connected to each other. Each of these graphs can then be solved for the minimum energy conformation of the residues in the graph. To accomplish this, each separate graph is broken up into its biconnected components, as shown in Figure 4(a). Biconnected components are cycles or nested cycles or bridges consisting of two nodes connected by an edge. Two biconnected components share a node called an articulation point, which when removed from the graph breaks the graph into two (or more) connected subgraphs. The global minimum of the energy can be found by beginning on the outside of the graph with biconnected components that have only one articulation point. For each rotamer of the articulation point, the minimum energy of the other rotamers is found and stored with the rotamer of the articulation point. Then the biconnected graph is "collapsed" onto the articulation point residue. This residue now contains information on all the residues in the biconnected component. The procedure continues to collapse biconnected components, until a single component is left, as shown in Figure 4.

Recently, two papers [111,236] have appeared that extend the graph theory algorithm further so that the smallest groups that need to be searched are much smaller than biconnected components. In this method, some nodes can be removed from the graph by collapsing a node or nodes onto an *edge*. This is shown in Figure 4(b), in which a single node that has two neighbors in the graph is collapsed onto an edge between the two neighbors. The new energy of each rotamer pair for residues i and j is now:

$$E_{\text{pair}}^{\text{new}}(r_i, r_j) = E_{\text{pair}}^{\text{old}}(r_i, r_j) + \min_{r_k} \left\{ E_{\text{self}}(r_k) + E_{\text{pair}}(r_i, r_k) + E_{\text{pair}}(r_j, r_k) \right\} \quad (9)$$

The size of the smallest group that must be solved combinatorially is called the tree width and is related to the size of the largest group of side chains that are all mutually connected to each other.

In most methods, the search is over a well-defined set of rotamers for each residue. As described above, these represent local minima on the side-chain conformational potential energy map. In several methods, however, nonrotamer positions are sampled. Summers and Karplus used CHARMM to calculate potential energy maps for side chains based on 10° grids [203,204]. Dunbrack and Karplus used CHARMM to minimize the energy of rotamers

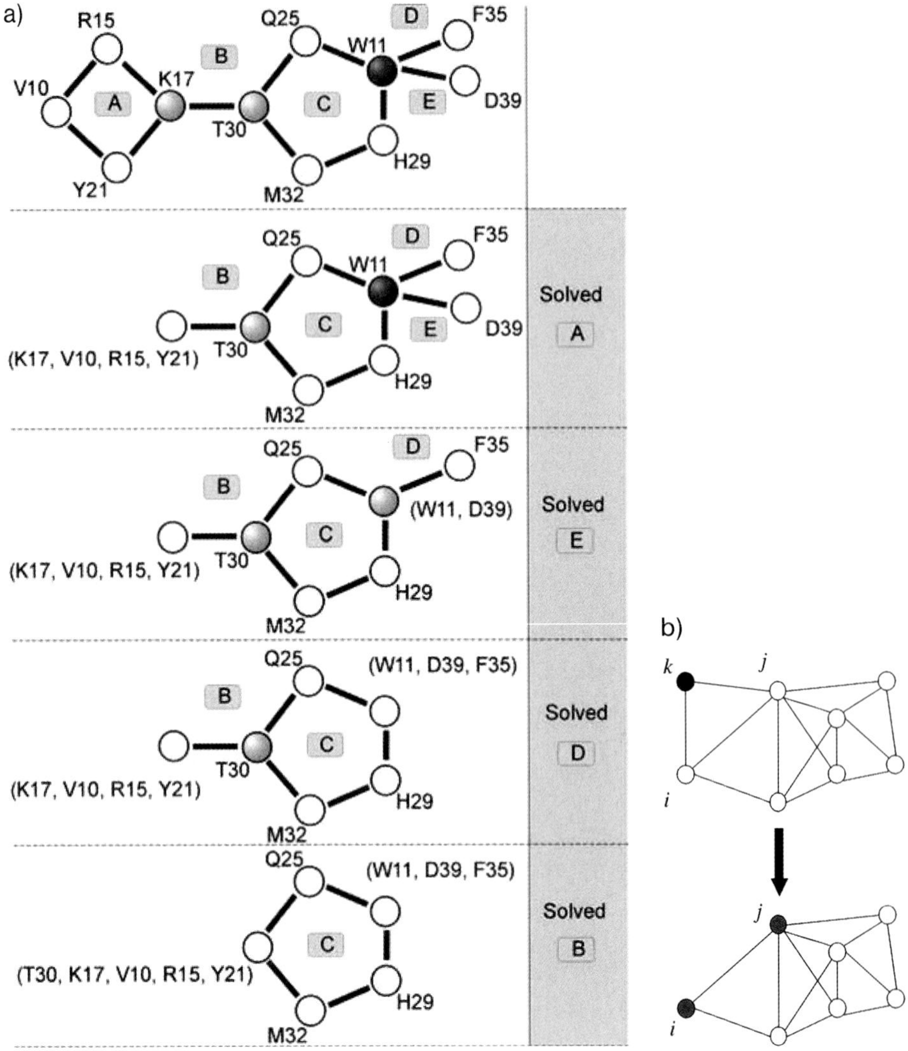

Figure 4 (a) Graph algorithm used in SCWRL3.0, solving a cluster using biconnected components. The minimum energy configuration of the cluster shown in Figure 1 is identified by stepwise solution of biconnected components. Each biconnected component is solved as shown in the right margin and the collapsed component is shown as superresidues in curly brackets. (b) Collapsing a node onto an edge.

from canonical starting conformations (–60°, 180° and +60°) [51]. Vasquez also used energy minimization [213], while Lee and Subbiah used a search over 10° increments in dihedral angles with a simple van der Waals term and a 3-fold alkane potential on side-chain dihedrals [112]. Mendes and coworkers [133, 134] used a mean-field method to sample from Gaussian distributions about the conformations in the rotamer library of Tuffery and coworkers [212].

3.2.4 Available Programs for Side-chain Prediction

While many methods for side-chain prediction have been presented over the years, only a small number of programs are publicly available at this time. Information on obtaining these programs is given in Table 2. We define "available" as either being downloadable (in source or executable form or both) from the Internet or able to be run from a webserver. Some authors will also provide their programs on request, but these programs do not generally have documentation nor are they designed for general use. They are not listed in Table 2.

3.3 Loop Modeling

3.3.1 Input Information

In stepwise construction methods, backbone segments that differ in length between the template and target (according to the sequence alignment) need to be rebuilt. In some situations, even when the sequence length of a coil segment is maintained, it may be necessary to consider alternative conformations to accommodate larger side chains or residues with differing backbone conformational requirements, Gly $\leftrightarrow$ non-Gly or Pro $\leftrightarrow$ non-Pro mutations. Most such loop construction methods have been tested only on native structures from which the loop to be built has been removed. However, the reality in homology modeling is more complicated, requiring several choices to be made in building the complete structure. These include how much of the template structure to remove before loop building, whether to model all side chains of the core before rebuilding the loops, and whether to rebuild multiple loops simultaneously or serially.

Deciding how much of the template structure to remove before loop building depends on examination of the sequence alignment and the template structure [114, 230]. Sequence alignments with insertions and deletions are usually not unambiguous. Most sequence alignment methods ignorant of structure will not juxtapose a gap in one sequence immediately adjacent to a gap in another sequence, i.e. they will produce an alignment that looks like this alignment:

```
AGVEPMENYKLS
SG---LDDFKLT
```

rather than like this one:

```
AGVEPMEN---YKLS
SGL-----LDDFKLT
```

However, the latter alignment is probably more realistic [1], indicating that a five-amino-acid loop in the first sequence and structure is to be replaced with a three-amino-acid loop in the second sequence. The customary practice is to remove the whole segment between two conserved secondary structures units. Even with this practice, ambiguity remains, since the ends of secondary structures, especially α-helices, are not well determined. If loop-building methods were accurate, then removing more of the segment would be a good idea. However, long loops (longer than seven amino acids) are difficult to rebuild accurately and hence there is cause to preserve as much of the starting structure as possible. Once the backbone has been borrowed from the template in stepwise modeling, one has to decide the order of building the core side chains, the backbone of loops to be built and their side chains. They may be built sequentially or allowed to vary simultaneously. Side chains from the core may guide the building of the loop, but at the same time may hinder correct placement. It is certainly the case that in the final structure there must be a reasonably low-energy conformation that can accommodate all loops and side chains simultaneously. Different authors have made different choices, and there has been little attempt to vary the procedure while keeping the search algorithm and potential energy function used fixed.

3.3.2 Loop Conformational Analysis

Loop structure prediction is always based in one way or another on an understanding of loop conformations in experimentally determined structures. Loop conformational analysis has been performed on a number of levels, ranging from classification of loops into a number of distinct types to statistical analysis of backbone dihedral angles. Loop classification schemes have usually been restricted to loops of a particular size range: short loops of one to four residues, medium loops of five to eight residues and long loops of nine residues or longer.

Thornton and coworkers have classified β-turns, which are short loops of two to five residues that connect two antiparallel β-sheet strands [195, 196, 226, 227]. These loops occur in a limited number of conformations that depend on the sequence of the loop, especially on the presence of glycine and proline residues at specific positions. The backbone conformation can be characterized by the conformations of each amino acid in terms of regions of the Ramachandran map occupied (usually defined as α_R, β_P, β_E, γ_R, α_L and γ_L) [227]. Usually one or more positions in the loops require an α_L conformation and therefore a glycine, asparagine or aspartic acid residue. One useful aspect of this analysis is that if a residue varies at certain positions or there are short insertions at certain positions, the effect on the loop can be predicted [196] since the number of possibilities for each length class is small. The programs BTPRED [192] and BHAIRPRED [103] are available (see Table 3) to predict the locations of specific types of β-turns from protein sequences and secondary structure predictions. Single-amino-acid changes tend to maintain the loop conformations, except when Pro residues substitute for residues with $\phi > 0°$, while insertions change the class of the loop.

Table 3 Publicly available loop conformation prediction programs

Program	Availability	Website
Rapper	web, download	http://raven.bioc.cam.ac.uk
ModLoop	web	http://alto.compbio.ucsf.edu/modloop//modloop
Loopy	download	http://honiglab.cpmc.columbia.edu/programs/loop
PLOP	download	http://francisco.compbio.ucsf.edu/~jacobson/plop_manual/plop_overview
MODELLER	download	http://salilab.org/modeller

In recent years with a larger number of structures available, medium-length loops have also been classified [38, 44, 56, 69, 104, 115, 135, 147, 149, 150, 229] by their patterns of backbone conformation residue by residue (α_R, β_P, etc.). A number of regularly occurring classes have been found, depending on length, type of secondary structure being connected and sequence. These classes cover many but by no means all of the loops seen in non-β turn contexts.

Longer loops (with more than eight amino acids) have been investigated by Martin and coworkers [131] and Ring and coworkers [165]. Martin and coworkers found that long loops fall into two classes: those that connect spatially adjacent secondary structures and those that connect secondary structures separated by some distance. Ring and coworkers provided a useful classification of longer loops as either strap (long extended loops), Ω loops (similar to those described by Leszczynski [115] and Pal and coworkers [150]), which resemble the Greek letter, and ζ loops, which are nonplanar and have a zigzag appearance. The different loop types were found to have different distributions of virtual C_α–C_α–C_α–C_α dihedrals to accommodate their shapes.

A number of groups have updated the Ramachandran propensities of the 20 amino acids. Swindells and coworkers [209] have calculated the intrinsic ϕ,ψ propensities of the 20 amino acids from the coil regions of 85 protein structures. The distribution for coil regions is quite different than for the regular secondary structure regions, with a large increase in β_P and α_L conformations, and much more diverse conformations in the β_E and α_R regions. Their results also indicate that the 18 non-Gly,Pro amino acid type are in fact quite different from each other in terms of their Ramachandran distributions, despite the fact that they are often treated as identically distributed in prediction methods [25, 57]. Their analysis was divided into the main broad regions of the Ramachandran map, ignoring the α_L region. The results are intriguing, in that the probability distributions are distinct enough even when calculated from a relatively small protein dataset. More recently Lovell and coworkers [124] and Anderson and coworkers [4] have produced new Ramachandran maps based on stricter criteria for inclusion of amino acids based on resolution, R-factors and B-factors, as well as data smoothing techniques that remove outliers and unpopulated parts of the Ramachandran map. Their results indicate that a stricter adherence to "allowed" regions is called for, since nearly all residues in disallowed regions are based on poor electron density.

3.3.3 Loop Prediction Methods

Loop prediction methods can be analyzed for a number of important factors in determining their usefulness: (i) method of backbone construction, (ii) what range of lengths are possible, (iii) how widely is the conformational space searched, (iv) how are side chains added, (v) how are the conformations scored (i.e., the potential energy function) and (vi) how much has the method been tested (length, number, self/nonself).

The most common approach to loop modeling involves using "spare parts" from other (unrelated) protein structures [10, 32, 60, 61, 63, 72, 81, 91, 96, 114, 135, 165, 168, 172, 205, 207, 217, 230, 231]. These database methods begin by measuring the orientation and separation of the backbone segments flanking the region to be modeled, and then search the PDB for segments of the same length that span a region of similar size and orientation. This work was pioneered by Jones and Thirup [91]. They defined a procedure in which C_α–C_α distances were measured among six residues, three on either side of a backbone segment to be constructed. These 15 C_α–C_α distances were used to search structures in the PDB for segments with similar C_α–C_α distances and the appropriate number of intervening residues. Other authors have used the same method for locating potential database candidates for the loop to be constructed [60, 96, 205, 217]. The fragment selection method used in Rosetta *ab initio* modeling [197] is based at least in part on the database approach to loop modeling, and is used in Rosetta for loop construction in homology modeling [167]. In recent years, as the size of the PDB has increased, database methods have continued to attract attention. With a larger database, recurring structural motifs have been classified for loop structures [44, 56, 104, 113, 135, 147, 172], including their sequence dependence.

Although many methods have been published, they have usually only been tested on a small number of loops, and then usually in the context of rebuilding loops onto their own backbones, rather than in the process of homology modeling. A recent exception is that of Fernandez-Fuentes and coworkers [61] who tested the ArchDB database [56] of loops as a predictive tool. They used a "jackknife" test that removed all loops from the same superfamily for each loop in a set of over 10 000 used to construct ArchDB.

The main alternative to database methods is *ab initio* construction of loops by random or exhaustive search mechanisms. These methods are quite varied in their generation and subsequent modification of loop structures to fit the environment of the fitted segment. The initial conformation may be random, starting from the N- or C-terminal anchor, so that the other end of the loop does not connect to the other anchor (the C- or N-terminal anchor, respectively). Such loops can then be closed using energy minimization that places some energetic constraints on a closed loop, or using loop closure methods, such as those based on inverse kinematics in robotics [19, 28, 35]. Other methods have built chains by sampling Ramachandran conformations randomly, keeping partial segments as long as they can complete the loop with the remaining residues to be built [64, 191, 198].

An alternative approach to the loop generation problem is to use a geometrically distorted loop that bridges the two anchors exactly and then to relax the structure into an undistorted protein-like structure. MODELLER starts loop modeling with a linear arrangement of the atoms in the loop, which is then relaxed into a protein-like conformation using energy minimization [65]. Zheng and coworkers used a scaling-relaxation method in which an initially generated or database loop is scaled in size until it fits the anchors [244–246]. This results in very short bond distances and unphysical connections to the anchors. From there, energy minimization is performed on the loop, slowly relaxing the scaling constant, until the loop is scaled back to full size.

One important aspect in the development of a prediction method based on random (or exhaustive) construction of backbone conformations is the free energy function used to discriminate among those conformations that successfully bridge the anchors. Fogolari and Tosatto have found that a free energy function including a molecular mechanics potential energy and a Poisson–Boltzmann solvent-accessible surface area solvation term was able to identify decoys from a large set that were close to the native structure [66]. Jacobson and coworkers recently used the OPLS (optimized potential for liquid simulations) molecular mechanics force field, with improved torsional energy parameters optimized to reproduce quantum-mechanical data and side-chain prediction [87], in combination with a surface-generalized Born/nonpolar (SGB/NP) hydration free energy model [68]. Their search method generated one residue at a time from a 5° resolution backbone model with steric and side-chain checks, from both ends of the loop, followed by clustering and energy minimizations of cluster representatives. Their method was tested on a large set of 833 loops with excellent results for loops up to 12 residues in length.

3.3.4 **Available Programs**

Very few loop modeling programs *per se* are publicly available, although loop modeling is integral to more complete modeling programs. A list of available loop modeling programs is given in Table 3. Some programs that do complete modeling but can be used for loop modeling without further refinement are listed (e.g. MODELLER).

3.4 Methods for Complete Modeling

Homology modeling is a complex process. Automated protocols that begin with a sequence and produce a complete model are few, and the resulting models should be examined with great care (as of course should all models). However, these methods usually allow for (and indeed recommend) some manual intervention in the choice of template structure or structures and in the sequence alignment. In these steps, manual intervention is likely to have important consequences. Later stages of modeling (actual building of the structure) are more easily automated and there are not usually obvious manual adjustments to make.

There are several publicly available programs available for homology modeling that are intended to make complete models from input sequences. These include MODELLER [174, 175, 180, 181], RAMP [176–178] and MolIDE [29]. There are also several webservers that provide homology modeling services, including SWISS-MODEL [74, 153, 154], Esypred [105] and 3D-JIGSAW [11]. Program availability is given in Table 4. Some of these programs provide only BLAST/PSI-BLAST searching followed by model-building with MODELLER (e.g. EsyPRED). We describe some of these programs.

Table 4 Publicly available comparative modeling programs

Program	Availability	Website
3d-JIGSAW	web	http://www.bmm.icnet.uk/servers/3djigsaw
CPHmodels	web	http://www.cbs.dtu.dk/services/CPHmodels
EsyPred	web	http://www.fundp.ac.be/urbm/bioinfo/esypred
FAMS	E-mail server	http://www.pharm.kitasato-u.ac.jp/fams
Geno3D	web	http://geno3d-pbil.ibcp.fr
MODELLER	download	http://salilab.org/modeller
ModWeb	web	http://salilab.org/modweb
Modzinger	web	http://peyo.ulb.ac.be/mz/index
nest	download	http://honiglab.cpmc.columbia.edu/programs/nest
parmodel	web	http://laboheme.df.ibilce.unesp.br/cluster/parmodel_mpi
Robetta	web	http://robetta.bakerlab.org
SDSC	web	http://cl.sdsc.edu/hm
SWISS-MODEL	web	http://swissmodel.expasy.org//SWISS-MODEL

3.4.1 MODELLER

MODELLER takes as input a protein sequence and a sequence alignment to the sequence(s) of known structure(s), and produces a comparative model. The program uses the input structure(s) to construct constraints on atomic distances, dihedral angles, etc., that when combined with statistical distributions derived from many homologous structure pairs in the PDB form a conditional probability distribution function for the degrees of freedom of the protein.

For instance, a probability function for the backbone dihedrals of a particular residue to be built in the model can be derived by combining information in the known structure (given the alignment) and information about the amino acid type's Ramachandran distribution in the PDB. The number of constraints is very large; for a protein of 100 residues there may be as many as 20 000 constraints. The constraints are combined with the CHARMM force field to form a function to be optimized. This function is optimized using conjugate gradient minimization and molecular dynamics with simulated annealing.

3.4.2 **MolIDE: A Graphical User Interface for Modeling**

MolIDE (Molecular Interactive Design Environment) is an open-source, extensible graphical user interface for homology modeling [29]. MolIDE provides a graphical interface for running sequence database searches with PSI-BLAST, searches of the PDB, secondary structure prediction, manual alignment editing, and running loop and side-chain prediction programs. One of MolIDE's main benefits is allowing a user to edit a sequence–structure alignment and to view the positions of insertions and deletions within the template structure in real time. MolIDE also allows manual choice of anchor residues for loop modeling with the assistance of a graphical view of the template protein structure. MolIDE runs on the Windows and Linux operating systems. The use of MolIDE will be illustrated in the next section with an example of comparative modeling of a protein of biological interest.

3.4.3 **RAMP and PROTINFO**

Samudrala and Moult described a method for "handling context sensitivity" of protein structure prediction, i.e. simultaneous loop and side-chain modeling, using a graph theory method [178, 179] and an all-atom distance-dependent statistical potential energy function [176]. These methods are also implemented in the PROTINFO webserver listed in Table 4.

3.4.4 **SWISS-MODEL**

SWISS-MODEL is intended to be a complete modeling procedure accessible via a web server that accepts the sequence to be modeled and then delivers the model by electronic mail [74, 154]. In contrast to MODELLER, SWISS-MODEL follows the standard protocol of homolog identification, sequence alignment, determining the core backbone, and modeling loops and side chains. SWISS-MODEL will search a sequence database of proteins in the PDB with BLAST, and will attempt to build a model for any PDB hits with p-values less than 10^{-5} and at least 30% sequence identity to the target. SWISS-MODEL allows for user intervention by specifying the template(s) and alignments to be used.

If more than one structure is found, the structures will be superimposed on the template structure closest in sequence identity to the target.

SWISS-MODEL determines the core backbone from the alignment of the target sequence to the template sequence(s) by averaging the structures according to their local degree of sequence identity with the target sequence. The program builds new segments of backbone for loop regions by a database scan of the PDB using anchors of four C_α atoms on each end. This method is used to build only the C_α atoms and the backbone is completed with a search of pentapeptide segments in the PDB that fit the C_α trace of the loop. Side chains are now built for those residues without information in the template structure by using the most common (backbone-independent) rotamer for that residue type. If a side chain can not be placed without steric overlaps, another rotamer is used. Some additional refinement is performed with energy minimization with the GROMOS [75] program.

4 Results

4.1 Range of Targets

A very large number of homology models have been built over the years by many authors. Recent targets have included proteins of significant interest in biology and medicine 10- [2,9,16,20,26,27,33,36,58,67,77,128,129,145,162, 171,206,216,242]. Several databases of homology models are available on the Internet, including ModBase [158], FAMSBASE [237] and the SWISS-MODEL repository [100]. Their websites are given in Table 4.

4.2 Example: Protein Kinase STK11/LKB1

The protein kinase STK11 is frequently mutated in human cancers and mutations in this gene are strongly associated with Peutz–Jeghers syndrome [79, 89]. Patients with Peutz–Jeghers syndrome often develop dark spots on the lips and inside the mouth as well as near the eyes and nostrils. These patients also develop polyps in the stomach and intestine, and are very susceptible to cancers of the breast, colon, pancreas, stomach and ovary [188]. This disease is inherited in an autosomal dominant manner, so that a mutation in a single copy of the gene is enough to confer risk [110].

One important use of homology modeling is to understand how missense mutations may lead to disease. In general, missense mutations that have deleterious effects lead to amino changes that either affect stability or dynamics of a folded protein, or affect interactions of the protein product with other molecules, including other proteins, DNA or ligands. While most of the mutations associated with the disease lead to a truncated protein, many missense mutations have also been linked to Peutz–Jeghers syndrome [110]. It is therefore of interest to build a model of the STK11 kinase domain and examine the location and likely effects of disease-associated mutations.

As described above, MolIDE [29] assists a user in modeling a protein by providing a graphical interface to the steps involved in basic homology modeling, including sequence database searching, alignment editing, and loop and side-chain modeling. We obtained the sequence of STK11 from the NCBI website in FASTA format, as shown in the middle panel of Figure 5. Once this sequence is input into MolIDE, the user runs PSI-BLAST from the Tools menu. PSI-BLAST is set up to run several rounds of search against the nonredundant protein database from NCBI. The version of PSI-BLAST distributed with MolIDE has been modified to output a profile matrix with a unique name after each iteration of the search so that each matrix can then be used to search the PDB sequence database included with MolIDE (G. Wang and R.L. Dunbrack, Jr., unpublished). These profiles are also used by the secondary structure prediction program PSIPRED [90], included with MolIDE. The secondary structure predictions are shown in the lower panel of Figure 5, where the red and green colors indicate α-helices and β-sheet strands, respectively, and the intensity of the color represents the confidence level of the prediction. As the profile includes more and more sequences remotely related to the query, the secondary structure prediction also changes. For some proteins, the prediction gets better as the signal from the multiple alignment becomes stronger, while for others the prediction may worsen if many sequences with variations in secondary structure (longer loops, shorter or longer secondary structure elements) get aligned or even misaligned to the profile.

As STK11 is a kinase, we have a large variety of structures in the PDB that can be used as a template. It is important to make a good choice, since the quality of the model will depend on the template or templates used. Currently, MolIDE does not model from multiple templates, so we need to select one structure. However, we could make a number of models based on different templates and compare them. A list of hits from the PDB for STK11 is shown in the upper panel of Figure 5. This list was generated from a PSI-BLAST search using the profile generated from the first round of PSI-BLAST on the nonredundant database. Information about each template, including PDB code, experiment type, resolution, E-value, sequence identity, length of template, starting positions of alignment and length of alignment, is given. Template choice is facilitated in this table by sorting based on any of the categories by clicking on the column header. Three different sorts of the table are shown in Figure 6, sorted by end query residue of the alignment (top), resolution (middle) and percent gaps in the alignment (bottom).

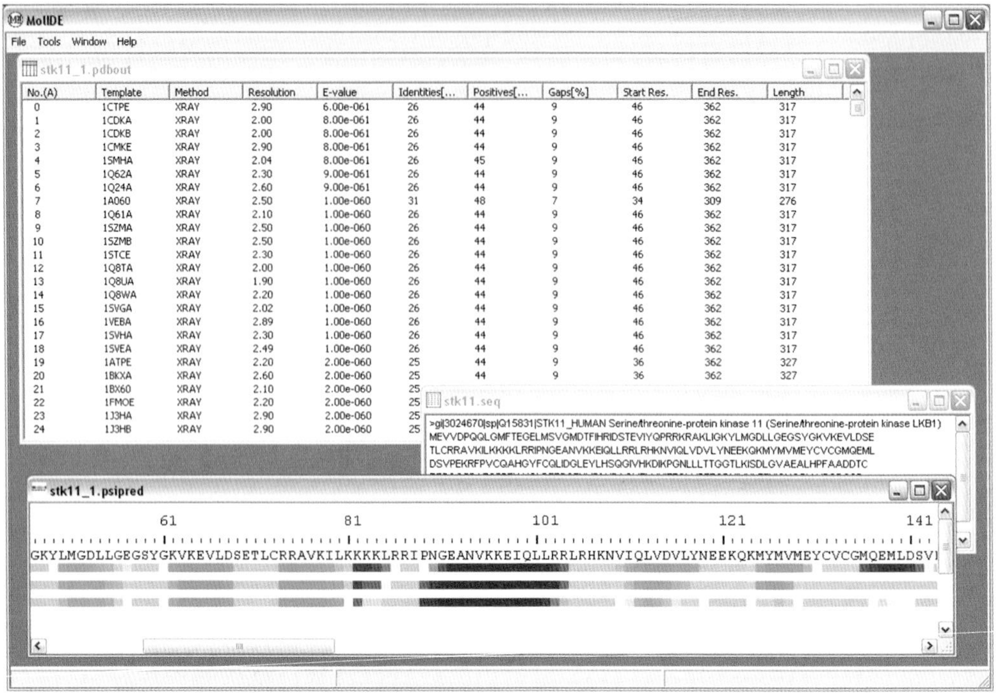

Figure 5 Screenshot of MolIDE for modeling STK11. Top: hits from a search of the PDB for protein target sequence STK11. Middle: sequence of STK11. Bottom: secondary structure prediction for STK11 using PSIPRED after each of three rounds of PSI-BLAST. Helix predictions are in red and sheet strand predictions are in green, and the intensity of the color is proportional to the confidence levels produced by PSIPRED.

The usefulness of a template depends on many factors. It is not necessarily the case that one should use the highest-sequence-identity hit or the best *E*-value. The number and location of gaps in the alignment should also be considered, as should the presence of desirable ligands and structure quality. For STK11, the majority of template alignments end around residue 310, which is the end of the kinase domain, but obviously it would be useful to model the region C-terminal to the kinase domain (residues 310–433).

By double-clicking on the PDB code, MolIDE opens up a window with the alignment as shown for that PDB entry. This window contains the alignment at the bottom and a rotatable view of the backbone of the template in the top part of the window (Figure 7). Conserved amino acids are indicated between the query (top) and hit (bottom) sequences. Gaps in the alignment are indicated by blue squares as are residues in the template that are missing in the coordinates due to poor electron density. Positions in the template structure where insertions need to be modeled, because the query sequence is longer in

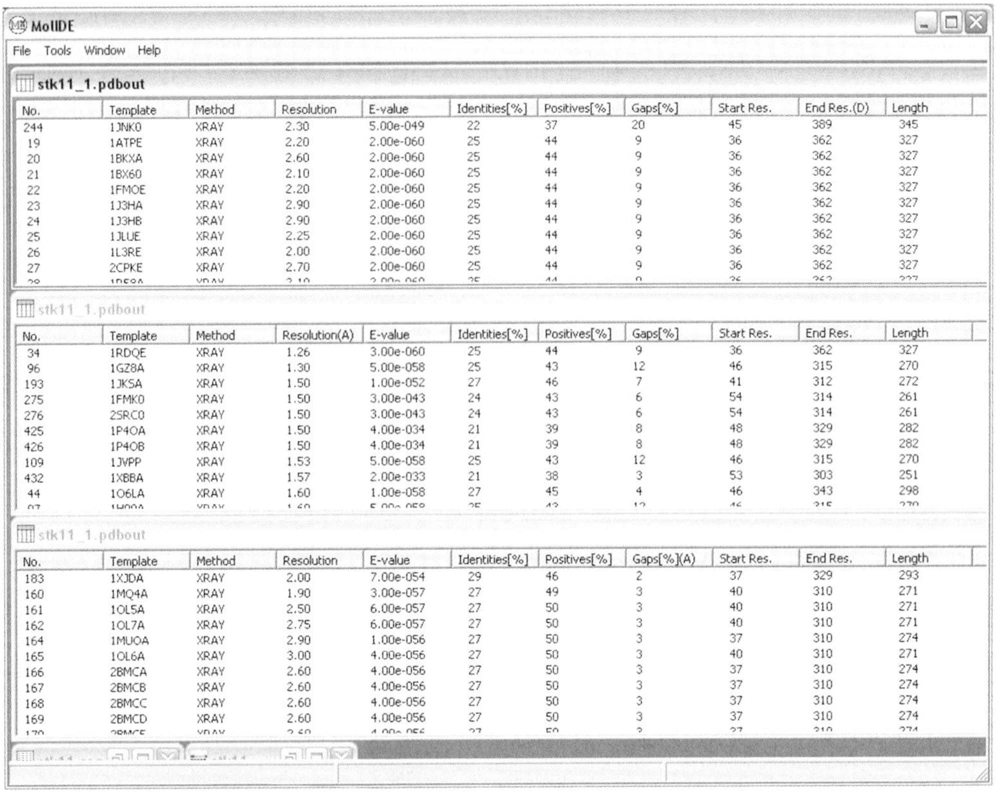
MolIDE
File Tools Window Help
stk11_1.pdbout

No.	Template	Method	Resolution	E-value	Identities[%]	Positives[%]	Gaps[%]	Start Res.	End Res.(D)	Length
244	1JNKO	XRAY	2.30	5.00e-049	22	37	20	45	389	345
19	1ATPE	XRAY	2.20	2.00e-060	25	44	9	36	362	327
20	1BKXA	XRAY	2.60	2.00e-060	25	44	9	36	362	327
21	1BX60	XRAY	2.10	2.00e-060	25	44	9	36	362	327
22	1FMOE	XRAY	2.20	2.00e-060	25	44	9	36	362	327
23	1J3HA	XRAY	2.90	2.00e-060	25	44	9	36	362	327
24	1J3HB	XRAY	2.90	2.00e-060	25	44	9	36	362	327
25	1JLUE	XRAY	2.25	2.00e-060	25	44	9	36	362	327
26	1L3RE	XRAY	2.00	2.00e-060	25	44	9	36	362	327
27	2CPKE	XRAY	2.70	2.00e-060	25	44	9	36	362	327

stk11_1.pdbout

No.	Template	Method	Resolution(A)	E-value	Identities[%]	Positives[%]	Gaps[%]	Start Res.	End Res.	Length
34	1RDQE	XRAY	1.26	3.00e-060	25	44	9	36	362	327
96	1GZ8A	XRAY	1.30	5.00e-058	25	43	12	46	315	270
193	1JKSA	XRAY	1.50	1.00e-052	27	46	7	41	312	272
275	1FMKO	XRAY	1.50	3.00e-043	24	43	6	54	314	261
276	2SRCO	XRAY	1.50	3.00e-043	24	43	6	54	314	261
425	1P4OA	XRAY	1.50	4.00e-034	21	39	8	48	329	282
426	1P4OB	XRAY	1.50	4.00e-034	21	39	8	48	329	282
109	1JVPP	XRAY	1.53	5.00e-058	25	43	12	46	315	270
432	1XBBA	XRAY	1.57	2.00e-033	21	38	3	53	303	251
44	1O6LA	XRAY	1.60	1.00e-058	27	45	4	46	343	298

stk11_1.pdbout

No.	Template	Method	Resolution	E-value	Identities[%]	Positives[%]	Gaps[%](A)	Start Res.	End Res.	Length
183	1XJDA	XRAY	2.00	7.00e-054	29	46	2	37	329	293
160	1MQ4A	XRAY	1.90	3.00e-057	27	49	3	40	310	271
161	1OL5A	XRAY	2.50	6.00e-057	27	50	3	40	310	271
162	1OL7A	XRAY	2.75	6.00e-057	27	50	3	40	310	271
164	1MUOA	XRAY	2.90	1.00e-056	27	50	3	37	310	274
165	1OL6A	XRAY	3.00	4.00e-056	27	50	3	40	310	271
166	2BMCA	XRAY	2.60	4.00e-056	27	50	3	37	310	274
167	2BMCB	XRAY	2.60	4.00e-056	27	50	3	37	310	274
168	2BMCC	XRAY	2.60	4.00e-056	27	50	3	37	310	274
169	2BMCD	XRAY	2.60	4.00e-056	27	50	3	37	310	274

Figure 6 Choosing templates for modeling STK11. Screenshot of MolIDE showing PDB hits sorted by end residue (top), resolution (middle) and percent gaps in alignment (bottom).

that region, are marked with yellow spheres on the structure view. Deletions from the structure are shown on the protein structure as red spheres.

Figure 7 shows the C-terminal end of the alignment for template 1RDQ (chain E) [238]. It is fairly clear from this view that the extension of the alignment beyond residue 316 of the query shows very little similarity in sequence with the template and includes a very large insertion that would need to be modeled. It is likely that this region is incorrectly aligned. This often happens with PSI-BLAST type alignments, such that alignments extend beyond conserved domains due to chance similarities, especially in regions without significant regular secondary structure.

After examining a number of the other templates that extend well beyond residue 310, it was clear that none of them gave a very good alignment for the C-terminal portion of STK11. While a large number of the templates share similar sequence identity to STK11 on the order of 21–27%, some of them contain substantially fewer gaps in the alignment than others. Sorting by

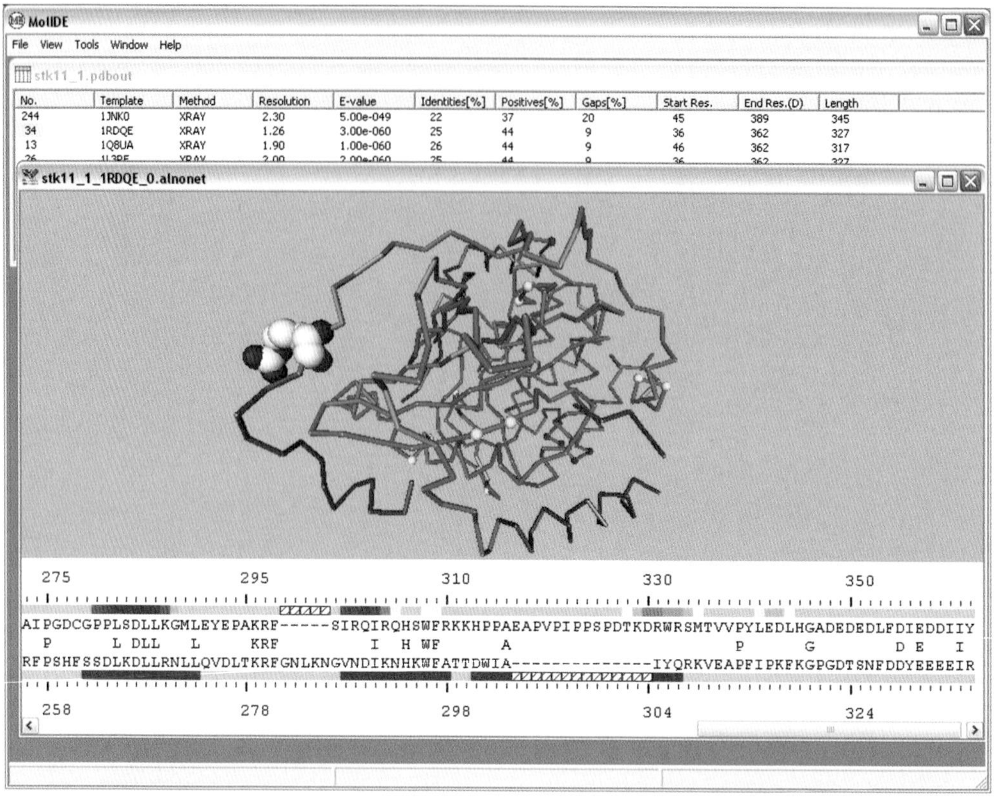

Figure 7 Template 1RDQ (chain E) as template for STK11. The end portion of the alignment is shown, indicating that the residues after STK11 residue 310 are poorly aligned with little sequence similarity and a large gap. Gaps in the alignment are marked with blue squares, and conserved residues are marked between the query (top) and hit (bottom) sequences. The predicted secondary structure is above the query (red = helix; green = sheet) and the experimental secondary structure of the hit is below the template sequence. In the structure view, residues deleted from the structure are indicated with red balls and points of insertion are marked by pairs of yellow balls. The aligned portion of the template is in green and the unaligned portions are in gray. The last residue of the aligned portion is in spacefill representation (Tyr336).

percent gaps, we find a group of templates that have 2–3% of the alignment as gaps for residues 40–310, as shown in the bottom panel of Figure 6. We chose as template PDB entry 1MQ4, a 1.9-Å structure of aurora-A kinase [143], since this structure was the highest resolution of these and contained ADP in the active site of the kinase.

One of the benefits of MolIDE in user-assisted homology modeling is the ability to edit the sequence with the assistance of a graphical view of the protein with the locations of insertions and deletions. For instance, in Figure 8 the alignments before (left screenshot) and after (right screenshot) editing are shown. In the left figure, the two-residue insertion (residues 123–124 with

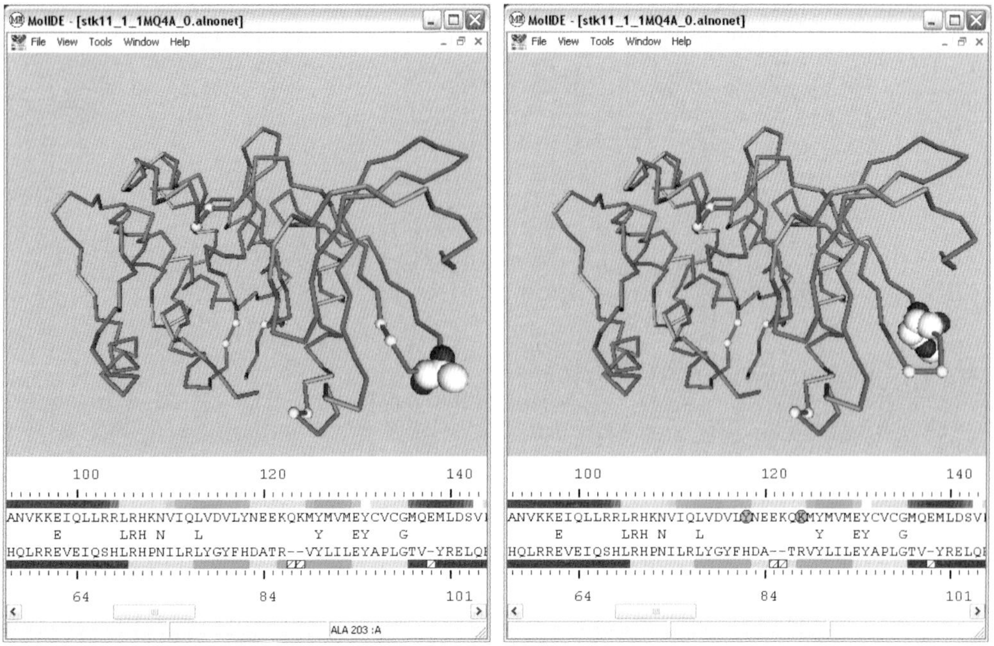

Figure 8 Manual editing of sequence alignments based on the structure view of template 1MQ4 for STK11. The alignment before editing is shown at left and after editing at right. The spacefilled residues mark the loop being edited (Ala84 left and His82 right). Note that the positions of the yellow balls marking the position of the insertion move as the alignment is edited. The anchor positions for loop modeling set manually are indicated on the right (Tyr118 and Lys124 of STK11).

sequence QK) occurs inside a β-sheet strand. Ala84 in the middle of the neighboring loop is shown in spacefill representation to mark the location on the structure. After editing the alignment by "ctrl"-clicking and "shift"-clicking to delete the gap and to create it in another location, the alignment appears as it does in the right side of the figure (with the last residue of the sheet strand, His82, in spacefill this time).

Once the sequence alignment is edited, e.g. by moving insertions and deletions into the middle of loop regions as described above, loop and side-chain modeling commands can be called from the menus. Our side-chain modeling program SCWRL is integrated into MolIDE and Loopy is used for loop modeling [235]. To model loops, the positions of left and right anchor residues are set by pointing and "right"-clicking on the query sequence and then calling Loopy from the Tools menu. The anchors for the loop consisting of residues 118–123 are shown in the right screenshot in Figure 8.

After modeling each of the five insertions and adding the coordinates of the ADP and magnesium ions, we can view the structure superposed on its template as shown in Figure 9. MolIDE is not set up to model in the presence

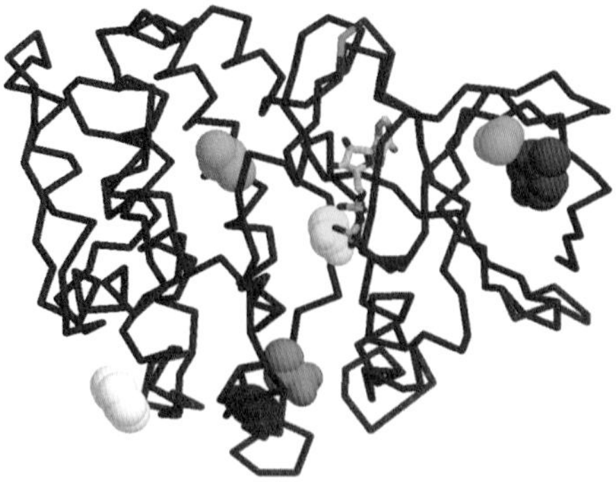

Figure 9 Model of STK11. The template backbone is in blue and the modeled loops for STK11 are in red. The ADP bound in the active site is in stick figure and CPK coloring (carbon = gray; oxygen = red; nitrogen = blue; phosphorus = orange). Several mutations associated with development of cancer are shown in spacefill representation, including Tyr49 (magenta), Val66 (cyan), Leu160 (orange), Asp194 (yellow), Glu199 (violet), Asp208 (red) and Phe231 (white). Gly135 is marked on the backbone in green.

of the ligand, although this can be accomplished for side chains with SCWRL outside of MolIDE. Most of the insertions are some distance away from the active site and four of these are relatively close in space to each other, all on the bottom face of the protein as oriented in Figure 9.

The positions of some missense mutations associated with cancer are marked with spacefill on the structure and three of these mutations are located in the modeled loops at the bottom of the protein. Another one is in the modeled loop at the top. Three mutations are buried in the hydrophobic core, of which two are in the N-terminal domain and one is in the C-terminal domain (colored in cyan, magenta and orange, respectively). These mutations are likely to disturb the hydrophobic cores of these regions, leading to instability of the folded structure. One additional mutation, in yellow spheres, is an Asp residue that binds the magnesium ions which stabilize the binding of ATP to the kinase. Loss of this Asp is likely to result in loss of magnesium and inability to bind ATP. STK11 binds a number of other proteins, and it is possible that one of these proteins binds to the modified loops at the bottom of the structure and that mutations in these loops leads to lack of binding of important interactors of STK11. Mutations which affect binding of the STRAD protein have been analyzed using a homology model of STK11 by Boudeau and coworkers [21]. While STRAD is homologous to kinase domains, none of the available dimeric kinase templates appears to have a binding interface consistent with these mutations, indicating that perhaps STRAD does not bind to STK11 in a manner similar to existing dimer interfaces of kinases.

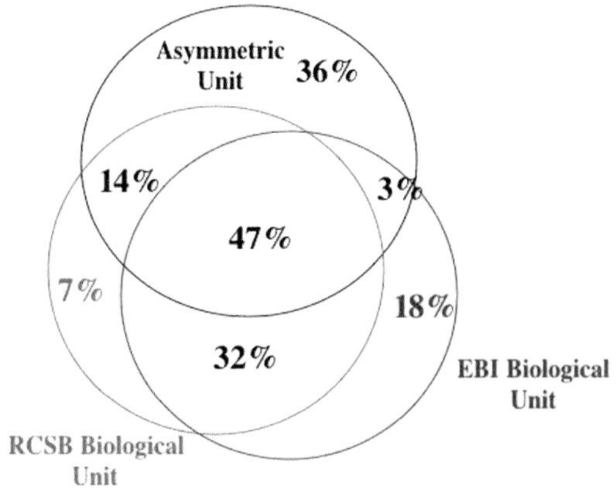

Figure 10 Venn diagram of similarities among asymmetric units, RCSB biological units and EBI (PQS) biological units. Each circle represents 24 000 entries available from both RCSN and PQS sites. Areas are only approximations to percentages marked in each overlapping or nonoverlapping region. For instance, 36% of the 24 000 entries have an asymmetric unit that is different from both RCSB and PQS.

4.3 The Importance of Protein Interactions

The STK11 example points out the importance of incorporating information on protein interactions in homology modeling. As described above, the Research Collaboratory for Structural Bioinformatics (RCSB) provides structures of the asymmetric unit, rather than the biological unit for crystal structures. RCSB also provides separate files that contain the proposed biological unit(s) for each structure, which may be larger or smaller than the asymmetric unit. To quantify this issue, we compared the asymmetric units and the biological units as provided by both the RCSB and the EBI/macromolecular structure database (MSD) [80]. For 23 418 structures available in PQS, Figure 10 shows the similarity of these three sets of units. Figure 10 shows that 53% of asymmetric units are different from either the RCSB or the PQS biological unit or both. This indicates that the standard entry from the PDB is not the biological unit at least half of the time and that the other two sources should be consulted. In addition, RCSB and PQS do not agree on biological units 21% of the time. Unfortunately, there are no automated ways to model homo-multimers, other than to model the sequence on each chain of a known multimer structure. SWISS-MODEL does provide a way to combine models made from different chains of a template biological unit file from PQS.

To illustrate the importance of these interactions, we investigated the large set of mutations in Lac Repressor investigated by Miller and colleagues [130, 148, 202], These authors presented functional data on 4042 mutations of the *Escherichia coil* Lac repressor. The Lac repressor function was evaluated *in vivo* by observing expression levels of β-galactosidase with and without allosteric induction by isopropyl-β-D-galactoside (IPTG). Thus mutations that reduce stability of the Lac repressor monomer, the affinity of monomers in the tetramer, as well as those that affect binding to DNA will produce a visible blue phenotype by the expression of β-galactosidase. Mutations that affect binding of IPTG will not be inducible and thus remain as white colonies, even in the presence of IPTG. This is an ideal system for identifying parameters that can be used to distinguish missense mutations that may cause functional changes in proteins from those that probably would not.

We used two pieces of information to analyze these mutations: (i) the location of residues in the structure of Lac tetramer bound to DNA and (ii) the log-odds scores in a position-specific scoring matrix (PSSM) generated from a multiple sequence alignment of repressor sequences. The PSSM includes two pieces of information: whether a particular site is well conserved in proteins related to Lac repressor and whether a proposed mutation is very different in physical character to residues at that position in proteins related to Lac repressor.

We defined four categories for location of an amino acid: face, buried, edge or surface depending on the value of the relative surface accessibility of a side-chain in the Lac repressor tetramer/DNA structure. Face residues were those that had reduced accessibility of their side-chains in the full complex compared to the Lac repressor monomer alone. Buried residues had less than 5% surface accessibility of their side-chains in the monomers. Edge residues were those with 5–30% accessibility and surface were all others. Surface residues are therefore those that are both on the surface and not in any binding site. The results are shown in Figure 11.

The data indicate that dissimilarity to amino acids in the repressor family as well as a location either buried in the hydrophobic core or in interfaces is sufficient to distinguish levels of "risk", i.e. a mutation with functional consequences. For each category of PSSM log-odds except PSSM = –4, face mutations are more likely to be deleterious than buried mutations and these are much more likely to be deleterious than edge or surface mutations. Interface residues do not tolerate even conservative mutations, so that even for PSSM values of 1 and 2, the proportion of deleterious mutations is 30% or higher. Mutations on the surface but not in an interface are very tolerant to mutations, with less than 1% of 743 mutations on the surface with PSSM of –2 or above having a negative phenotype.

The data in Figure 11 indicate the importance of modeling the biological unit, since all of the mutations in the "face" region would be considered surface or edge residues if the monomer was not present in a dimer complex with DNA.

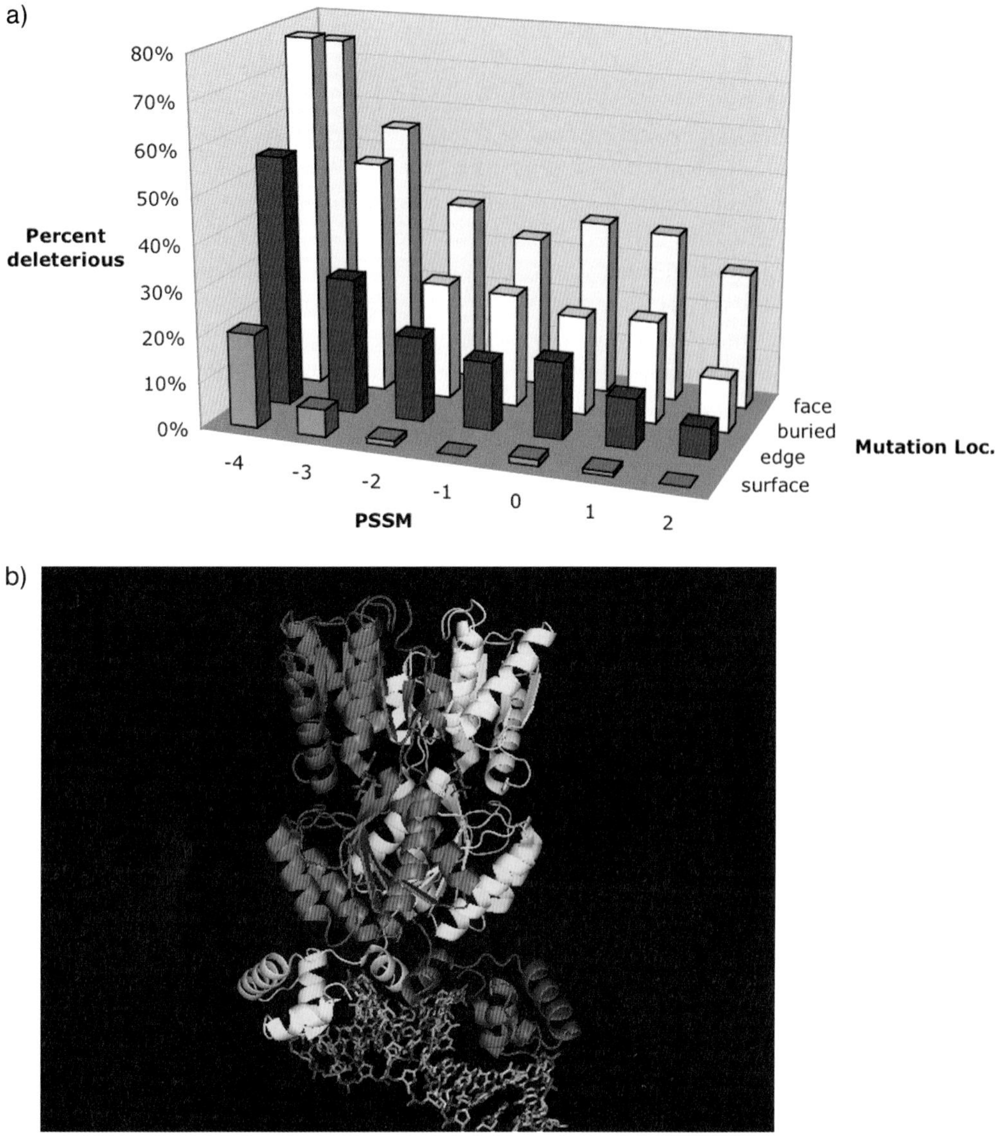

Figure 11 (a) Rates of deleterious mutations based on PSSM score and physical location for 4042 Lac repressor mutants. Face residues are those with lower surface accessibility in the Lac repressor dimer/DNA complex than in the Lac repressor monomer structure. Buried residues have less than 5% surface accessibility in the Lac repressor monomer. Surface residues have greater than 30% accessibility in the Lac repressor monomer or dimer/DNA complex, and "edge" residues have between 5 and 30% accessibility. PSSM is the log-odds score of finding the mutation at each location of the Lac repressor sequence based on a multiple alignment of homologous sequences. (b) Crystal structure of the Lac repressor dimer bound to DNA.

5 Strengths and Limitations

The strengths of homology modeling are based on the insights provided for protein function, structure and evolution that would not be available in the absence of an experimental structure. In many situations, a model built by homology is sufficient for interpreting a great deal of experimental information and will provide enough information for designing new experiments. Homology modeling may also provide functional information beyond the identification of homologous sequences to the target, i.e. a model may serve to distinguish orthologous and paralogous relationships.

The limitations are due to decreasing accuracy as the evolutionary distance between target and template increases. Alignment becomes more uncertain, insertions and deletions more frequent, and even secondary structural units may be of different lengths, numbers, and positions in very remote homologs. Predicting the locations of secondary structure units that are not present in the template structure is a difficult problem and there has been little attention paid to this problem.

The limitations of homology modeling also arise when we have insufficient information to build a model for an entire protein. For instance, we may be able to model one or more domains of a multi-domain protein or a multisubunit complex, but it may not be possible to predict the relative organization of the domains or subunits within the full protein. This remains a challenge for further research. And we are of course limited by structures present in the PDB, which are almost exclusively soluble proteins. Up to 30% of some genomes are membrane proteins, which are at present difficult to model because of the small number of membrane proteins of known structure. Recent structures [119, 146] of the G-protein-coupled receptor (GPCR) rhodopsin at higher resolution than previous structures [151] create new opportunities to model many of these membrane proteins more accurately. A number of GPCRs have been modeled on the bovine rhodopsin structures [33, 58, 136, 144, 152, 243]. In addition, recent structures of bacterial ATP-binding cassette (ABC) transporters at various stages of the transport process [45, 122, 164, 186] also provide opportunities for modeling of a large number of human ABC transporters implicated in drug resistance, such as P-glycoprotein and the multidrug resistance protein (MRP) proteins [160, 190].

Another problem is the quality of data in sequencing and structure determination. There are substantial errors in determining protein sequences from genome sequences, either because of errors in the DNA sequence or in locating exons in eukaryotic genomic DNA [199]. Over 50% of X-ray structures are solved at relatively low resolution, levels of greater than 2.0 Å. Despite progress in determining protein structures by NMR, these structures are of lower resolution than high-quality X-ray structures. While high-throughput structure determination will be of great value to modeling by homology, one concern is the quality of structure determination when the function of the proteins being determined is unknown.

6 Validation

Validation for homology modeling is available in two distinct ways: (i) the prediction rates for each method based on the prediction of known structures given information from other structures and (ii) criteria used to judge each model individually. Most structure prediction method papers have included predictions of known structure, serving as test sets of their accuracy. However, in many cases the number of test cases is inadequate (see Ref. [48]). It is also very easy to select test structures that behave particularly well for a given method and many methods do not stand up to scrutiny of large test sets performed by other researchers. Test sets vary in number of test cases as well as whether predictions of loops or side-chains are performed by building replacements on the template structure scaffold, or in real homology modeling situations where the loops/side-chains are built on nonself scaffolds. The realistic case is more difficult to perform in a comprehensive way, since it requires many sequence–structure alignments to provide the input information on which models are to be built. Another problem is that each method is judged using widely varying criteria, and so no head-to-head comparison is possible from the published papers. The problem of biased test sets and subsequent development of larger benchmarks has a long history in the secondary structure prediction field [170, 239].

While sequence alignment methods have been extensively benchmarked [184], programs that build coordinates from alignments, including the backbone, loops and side chains, have not been extensively compared to one another in large-scale tests. Recently, however, Wallner and Elofsson [220] compared several programs that build coordinates from templates given template-target alignments, including MODELLER [174], SegMod/ENCAD [116], SWISS-MODEL [189], 3D-JIGSAW [11], nest [234, 235], Builder [96, 97] and SCWRL (for side chains without modification of the template backbone) [30]. They found that three of the programs, MODELLER, nest and SegMod/ENCAD, perform better than the others. In particular, SegMod is a very old program and still performs as well as much more recent programs. They also observed that none of the homology modeling programs builds side chains as well as SCWRL.

6.1 The CASP Meeting

Another forum for testing homology modeling methods has been the ongoing series of CASP meetings organized by John Moult and colleagues [138–141, 214,215,240]. In the spring and summer before each meeting held in December 1994, 1996, 1998, 2000, 2002 and 2004, sequences of proteins whose structure was under active experimental determination by NMR or X-ray crystallography were distributed via the Internet. Anyone can submit structure predictions at various levels of detail (secondary structure predictions, sequence alignments to structures and full 3-D coordinates) before specific expiration dates for each target sequence. The models are evaluated via a number of computer programs written for the purpose, and then assessed by experts in each field, including comparative modeling, fold recognition and *ab initio* structure prediction. The organizers then invite predictors whose predictions are outstanding to present their methods and results at the meeting, and to describe their work in a special issue of the journal *Proteins*, published in the following year.

Ordinarily when protein structure prediction methods are developed, they are tested on sets of protein structures where the answer is known. Unfortunately, it is easy to select targets, even subconsciously, for which a particular method under development may work well. Also, it is easy to optimize parameters for a small test set that do not work as well for larger test sets. While the number of prediction targets in CASP is limited to numbers on the order of 10–20 per category, these numbers are still higher than many of the test sets used in testing new methods under development.

6.2 Protein Health

A number of programs have been developed to ascertain the quality of experimentally determined structures and these can be used to determine whether a protein model obeys appropriate stereochemical rules. The two most popular programs are ProCheck and WhatCheck [219]. Recently, the Richardson group has developed MolProbity, which seeks to identify a number of features in protein structures that are statistically unlikely, when compared to a manually curated set of very high-resolution structures [39]. This is a webserver that reports bad rotamer conformations, close contacts, flipped amide side-chains and other potential errors in structures. Although this site is more geared to analysis of new experimental structures, it can also be used on homology models to identify steric clashes or poorly modeled regions of proteins.

These programs check bond lengths and angles, dihedral angles, planarity of sp^2 groups, nonbonded atomic distances, disulfide bonds and other characteristics of protein structures. One of the more useful checks is to see whether backbone geometries are in acceptable regions of the Ramachandran map. Backbone conformations in the forbidden regions are very likely to be incorrect. It should be noted once again that correct geometry is no guarantee of correct structure prediction. In some cases, it may be better to tolerate a few steric conflicts or bad dihedral angles, rather than to minimize the structure's energy. While the geometry may look better, the final structure may be further away from the true structure (if it were known) than the unminimized structure. Chapter 11 discusses the problem of health of protein models and describes the respective Model Quality Assessment Programs (MQAPs).

Acknowledgements

This work was funded in part by grants R01-HG02302 and CA06927 from the National Institutes of Health. I thank Guoli Wang for providing the genome analysis in Figure 2, and Adrian Canutescu for assistance with MolIDE and Figure 3.

References

1 Alexandrov, N. N. and R. Luethy. 1998. Alignment algorithm for homology modeling and threading. Protein Sci. **7**: 254–8.

2 Allorge, D., D. Breant, J. Harlow, *et al.* 2005. Functional analysis of CYP2D6.31 variant: homology modeling suggests possible disruption of redox partner interaction by Arg440His substitution. Proteins **59**: 339–46.

3 Altschul, S. F., T. L. Madden, A. A. Schäffer, J. Zhang, Z. Zhang, W. Miller and D. J. Lipman. 1997. Gapped BLAST and PSI-BLAST: a new generation of database programs. Nucleic Acids Res. **25**: 3389–402.

4 Anderson, R. J., Z. Weng, R. K. Campbell and X. Jiang. 2005. Main-chain conformational tendencies of amino acids. Proteins **60**: 697–89.

5 Andreeva, A., D. Howorth, S. E. Brenner, T. J. Hubbard, C. Chothia and A. G. Murzin. 2004. SCOP database in 2004: refinements integrate structure and sequence family data. Nucleic Acids Res. **32**: D226–9.

6 Bailey, T. L. and M. Gribskov. 1996. The megaprior heuristic for discovering protein sequence patterns. Proc. ISMB **4**: 15–24.

7 Bairoch, A., R. Apweiler, C. H. Wu, *et al.* 2005. The Universal Protein Resource (UniProt). Nucleic Acids Res. **33**: D154–9.

8 Baldi, P., Y. Chauvin, T. Hunkapiller and M. A. McClure. 1994. Hidden Markov models of biological primary sequence information. Proc. Natl Acad. Sci. USA **91**: 1059–63.

9 Barre, A., J. P. Borges, R. Culerrier and P. Rouge. 2005. Homology modelling of the major peanut allergen Ara h 2 and surface mapping of IgE-binding epitopes. Immunol Lett. **100**: 153–8.

10 Bates, P. A., R. M. Jackson and M. J. Sternberg. 1997. Model building by comparison: a combination of expert knowledge and computer automation. Proteins **Suppl. 1**: 59–67.

11 Bates, P. A., L. A. Kelley, R. M. MacCallum and M. J. Sternberg. 2001. Enhancement of protein modeling by human intervention in applying the automatic programs 3D-JIGSAW and 3D-PSSM. Proteins **Suppl. 5**: 39–46.

12 Benedetti, E., G. Morelli, G. Nemethy and H. A. Scheraga. 1983. Statistical and energetic analysis of sidechain conformations in oligopeptides. Int. J. Peptide Protein Res. **22**: 1–15.

13 Benson, D. A., I. Karsch-Mizrachi, D. J. Lipman, J. Ostell and D. L. Wheeler. 2005. GenBank. Nucleic Acids Res. **33**: D34–8.

14 Berman, H. M., J. Westbrook, Z. Feng, G. Gilliland, T. N. Bhat, H. Weissig, I. N. Shindyalov and P. E. Bourne. 2000. The Protein Data Bank. Nucleic Acids Res. **28**: 235–42.

15 Berman, H. M. and J. D. Westbrook. 2004. The impact of structural genomics on the protein data bank. Am. J. Pharmacogenomics **4**: 247–52.

16 Bertaccini, E. J., J. Shapiro, D. L. Brutlag and J. R. Trudell. 2005. Homology modeling of a human glycine alpha 1 receptor reveals a plausible anesthetic binding site. J. Chem. Inf. Model. **45**: 128–35.

17 Bhat, T. N., V. Sasisekharan and M. Vijayan. 1979. An analysis of sidechain conformation in proteins. Int. J. Peptide Protein Res. **13**: 170–84.

18 Blundell, T. L., B. L. Sibanda, M. J. E. Sternberg and J. M. Thornton. 1987. Knowledge-based prediction of protein structures and the design of novel molecules. Nature **326**: 347–52.

19 Boomsma, W. and T. Hamelryck. 2005. Full cyclic coordinate descent: solving the protein loop closure problem in Calpha space. BMC Bioinformatics **6**: 159.

20 Borysenko, C. W., W. F. Furey and H. C. Blair. 2005. Comparative modeling of TNFRSF25 (DR3) predicts receptor destabilization by a mutation linked to rheumatoid arthritis. Biochem. Biophys. Res. Commun. **328**: 794–9.

21 Boudeau, J., J. W. Scott, N. Resta, *et al.* 2004. Analysis of the LKB1–STRAD–MO25 complex. J. Cell Sci. **117**: 6365–75.

22 Bower, M. J., F. E. Cohen and R. L. Dunbrack, Jr. 1997. Prediction of protein side-chain rotamers from a backbone-dependent rotamer library: a new homology modeling tool. J. Mol. Biol. **267**: 1268–82.

23 Brown, M., R. Hughey, A. Krogh, I. S. Mian, K. Sjolander and D. Haussler. 1993. Using Dirichlet mixture priors to derive hidden Markov models for protein families. Proc. ISMB **1**: 47–55.

24 Browne, W. J., A. C. North and D. C. Phillips. 1969. A possible three-dimensional structure of bovine alpha-lactalbumin based on that of hen's egg-white lysozyme. J. Mol. Biol. **42**: 65–86.

25 Bruccoleri, R. E. and M. Karplus. 1987. Prediction of the folding of short polypeptide segments by uniform conformational sampling. Biopolymers **26**: 137–68.

26 Caffrey, C. R., L. Placha, C. Barinka, *et al.* 2005. Homology modeling and SAR analysis of *Schistosoma japonicum* cathepsin D (SjCD) with statin inhibitors identify a unique active site steric barrier with potential for the design of specific inhibitors. Biol. Chem. **386**: 339–49.

27 Campillo, N. E., J. Antonio Paez, L. Lagartera and A. Gonzalez. 2005. Homology modelling and active-site-mutagenesis study of the catalytic domain of the pneumococcal phosphorylcholine esterase. Bioorg. Med. Chem. **13**: 6404–13.

28 Canutescu, A. A. and R. L. Dunbrack, Jr. 2003. Cyclic coordinate descent: a robotics algorithm for protein loop closure. Protein Sci. **12**: 963–72.

29 Canutescu, A. A. and R. L. Dunbrack, Jr. 2005. MollDE: a homology modeling framework you can click with. Bioinformatics **21**: 2914–6.

30 Canutescu, A. A., A. A. Shelenkov and R. L. Dunbrack, Jr. 2003. A graph-theory algorithm for rapid protein side-chain prediction. Protein Sci. **12**: 2001–14.

31 Chothia, C. and A. M. Lesk. 1986. The relation between the divergence of sequence and structure in proteins. EMBO J. **5**: 823–6.

32 Chothia, C., A. M. Lesk, A. Tramontano, *et al.* 1989. Conformations of immunoglobulin hypervariable regions. Nature **342**: 877–83.

33 COSTANZI, S., L. MAMEDOVA, Z. G. GAO AND K. A. JACOBSON. 2004. Architecture of P2Y nucleotide receptors: structural comparison based on sequence analysis, mutagenesis, and homology modeling. J. Med. Chem. **47**: 5393–404.

34 COUCH, F. J. AND B. L. WEBER. 1996. Mutations and polymorphisms in the familial early-onset breast cancer (BRCA1) gene. Breast Cancer Information Core. Hum. Mutat. **8**: 8–18.

35 COUTSIAS, E. A., C. SEOK, M. P. JACOBSON AND K. A. DILL. 2004. A kinematic view of loop closure. J. Comput. Chem. **25**: 510–28.

36 DA SILVA, C. H., I. CARVALHO AND C. A. TAFT. 2005. Homology modeling and molecular interaction field studies of alpha-glucosidases as a guide to structure-based design of novel proposed anti-HIV inhibitors. J. Comput. Aided Mol. Des. **19**: 83–92.

37 DAHIYAT, B. I., C. A. SARISKY AND S. L. MAYO. 1997. De novo protein design: towards fully automated sequence selection. J. Mol. Biol. **273**: 789–96.

38 DASGUPTA, B., L. PAL, G. BASU AND P. CHAKRABARTI. 2004. Expanded turn conformations: characterization and sequence–structure correspondence in alpha-turns with implications in helix folding. Proteins **55**: 305–15.

39 DAVIS, I. W., L. W. MURRAY, J. S. RICHARDSON AND D. C. RICHARDSON. 2004. MOLPROBITY: structure validation and all-atom contact analysis for nucleic acids and their complexes. Nucleic Acids Res. **32**: W615–9.

40 DE MAEYER, M., J. DESMET AND I. LASTERS. 2000. The dead-end elimination theorem: mathematical aspects, implementation, optimizations, evaluation, and performance. Methods Mol. Biol. **143**: 265–304.

41 DESMET, J., M. DE MAEYER, B. HAZES AND I. LASTERS. 1992. The dead-end elimination theorem and its use in protein sidechain positioning. Nature **356**: 539–42.

42 DESMET, J., M. DE MAEYER AND I. LASTERS. 1997. Theoretical and algorithmical optimization of the dead-end elimination theorem. Pac. Symp. Biocomput. 1997: 122–33.

43 DODGE, C., R. SCHNEIDER AND C. SANDER. 1998. The HSSP database of protein structure–sequence alignments and family profiles. Nucleic Acids Res. **26**: 313–5.

44 DONATE, L. E., S. D. RUFINO, L. H. J. CANARD AND T. L. BLUNDELL. 1996. Conformational analysis and clustering of short and medium size loops connecting regular secondary structures: a database for modeling and prediction. Protein Sci. **5**: 2600–16.

45 DONG, J., G. YANG AND H. S. MCHAOURAB. 2005. Structural basis of energy transduction in the transport cycle of MsbA. Science **308**: 1023–8.

46 DORIT, R. L., L. SCHOENBACH AND W. GILBERT. 1990. How big is the universe of exons? Science **250**: 1377–82.

47 DUNBRACK, R. L., JR. 1999. Comparative modeling of CASP3 targets using PSI-BLAST and SCWRL. Proteins **Suppl. 3**: 81–7.

48 DUNBRACK, R. L., JR. 2001. Protein structure prediction in biology and medicine. In LENGAUER, T. (ed.), *Bioinformatics and Drug Design*. Wiley, New York, NY: 145–235.

49 DUNBRACK, R. L., JR. 2002. Rotamer libraries in the 21st century. Curr. Opin. Struct. Biol. **12**: 431–40.

50 DUNBRACK, R. L., JR. AND F. E. COHEN. 1997. Bayesian statistical analysis of protein sidechain rotamer preferences. Protein Sci. **6**: 1661–81.

51 DUNBRACK, R. L., JR. AND M. KARPLUS. 1993. Backbone-dependent rotamer library for proteins. Application to side-chain prediction. J. Mol. Biol. **230**: 543–74.

52 DUNBRACK, R. L., JR. AND M. KARPLUS. 1994. Conformational analysis of the backbone-dependent rotamer preferences of protein sidechains. Nat. Struct. Biol. **1**: 334–40.

53 EDDY, S. R. 1996. Hidden Markov models. Curr. Opin. Struct. Biol. **6**: 361–5.

54 EDDY, S. R., G. MITCHISON AND R. DURBIN. 1995. Maximum discrimination

hidden Markov models of sequence consensus. J. Comput. Biol. **2**: 9–23.

55 Edgar, R. C. 2004. MUSCLE: multiple sequence alignment with high accuracy and high throughput. Nucleic Acids Res. **32**: 1792–7.

56 Espadaler, J., N. Fernandez-Fuentes, A. Hermoso, E. Querol, F. X. Aviles, M. J. Sternberg and B. Oliva. 2004. ArchDB: automated protein loop classification as a tool for structural genomics. Nucleic Acids Res. **32**: D185–8.

57 Evans, J. S., S. I. Chan and W. A. Goddard, 3rd. 1995. Prediction of polyelectrolyte polypeptide structures using Monte Carlo conformational search methods with implicit solvation modeling. Protein Sci. **4**: 2019–31.

58 Evers, A. and T. Klabunde. 2005. Structure-based drug discovery using GPCR homology modeling: successful virtual screening for antagonists of the alpha1A adrenergic receptor. J. Med. Chem. **48**: 1088–97.

59 Fan, H. and A. E. Mark. 2004. Refinement of homology-based protein structures by molecular dynamics simulation techniques. Protein Sci. **13**: 211–20.

60 Fechteler, T., U. Dengler and D. Schomburg. 1995. Prediction of protein three-dimensional structures in insertion and deletion regions: a procedure for searching data bases of representative protein fragments using geometric scoring criteria. J. Mol. Biol. **253**: 114–31.

61 Fernandez-Fuentes, N., E. Querol, F. X. Aviles, M. J. Sternberg and B. Oliva. 2005. Prediction of the conformation and geometry of loops in globular proteins: testing ArchDB, a structural classification of loops. Proteins **60**: 746–57.

62 Fetrow, J. S., A. Godzik and J. Skolnick. 1998. Functional analysis of the *Escherichia coli* genome using the sequence-to-structure-to-function paradigm: identification of proteins exhibiting the glutaredoxin/thioredoxin disulfide oxidoreductase activity. J. Mol. Biol. **282**: 703–11.

63 Fidelis, K., P. S. Stern, D. Bacon and J. Moult. 1994. Comparison of systematic search and database methods for constructing segments of protein structures. Protein Eng. **7**: 953–60.

64 Fine, R. M., H. Wang, P. S. Shenkin, D. L. Yarmush and C. Levinthal. 1986. Predicting antibody hypervariable loop conformations. II: minimization and molecular dynamics studies of MCPC603 from many randomly generated loop conformations. Proteins **1**: 342–62.

65 Fiser, A., R. K. Do and A. Sali. 2000. Modeling of loops in protein structures. Protein Sci. **9**: 1753–73.

66 Fogolari, F. and S. C. Tosatto. 2005. Application of MM/PBSA colony free energy to loop decoy discrimination: toward correlation between energy and root mean square deviation. Protein Sci. **14**: 889–901.

67 Franca, T. C., P. G. Pascutti, T. C. Ramalho and J. D. Figueroa-Villar. 2005. A three-dimensional structure of *Plasmodium falciparum* serine hydroxymethyltransferase in complex with glycine and 5-formyl-tetrahydrofolate. Homology modeling and molecular dynamics. Biophys. Chem. **115**: 1–10.

68 Gallicchio, E., L. Y. Zhang and R. M. Levy. 2002. The SGB/NP hydration free energy model based on the surface generalized born solvent reaction field and novel nonpolar hydration free energy estimators. J. Comput. Chem. **23**: 517–29.

69 Geetha, V. and P. J. Munson. 1997. Linkers of secondary structures in proteins. Protein Sci. **6**: 2538–47.

70 Goldstein, R. F. 1994. Efficient rotamer elimination applied to protein side-chains and related spin glasses. Biophys. J. **66**: 1335–40.

71 Gordon, D. B. and S. L. Mayo. 1999. Branch-and-terminate: a combinatorial optimization algorithm for protein design. Struct. Fold. Des. **7**: 1089–98.

72 Greer, J. 1990. Comparative modeling methods: application to the family of the mammalian serine proteases. Proteins **7**: 317–34.

73 Greer, J. 1980. Model for haptoglobin heavy chain based upon structural homology. Proc. Natl Acad. Sci. USA 77: 3393–7.

74 Guex, N. and M. C. Peitsch. 1997. SWISS-MODEL and the Swiss-PdbViewer: an environment for comparative protein modeling. Electrophoresis **18**: 2714–23.

75 Gunsteren, W. F. v., P. H. Hünenberger, A. E. Mark, P. E. Smith and I. G. Tironi. 1995. Computer simulation of protein motion. Comp. Phys. Commun. **91**: 305–19.

76 Hamosh, A., A. F. Scott, J. Amberger, D. Valle and V. A. McKusick. 2000. Online Mendelian Inheritance in Man (OMIM). Hum. Mutat. **15**: 57–61.

77 Hamza, A., H. Cho, H. H. Tai and C. G. Zhan. 2005. Understanding human 15-hydroxyprostaglandin dehydrogenase binding with NAD^+ and PGE_2 by homology modeling, docking and molecular dynamics simulation. Bioorg. Med. Chem. **13**: 4544–51.

78 Havel, T. F. and M. E. Snow. 1991. A new method for building protein conformations from sequence alignments with homologues of known structure. J. Mol. Biol. **217**: 1–7.

79 Hemminki, A., D. Markie, I. Tomlinson, *et al.* 1998. A serine/threonine kinase gene defective in Peutz–Jeghers syndrome. Nature **391**: 184–7.

80 Henrick, K. and J. M. Thornton. 1998. PQS: a protein quaternary structure file server. Trends Biochem. Sci. **23**: 358–61.

81 Heuser, P., G. Wohlfahrt and D. Schomburg. 2004. Efficient methods for filtering and ranking fragments for the prediction of structurally variable regions in proteins. Proteins **54**: 583–95.

82 Holm, L. and C. Sander. 1995. Dali: a network tool for protein structure comparison. Trends Biochem. Sci. **20**: 478–80.

83 Holm, L. and C. Sander. 1992. Fast and simple Monte Carlo algorithm for side chain optimization in proteins: application to model building by homology. Proteins **14**: 213–23.

84 Holm, L. and C. Sander. 1996. Mapping the protein universe. Science **273**: 595–603.

85 Huynen, M., T. Doerks, F. Eisenhaber, C. Orengo, S. Sunyaev, Y. Yuan and P. Bork. 1998. Homology-based fold predictions for *Mycoplasma genitalium* proteins. J. Mol. Biol. **280**: 323–6.

86 Hwang, J. K. and W. F. Liao. 1995. Side-chain prediction by neural networks and simulated annealing optimization. Protein Eng. **8**: 363–70.

87 Jacobson, M. P., D. L. Pincus, C. S. Rapp, T. J. Day, B. Honig, D. E. Shaw and R. A. Friesner. 2004. A hierarchical approach to all-atom protein loop prediction. Proteins **55**: 351–67.

88 Janin, J., S. Wodak, M. Levitt and B. Maigret. 1978. Conformations of amino acid side-chains in proteins. J. Mol. Biol. **125**: 357–86.

89 Jenne, D. E., H. Reimann, J. Nezu, *et al.* 1998. Peutz–Jeghers syndrome is caused by mutations in a novel serine threonine kinase. Nat. Genet. **18**: 38–43.

90 Jones, D. T. 1999. Protein secondary structure prediction based on position-specific scoring matrices. J. Mol. Biol. **292**: 195–202.

91 Jones, T. A. and S. Thirup. 1986. Using known substructures in protein model building and crystallography. EMBO J. **5**: 819–22.

92 Karchin, R., M. Diekhans, L. Kelly, D. J. Thomas, U. Pieper, N. Eswar, D. Haussler and A. Sali. 2005. LS-SNP: large-scale annotation of coding non-synonymous SNPs based on multiple information sources. Bioinformatics **21**: 2814–20.

93 Karplus, K., C. Barrett and R. Hughey. 1998. Hidden Markov models for detecting remote protein homologies. Bioinformatics **14**: 846–56.

94 Karplus, M. and R. G. Parr. 1963. An approach to the internal rotation problem. J. Chem. Phys. **38**: 1547–52.

95 Kim, S. H., D. H. Shin, I. G. Choi, U. Schulze-Gahmen, S. Chen and R. Kim. 2003. Structure-based functional inference in structural genomics. J. Struct. Funct. Genomics **4**: 129–35.

96 Koehl, P. and M. Delarue. 1994. Application of a self-consistent mean field theory to predict protein side-chains conformation and estimate their conformational entropy. J. Mol. Biol. **239**: 249–75.

97 Koehl, P. and M. Delarue. 1995. A self consistent mean field approach to simultaneous gap closure and side-chain positioning in homology modelling. Nat. Struct. Biol. **2**: 163–70.

98 Koehl, P. and M. Levitt. 1999. A brighter future for protein structure prediction. Nat. Struct. Biol. **6**: 108–11.

99 Kono, H. and J. Doi. 1994. Energy minimization method using automata network for sequence and side-chain conformation prediction from given backbone geometry. Proteins **19**: 244–55.

100 Kopp, J. and T. Schwede. 2004. The SWISS-MODEL Repository of annotated three-dimensional protein structure homology models. Nucleic Acids Res. **32**: D230–4.

101 Kraus, J. P., M. Janosik, V. Kozich, *et al.* 1999. Cystathionine beta-synthase mutations in homocystinuria. Hum. Mutat. **13**: 362–75.

102 Krawczak, M., E. V. Ball, I. Fenton, P. D. Stenson, S. Abeysinghe, N. Thomas and D. N. Cooper. 2000. Human gene mutation database – a biomedical information and research resource. Hum. Mutat. **15**: 45–51.

103 Kumar, M., M. Bhasin, N. K. Natt and G. P. Raghava. 2005. BhairPred: prediction of beta-hairpins in a protein from multiple alignment information using ANN and SVM techniques. Nucleic Acids Res. **33**: W154–9.

104 Kwasigroch, J., J. Chomilier and J. Mornon. 1996. A global taxonomy of loops in globular proteins. J. Mol. Biol. **259**: 855–72.

105 Lambert, C., N. Leonard, X. De Bolle and E. Depiereux. 2002. ESyPred3D: prediction of proteins 3D structures. Bioinformatics **18**: 1250–6.

106 Laskowski, R. A., J. D. Watson and J. M. Thornton. 2005. ProFunc: a server for predicting protein function from 3D structure. Nucleic Acids Res. **33**: W89–93.

107 Lasters, I., M. De Maeyer and J. Desmet. 1995. Enhanced dead-end elimination in the search for the global minimum energy conformation of a collection of protein side chains. Protein Eng. **8**: 815–22.

108 Lasters, I. and J. Desmet. 1993. The fuzzy-end elimination theorem: correctly implementing the sidechain placement algorithm based on the dead-end elimination theorem. Protein Eng. **6**: 717–22.

109 Laughton, C. A. 1994. Prediction of protein sidechain conformations from local three-dimensional homology relationships. J. Mol. Biol. **235**: 1088–97.

110 Launonen, V. 2005. Mutations in the human LKB1/STK11 gene. Hum. Mutat. **26**: 291–7.

111 Leaver-Fay, A., B. Kuhlman and J. Snoeyink. 2005. An adaptive dynamic programming algorithm for the side chain placement problem. Pac. Symp. Biocomput.: 16–27.

112 Lee, C. and S. Subbiah. 1991. Prediction of protein side-chain conformation by packing optimization. J. Mol. Biol. **217**: 373–88.

113 Lessel, U. and D. Schomburg. 1997. Creation and characterization of a new, non-redundant fragment data bank. Protein Eng. **10**: 659–64.

114 Lessel, U. and D. Schomburg. 1999. Importance of anchor group positioning in protein loop prediction. Proteins **37**: 56–64.

115 Leszczynski, J. F. and G. D. Rose. 1986. Loops in globular proteins: a novel category of secondary structure. Science **234**: 849–55.

116 Levitt, M. 1992. Accurate modeling of protein conformation by automatic segment matching. J. Mol. Biol. **226**: 507–33.

117 Levitt, M., M. Gerstein, E. Huang, S. Subbiah and J. Tsai. 1997. Protein folding: the endgame. Annu. Rev. Biochem. **66**: 549–79.

118 Li, H., R. Tejero, D. Monleon, D. Bassolino-Klimas, C. Abate-Shen, R. E. Bruccoleri and G. T. Montelione. 1997. Homology modeling using simulated annealing

of restrained molecular dynamics and conformational search calculations with CONGEN: application in predicting the three-dimensional structure of murine homeodomain Msx-1. Protein Sci. **6**: 956–70.

119 Li, J., P. C. Edwards, M. Burghammer, C. Villa and G. F. Schertler. 2004. Structure of bovine rhodopsin in a trigonal crystal form. J. Mol. Biol. **343**: 1409–38.

120 Liang, S. and N. V. Grishin. 2002. Side-chain modeling with an optimized scoring function. Protein Sci. **11**: 322–31.

121 Lichtarge, O., H. R. Bourne and F. E. Cohen. 1996. An evolutionary trace method defines binding surfaces common to protein families. J. Mol. Biol. **257**: 342–58.

122 Locher, K. P., A. T. Lee and D. C. Rees. 2002. The *E. coli* BtuCD structure: a framework for ABC transporter architecture and mechanism. Science **296**: 1091–8.

123 Looger, L. L. and H. W. Hellinga. 2001. Generalized dead-end elimination algorithms make large-scale protein side-chain structure prediction tractable: implications for protein design and structural genomics. J. Mol. Biol. **307**: 429–45.

124 Lovell, S. C., I. W. Davis, W. B. Arendall, 3rd, P. I. de Bakker, J. M. Word, M. G. Prisant, J. S. Richardson and D. C. Richardson. 2003. Structure validation by Calpha geometry: phi,psi and Cbeta deviation. Proteins **50**: 437–50.

125 Lovell, S. C., J. M. Word, J. S. Richardson and D. C. Richardson. 2000. The penultimate rotamer library. Proteins **40**: 389–408.

126 Lundstrom, K. 2004. Structural genomics on membrane proteins: mini review. Comb. Chem. High Throughput Screen. **7**: 431–9.

127 MacKerell, A. D., Jr., D. Bashford, M. Bellott, *et al.* 1998. All-atom empirical potential for molecular modeling and dynamics studies of proteins. J. Phys. Chem. B**102**: 3586–616.

128 Marabotti, A. and A. M. Facchiano. 2005. Homology modeling studies on human galactose-1-phosphate uridylyltransferase and on its galactosemia-related mutant Q188R provide an explanation of molecular effects of the mutation on homo- and heterodimers. J. Med. Chem. **48**: 773–9.

129 Marinelli, L., K. E. Gottschalk, A. Meyer, E. Novellino and H. Kessler. 2004. Human integrin alphavbeta5: homology modeling and ligand binding. J. Med. Chem. **47**: 4166–77.

130 Markiewicz, P., L. G. Kleina, C. Cruz, S. Ehret and J. H. Miller. 1994. Genetic studies of the lac repressor. XIV. Analysis of 4000 altered *Escherichia coli* lac repressors reveals essential and non-essential residues, as well as "spacers" which do not require a specific sequence. J. Mol. Biol. **240**: 421–33.

131 Martin, A. C., K. Toda, H. J. Stirk and J. M. Thornton. 1995. Long loops in proteins. Protein Eng. **8**: 1093–101.

132 McGregor, M. J., S. A. Islam and M. J. E. Sternberg. 1987. Analysis of the relationship between sidechain conformation and secondary structure in globular proteins. J. Mol. Biol. **198**: 295–310.

133 Mendes, J., A. M. Baptista, M. A. Carrondo and C. M. Soares. 1999. Improved modeling of side-chains in proteins with rotamer-based methods: a flexible rotamer model. Proteins **37**: 530–43.

134 Mendes, J., C. M. Soares and M. A. Carrondo. 1999. Improvement of side-chain modeling in proteins with the self-consistent mean field theory method based on an analysis of the factors influencing prediction. Biopolymers **50**: 111–31.

135 Michalsky, E., A. Goede and R. Preissner. 2003. Loops In Proteins (LIP) – a comprehensive loop database for homology modelling. Protein Eng. **16**: 979–85.

136 Miedlich, S. U., L. Gama, K. Seuwen, R. M. Wolf and G. E. Breitwieser. 2004. Homology modeling of the transmembrane domain of the human

calcium sensing receptor and localization of an allosteric binding site. J. Biol. Chem. **279**: 7254–63.

137 MISURA, K. M., A. V. MOROZOV AND D. BAKER. 2004. Analysis of anisotropic side-chain packing in proteins and application to high-resolution structure prediction. J. Mol. Biol. **342**: 651–64.

138 MOULT, J. 1996. The current state of the art in protein structure prediction. Curr. Opin. Biotechnol. **7**: 422–27.

139 MOULT, J. 1999. Predicting protein three-dimensional structure. Curr. Opin. Biotechnol. **10**: 583–8.

140 MOULT, J., T. HUBBARD, S. H. BRYANT, K. FIDELIS AND J. T. PEDERSEN. 1997. Critical assessment of methods of protein structure prediction (CASP): round II. Proteins **Suppl. 1**: 2–6.

141 MOULT, J., T. HUBBARD, K. FIDELIS AND J. T. PEDERSEN. 1999. Critical assessment of methods of protein structure prediction (CASP): round III. Proteins **Suppl. 3**: 2–6.

142 NAJMANOVICH, R. J., J. W. TORRANCE AND J. M. THORNTON. 2005. Prediction of protein function from structure: insights from methods for the detection of local structural similarities. Biotechniques **38**: 847, 849, 851.

143 NOWAKOWSKI, J., C. N. CRONIN, D. E. MCREE, *et al.* 2002. Structures of the cancer-related Aurora-A, FAK, and EphA2 protein kinases from nanovolume crystallography. Structure (Camb.) **10**: 1659–67.

144 NUNEZ MIGUEL, R., J. SANDERS, J. JEFFREYS, *et al.* 2004. Analysis of the thyrotropin receptor-thyrotropin interaction by comparative modeling. Thyroid **14**: 991–1011.

145 O'CONNELL, N. M., R. E. SAUNDERS, C. A. LEE, D. J. PERRY AND S. J. PERKINS. 2005. Structural interpretation of 42 mutations causing factor XI deficiency using homology modeling. J. Thromb. Haemost. **3**: 127–38.

146 OKADA, T., M. SUGIHARA, A. N. BONDAR, M. ELSTNER, P. ENTEL AND V. BUSS. 2004. The retinal conformation and its environment in rhodopsin in light of a new 2.2 Å crystal structure. J. Mol. Biol. **342**: 571–83.

147 OLIVA, B., P. A. BATES, E. QUEROL, F. X. AVILÉS AND M. J. E. STERNBERG. 1997. An automated classification of the structure of protein loops. J. Mol. Biol. **266**: 814–30.

148 PACE, H. C., M. A. KERCHER, P. LU, P. MARKIEWICZ, J. H. MILLER, G. CHANG AND M. LEWIS. 1997. Lac repressor genetic map in real space. Trends Biochem.Sci. **22**: 334–9.

149 PAL, L., B. DASGUPTA AND P. CHAKRABARTI. 2005. 3_{10}-Helix adjoining alpha-helix and beta-strand: sequence and structural features and their conservation. Biopolymers **78**: 147–62.

150 PAL, M. AND S. DASGUPTA. 2003. The nature of the turn in omega loops of proteins. Proteins **51**: 591–606.

151 PALCZEWSKI, K., T. KUMASAKA, T. HORI, *et al.* 2000. Crystal structure of rhodopsin: a G protein-coupled receptor. Science **289**: 739–45.

152 PEDRETTI, A., M. ELENA SILVA, L. VILLA AND G. VISTOLI. 2004. Binding site analysis of full-length alpha1a adrenergic receptor using homology modeling and molecular docking. Biochem. Biophys. Res. Commun. **319**: 493–500.

153 PEITSCH, M. C. 1997. Large scale protein modelling and model repository. Proc. ISMB **5**: 234–6.

154 PEITSCH, M. C. 1996. ProMod and Swiss-Model: Internet-based tools for automated comparative protein modelling. Biochem. Soc. Trans. **24**: 274–9.

155 PEITSCH, M. C., M. R. WILKINS, L. TONELLA, J. C. SANCHEZ, R. D. APPEL AND D. F. HOCHSTRASSER. 1997. Large-scale protein modelling and integration with the SWISS-PROT and SWISS-2DPAGE databases: the example of *Escherichia coli*. Electrophoresis **18**: 498–501.

156 PERUTZ, M. F., J. C. KENDREW AND H. C. WATSON. 1965. Structure and function of haemoglobin. J. Mol. Biol. **13**: 669–78.

157 PETERSON, R. W., P. L. DUTTON AND A. J. WAND. 2004. Improved side-chain prediction accuracy using an *ab initio* potential energy function and a very large rotamer library. Protein Sci. **13**: 735–51.

158 PIEPER, U., N. ESWAR, A. C. STUART, V. A. ILYIN AND A. SALI. 2002. MODBASE,

a database of annotated comparative protein structure models. Nucleic Acids Res. **30**: 255–9.

159 PIERCE, N. A., J. A. SPRIET, J. DESMET AND S. L. MAYO. 1999. Conformational splitting: a more powerful criterion for dead-end elimination. J. Comp. Chem. **21**: 999–1009.

160 PLEBAN, K., A. MACCHIARULO, G. COSTANTINO, R. PELLICCIARI, P. CHIBA AND G. F. ECKER. 2004. Homology model of the multidrug transporter LmrA from *Lactococcus lactis*. Bioorg. Med. Chem. Lett. **14**: 5823–6.

161 PONDER, J. W. AND F. M. RICHARDS. 1987. Tertiary templates for proteins: Use of packing criteria in the enumeration of allowed sequences for different structural classes. J. Mol. Biol. **193**: 775–92.

162 PURTA, E., F. VAN VLIET, C. TRICOT, L. G. DE BIE, M. FEDER, K. SKOWRONEK, L. DROOGMANS AND J. M. BUJNICKI. 2005. Sequence–structure–function relationships of a tRNA (m7G46) methyltransferase studied by homology modeling and site-directed mutagenesis. Proteins **59**: 482–8.

163 QUEVILLON-CHERUEL, S., B. COLLINET, C. Z. ZHOU, *et al.* 2003. A structural genomics initiative on yeast proteins. J. Synchrotron Radiat. **10**: 4–8.

164 REYES, C. L. AND G. CHANG. 2005. Structure of the ABC transporter MsbA in complex with ADP·vanadate and lipopolysaccharide. Science **308**: 1028–31.

165 RING, C. S., D. G. KNELLER, R. LANGRIDGE AND F. E. COHEN. 1992. Taxonomy and conformational analysis of loops in proteins [published erratum appears in J. Mol. Biol. 1992; **227**(3): 977]. J. Mol. Biol. **224**: 685–99.

166 RING, C. S., E. SUN, J. H. MCKERROW, G. K. LEE, P. J. ROSENTHAL, I. D. KUNTZ AND F. E. COHEN. 1993. Structure-based inhibitor design by using protein models for the development of antiparasitic agents. Proc. Natl. Acad. Sci. USA **90**: 3583–7.

167 ROHL, C. A., C. E. STRAUSS, D. CHIVIAN AND D. BAKER. 2004. Modeling structurally variable regions in homologous proteins with rosetta. Proteins **55**: 656–77.

168 ROOMAN, M. J. AND S. J. WODAK. 1991. Weak correlation between predictive power of individual sequence patterns and overall prediction accuracy in proteins. Proteins **9**: 69–78.

169 ROST, B. 1999. Twilight zone of protein sequence alignments. Protein Eng. **12**: 85–94.

170 ROST, B., C. SANDER AND R. SCHNEIDER. 1994. Redefining the goals of structure prediction. J. Mol. Biol. **235**: 13–26.

171 ROY, S. AND S. SEN. 2005. Homology modeling based solution structure of Hoxc8–DNA complex: role of context bases outside TAAT stretch. J. Biomol. Struct. Dyn. **22**: 707–18.

172 RUFINO, S. D., L. E. DONATE, L. H. J. CANARD AND T. L. BLUNDELL. 1997. Predicting the conformational class of short and medium size loops connecting regular secondary structures: application to comparative modeling. J. Mol. Biol. **267**: 352–67.

173 SAHASRABUDHE, P. V., R. TEJERO, S. KITAO, Y. FURUICHI AND G. T. MONTELIONE. 1998. Homology modeling of an RNP domain from a human RNA-binding protein: homology-constrained energy optimization provides a criterion for distinguishing potential sequence alignments. Proteins **33**: 558–66.

174 SALI, A. AND T. L. BLUNDELL. 1993. Comparative protein modelling by satisfaction of spatial restraints. J. Mol. Biol. **234**: 779–815.

175 SALI, A. AND J. P. OVERINGTON. 1994. Derivation of rules for comparative protein modeling from a database of protein structure alignments. Protein Sci. **3**: 1582–96.

176 SAMUDRALA, R. AND J. MOULT. 1998. An all-atom distance-dependent conditional probability discriminatory function for protein structure prediction. J. Mol. Biol. **275**: 895–916.

177 SAMUDRALA, R. AND J. MOULT. 1998. Determinants of side chain conformational preferences in protein structures. Protein Eng. **11**: 991–7.

178 SAMUDRALA, R. AND J. MOULT. 1998. A graph-theoretic algorithm for comparative modeling of protein structure. J. Mol. Biol. **279**: 287–302.

179 SAMUDRALA, R. AND J. MOULT. 1997. Handling context-sensitivity in protein structures using graph theory: bona fide prediction. Proteins **Suppl. 1**: 43–9.

180 SANCHEZ, R. AND A. SALI. 1997. Evaluation of comparative protein structure modeling by MODELLER-3. Proteins **Suppl. 1**: 50–8.

181 SANCHEZ, R. AND A. SALI. 1998. Large-scale protein structure modeling of the Saccharomyces cerevisiae genome. Proc. Natl Acad. Sci. USA **95**: 13597–602.

182 SASISEKHARAN, V. AND P. K. PONNUSWAMY. 1970. Backbone and sidechain conformations of amino acids and amino acid residues in peptides. Biopolymers **9**: 1249–56.

183 SASISEKHARAN, V. AND P. K. PONNUSWAMY. 1971. Studies on the conformation of amino acids. X. Conformations of norvalyl, leucyl, aromatic side groups in a dipeptide unit. Biopolymers **10**: 583–92.

184 SAUDER, J. M., J. W. ARTHUR AND R. L. DUNBRACK, JR. 2000. Large-scale comparison of protein sequence alignment algorithms with structure alignments. Proteins **40**: 6–22.

185 SCHIFFER, C. A., J. W. CALDWELL, P. A. KOLLMAN AND R. M. STROUD. 1990. Prediction of homologous protein structures based on conformational searches and energetics. Proteins **8**: 30–43.

186 SCHMITT, L. 2002. The first view of an ABC transporter: the X-ray crystal structure of MsbA from *E. coli*. ChemBiochem. **3**: 161–5.

187 SCHRAUBER, H., F. EISENHABER AND P. ARGOS. 1993. Rotamers: To be or not to be? An analysis of amino acid sidechain conformations in globular proteins. J. Mol. Biol. **230**: 592–612.

188 SCHUMACHER, V., T. VOGEL, B. LEUBE, C. DRIEMEL, T. GOECKE, G. MOESLEIN AND B. ROYER-POKORA. 2005. Gene symbol: STK11. Disease: Peutz–Jeghers syndrome. Hum. Genet. **116**: 541.

189 SCHWEDE, T., J. KOPP, N. GUEX AND M. C. PEITSCH. 2003. SWISS-MODEL: an automated protein homology-modeling server. Nucleic Acids Res. **31**: 3381–5.

190 SEIGNEURET, M. AND A. GARNIER-SUILLEROT. 2003. A structural model for the open conformation of the mdr1 P-glycoprotein based on the MsbA crystal structure. J. Biol. Chem. **278**: 30115–24.

191 SHENKIN, P. S., D. L. YARMUSH, R. M. FINE, H. J. WANG AND C. LEVINTHAL. 1987. Predicting antibody hypervariable loop conformation. I. Ensembles of random conformations for ringlike structures. Biopolymers **26**: 2053–85.

192 SHEPHERD, A. J., D. GORSE AND J. M. THORNTON. 1999. Prediction of the location and type of beta-turns in proteins using neural networks. Protein Sci. **8**: 1045–55.

193 SHETTY, R. P., P. I. DE BAKKER, M. A. DEPRISTO AND T. L. BLUNDELL. 2003. Advantages of fine-grained side chain conformer libraries. Protein Eng. **16**: 963–9.

194 SHINDYALOV, I. N. AND P. E. BOURNE. 1998. Protein structure alignment by incremental combinatorial extension (CE) of the optimal path. Protein Eng. **11**: 739–47.

195 SIBANDA, B. L. AND J. M. THORNTON. 1985. Beta-hairpin families in globular proteins. Nature **316**: 170–4.

196 SIBANDA, B. L. AND J. M. THORNTON. 1991. Conformation of beta hairpins in protein structures: classification and diversity in homologous structures. Methods Enzymol. **202**: 59–82.

197 SIMONS, K. T., C. KOOPERBERG, E. HUANG AND D. BAKER. 1997. Assembly of protein tertiary structures from fragments with similar local sequences using simulated annealing and Bayesian scoring functions. J. Mol. Biol. **268**: 209–25.

198 SOWDHAMINI, R., S. D. RUFINO AND T. L. BLUNDELL. 1996. A database of globular protein structural domains: clustering of representative family members into similar folds. Fold. Des. **1**: 209–20.

199 STATES, D. J. AND D. BOTSTEIN. 1991. Molecular sequence accuracy and the

analysis of protein coding regions. Proc. Natl Acad. Sci. USA **88**: 5518–22.

200 Stenson, P. D., E. V. Ball, M. Mort, *et al.* 2003. Human Gene Mutation Database (HGMD): 2003 update. Hum. Mutat. **21**: 577–81.

201 Strausberg, R. L., K. H. Buetow, M. R. Emmert-Buck and R. D. Klausner. 2000. The cancer genome anatomy project: building an annotated gene index. Trends Genet. **16**: 103–6.

202 Suckow, J., P. Markiewicz, L. G. Kleina, J. Miller, B. Kisters-Woike and B. Muller-Hill. 1996. Genetic studies of the Lac repressor. XV: 4000 single amino acid substitutions and analysis of the resulting phenotypes on the basis of the protein structure. J. Mol. Biol. **261**: 509–23.

203 Summers, N. L., W. D. Carlson and M. Karplus. 1987. Analysis of sidechain orientations in homologous proteins. J. Mol. Biol. **196**: 175–98.

204 Summers, N. L. and M. Karplus. 1989. Construction of side-chains in homology modelling: Application to the C-terminal lobe of rhizopuspepsin. J. Mol. Biol. **210**: 785–811.

205 Summers, N. L. and M. Karplus. 1990. Modeling of globular proteins: A distance-based search procedure for the construction of insertion/deletion regions and Pro ↔ non-Pro mutations. J. Mol. Biol. **216**: 991–1016.

206 Sun, M., Z. Li, Y. Zhang, Q. Zheng and C. C. Sun. 2005. Homology modeling and docking study of cyclin-dependent kinase (CDK) 10. Bioorg. Med. Chem. Lett. **15**: 2851–6.

207 Sutcliffe, M. J., I. Haneef, D. Carney and T. L. Blundell. 1987. Knowledge based modeling of homologous proteins, part I: three-dimensional frameworks derived from the simultaneous superposition of multiple structures. Protein Eng. **5**: 377–84.

208 Sutcliffe, M. J., F. R. Hayes and T. L. Blundell. 1987. Knowledge based modeling of homologous proteins, Part II: rules for the conformations of substituted sidechains. Protein Eng. **1**: 385–92.

209 Swindells, M. B., M. W. MacArthur and J. M. Thornton. 1995. Intrinsic ϕ,ψ propensities of amino acids, derived from the coil regions of known structures. Nat. Struct. Biol. **2**: 596–603.

210 Terwilliger, T. C., M. S. Park, G. S. Waldo, *et al.* 2003. The TB structural genomics consortium: a resource for *Mycobacterium tuberculosis* biology. Tuberculosis (Edinb.) **83**: 223–49.

211 Thompson, J. D., D. G. Higgins and T. J. Gibson. 1994. CLUSTAL W: improving the sensitivity of progressive multiple sequence alignment through sequence weighting, position-specific gap penalties and weight matrix choice. Nucleic Acids Res. **22**: 4673–80.

212 Tuffery, P., C. Etchebest, S. Hazout and R. Lavery. 1991. A new approach to the rapid determination of protein side chain conformations. J. Biomol. Struct. Dyn. **8**: 1267–89.

213 Vasquez, M. 1995. An evaluation of discrete and continuum search techniques for conformational analysis of side-chains in proteins. Biopolymers **36**: 53–70.

214 Venclovas, C., A. Zemla, K. Fidelis and J. Moult. 1997. Criteria for evaluating protein structures derived from comparative modeling. Proteins **Suppl. 1**: 7–13.

215 Venclovas, C., A. Zemla, K. Fidelis and J. Moult. 1999. Some measures of comparative performance in the three CASPs. Proteins **Suppl. 3**: 231–7.

216 Vijayasri, S. and S. Agrawal. 2005. Domain-based homology modeling and mapping of the conformational epitopes of envelope glycoprotein of West Nile virus. J. Mol. Model. **11**: 248–55 [Online].

217 Vlijmen, H. W. T. v. and M. Karplus. 1997. PDB-based protein loop prediction: Parameters for selection and methods for optimization. J. Mol. Biol. **267**: 975–1001.

218 Voigt, C. A., D. B. Gordon and S. L. Mayo. 2000. Trading accuracy for speed: A quantitative comparison of search algorithms in protein sequence design. J. Mol. Biol. **299**: 789–803.

219 Vriend, G. 1990. WHAT IF: a molecular modeling and drug design program. J. Mol. Graph. **8**: 52–56.

220 WALLNER, B. AND A. ELOFSSON. 2005. All are not equal: a benchmark of different homology modeling programs. Protein Sci. **14**: 1315–27.

221 WANG, G., J. M. SAUDER AND R. L. DUNBRACK JR. 2005. Comparative modeling in structural genomics. In SUNDSTROM, M., M. NORIN AND A. EDWARDS (eds.), *Structural Proteomics and High Throughput Structural Biology*. CRC Press, New York, NY: 109–36.

222 WATSON, J. D., R. A. LASKOWSKI AND J. M. THORNTON. 2005. Predicting protein function from sequence and structural data. Curr Opin Struct Biol **15**: 275–84.

223 WEBER, I. T. 1990. Evaluation of homology modeling of HIV protease. Proteins **7**: 172–84.

224 WEBER, I. T., M. MILLER, M. JASKOLSKI, J. LEIS, A. M. SKALKA AND A. WLODAWER. 1989. Molecular modeling of the HIV-1 protease and its substrate binding site. Science **243**: 928–31.

225 WHISSTOCK, J. C. AND A. M. LESK. 2003. Prediction of protein function from protein sequence and structure. Q. Rev. Biophys. **36**: 307–40.

226 WILMOT, C. M. AND J. M. THORNTON. 1988. Analysis and prediction of the different types of beta-turn in proteins. J. Mol. Biol. **203**: 221–32.

227 WILMOT, C. M. AND J. M. THORNTON. 1990. Beta-turns and their distortions: a proposed new nomenclature. Protein Eng. **3**: 479–93.

228 WILSON, C., L. GREGORET AND D. AGARD. 1993. Modeling sidechain conformation for homologous proteins using an energy-based rotamer search. J. Mol. Biol. **229**: 996–1006.

229 WINTJENS, R. T., M. J. ROOMAN AND S. J. WODAK. 1996. Automatic classification and analysis of alpha alpha-turn motifs in proteins. J. Mol. Biol. **255**: 235–53.

230 WOHLFAHRT, G., V. HANGOC AND D. SCHOMBURG. 2002. Positioning of anchor groups in protein loop prediction: the importance of solvent accessibility and secondary structure elements. Proteins **47**: 370–8.

231 WOJCIK, J., J. P. MORNON AND J. CHOMILIER. 1999. New efficient statistical sequence-dependent structure prediction of short to medium-sized protein loops based on an exhaustive loop classification. J. Mol. Biol. **289**: 1469–90.

232 WOLF, Y. I., S. E. BRENNER, P. A. BASH AND E. V. KOONIN. 1999. Distribution of protein folds in the three superkingdoms of life. Genome Res. **9**: 17–26.

233 WOLF, Y. I., N. V. GRISHIN AND E. V. KOONIN. 2000. Estimating the number of protein folds and families from complete genome data. J. Mol. Biol. **299**: 897–905.

234 XIANG, Z. AND B. HONIG. 2001. Extending the accuracy limits of prediction for side-chain conformations. J. Mol. Biol. **311**: 421–30.

235 XIANG, Z., C. S. SOTO AND B. HONIG. 2002. Evaluating conformational free energies: the colony energy and its application to the problem of protein loop prediction. Proc. Natl. Acad. Sci. USA **99**: 7432–7.

236 XU, J. 2005. Rapid protein side-chain packing via tree decomposition. Proc. RECOMB **9**: 423.

237 YAMAGUCHI, A., M. IWADATE, E. SUZUKI, K. YURA, S. KAWAKITA, H. UMEYAMA AND M. GO. 2003. Enlarged FAMSBASE: protein 3D structure models of genome sequences for 41 species. Nucleic Acids Res. **31**: 463–8.

238 YANG, J., L. F. TEN EYCK, N. H. XUONG AND S. S. TAYLOR. 2004. Crystal structure of a cAMP-dependent protein kinase mutant at 1.26A: new insights into the catalytic mechanism. J. Mol. Biol. **336**: 473–87.

239 ZEMLA, A., C. VENCLOVAS, K. FIDELIS AND B. ROST. 1999. A modified definition of Sov, a segment-based measure for protein secondary structure prediction assessment. Proteins **34**: 220–3.

240 ZEMLA, A., C. VENCLOVAS, J. MOULT AND K. FIDELIS. 1999. Processing and analysis of CASP3 protein structure predictions. Proteins **Suppl. 3**: 22–9.

241 ZHANG, C. AND S. H. KIM. 2003. Overview of structural genomics: from structure to function. Curr. Opin. Chem. Biol. **7**: 28–32.

242 ZHANG, P., J. XIE, G. YI, C. ZHANG AND R. ZHOU. 2005. *De novo* RNA synthesis and homology modeling of the classical

swine fever virus RNA polymerase. Virus Res. **112**: 9–23.

243 ZHANG, Y., Y. Y. SHAM, R. RAJAMANI, J. GAO AND P. S. PORTOGHESE. 2005. Homology modeling and molecular dynamics simulations of the mu opioid receptor in a membrane-aqueous system. ChemBiochem. **6**: 853–9.

244 ZHENG, Q. AND D. J. KYLE. 1996. Accuracy and reliability of the scaling-relaxation method for loop closure: an evaluation based on extensive and multiple copy conformational samplings. Proteins **24**: 209–17.

245 ZHENG, Q., R. ROSENFELD, C. DELISI AND D. J. KYLE. 1994. Multiple copy sampling in protein loop modeling: computational efficiency and sensitivity to dihedral angle perturbations. Protein Sci. **3**: 493–506.

246 ZHENG, Q., R. ROSENFELD, S. VAJDA AND C. DELISI. 1993. Determining protein loop conformation using scaling-relaxation techniques. Protein Sci. **2**: 1242–8.

247 ZOLLER, B. AND B. DAHLBACK. 1994. Linkage between inherited resistance to activated protein C and factor V gene mutation in venous thrombosis. Lancet **343**: 1536–8.

11
Protein Fold Recognition Based on Distant Homologs

Ingolf Sommer

1 Introduction

In the 1960s Anfinsen showed with a rather simple experiment that for many proteins the sequence is the sole determinant for the three-dimensional (3-D) structure [7, 8]. A denaturating substance was added to the solution of a protein, resulting in the loss of native protein structure and function. After removal of the denaturing substance, proteins recovered the functional activity (an enzymatic reaction in Anfinsen's experiment). Thereby, it was concluded that the protein managed to refold itself in the absence of any other agents.

Later, it became evident that some proteins need other proteins to fold and some proteins or parts of proteins remain unfolded (see Chapter 9). Still, the Anfinsen principle has been a guiding force for much research that aims at understanding mechanisms for determining the 3-D structure of proteins given their sequence. Although great improvements of these methods have been achieved, the protein structure prediction problem is still unsolved, in general. The folding process that determines the structure is not known to enough detail to serve as a basis for modeling. Instead, prediction methods have to rely on heuristic inductive inferences.

One very successful approach to the prediction of protein structures today models the protein structure based on another structurally resolved protein as a structural template. We call this approach template-based modeling. The protein whose structure is to be predicted is called the target. In addition to the target sequence, the method requires the input of a database of resolved protein structures, the so-called template structures. Rather than modeling the protein structure *de novo*, we repeatedly ask the question whether the protein structure of the target sequence is similar to a template structure. This leaves us with a set of candidates for templates after which we can model the structure of the target sequence.

This chapter deals with the question how likely it is that a target protein sequence attains a structure similar to a given template structure. In principle

Bioinformatics - From Genomes to Therapies Vol. 1. Edited by Thomas Lengauer

ISBN: 978-3-527-31278-8

one can compare the sequence itself to the sequences of the template structures (sequence–sequence comparison, Section 3.1), additionally take evolutionary information into account (profile methods, Sections 3.2 and 3.3) or thread the sequence onto the given 3-D template structure taking physico-chemical properties of the template structure into account (Section 4). The problem of identifying suitable templates typically becomes harder the more distant the target sequence is related to its most similar sequences in the template database. More similar sequences are easier to identify when looking at sequence information only; typically, they also have a more similar structure within a chemically more similar environment.

Traditionally, protein structure prediction has been divided into homology modeling (also called comparative modeling; see Chapter 10), fold recognition (this chapter) and prediction of novel folds (Chapter 12) [97]. In homology modeling, closely homologous templates are available affording very precise models for the protein structure. In fold recognition, identifying a suitable template becomes a challenge. Once a template is obtained, a prediction of the 3-D arrangement can be made, whereas a constructed full-atom model is not reliable, in general. In contrast, in the new fold category no suitable template is available and fragment assembly or *de novo* methods need to be applied.

The basic pipeline exercised is identical in the first two categories, in principle. Thus, today homology modeling and fold recognition are merging. The focus in homology modeling is more on obtaining detailed models with high resolution. The focus of fold recognition is more on the identification of suitable templates.

2 Overview of Template-based Modeling

2.1 Key Steps in Template-based Modeling

The input to template-based modeling is a target sequence and a database of previously resolved template structures. The output is a 3-D model for the target sequence, constructed according to the template.

2.1.1 Identifying Templates

How do we decide how likely the sequence of a target protein attains a given template structure? We perform a pairwise alignment of the target sequence with the sequences of the template structures. Here, different techniques can be used: the target sequence can be aligned with the amino acid sequences of the templates. If evolutionary information on the target sequence is available (see Chapters 3 and 4), a multiple alignment of related sequences can be used to construct a position-specific scoring matrix (PSSM or, similarly, a frequency profile, see Section 3.2.1). A PSSM represents the preferences of a residue in the target sequence to be matched with a residue in the template structure (Section 3.2.1). This kind of evolutionary information can also be used on the template side or on both target and template sides. One can enhance this approach by introducing additional information, e.g. stemming from secondary structure predictions (Section 5.1.1). Sequence–structure alignment methods using the 3-D structure information on the template sides during the template identification process are presented in Section 4.

2.1.2 **Assessing Significance**

The score of the alignment tells us the propensity of the target sequence to attain the template structure. Ideally, the template that can be aligned to the target sequence with the highest alignment score should provide the structural model for the target protein. However, since alignment scores reflect structural preference only inadequately, this model selection procedure is fallible. Thus, the score has to be accompanied with a confidence value that rates how much we can trust the prediction. Often, a confidence score is based on theoretical statistical significance which rates how unlikely it is to obtain the alignment by chance. In addition, empirical choices of confidence values have also proven effective (Section 6.1).

2.1.3 **Model Building**

Aligning the target sequence to a template protein is only the first step of producing a full-atom protein structure model for the target protein. While it is the critical step for the similarity range of below about 40% sequence identity between target and template sequence, it is less difficult in high similarity ranges (see Chapter 10). An incorrect alignment invariably leads to a wrong protein model.

The alignment maps residues of the target onto the template. An initial model is constructed by copying coordinates of the template structure and changing the template residue types according to the target sequence. This model only provides a part of the structure of the target protein's backbone. Gaps in the alignment represent parts of the target sequence that we cannot map onto the template structure (if gaps occur in the template sequence) or tears and rips in the backbone model of the target sequence (if gaps occur in the target sequence). The former gaps mostly coincide with loops in the target protein that have no counterpart in the template structure. These loops have to be modeled in a separate loop modeling step by inserting loop fragments from a database of protein structures, or by using energy optimization methods. Rips in the backbone of the target protein have to be mended and, finally, the sidechains of the target protein have to be attached to the backbone. Here, the variants are to use a database of side chain rotamers or to do energy optimization. Similarity-based protein modeling tools combine these steps in different ways and using different algorithmic procedures (see Chapter 10) [32,129,133].

2.1.4 Evaluation

The performance of fold recognition methods is typically assessed by benchmarks [20,37,71,171]. Here, the objective is to retrieve the template structure that is most similar to the structure of the target protein. The performance can be quantified in terms of the number of correctly assigned folds or, in a more detailed fashion, by rating the quality of the alignment, on which the fold assignment is based. The accuracy of protein sequence–structure alignment methods depends highly on the sequence similarity between the target and the template protein. It is low in the case of low sequence similarity and high in the case of high sequence similarity. Over the whole protein structure database we can today achieve an accuracy of above 70% correctly assigned folds [165]. Starting 1994, the Critical Assessment of Protein Structure Prediction (CASP) experiment, a biannual blind test for protein structure prediction methods was established to measure progress in the field [98].

While similarity-based modeling is quite successful, this approach cannot discover yet unseen protein structures. Rather, it can only rediscover structures that have been seen before as attained by different protein sequences. As we can only assume to have uncovered about one half of all protein folds used by nature [72,178], this approach has strong limitations.

2.2 Template Databases

Typically the structures serving as templates are experimentally determined by X-ray crystallography or nuclear magnetic resonance (NMR) spectroscopy.

X-ray crystallography determines structures of macromolecules by analyzing their diffraction patterns when irradiated by X-rays. In order to obtain a diffraction pattern a protein has to be crystallized. During crystallization many instances of the same protein are symmetrically arranged along a lattice. Even when crystallized, proteins frequently display biological activity indicating that the crystallization process captured them in the biologically active form. Some proteins are very hard to crystallize (e.g. membrane proteins). When the protein crystal is properly irradiated with X-rays, diffraction patterns result that are captured on film or digital media. From these patterns the electron density of the protein is computed and from the density information the atomic structure of the protein is derived [13,14,120].

NMR spectroscopy is the second prominent method for structure determination. It can be applied to smaller proteins (up to around 40 kDa) in highly concentrated solution. Some atomic nuclei, e.g. protons in hydrogen, display an intrinsic magnetic property called spin. By applying an external varying magnetic field the state of the spin may change, provided resonance takes place. A resonance spectrum can be obtained. The surrounding local environment of nuclei may cause a shift of the peaks in the resonance spectrum. Peaks in the spectrum are associated with residues in the protein sequence in a sequential assignment step and a list of distance constraints can be deduced. Applying distance geometry techniques to these data, atom coordinates can be estimated [13, 14, 162].

The most frequently used public resource for coordinates of protein structures determined with NMR spectroscopy or X-ray crystallography is the Protein Data Bank (PDB). Quality of structures deposited in the PDB can be judged with a number of programs (Procheck [94], Whatcheck [55]). The ASTRAL compendium for structure and sequence analysis conveniently combines the output of several of these programs with additional manual annotation into a joint AEROSPACI score [16, 24].

Most proteins are globular. Larger proteins often fold into several independent folding units or domains. Domains are compact regions of structure often capable of folding on their own in aqueous solution. Domains can be defined as folding units, as units of structural similarity, or as evolutionary units [115]. Several resources describe the domain composition within proteins and classify the identified domains hierarchically: most prominent are SCOP, CATH and FSSP. SCOP, (structural classification of proteins [99]) is a human-curated database organized hierarchically into classes, folds, superfamilies and families. CATH (class, architecture, topology and homologous superfamily [106]) is a semi-automated procedure. The FSSP (families of structurally similar proteins [54]) relies on a fully automatic procedure.

Since domains are recurring folding units, often domains are chosen as templates for structure prediction. For the construction of template databases, reasonably different templates with high quality structures are favored. Representative sets of templates can be constructed limiting the maximal percentage of sequence identity while choosing structures with a high SPACI score for example. Choosing SCOP domains with at most 40% (95%) sequence identity results in a set of 7290 (12 065) structures for SCOP version 1.69 of July 2005.

3 Sequence-based Methods for Identifying Templates

3.1 Sequence–Sequence Comparison Methods

The simplest method for assigning a fold to a target sequence is to compare the target sequence to sequences of proteins with known structures. In order to compare the sequences they are aligned (see Chapter 3, [34]).

Scoring the similarity of an individual pair of amino acids amounts to comparing the likelihoods of generating this pair from two alternative stochastic models. One model, the model of related amino acids, describes the distribution of amino acid pairs originating from related positions in pairwise alignments of homologous sequences. The second model is a null model which describes the distribution of unrelated amino acid pairs. The amino acid pair is denoted by a pair of random variables (X, Y) both with values in $\{1, \ldots, 20\}$. The distribution of (X, Y) under the related model is defined as $p_{\mathrm{rel}}(i, j)$. Note that the background distribution of amino acids can be computed as $p_i = \sum_{j=1}^{20} p_{\mathrm{rel}}(i, j)$. The log-likelihood ratios of all pairs of amino acids are stored in an amino acid similarity matrix: $M_{i,j} = \lambda \log\left(\frac{p_{\mathrm{rel}}(i,j)}{p_i p_j}\right)$, with constant scaling factors λ. Different factors were introduced by different authors: Dayhoff and coworkers [29] use $\lambda = 10/\log 10$, the BLOSUM series [51] uses $\lambda = 2/\log 2$.

Introducing an additional penalty for inserting and extending gaps, two sequences of unequal length can be aligned. For arbitrary sequences, the Needleman–Wunsch algorithm [100], aligns two sequences using dynamic programming. It is appropriate, when sequences are expected to be similar from beginning to end, computing a so-called *global* alignment. In cases where only limited patches of the sequences share similarity, *local* alignment is used, which is computed using the Smith–Waterman algorithm [145]. When searching for similarity of a complete subsequence within a sequence, as is the case for example when identifying domains within longer sequences, *free-shift* alignments are appropriate (domain identification is also discussed in the context of Chapter 12).

The runtime of alignment algorithms is measured in terms of (sum of the) length of the aligned sequences. Using the Needleman–Wunsch algorithm, alignments can be computed in cubic time for general gap cost functions. For affine gap-cost functions, alignments can be computed in quadratic time using the Gotoh algorithm [48]. Since these dynamic programming programming algorithm can be slow for high-throughput applications fast heuristics like BLAST [5] and FASTA [113] were developed which approximate the optimal alignment algorithms.

The parameters of sequence alignment methods have to be calibrated for the intended application, e.g. Pearson details a protocol for searching genomes with these methods [111]. The major disadvantage of these pairwise sequence comparison methods is that conserved and variable positions are treated indifferently and contribute equally to the final alignment score. This limits the ability to identify distant homologs.

Sequence–sequence comparison methods like the Smith–Waterman algorithm [145] and the BLAST [5] or FASTA [113] tools can assign a fold to approximately 20–30% of the proteins coded by genes in a microbial genome [45, 173].

3.2 Frequency Profile Methods

Exploiting evolutionary information in addition to the plain sequence information, frequency profile methods are powerful tools for detecting distant relationships between amino acid sequences, often picking up signals even when other methods fail [170]. In this section we define frequency profiles, and describe why one would use them, how they are generated and different ways of comparing profiles to sequences and profiles to other profiles.

3.2.1 Definition of a Frequency Profile and PSSM

A frequency profile (or profile for short) is a sequence of position-specific frequencies of amino acids, which can be used for sequence alignments instead of a sequence containing individual amino acids at each position [49, 150].

Starting with an alignment of several related sequences (a multiple sequence alignment), for each position in that multiple alignment the occurrences of the types of amino acids are counted (for an example, see Figures 1 and 2). This yields an estimate of the likelihood of different types of amino acids occurring at different positions along the sequence. By counting gap characters in an alignment column, information on the likelihood of insertions at individual positions along the sequence can be gathered. Before counting, each sequence can be given a weight, which is a useful debiasing procedure if several of the sequences are very similar [49, 52, 91, 134].

Formally, a frequency profile matrix $(F_{p,a})$ is composed of at least 20 columns and N rows. The row p corresponds to the sequence position in the multiple alignment of length N and the column a to the type of amino acid. The first 20 columns of each row specify the relative frequencies of the types of amino acids at that position [123, 167, 185]. Some scoring schemes require additional columns containing penalties for insertions or deletions at that position [49], or unexpected characters in the sequences [18].

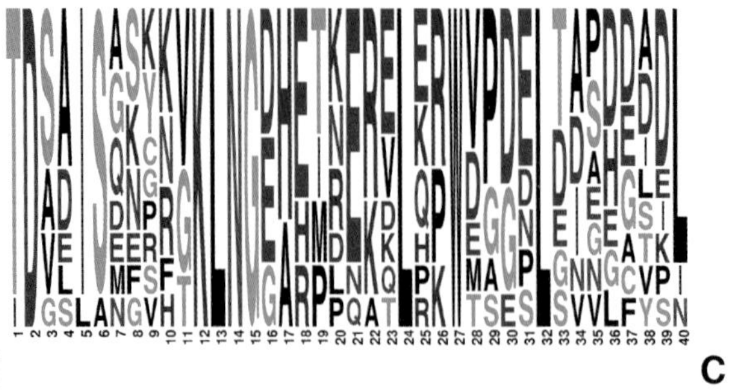

Figure 1 Sequence logo corresponding to the first 40 sequence positions of Pfam [11] multiple alignment of the Ataxin-2 N-terminal region. The figure is produced with WebLogo [28]. The overall height of the stack at each sequence position indicates the sequence conservation at that position, while the height of symbols within the stack indicates the relative frequency of each amino acid at that position.

	1	2	3	4	5	6	7	8	9	10	11	12	13	14	15	16	17	18	19	20
A			0.2	0.5		0.1	0.2										0.3			
C									0.1											
D		1.0		0.2			0.1									0.4				0.1
E				0.1			0.1	0.1								0.4		0.6		
F								0.1		0.1										
G			0.1				0.2	0.1	0.1		0.3				1.0	0.2				
H										0.1							0.7	0.2		
I	0.1				0.9														0.2	
K								0.2	0.2	0.3		1.0								0.3
L				0.1	0.1								1.0							0.1
M							0.1												0.2	
N							0.1	0.2		0.2				1.0						0.2
P									0.1										0.2	0.1
Q							0.2													
R									0.1	0.2								0.2		0.2
S			0.5	0.1		0.9		0.3	0.1											
T	0.9										0.2								0.4	
V			0.2						0.1		0.5									
W																				
Y									0.2											

Figure 2 First 20 sequence positions of the frequency profile matrix corresponding to the sequence logo depicted in Figure 1. All sequences are weighted identically.

Originally, Gribskov [49] combined the frequency profile matrix directly with amino acid substitution matrices [29,51], yielding

$$(S_{p,a}) = \left(\sum_{b=1}^{20} F_{p,b} \cdot M_{a,b} \right),$$

where $(M_{a,b})$ is Dayhoff's substitution matrix. This construct is referred to as a PSSM. In contrast to a frequency profile, a PSSM is a frequency profile multiplied with substitution preferences.

Other authors also use the word *profile* in the context of structure-based template identification methods to describe the sequence-based information extracted from protein structures, as discussed in Sections 4 and 4.3, in particular.

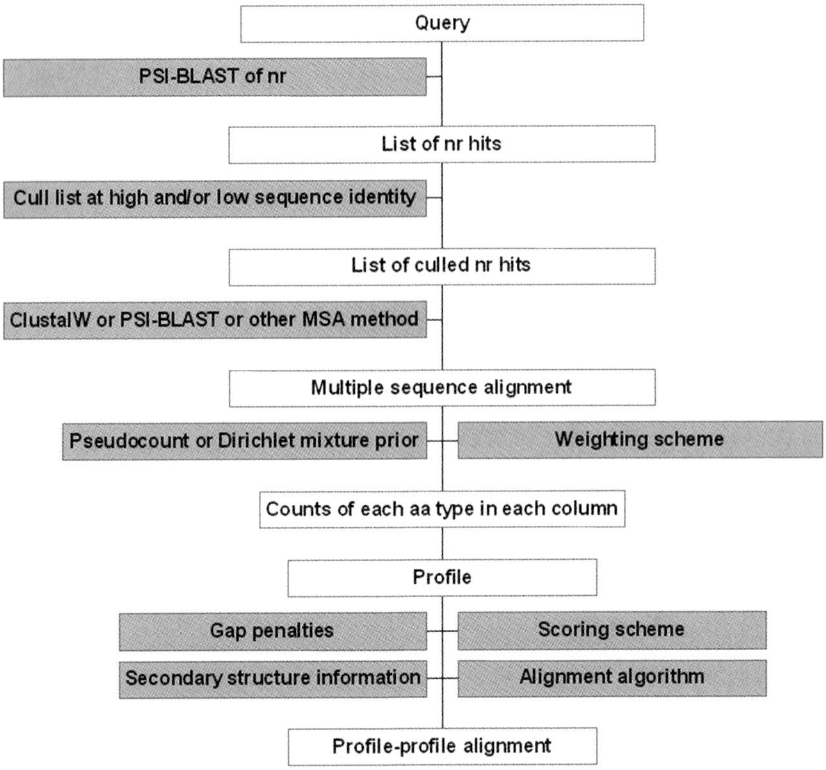

Figure 3 Illustration of the generation and application of profile alignments (from Wang and Dunbrack [171], reprinted with permission of Cold Spring Harbor Laboratory Press and the authors).

3.2.2 Generating Frequency Profiles

For constructing a profile one needs a multiple alignment of related sequences. Such alignment can be readily available, e.g. as in Pfam [11]. If not, one can start generating a profile from a single sequence by searching related sequences and multiply aligning them. Once a few sequences are found, a profile can be constructed from them and employed to search more sequences. This iterative approach for searching homologs is implemented in the popular PSI-BLAST program [6, 131], for example. In practice, the number of sequences identified by this search matters. This number is controlled by the BLAST parameters "number of iterations" and "*E*-value" for inclusion into the PSSM. The multiple alignment can be taken directly from the BLAST output, iteratively optimized [136], or improved with tools for multiple sequence alignment like ClustalW [156] or T-Coffee [101, 107]. The process of profile generation is illustrated in Figure 3.

If one particular sequence is the seed for searching further sequences, the multiple sequence alignment can be cropped by deleting the columns which contain gaps in the master sequence (master–slave alignment in BLAST terminology). Such an alignment can be used to construct a profile specifically reflecting that sequence. In the matrix notation above, this results in a frequency profile matrix F with rows eliminated until the N remaining rows correspond to the N residues of the master sequence.

Profiles calculated from multiple alignments that originate from similarity searches are subject to a bias introduced by the composition of the sequence databases that are searched [3]. An example is a case such that a search yields a large number of hits of mammalian origin and only few distantly related plant sequences. If a frequency profile from such a multiple alignment were calculated by simply taking the relative frequencies of the amino acids, this would reflect a strong emphasis on the mammalian sequences. This effect is not desirable, since the goal is to present all sequences of the family in an unbiased manner in the profile. To tackle this problem, it is generally assumed that only a fraction of the sequences of the family to be modeled is available in the databases and many sequences have not been observed, so far. A number of methods has been developed to estimate the size of the sequence families from the available sample. The most important are Dirichlet mixture models [17], pseudocounts [153], minimal-risk estimation [180] and sequence weighting models [52,74,89,123,125,134,150]. For a discussion of sequence weighting models, see Ref. [171].

3.2.3 Scoring Frequency Profiles

Frequency profiles can be aligned using the same algorithms as in sequence alignment. Whereas in plain sequence alignments two individual amino acids are matched, here profile vectors are compared. Profile vectors can be matched to individual amino acids or to other profile vectors. Different schoring schemes exist for both approaches .

3.2.4 Scoring Profiles Against Sequences

Let α be a row-vector at a certain sequence position p in a profile F and let $a_i = F(p,i)$ be the frequency observed for amino acid number i at position p. In the following, we will investigate ways of scoring α against an individual amino acid of type j (profile–sequence), or against another row-vector β from another profile F' (profile–profile).

The average score was the first profile–sequence score used in bioinformatics [22,49]. It is defined as

$$\text{score}_{\text{avg}}(\alpha, j) = \sum_{i=1}^{20} \alpha_i M_{i,j}.$$

Its basic idea is to compute the expected value of the sequence–sequence score under the profile distribution. Later, we will see an extension of this idea to profile–profile scoring. This scoring system has several drawbacks which led to further development.

The previously mentioned iterative PSI-BLAST program [6] starts with a round of BLAST using sequence–sequence alignment techniques to identify related sequences. From the sequences found, a profile is constructed which is used to identify further sequences using essentially the average scoring scheme. No position-specific gap costs are used, instead for each iteration the same gap costs that are used in the initial BLAST run are applied. In contrast to PSI-BLAST, the related tool IMPALA [131] implements the opposite sequence–profile direction and compares a single target sequence to a database of PSSMs previously generated with PSI-BLAST. This method can be used to quickly compare a sequence to a database of template structures with precomputed PSSMs.

Log-likelihood profile–sequence scoring (e.g. Ref. [27]):

$$\mathrm{score}_{\mathrm{lq}}(\alpha, j) = \log \frac{\alpha_j}{p_j},$$

is an optimal (in the sense that the likelihood ratio guarantees the lowest error of type II of all tests at the same level of significance) test statistic according to the Neyman–Pearson lemma from statistical test theory for deciding whether the amino acid j is a sample from the distribution α or rather from the background distribution p. The values summed up over all aligned positions provide a direct measure of the likelihood of the amino acid sequence being a sample from the profile.

Evolutionary profile–sequence scoring [27,50] is defined as

$$\mathrm{score}_{\mathrm{ev}}(\alpha, j) = \log \sum_{i=1}^{20} \alpha_i \frac{p_{\mathrm{rel}}(i, j)}{p_i p_j}.$$

This score summed up over all aligned positions in an alignment is an optimal means of deciding whether the sequence occurs by chance or is more likely to be the result of sampling from the profile having undergone evolutionary transition.

3.2.5 **Scoring Profiles against Profiles**

Since profiles have been invented, several ways of comparing profiles to profiles have been developed and tested. These scoring schemes perform differently in their abilities for searching templates and in the quality of the alignments produced.

A simple, fast and heuristic way of comparing two profile vectors independently of any substitution matrix is the dot product scoring, which is the summation of the products of the frequencies per type of amino acid [117,123,124]:

$$\text{score}_{\text{dotprod}}(\alpha, \beta) = \sum_{i=1}^{20} \alpha_i \beta_i.$$

Average or cross-product profile–profile scoring is a straightforward extension of the approach used in profile–sequence scoring. The products of the frequencies are multiplied with the corresponding log–odds elements of the substitution matrix:

$$\text{score}_{\text{avg}}(\alpha, \beta) = \sum_{i=1}^{20} \sum_{j=1}^{20} \alpha_i \beta_j \log \frac{p_{\text{rel}}(i, j)}{p_i p_j}.$$

Though not called profile–profile alignment, this approach has been used in ClustalW [155, 156], where two multiple alignments are aligned using the average over all pairwise scores between residues.

The log-average scoring multiplies the products of the frequencies by the corresponding probabilities in the substitution matrix and then takes the logarithm [165, 167]:

$$\text{score}_{\text{logavg}}(\alpha, \beta) = \log \sum_{i=1}^{20} \sum_{j=1}^{20} \alpha_i \beta_j \frac{p_{\text{rel}}(i, j)}{p_i p_j}.$$

This scoring is symmetric and scores zero against the background distribution. For the special case of profile–sequence alignment, i.e. for one of the profiles corresponding to just one sequence, log-average scoring reduces to score_{ev}. For the special case of both profiles corresponding to one sequence, log-average scoring reduces to the scoring in normal sequence–sequence alignment.

The Jensen–Shannon divergence $D_\lambda^{\text{JS}}(\alpha, \beta) = \lambda D^{\text{KL}}(\alpha, \gamma) + (1 - \lambda) D^{\text{KL}}(\beta, \gamma)$, with $\gamma = \lambda\alpha + (1 - \lambda)\beta$ for $0 \leq \lambda \leq 1$ is based on the information theoretic Kullback–Leibler distance $D^{\text{KL}}(\alpha, \beta) = \sum_{k=1}^{20} \alpha_k \log_2 \frac{\alpha_k}{\beta_k}$. With p as amino acid background distribution, Yona and Levitt developed the score [185]:

$$\text{score}_{\text{JensenShannon}}(\alpha, \beta) = \frac{1}{2}(1 - D^{\text{JS}}(\alpha, \beta))(1 + D^{\text{JS}}(\gamma, p)).$$

Given two profile vectors α and β, the corresponding row vectors of the PSI-BLAST PSSMs S and T, and the effective number of observations n and m, the Panchenko score [108] is computed as:

$$\text{score}_{\text{Panchenko}}((\alpha, S, m), (\beta, T, n)) = \frac{m\, \alpha \cdot T + n\, \beta \cdot S}{n + m}.$$

The LogOddsMultin score used in the COMPASS tool for comparison of multiple protein alignments [125] is an extension of the scoring that PSI-BLAST uses:

$$\text{score}_{\text{LogOddsMultin}} = c_1 \sum_{i=1}^{20} n_i^{(1)} \log \frac{q_i^{(2)}}{p_i} + c_2 \sum_{i=1}^{20} n_i^{(2)} \log \frac{q_i^{(1)}}{p_i},$$

where $n_a^{(1)}$ and $n_a^{(2)}$ are the effective counts for each amino acid in columns 1 and 2, and where $c_1 = \frac{n^{(2)}-1}{n^{(1)}+n^{(2)}-2}$, $c_2 = \frac{n^{(1)}-1}{n^{(1)}+n^{(2)}-2} = 1 - c_1$ and $n^{(k)} = \sum_{j=1}^{20} n_j^{(k)}$.

These scoring schemes are extensively discussed and experimentally compared in Refs. [164, 171]. Both studies compare and analyze the searching abilities as well as the quality of the alignments produced with these scorings schemes. Alignment quality can be monitored with measures like Q_{Modeler} as fraction of correctly aligned positions in the profile–profile alignment [130], $Q_{\text{Developer}}$ as fraction of correctly aligned positions in the structural alignment [130], or similarly Q_{Combined} that penalizes sequence alignments that are either too long or too short [185]. Measuring alignment quality with Q_{Combined}, the above profile–profile scoring functions perform similarly, while displaying differences in Q_{Modeler} and $Q_{\text{Developer}}$ [171]. The Jensen–Shannon and LogOddsMultin functions produce shorter, more accurate alignments [171]. The Jensen–Shannon scoring produces better alignments for closely related sequences [164]. Of the scoring functions above, in terms of search specificity and sensitivity the LogOddsMultin and log-average perform significantly better [171].

3.3 Hidden Markov Models (HMMs)

3.3.1 Definition

HMMs (see Chapter 3 or Ref. [34]) can be regarded as a generalization of profiles and have become an important tool in fold recognition. Introduced in the late 1960s and 1970s, and popular in speech recognition [118], HMMs made it into computational biology in the late 1980s [26] and have been used as profile models since the mid-1990s [73, 151]. For an introduction to HMMs, see Chapter 3 or Refs. [9, 118]; for review articles in the context of computational biology, see Refs. [25, 35].

HMMs are probabilistic models that are applicable to signals, time series or linear sequences. An HMM is a system characterized by the following: It has a set of hidden states $S = \{S_1, S_2, \ldots, S_N\}$. The system is in state q_t at time t and has a number M of distinct observation symbols per state, i.e. a discrete alphabet $V = \{v_1, v_2, \ldots, v_M\}$. The system randomly evolves according to a state transition probability distribution matrix $A = (a_{ij})$, where $a_{ij} = P(q_{t+1} = S_j | q_t = S_i)$ for $1 \leq i, j \leq N$ and emits characters from the alphabet V according to an emission probability matrix $E = (e_{ik})$, $1 \leq i \leq N, 1 \leq k \leq M$. When the system is in a given state i, it has a probability a_{ij} of moving to state j and a probability e_{ik} of emitting letter v_k. Biological relevance is attached to the hidden states of the Markov model. Transition and emission probabilities depend on the current state only and not on the past. This property is called the first-order Markov assumption. Only the emitted symbols can be observed, not the underlying random walk from state to state.

3.3.2 Profile HMM Technology

When applying HMMs to model families of protein sequences one speaks of profile HMMs. In this case, mainly an alphabet of 20 amino acid letters is used, but also other alphabets exist such as 64-letter alphabets for codon triplets, three-letter alphabets (helix, sheet and coil) for secondary structure prediction and Cartesian products of these alphabets. If necessary, meta-characters such as gap symbols can be added to the alphabets, as well.

In a standard architecture for protein sequence HMMs there are three classes of states [9], besides the *start* and *end* state: the *match*, *insert* and *delete* states. Thus $S = \{start, m_1, \ldots, m_N, i_1, \ldots, i_{N+1}, d_1, \ldots, d_N, end\}$. Typically, the length of the model N is the average length of the sequences in the family. The *match* and *insert* states always emit amino acid symbols, whereas the delete states never do. The basic path of state transitions is *start* $\rightarrow m_1 \rightarrow m_2 \rightarrow \ldots \rightarrow m_N \rightarrow$ *end*. Each *match* state m_j has an outgoing edge to succeeding *match*, *delete*, and *insert* states, m_{j+1}, d_{j+1} and i_{j+1}, respectively. Each *delete* state d_j has outgoing edges to the succeeding *delete*, *match* and *insert* states d_{j+1}, m_{j+1} and i_{j+1}. The *insert* state i_j is connected to the succeeding *match* and *delete* states m_{j+1} and d_{j+1}, and has a loop to itself allowing for multiple insertions.

There are three typical questions associated with HMMs in this context [118]. Given an observation sequence of emission characters and a model, how likely is that sequence for the particular model? Given the observation, i.e. target, sequence and a model, what is the underlying sequence of states? How to adjust the model parameters so that the model best describes a multiple alignment of amino acid sequences?

With profile HMMs [73], sequence families can be characterized. For a profile HMM, the first question addresses the likelihood of a given amino acid sequence to be a member of the given family.

For computing the likelihood of a sequence being emitted by a model, the forward procedure [9, 118] is used. Probabilities of states are propagated through the model from the *start* to the *end* states. From the state probabilities, the probabilities of a certain letter being emitted at a certain time can then be computed.

The most likely path of states is computed with the Viterbi algorithm [9, 118]. The path consists of a sequence of matches, insertions and deletions, and thus corresponds to an alignment of the target sequence with the sequence family.

Commonly, with a given multiple alignment the parameters of the standard architecture are initialized prior to learning as follows: The *match* state is assigned to any column of the alignment that contains less than 50% gaps. *Delete* states are associated with columns that contain any gaps. Columns with more than 50% gaps are assigned to corresponding *insert* states. Emissions of *match* and *insert* states can be initialized from the frequency counts of the corresponding columns in the multiple alignment. These counts need to be regularized with Dirichlet distributions or Dirichlet mixtures in order to avoid emissions associated with zero counts [9]. In a subsequent *learning* phase, the HMM parameters are optimized. Various algorithms are available for that, including the expectation maximization (EM or Baum–Welch) algorithm and different generalizations of it as well as gradient descent methods [9, 10, 30, 118]. This learning process corresponds to the generation of a profile in the previous section on profile–profile scoring.

3.3.3 **HMMs in Fold Recognition**

HMMs have been used for a number of years in fold recognition [25, 35, 66–68] and have been extensively tested (e.g Ref. [110]). One approach [66] is to iteratively add homologous sequences to a HMM (like the PSI-BLAST approach does for sequence alignments). Similar to profile methods, HMMs can turn multiple sequence alignments into position-specific scoring systems suitable for searching databases for remotely homologous sequences [35].

3.3.4 **HMM–HMM Comparisons**

Similar to the scoring schemes described in the section on frequency profile scoring, there are several ways to score profile HMMs against sequences or against other profile HMMs. Whereas the standard HMM approach was to compare an HMM to one sequence, Lyngsø and coworkers developed an algorithm for the alignment of two HMMs based on the maximization of the coemission probabilities [90]. Recently, Edgar and Sjölander proposed an approach to align two multiple alignments by constructing a profile HMM from one of the alignments and aligning the other to that HMM [36]. Söding has generalized the log-likelihood score maximized in HMM sequence alignments to the case of HMM–HMM alignments [146].

3.4 Support Vector Machines (SVMs)

3.4.1 **Definition**

SVMs are a state-of-the-art machine learning method for classification problems. SVMs have been succesfully applied for fold recognition.

Conceptually, SVMs map the data points to be classified into a high-dimensional space called feature space, in which an optimally separating hyperplane is sought that separates the two classes of points to be distinguished by the (binary) classification [132]. Technically, the transformation into the high-dimensional space can be avoided and only inner products in that space, called kernels, need to be computed. While the SVM machinery is fairly standardized, kernel functions are highly problem specific. One problem for protein sequences is to map the sequences of typically differing lengths into a space with a constant dimensionality. Several kernel functions exist for the protein classification problem.

3.4.2 Various Kernels

Jaakkola and coworkers suggest the Fisher kernel function, which is specific to a protein family [57]. An HMM is trained from positive samples of the family. The so-called Fisher score is the gradient of the log likelihood score for an arbitrary sequence X with respect to the HMM parameters. This score maps the sequence X into a fixed length vector. The Fisher kernel function is then computed on the basis of Euclidean distances between the Fisher score vector for X and the score vectors for known positive and negative examples of the protein family.

Another example, the SVM-pairwise kernel, uses the Smith–Waterman algorithm to align a new protein sequence X against all n sequences in the training set. The feature vector corresponding to protein X is $F_X = f_{x1}, f_{x2}, \ldots, f_{xn}$, where f_{xi} is the E-value of the Smith–Waterman score between sequence X and the ith training set sequence [81].

There are numerous other kernels to treat protein sequences, e.g. mismatch kernels [79], string kernels [126] or motif kernels [12].

3.4.3 Experimental Assessment

Machine learning methods need sufficient training data, which imposes some restrictions on the experiments. The typical problem tackled by the machine learning community is the protein classification problem, as described by Jaakkola and coworkers in 1999 [56]. The protein classification problem is to predict the SCOP structural class of a protein given its amino acid sequence. Two sequences with the same superfamily are considered as related by homology and two domains from different folds are considered as unrelated. Proteins from different superfamilies within the same fold have an uncertain relationship, and are not considered in experiments. The classification question then is to decide whether a protein belongs to a certain superfamily.

In the standardized experimental setup defined by Jaakkola and coworkers, and typically used, for each family the protein domains outside the family, but within the same superfamily are taken as positive training samples. Positive test samples are the members of the family. Negative samples are taken from outside the fold to which the family belongs. The set of negative samples is randomly split into training and test samples.

Jaakkola suggests to use only families with at least five family members (positive test) and 10 superfamily members outside the family (positive train). Liao and coworkers [81] suggest using only families with at least 10 family members and five superfamily members outside the family. Both restrictions imply a drastic reduction of the number of families considered at all.

A distincion is made whether all the training data are labeled (supervised learning) or not (semi-supervised learning). The additional data in semi-supervised learning can help to better structure the space around the labeled points. Typically, semi-supervised learning is more expensive and experimentally performs better than supervised learning [177].

Despite the promising developments, it has to be clearly pointed out that SVMs operating directly on the amino acid sequence with kernels like the ones mentioned above currently are not actively used to identify suitable templates. The machine learning methodology requires a certain amount of data for training. Therefore, the protocols used for testing SVMs on the fold recognition problem vary slightly, but significantly, from protocols used to evaluate other current methods. For instance, in CASP6 (see Section 6.3) there was no prediction group relying on kernel methods for template identification.

4 Structure-based Methods for Identifying Templates

Often proteins share similar structure while showing very little (15% or less) sequence identity [62]. Sequence similarity is not necessary for structural similarity, instead extreme divergences of sequences are observed as well as convergent evolution where similar 3-D folds are adopted several times.

Starting in the 1990s, methods were developed for template identification based on structural information such as secondary structure, burial patterns or side-chain pair-interactions. These methods are often referred to as inverse folding, sequence–structure alignment or threading methods. Note, that the term "threading" is widely used to label any method which attempts to tackle the problem of aligning two protein sequences the structure of one of which is known. Originally invented by Jones and coworkers [62] and also frequently used is an alternative definition, according to which "threading" is the alignment of a sequence with a protein structure in 3-D without regard to the sequence associated with the structure [58].

Sequence–structure alignment methods should be able to recognize not only homologous proteins but also analogous proteins sharing the same fold [15, 62]. Most of these methods employ inverse Boltzmann statistics: the frequencies of observed findings are converted into a pseudo-energy function which is optimized in order to find a good sequence–structure alignment (see Section 4.1).

Basically there are two ways of scoring a sequence against a structure. One is to score the interactions of pairs of amino acids of the target sequence within the template structure using pair-interaction potentials as energy functions. Unfortunately, finding a global optimum for this kind of problem has been shown to be NP-hard [77]. There are several approaches to tackle this and we will present four conceptually important ones in Section 4.2.

Alternatively, locations of residues may be evaluated with respect to their placement inside the original template structure afforded by sequence alignments of target and template protein. In this case the global optimum sequence–structure alignment can be found using dynamic programming methods. Standard methods for this are reviewed in Section 4.3.

Today, hybrid algorithms, combining sequence frequency profile methods with sequence–structure methods, are common practice. Either profiles can be integrated into the threading process or methods are exercised separately and the results are merged afterwards. Reviews can be found in Refs. [59, 70, 149, 160]; however, we will first focus on the basic principles.

4.1 Boltzmann's Principle and Knowledge-based Potentials

In protein kinetics, the topography of the landscape of free energies is often described as a funnel [31]. Typically, the energy is assumed to decrease as the folding process proceeds. This organization of the energy landscape is not characteristic of random polypeptides, but is a result of evolution. A common assumption is that the native structure is the one with the lowest free energy. However, many factors contribute to the free energy of the system. Not all factors are known and neither is their interplay completely understood. Therefore, energy landscapes cannot be determined exactly [105].

In addition, in threading we are actually dealing with two structures, i.e. the one to be identified and the one serving as template. The environments of the two structures can be substantially different, making it difficult to apply the same detailed energy potentials to both structures. In spite of these problems, knowledge-based potentials have been employed successfully in threading [19]. Information on different levels of abstraction can be extracted from databases of known structures, with the help of inverse Boltzmann statistics converted into empirical energy potentials.

According to the law of Boltzmann [137, 187] a particular state x of a physical system in equilibrium is occupied with probability $f(x)$:

$$f(x) = \frac{1}{Z} e^{-\frac{E(x)}{kT}}, \quad \text{where} \quad Z = \int \cdots \int e^{-\frac{E(x)}{kT}} dx,$$

k is the Boltzmann's constant, T is the absolute temperature and Z is called the partition function. In a discrete state space, the integrals are replaced by sums. If the energies of all states x are known, the probability densities $f(x)$ can be computed.

Conversely, if the probability density functions f of a system are given [138], the energy can be calculated as:

$$E(x) = -kT \ln(f(x)) - kT \ln(Z).$$

This is frequently referred to as the inverse Boltzmann principle. Z cannot be evaluated by measuring the densities and therefore the energy can be only determined up to the additive constant $-kT \ln(Z)$.

Originally, Boltzmann's equation assumes that the system is in equilibrium. Also, the principle can only be applied to complete systems not to their parts. However, Finkelstein and coworkers showed that a Boltzmann-like distribution arises naturally from low energy conformations of random heteropolymers and similarly of proteins [38, 187]. This suggests that the Boltzmann model can be applied to derive empirical energy functions from ensembles of protein structures even though they are not systems in equilibrium.

4.2 Threading Using Pair-interaction Potentials

Different types of potentials are used in threading. The more accurate potentials rely on pair-interactions of residues (many-body interactions are not considered at all in threading due to their complexity). Most often, distances between pairs of residues, considering their amino acid side-chain types, are condensed into pair-interaction potentials using knowledge-based inverse Boltzmann approaches. Commonly used choices of interaction centers are the C_α atoms, the C_β atoms, the side-chain centers of mass, specially defined interaction centers or any side-chain atom [142].

Finding the globally optimum threading involving a pair-interaction scoring function is NP-hard if variable-length gaps and interactions between neighboring amino residues are allowed [77]. This means that, in order to find an optimal solution, an algorithm requires an amount of time that, in the worst case, is exponential in the size of the protein. Several strategies have been developed to tackle this.

Jones and coworkers [62] use a double dynamic programming algorithm in conjunction with potentials that do not require explicit modeling of all side-chain atoms. They use a set of knowledge-based potentials which are derived from a statistical analysis of known protein structures, according to the inverse Boltzmann principle [137]. For a given pair of atoms, a given residue sequence separation and a given interaction distance these potentials provide a measure of pseudo-energy, which relates to the probability of observing the proposed interaction in native protein structures. By providing different empirical potentials for different ranges of sequence separation, specific structural significance is conferred on each range. The short-range terms predominate in the matching of secondary structural elements. By threading a sequence segment onto the template of an α-helical conformation and evaluating the short-range potential terms, the probability of the sequence folding into an α-helix is evaluated. In a similar way, medium-range terms mediate the matching of super-secondary structural motifs, and the long-range terms the tertiary packing. Around each residue in turn, their algorithm uses dynamic programming as in sequence alignment to optimize the threading of the sequence onto the structure. It finally computes the best threading through the whole structure by means of a shortest-path algorithm.

Lathrop and Smith use core structural models to derive a branch and bound algorithm for threading [78]. A core structural model consists of several core segments. Each position of each core element is occupied by a single amino acid residue from the threaded sequence. Typically core segments correspond to secondary structure elements, i.e. helices or strands. They are connected by a set of loop regions. Neighboring positions are computed, where positions are defined to be neighbors if they contribute a pair-interaction term to the energy-score function. This often but not always requires that they lie close in space, that is make a noncovalent interaction. The use of core elements as larger building blocks reduces the problem size drastically.

Recursive dynamic programming (RDP) [154] is another approach to addressing the full threading problem using heuristics and without restriction to core elements. It is based on the "divide-and-conquer" paradigm and maps the target sequence onto the known backbone structure of a template protein in a stepwise fashion – a technique that is similar to computing local alignments but utilizing different cost functions. It starts by mapping parts of the target onto the template that show statistically significant similarity with the template sequence. After mapping, the template structure is modified in order to account for the mapped target residues. Then significant similarities between the yet unmapped parts of the target and the modified template are searched, and the resulting segments of the target are mapped onto the template. This recursive process of identifying segments in the target to be mapped onto the template and modifying the template is continued until no significant similarities between the remaining parts of target and template are found. Those parts which are left unmapped by the procedure are interpreted as gaps. The RDP method is robust in the sense that different local alignment methods can be used, several alternatives of mapping parts of the target onto the template can be handled and compared in the process, and the cost functions can be dynamically adapted to biological needs.

Xu and coworkers, in their RAPTOR (RApid Protein Threading by Operation Research technique) method, use a linear programming approach to do protein 3-D structure prediction via threading [181–183]. Based on the contact map graph of the protein 3-D structure template, the protein threading problem is formulated as a large-scale integer programming problem. This formulation is then relaxed to a linear programming problem, and solved by a branch-and-bound method. The final solution is globally optimal with respect to their energy functions. The energy function includes pairwise interaction preferences and allows variable gaps.

4.3 Threading using Frozen Approximation Algorithms

The alternative to using full pair-potentials in threading is to evaluate a target sequence with respect to the template structure's original native sequence. While threading the target onto the structure, the interaction partners in the potentials or a set of local environmental preferences are taken from the template protein. With these frozen approximation approaches [47,142], a globally optimum threading – of a problem with reduced complexity compared to the threading problem using full pair-interaction potentials – can be found using dynamic programming methods.

Bowie and coworkers [15] start with a known template structure and describe the environments of its residues by three types of properties: (i) the area of the residues buried in the protein and inaccessible to solvent, (ii) the fraction of side-chain area that is covered by polar atoms, and (iii) the local secondary structure. Based on these parameters each residue is categorized into an environment class. In this manner, a 3-D protein structure is converted into a 1-D string, like a sequence, which represents the environment class of each residue in the folded protein structure. With a sequence-alignment-like algorithm they then seek the most favorable alignment of a protein sequence to this environment string. An alignment column now aligns a residue in the target sequence with an environment class in the template structure. Using inverse Boltzmann statistics, knowledge-based scoring functions have been derived for this kind of match. Later the method was extended to also incorporate secondary structure predictions (helix, strand, coil) on the sequence side to be matched to the secondary structure of the templates [121].

Flöckner and Sippl base their method to determine a sequence to structure alignment on the Needleman–Wunsch algorithm [43,140]. While in sequence comparison the similarity of amino acids is measured directly, Flöckner and

Sippl evaluate the fitness of an amino acid of the query sequence within the template structure by using the energy field generated by the original template structure, while mutating that single residue to the type observed within the query sequence. For this mutated structure a knowledge-based energy function, composed of pairwise atom–atom interactions is evaluated for C_β–C_β interactions.

Also starting from known structures, Alexandrov and Zimmer [2] describe the environments of residues by counting the number of contacts that each amino acid makes in a structure. This information can be matched with sequence information by previously counting the so-called contact capacities, i.e. is the normalized number of contacts that each type of amino acid makes in an ensemble of proteins. Given the number of contact counts per sequence position in the structure and the number of counts a type of amino acid prefers, for each position in the structure's sequence certain types of amino acids are preferable. This information can be aligned to the target sequence with dynamic programming just as in sequence alignment.

5 Hybrid Methods and Recent Developments

5.1 Using Different Sources of Information

Secondary structure prediction and disorder prediction are methods for predicting additional structural features of amino acid sequences (see Chapter 9). Such methods can be used as stand-alone tools to learn more about a target sequence. Alternatively, they can be directly integrated into template identification methods by incorporating the additional sources of information into the scoring functions. In the context of this section, the focus is on the integration of these methods into sequence comparison.

5.1.1 Incorporating Secondary Structure Prediction into Frequency Profiles and HMMs

Often, one of the first steps in structure prediction is to predict the secondary structure of the target protein, that is to annotate each of a sequence's residues with a probability of being contained in a helix, coil or sheet (see Chapter 9). As a result, the field of secondary structure prediction has received considerable attention and is reaching a mature state [1,61,83,92,179].

Predicted secondary structure can be incorporated into other fold recognition methods, e.g. profile–profile, HMM or threading methods [39,88]. In the frequency profile case, the profile matrix with 20 columns for each type of amino acid is extended with an additional profile matrix with three columns for helix, coil and sheet. Both profile matrices are scored against other profile matrices of the same type and the results is merged to a joint score [46, 166]. Wang and Dunbrack state that incorporation of secondary structure information improves alignment accuracy slightly [171] and improves the search capabilities of the average score mentioned in Section 3.2.5 significantly.

Secondary structure information was also used to extend the HMM principle [64, 67]. Karchin and Karplus incorporated predicted local structure into so-called two-track profile HMMs. They did not rely on a simple helix–strand–coil definition of secondary structure, but experimented with a variety of local structure descriptions, and established which descriptions are most useful for improving fold recognition and alignment quality. On a test set of 1298 nonhomologous proteins, HMMs incorporating a three-letter alphabet improved fold recognition accuracy by 15% over HMMs using amino acids only. Comparing two-track HMMs to HMMs operating on amino acids only, on a difficult alignment test set of 200 protein pairs, Karchin found that HMMs with a six-letter secondary track improved alignment quality by 62%, relative to DALI [53] structural alignments.

5.1.2 Intrinsically Disordered Regions in Proteins

While the analysis and prediction of secondary structure is a matured subject, the interest in intrinsic disorder of proteins (see Chapter 9) has grown tremendously over the last couple of years [33, 63, 84, 102, 104, 119, 157, 158, 174].

Intrinsically disordered proteins do not fold into stable 3-D structure; instead, in solution they exist as an ensemble of interchanging conformations. There are examples of proteins which are disordered completely and others where only part of the amino acid sequence does not fold stably. In a recent study [103] based on six genomes, roughly 5% of proteins in bacteria, 7% in archea and 25% in eukaryotes were estimated to be mostly disordered. Ward and coworkers [174] predict an average of 2% of archaean, 4% of eubacteria and 33% of eukaryotic proteins to contain more than 30 residues of disorder. Thus, proteins with disordered regions seem to be especially common in eukaryotic cells. Often, proteins with disordered regions perform a function: they are involved in protein–protein [76, 82, 85] or protein–nucleic acid [148] interactions. Under certain conditions disordered fragments become structured during the process of interaction. In the electron density maps of X-ray crystallographic studies, disordered regions frequently do not appear; in NMR experiments, they appear highly flexible.

As stretches of amino acids that do not fold into a stable structure should not be predicted to have a structure, disorder prediction (see Chapter 9) has become important in structure prediction. Methods have been developed which annotate each residue of an amino acid sequence with a value of predicted disorder. These methods are based on machine learning techniques like neural nets or support vector machines and employ training sets of disordered regions to learn to discriminate ordered from disordered stretches of sequence [63, 84, 103, 104, 119, 168].

To the best of our knowledge such methods have not yet been incorporated into fold recognition methods as has been done with secondary structure prediction, but this should only be a matter of time.

5.1.3 Incorporating 3-D Structure into Frequency Profiles

Frequency profiles can be derived from sequences alone, as reviewed in Section 3.2.2. Alternatively, structure-based multiple alignments can be used for the generation of profiles. For this, multiple structure superposition is performed on the available structures to create a multiple alignment of their sequences which is then used to generate a frequency profile as described in Section 3.2.2 above [23, 69, 107, 114, 152, 184]. There are actually two aspects to this. First, using structure superposition to create multiple alignments even for cases where sequence alignment is not feasible. This potentially results in an increase of coverage at a loss of precision. Second, using available structures to generate more reliable seed alignments, increasing precision.

5.2 Combining Information

Recent contributions to the field of fold recognition have been integrative, collecting information from many sources. The GenTHREADER program [60, 93] for automatic fold recognition consists of a neural net which was trained to combine sequence alignment score, length information and energy potentials derived from threading into a single score representing the relationship between two proteins [60]. An improved version incorporates PSI-BLAST searches and also makes use of predicted secondary structure [93].

Another competitive example for the combination of information is the TASSER/PROSPECTOR suite of programs developed by Skolnick and coworkers. PROSPECTOR_3 is an iterative approach to search for diverse templates using a variety of pair potentials and scoring functions [143, 188]. Different frequency profiles are used in a first round to identify templates. These targets are further evaluated in subsequent evaluations of pair interactions. The scoring functions include a quasi-chemical based pair potential [141], a protein-specific, orientation-independent pair potential based on local sequence fragments [144] and a pair potential that depends on the orientation of the side-chains [144]. With the templates identified by PROSPECTOR, the TASSER program performs tertiary structure assembly via the rearrangement of continuous template fragments [189].

Another example is the SPARKS method (Sequence, secondary structure Profiles and Residue-level Knowledge-based energy Score) [191], in which Zhou and Zhou use a knowledge-based energy function for fold recognition.

Being a residue-level frozen approximation potential, the dynamic programming method can be used for alignment optimization. The potential contains a backbone torsion term, a buried surface term and a contact-energy term. With sequence profile and secondary structure information it is combined into a joint fold recognition method. Taking advantage of sequence and structure methods it was highly competitive in the latest CASP experiment.

Like this one, there are a number of other approaches integration information from programs running locally or collecting information from web services [166].

5.3 Meta-servers

Meta-servers collect and analyze results from individual web servers and combine them into a joint result. Benchmarking results obtained in the last years indicate that, on average, meta-servers are more accurate than individual methods.

One first successful attempt to benefit from a number of distributed information sources was the PCONS meta-server, which concentrates on a number of reliable servers at different locations and selects the most abundant fold among their high-scoring models [86]. It translates the confidence scores reported by each server into uniformly scaled values corresponding to the expected accuracy of each model. The translated scores as well as the similarity between initial models is used as input to a neural network in charge of the final selection.

Several other meta-methods followed soon afterwards. Current methods differ in how the initial models are compared, whether scores provided by the individual methods are used and how the final model is generated.

The 3D-Jury system uses the rationale that the high-scoring models which are produced by several servers are closer to the native structure than the single model with highest score. Thus, models occurring with higher than expected frequencies are are taken for the preferred conformation [44, 163].

The 3D-SHOTGUN meta-predictor employs techniques of so-called cooperative algorithms from computer vision. As input it takes the models with their confidence scores. The result is a hybrid model, which is spliced from fragments of the input models. It has the potential of covering a larger part of the native protein than any template structure alone. Thus, 3D-SHOTGUN entails the first fold recognition meta-predictor attempt to go beyond the simple selection of one of the input models [40].

6 Assessment of Models

Once a template is identified it can be assessed in a number of ways. (i) The significance of the selection of the template can be estimated. (ii) After a model has been constructed on the basis of the template, the quality of the 3-D model can be scored. (iii) If the native structure for which the prediction was made becomes known later, the quality of the model can be evaluated, in terms of how faithfully it represents the true structure. Thus, conclusions about the method producing the prediction can be drawn.

6.1 Estimating Significance of Sequence Hits

Essentially, methods for template identification compute a list of candidate templates for a target sequence. The method typically employs a scoring system or (virtual) energy function according to which the list is sorted and the maximum scoring candidate is selected as template. Before using a template to build a model the question how reliably the template is related to the target needs to be addressed.

The classic approach to this is to calculate the probability of obtaining a maximum score greater than the observed score assuming that the protein sequences compared with the scoring system are unrelated. This probability is called p-value. To compute it one needs to know the distribution of maximal scores for unrelated sequences for the particular scoring scheme. The E-value is a similar concept, additionally taking the size of the database of templates into account [5,6].

For some template identification methods the score distributions are known, thus scores can be readily converted into p-values (see Section 4 in Chapter 3). For other methods empirical confidence scores have been developed.

For optimal local gapless sequence alignments of independent random sequences the score distributions are known to be of an asymptotically extreme-value or Gumbel form [65]: $P(score > t) \approx 1 - e^{-Ke^{-\lambda t}}$, where λ depends only on the scoring system and K depends on the scoring system and the sequence lengths, such that the distribution reflects the fact that the chance of spurious high scores increases with sequence lengths. The dependence of the parameters on the scoring system and sequence lengths is known.

For local alignments with gaps of unrelated biological sequences no general theory is available. However, there is considerable evidence that the distribution is still of extreme-value form and parameters can be fitted experimentally [4,75,80,95,112,175]. Local alignments using frequency profile sequences were also shown to follow an extreme-value distribution [95].

For optimal global alignments, whether with plain sequences or sequence profiles, theoretically neither the family of distributions nor the dependence of the expected score (or of other parameters) on the sequence lengths is known. However, there exist approaches for experimentally fitting distributions to scores [109, 176].

For frozen approximation sequence–structure alignment the situation is similar to sequence alignment: the local threading scores of sample sequence–structure pairs often can be fitted to a Gumbel distribution [147]. In the global threading case the distribution is unknown.

For the methods for which no theory is available, heuristic approaches exist for estimating reliability of the predictions. One generic approach is to derive a function of target and template which is reasonably associated with reliability, and then statistically test it on a set of proteins to estimate the probability for an identified target being correct (e.g. Ref. [147]). One such target–template comparison function is, for example, the difference of the score of the target sequence aligned to the template and the reversed target sequence aligned to the template. This is much faster than repeatedly evaluating the score relative to that of a randomized sequence [66, 143].

For a more detailed discussion, see Ref. [34], where the classical statistical assessment of significance of scores is also compared to Bayesian approaches.

6.2 Scoring 3-D Model Quality: Model Quality Assessment Programs (MQAPs)

The template identification methods discussed above compute suggestions for templates. With a target-template alignment, a model can be computed. MQAPs serve to distinguish near-native structures (i.e. "good" predictions) from decoys (i.e. "bad" predictions). MQAPs are programs that receive as input a 3-D model of a protein structure and produce as output a real number representing the quality of the model (http://www.cs.bgu.ac.il/~dfischer/CAFASP4/mqap.html). MQAPs only use the model as input, not the native structure, and thus stand in contrast to methods that evaluate the quality of a model when comparing it to the native structure.

In contrast to scoring functions in sequence–structure alignment and to physical energy functions, MQAPs operate on an intermediate level – they are more flexible than a sequence–structure alignment function as the dynamic programming paradigm used in alignment imposes the requirement of prefix optimality which is not required in MQAPs. On the other hand, MQAP functions are not sensitive to ruptures in the sequences in contrast to physical energy functions. MQAPs target at scoring the quality of predicted models. Typically, MQAPs use one or more different statistical potentials, representing information coded in protein structures [87, 116, 139, 159]. For example, the FRST method uses a combination of pairwise, solvation, hydrogen bond and torsion angle potentials [159]. An intuitive test for a scoring function is picking the native structure among a set of decoys. A number of decoy sets have been made available [127] and are used for training.

In 2004, CAFASP4 [42] provided infrastructure to perform a fully automated blind test of MQAPs on the CASP target proteins and the structure predictions made during CASP. In this test the five following programs were rated as most reliable: FRST [159], Verify3-D [87], RAPDF [128], ProsaII [139] and ProQ [169].

Generating different alignments between target and template using various alignment methods and then employing an MQAP to pick out the best one can potentially lead to an improved overall alignment result.

6.3 Evaluation of Protein Structure Prediction: Critical Assessment of Techniques for Protein Structure Prediction

The performance of methods like the ones reviewed above needs assessment. The performance of sequence–structure alignment methods can be assessed either by testing their fold recognition performance [20, 37, 71] or by benchmarking their alignment quality [130, 171]. For assessing fold recognition performance, classification performance of different methods is tested versus a standard like the SCOP database for structure classification. For measuring alignment quality, the sequence–structure alignments produced by threading methods are compared to high-quality alignments as produced by structure superposition methods. CASP, a blind testing experiment, has had a large impact on the community and is therefore summarized in the following.

CASP was started in 1994 [98]. The idea was to establish a clearinghouse between experimental and predicted protein structure. Protein sequences whose structure is currently being analyzed experimentally are made available to structure prediction groups. Structural models are predicted by a number of participating groups and submitted to the CASP organizers before the release of the crystal structures. After the release of the crystal structures, the predicted models are compared with them by a number of human experts – the CASP assessors. Typically, the structures are categorized according to their difficulty into homology modeling, fold recognition and *de novo* targets. CASP has been held biannually. The number of targets has been growing from 33 (CASP1 in 1994) to 42, (CASP2), 43 (CASP3), 43 (CASP4), 67 (CASP5) and 64 (CASP6 in 2004), as well as the number of participating prediction groups from 35 (CASP1) to 152 (CASP2) to 201 human groups and 65 servers in CASP6.

Within the original setup of CASP humans can interact with computer programs to generate protein models. In this setting, it is hard to discriminate between the contribution of the human and the program, respectively. This issue was resolved by introducing CAFASP [41], where the additional FA in the acronym stands for fully automated. In CAFASP, server programs are directly contacted by the clearinghouse and have to respond within a short timeframe (usually 48 h) and without human intervention.

In order to perform the comparisons of predicted models and experimental structures, different assessors have chosen different approaches. Some completely relied on manual inspection (which is a tremendous amount of work). Due to the number of participating groups, computerized preprocessing of the results is becoming the standard procedure. Different distance measures for measuring the fractional similarity between the experimental and the predicted structure have been employed. Currently the GDT_TS score is used for CASP and the MaxSub score in CAFASP. The GDT_TS score is computed by the LGA program which simultaneously optimizes for local structure and global RMSD superposition using different cutoffs [186]. MaxSub aims at identifying the largest subset of C_α atoms of a model that superimpose "well" over the experimental structure [135]. Both measures produce a single normalized score that represents the quality of the model. Both measures differ in details, but correlate on concrete examples [190].

CASP5 (2002) results for fold recognition methods are summarized in Ref. [70], results of CASP6 (2004) are in press at the time of writing [96, 161, 172]. While there is an obvious danger of overtraining to the experiment, the CASP community has been eager to pick up new trends and find ways to evaluate them. Examples are the evaluation of predicted intrinsic disorder at CASP5 and CASP6 or the assessment of prediction of domain boundaries and model quality assessment programs at CASP6.

Continuous experiments, LiveBench and EVA, are performed by weekly extracting newly published structures from the PDB and submitting them to automated servers. Based on automated measures like MaxSub, the quality of the predictions of participating servers can be measured online [21, 122].

7 Programs and Web Resources

The web resources for protein structure prediction are extensive. Since search engines have a tendency to be more up-to-date and practical than link lists in books, we give an overview and further pointers here only. Good starting points for fold recognition via the Internet are meta-servers and the CASP experiment. Meta-servers provide functionality for unproblematically exercising a number of servers simultaneously in order to perform structure predictions. The CASP experiment provides a list of the best performing methods today; although some of these might not be as easily accessible or very compute intensive to exercise.

Bioinfo/3D-Jury	http://bioinfo.pl/meta
PredictProtein	http://www.embl-heidelberg.de/predictprotein/predictprotein
CASP	http://predictioncenter.org
CAFASP	http://www.cs.bgu.ac.il/~dfischer/CAFASP4

For high-throughput experiments as well as confidentiality reasons one may want to install fold recognition software locally. Unfortunately, this software tends to depend on a number of libraries (templates, profiles, potentials, motifs, etc.) which need to be up-to-date in order to be performant and which have a tendency to be cumbersome to install. An obvious starting point for downloadable software is PSI-BLAST. Some HMM software is freely available. Most profile software is available through contacting the respective authors only. URLs of BLAST and two HMM libraries are:

BLAST	http://www.ncbi.nlm.nih.gov/BLAST
HMMer	http://hmmer.wustl.edu
SAM	http://www.cse.ucsc.edu/compbio/sam.html

Links to key databases are:

PDB	http://www.rcsb.org/pdb
SCOP	http://scop.mrc-lmb.cam.ac.uk/scop
Entrez DB Collection	http://www.ncbi.nlm.nih.gov/gquery/gquery.fcgi

Also Kevin Karplus provides an up to date link list, related to this chapter:

http://www.soe.ucsc.edu/~karplus/compbio_pages

Acknowledgments

This work was funded in part by grant Le 491/14 of the Deutsche Forschungsgemeinschaft. I thank my colleagues at the Max-Planck-Institut for Informatics for their support. I am particularly grateful to Mario Albrecht, Francisco S. Domingues, Andreas Kaemper, Lars Kunert, Oliver Sander and Thomas Lengauer for critically reading the manuscript and improving it considerably.

References

1 Albrecht, M., S. C. E. Tosatto, T. Lengauer, and G. Valle. 2003. Simple consensus procedures are effective and sufficient in secondary structure prediction. Protein Eng. **16**: 459–62.

2 Alexandrov, N., R. Nussinov, and R. Zimmer. 1996. Fast protein fold recognition via sequence to structure alignment and contact capacity potentials. Pac. Symp. Biocomput. **1**: 53–72.

3 Altschul, S., R. Caroll, and D. Lipman. 1989. Weights for data related by a tree. J. Mol. Biol. **207**: 647–53.

4 Altschul, S. and W. Gish. 1996. Local alignment statistics. Methods Enzymol. **266**: 460–80.

5 Altschul, S. F., W. Gish, W. Miller, E. W. Myers, and D. J. Lipman. 1990. Basic local alignment search tool. J. Mol. Biol. **215**: 403–10.

6 ALTSCHUL, S. F., T. L. MADDEN, A. A. SCHÄFFER, J. ZHANG, Z. ZHANG, W. MILLER, AND D. J. LIPMAN. 1997. Gapped BLAST and PSI-BLAST: a new generation of protein database search programs. Nucleic Acids Res. **25**: 3389–402.

7 ANFINSEN, C. 1973. Principles that govern the folding of protein chains. Science **181**: 223–30.

8 ANFINSEN, C. B., E. HABER, M. SELA, AND F. H. WHITE. 1961. The kinetics of formation of native ribonuclease during oxidation of the reduced polypeptide chain. Proc. Natl. Acad. Sci. USA **47**: 1309–14.

9 BALDI, P. AND S. BRUNAK. 2001. *Bioinformatics: The Machine Learning Approach*. MIT Press, Cambridge, MA.

10 BALDI, P. AND Y. CHAUVIN. 1994. Smooth on-line learning algorithms for hidden markov models. Neural Comput. **6**: 305–16.

11 BATEMAN, A., L. COIN, R. DURBIN, ET AL. 2004. The Pfam protein families database. Nucleic Acids Res. **32**: D138–41.

12 BEN-HUR, A. AND D. BRUTLAG. 2003. Remote homology detection: a motif based approach. Bioinformatics **19 (Suppl. 1)**: i26–33.

13 BERG, J., J. TYMOCZKO, AND L. STRYER. 2001. *Biochemistry*. Freeman, San Francisco, CA.

14 BOURNE, P. E. AND H. WEISSIG. 2003. *Structural Bioinformatics*. Wiley, New York, NY.

15 BOWIE, J., R. LUTHY, AND D. EISENBERG. 1991. A method to identify protein sequences that fold into a known three-dimensional structure. Science **253**: 164–70.

16 BRENNER, S., P. KOEHL, AND M. LEVITT. 2000. The ASTRAL compendium for protein structure and sequence analysis. Nucleic Acids Res. **28**: 254–6.

17 BROWN, M. P., R. HUGHEY, A. KROGH, I. S. MIAN, K. SJÖLANDER, AND D. HAUSSLER. 1993. Using Dirichlet mixture priors to derive hidden Markov models for protein families. Proc. ISMB **1**: 47–55.

18 BUCHER, P. AND A. BAIROCH. 1994. A generalized profile syntax for biomolecular sequence motifs and its function in automatic sequence interpretation. Proc. ISMB **2**: 53–61.

19 BUCHETE, N.-V., J. E. STRAUB, AND D. THIRUMALAI. 2004. Development of novel statistical potentials for protein fold recognition. Curr. Opin. Struct. Biol. **14**: 225–32.

20 BUJNICKI, J., A. ELOFSSON, D. FISCHER, AND L. RYCHLEWSKI. (2001a). LiveBench-2: large-scale automated evaluation of protein structure prediction servers. Proteins **45 (Suppl. 5)**: 184–91.

21 BUJNICKI, J. M., A. ELOFSSON, D. FISCHER, AND L. RYCHLEWSKI. (2001b). LiveBench-1: continuous benchmarking of protein structure prediction servers. Protein Sci. **10**: 352–61.

22 CARILLO, H. AND D. LIPMAN. 1988. The multiple sequence alignment in biology. SIAM J. Appl. Math. **48**: 1073–82.

23 CASBON, J. AND M. A. S. SAQI. 2005. S4: structure-based sequence alignments of SCOP superfamilies. Nucleic Acids Res. **33**: D219–22.

24 CHANDONIA, J.-M., G. HON, N. S. WALKER, L. L. CONTE, P. KOEHL, M. LEVITT, AND S. E. BRENNER. 2004. The ASTRAL Compendium in 2004. Nucleic Acids Res. **32**: D189–92.

25 CHOO, K. H., J. C. TONG, AND L. ZHANG. 2004. Recent applications of Hidden Markov Models in computational biology. Genomics Proteomics Bioinf. **2**: 84–96.

26 CHURCHILL, G. A. 1989. Stochastic models for heterogeneous DNA sequences. Bull. Math. Biol. **51**: 79–94.

27 CLAVERIE, J.-M. 1994. Some useful statistical properties of position-weight matrices. Comput. Chem. **18**: 287–94.

28 CROOKS, G. E., G. HON, J.-M. CHANDONIA, AND S. E. BRENNER. 2004. WebLogo: a sequence logo generator. Genome Res. **14**: 1188–90.

29 DAYHOFF, M. O., R. M. SCHWARTZ, AND B. C. ORCUTT. 1978. A model of evolutionary change in proteins. In DAYHOFF M. O. (ed.) *Atlas of Protein Sequence and Structure*, Volume 5,

Chapter 22, pp. 345–352. National Biomedical Research Foundation.

30 DEMPSTER, A., N. LAIRD, AND D. RUBIN. 1977. Maximum likelihood from incomplete data via the em algorithm. J. R. Stat. Soc. B **39**: 1–38.

31 DINNER, A. R., A. SALI, L. J. SMITH, C. M. DOBSON, AND M. KARPLUS. 2000. Understanding protein folding via free-energy surfaces from theory and experiment. Trends Biochem. Sci. **25**: 331–9.

32 DUNBRACK, R. 1999. Comparative modeling of CASP3 targets using PSI-BLAST and SCWRL. Proteins **37 (Suppl. 3)**: 81–7.

33 DUNKER, A. K. AND Z. OBRADOVIC. 2001. The protein trinity – linking function and disorder. Nat. Biotechnol. **19**: 805–6.

34 DURBIN, R., S. EDDY, A. KROGH, AND G. MITCHISON. 1998. *Biological Sequence Analysis*. Cambridge University Press, Cambridge.

35 EDDY, S. R. 1998. Profile hidden Markov models. Bioinformatics **14**: 755–63.

36 EDGAR, R. C. AND K. SJÖLANDER. 2004. COACH: profile–profile alignment of protein families using hidden Markov models. Bioinformatics **20**: 1309–18.

37 EYRICH, V. A., D. PRZYBYLSKI, I. Y. Y. KOH, O. GRANA, F. PAZOS, A. VALENCIA, AND B. ROST. 2003. CAFASP3 in the spotlight of EVA. Proteins **53 (Suppl. 6)**: 548–60.

38 FINKELSTEIN, A. V., A. Y. BADRETDINOV, AND A. M. GUTIN. 1995. Why do protein architectures have Boltzmann-like statistics? Proteins **23**: 142–50.

39 FISCHEL-GHODSIAN, F., G. MATHIOWITZ, AND T. SMITH. 1990. Alignment of protein sequences using secondary structure: a modified dynamic programming method. Protein Eng. **3**: 577–81.

40 FISCHER, D. 2003. 3D-SHOTGUN: a novel, cooperative, fold-recognition meta-predictor. Proteins **51**: 434–41.

41 FISCHER, D., C. BARRET, K. BRYSON, ET AL. 1999. CAFASP-1: critical assessment of fully automated structure prediction methods. Proteins **37 (Suppl. 3)**: 209–17.

42 FISCHER, D., L. RYCHLEWSKI, R. L. DUNBRACK, A. R. ORTIZ, AND A. ELOFSSON. 2003. CAFASP3: the third critical assessment of fully automated structure prediction methods. Proteins **53 Suppl. 6**: 503–16.

43 FLÖCKNER, H., M. BRAXENTHALER, P. LACKNER, M. JARITZ, M. ORTNER, AND M. J. SIPPL. 1995. Progress in fold recognition. Proteins **23**: 376–86.

44 GINALSKI, K., A. ELOFSSON, D. FISCHER, AND L. RYCHLEWSKI. 2003. 3D-Jury: a simple approach to improve protein structure predictions. Bioinformatics **19**: 1015–8.

45 GINALSKI, K., N. V. GRISHIN, A. GODZIK, AND L. RYCHLEWSKI. 2005. Practical lessons from protein structure prediction. Nucleic Acids Res. **33**: 1874–91.

46 GINALSKI, K., J. PAS, L. S. WYRWICZ, M. VON GROTTHUSS, J. M. BUJNICKI, AND L. RYCHLEWSKI. 2003. ORFeus: detection of distant homology using sequence profiles and predicted secondary structure. Nucleic Acids Res. **31**: 3804–7.

47 GODZIK, A., A. KOLINSKI, AND J. SKOLNICK. 1992. Topology fingerprint approach to the inverse protein folding problem. J. Mol. Biol. **227**: 227–38.

48 GOTOH, O. 1982. An improved algorithm for matching biological sequences. J. Mol. Biol. **162**: 705–8.

49 GRIBSKOV, M., A. MCLACHLAN, AND D. EISENBERG. 1987. Profile analysis: detection of distantly related proteins. Proc. Natl Acad. Sci. USA **84**: 4355–8.

50 GRIBSKOV, M. AND S. VERETNIK. 1996. Identification of sequence patterns with profile analysis. Methods Enzymol. **266**: 198–212.

51 HENIKOFF, S. AND J. G. HENIKOFF. 1992. Amino acid substitution matrices from protein blocks. Proc. Natl Acad. Sci. USA **89**: 10 915–9.

52 HENIKOFF, S. AND J. G. HENIKOFF. 1994. Position-based sequence weights. J. Mol. Biol. **243**: 574–8.

53 HOLM, L. AND C. SANDER. 1992. Evaluation of protein models by atomic solvation preference. J. Mol. Biol. **225**: 93–105.

54 HOLM, L. AND C. SANDER. 1996. Mapping the protein universe. Science **273**: 595–603.

55 HOOFT, R. W., C. SANDER, M. SCHARF, AND G. VRIEND. 1996. The PDBFINDER database: a summary of PDB, DSSP and HSSP information with added value. Comput. Appl. Biosci. **12**: 525–9.

56 JAAKKOLA, T., M. DIEKHANS, AND D. HAUSSLER. 1999. Using the Fisher kernel method to detect remote protein homologies. Proc. ISMB **7**: 149–58.

57 JAAKKOLA, T., M. DIEKHANS, AND D. HAUSSLER. 2000. A discriminative framework for detecting remote protein homologies. J. Comput. Biol. **7**: 95–114.

58 JONES, D., R. MILLER, AND J. THORNTON. 1995. Successful protein fold recognition by optimal sequence threading validated by rigorous blind testing. Proteins **23**: 387–97.

59 JONES, D. T. 1997. Progress in protein structure prediction. Curr. Opin. Struct. Biol. **7**: 377–87.

60 JONES, D. T. (1999a). GenTHREADER: an efficient and reliable protein fold recognition method for genomic sequences. J. Mol. Biol. **287**: 797–815.

61 JONES, D. T. (1999b). Protein secondary structure prediction based on position-specific scoring matrices. J. Mol. Biol. **292**: 195–202.

62 JONES, D. T., W. R. TAYLOR, AND J. M. THORNTON. 1992. A new approach to protein fold recognition. Nature **358**: 86–9.

63 JONES, D. T. AND J. J. WARD. 2003. Prediction of disordered regions in proteins from position specific score matrices. Proteins **53 (Suppl. 6)**: 573–8.

64 KARCHIN, R., M. CLINE, Y. MANDEL-GUTFREUND, AND K. KARPLUS. 2003. Hidden Markov models that use predicted local structure for fold recognition: alphabets of backbone geometry. Proteins **51**: 504–14.

65 KARLIN, S. AND S. ALTSCHUL. 1990. Methods for assessing the statistical significance of molecular sequence features by using general scoring schemes. Proc. Natl Acad. Sci. USA **87**: 2264–8.

66 KARPLUS, K., C. BARRETT, AND R. HUGHEY. 1998. Hidden Markov models for detecting remote protein homologies. Bioinformatics **14**: 846–56.

67 KARPLUS, K., R. KARCHIN, J. DRAPER, J. CASPER, Y. MANDEL-GUTFREUND, M. DIEKHANS, AND R. HUGHEY. 2003. Combining local-structure, fold-recognition, and new fold methods for protein structure prediction. Proteins **53 (Suppl. 6)**: 491–6.

68 KARPLUS, K., K. SJÖLANDER, C. BARRETT, M. CLINE, D. HAUSSLER, R. HUGHEY, L. HOLM, AND C. SANDER. 1997. Predicting protein structure using hidden Markov models. Proteins **29 (Suppl. 1)**: 134–9.

69 KELLEY, L., R. MACCALLUM, AND M. STERNBERG. 2000. Enhanced genome annotation using structural profiles in the program 3D-PSSM. J. Mol. Biol. **299**: 499–520.

70 KINCH, L. N., J. O. WRABL, S. S. KRISHNA, ET AL. 2003. CASP5 assessment of fold recognition target predictions. Proteins **53 (Suppl. 6)**: 395–409.

71 KOH, I. Y. Y., V. A. EYRICH, M. A. MARTI-RENOM, ET AL. 2003. EVA: Evaluation of protein structure prediction servers. Nucleic Acids Res. **31**: 3311–5.

72 KOONIN, E. V., Y. I. WOLF, AND G. P. KAREV. 2002. The structure of the protein universe and genome evolution. Nature **420**: 218–23.

73 KROGH, A., M. BROWN, I. S. MIAN, K. SJÖLANDER, AND D. HAUSSLER. 1994. Hidden Markov models in computational biology. Applications to protein modeling. J. Mol. Biol. **235**: 1501–31.

74 KROGH, A. AND G. MITCHISON. 1995. Maximum entropy weighting of aligned sequences of protein or dna. Proc. ISMB **3**: 215–21.

75 KSCHISCHO, M., M. LÄSSIG, AND Y.-K. YU. 2005. Toward an accurate statistics of gapped alignments. Bull. Math. Biol. **67**: 169–91.

76 LACY, E. R., I. FILIPPOV, W. S. LEWIS, S. OTIENO, L. XIAO, S. WEISS, L. HENGST, AND R. W. KRIWACKI. 2004. p27 binds cyclin–CDK complexes through

a sequential mechanism involving binding-induced protein folding. Nat. Struct. Mol. Biol. **11**: 358–64.

77 Lathrop, R. H. 1994. The protein threading problem with sequence amino acid interaction preferences is NP-complete. Protein Eng. **7**: 1059–68.

78 Lathrop, R. H. and T. F. Smith. 1996. Global optimum protein threading with gapped alignment and empirical pair score functions. J. Mol. Biol. **255**: 641–65.

79 Leslie, C. S., E. Eskin, A. Cohen, J. Weston, and W. S. Noble. 2004. Mismatch string kernels for discriminative protein classification. Bioinformatics **20**: 467–76.

80 Levitt, M. and M. Gerstein. 1998. A unified statistical framework for sequence comparison and structure comparison. Proc. Natl Acad. Sci. USA **95**: 5913–20.

81 Liao, L. and W. S. Noble. 2003. Combining pairwise sequence similarity and support vector machines for detecting remote protein evolutionary and structural relationships. J. Comput. Biol. **10**: 857–68.

82 Liddington, R. C. 2004. Structural basis of protein–protein interactions. Methods Mol. Biol. **261**: 3–14.

83 Lin, K., V. A. Simossis, W. R. Taylor, and J. Heringa. 2005. A simple and fast secondary structure prediction method using hidden neural networks. Bioinformatics **21**: 152–9.

84 Linding, R., L. J. Jensen, F. Diella, P. Bork, T. J. Gibson, and R. B. Russell. 2003. Protein disorder prediction: implications for structural proteomics. Structure (Camb.) **11**: 1453–9.

85 Liu, D., R. Ishima, K. I. Tong, et al. 1998. Solution structure of a TBP–TAF(II)230 complex: protein mimicry of the minor groove surface of the TATA box unwound by TBP. Cell **94**: 573–83.

86 Lundström, J., L. Rychlewski, J. Bujnicki, and A. Elofsson. 2001. Pcons: a neural-network-based consensus predictor that improves fold recognition. Protein Sci. **10**: 2354–62.

87 Lüthy, R., J. Bowie, and D. Eisenberg. 1992. Assessment of protein models with three-dimensional profiles. Nature **356**: 83–5.

88 Lüthy, R., A. McLachlan, and D. Eisenberg. 1991. Secondary structure-based profiles: use of structure-conserving scoring tables in searching protein sequence databases for structural similarities. Proteins **10**: 229–39.

89 Lüthy, R., I. Xenarios, and P. Bucher. 1994. Improving the sensitivity of the sequence profile method. Protein Sci. **3**: 139–46.

90 Lyngsø, R. B., C. N. Pedersen, and H. Nielsen. 1999. Metrics and similarity measures for hidden Markov models. Proc. ISMB **7**: 178–86.

91 Marti-Renom, M. A., M. Madhusudhan, and A. Sali. 2004. Alignment of protein sequences by their profiles. Protein Sci. **13**: 1071–87.

92 McGuffin, L. J., K. Bryson, and D. T. Jones. 2000. The PSIPRED protein structure prediction server. Bioinformatics **16**: 404–5.

93 McGuffin, L. J. and D. T. Jones. 2003. Improvement of the GenTHREADER method for genomic fold recognition. Bioinformatics **19**: 874–81.

94 Morris, A. L., M. W. MacArthur, E. G. Hutchinson, and J. M. Thornton (1992, Apr). Stereochemical quality of protein structure coordinates. Proteins **12**: 345–64.

95 Mott, R. 2000. Accurate formula for *p*-values of gapped local sequence and profile alignments. J. Mol. Biol. **300**: 649–59.

96 Moult, J., K. Fidelis, A. Tramontano, B. Rost, and T. Hubbard. 2005. Critical assessment of methods of protein structure prediction (CASP) – round VI. Proteins **61 (Suppl. 7)**: 3–7.

97 Moult, J., K. Fidelis, A. Zemla, and T. Hubbard. 2003. Critical assessment of methods of protein structure prediction (CASP) – round V. Proteins **53 (Suppl. 6)**: 334–9.

98 Moult, J., J. T. Pedersen, R. Judson, and K. Fidelis. 1995. A large-scale experiment to assess protein structure prediction methods. Proteins **23**: ii–v.

99 Murzin, A. G., S. E. Brenner, T. Hubbard, and C. Chothia. 1995. SCOP: a structural classification of

proteins database for the investigation of sequences and structures. J. Mol. Biol. **247**: 536–40.

100 Needleman, S. B. and C. D. Wunsch. 1970. A general method applicable to the search for similarities in the amino acid sequence of two proteins. J. Mol. Biol. **48**: 443–53.

101 Notredame, C., D. Higgins, and J. Heringa. 2000. T-Coffee: A novel method for fast and accurate multiple sequence alignment. J. Mol. Biol. **302**: 205–17.

102 Obradovic, Z., K. Peng, S. Vucetic, P. Radivojac, C. J. Brown, and A. K. Dunker. 2003. Predicting intrinsic disorder from amino acid sequence. Proteins **53 (Suppl. 6)**: 566–72.

103 Oldfield, C. J., Y. Cheng, M. S. Cortese, C. J. Brown, V. N. Uversky, and A. K. Dunker. 2005. Comparing and combining predictors of mostly disordered proteins. Biochemistry **44**: 1989–2000.

104 Oldfield, C. J., E. L. Ulrich, Y. Cheng, A. K. Dunker, and J. L. Markley. (2005). Addressing the intrinsic disorder bottleneck in structural proteomics. Proteins **59**: 444–53.

105 Onuchic, J. N. and P. G. Wolynes. 2004. Theory of protein folding. Curr. Opin. Struct. Biol. **14**: 70–5.

106 Orengo, C. A., A. D. Michie, S. Jones, D. T. Jones, M. B. Swindells, and J. M. Thornton. 1997. CATH – a hierarchic classification of protein domain structures. Structure **5**: 1093–108.

107 O'Sullivan, O., K. Suhre, C. Abergel, D. G. Higgins, and C. Notredame. (2004). 3DCoffee: combining protein sequences and structures within multiple sequence alignments. J. Mol. Biol. **340**: 385–95.

108 Panchenko, A. R. 2003. Finding weak similarities between proteins by sequence profile comparison. Nucleic Acids Res. **31**: 683–9.

109 Pang, H., J. Tang, S.-S. Chen, and S. Tao. 2005. Statistical distributions of optimal global alignment scores of random protein sequences. BMC Bioinformatics **6**: 257.

110 Park, J., K. Karplus, C. Barrett, R. Hughey, D. Haussler, T. Hubbard, and C. Chothia. 1998. Sequence comparisons using multiple sequences detect three times as many remote homologues as pairwise methods. J. Mol. Biol. **284**: 1201–10.

111 Pearson, W. 1996. Effective protein sequence comparison. Methods Enzymol. **266**: 227–58.

112 Pearson, W. R. 1998. Empirical statistical estimates for sequence similarity searches. J. Mol. Biol. **276**: 71–84.

113 Pearson, W. R. and D. J. Lipman. 1988. Improved tools for biological sequence comparison. Proc. Natl Acad. Sci. USA **85**: 2444–8.

114 Petrey, D., Z. Xiang, C. L. Tang, et al. 2003. Using multiple structure alignments, fast model building, and energetic analysis in fold recognition and homology modeling. Proteins **53 (Suppl. 6)**: 430–5.

115 Petsko, G. A. and D. Ringe. 2003. *Protein Structure and Function*. New Science Press, London.

116 Pettitt, C. S., L. J. McGuffin, and D. T. Jones. 2005. Improving sequence-based fold recognition by using 3D model quality assessment. Bioinformatics **21**: 3509–15.

117 Pietrokovski, S. 1996. Searching databases of conserved sequence regions by aligning protein multiple-alignments. Nucleic Acids Res. **24**: 3836–45.

118 Rabiner, L. 1989. A tutorial on hidden markov models and selected applications in speech recognition. Proc. IEEE **77**: 257–86.

119 Radivojac, P., Z. Obradovic, D. K. Smith, G. Zhu, S. Vucetic, C. J. Brown, J. D. Lawson, and A. K. Dunker. 2004. Protein flexibility and intrinsic disorder. Protein Sci. **13**: 71–80.

120 Rhodes, G. 2004. *Crystallography Made Crystal Clear*. Academic Press, New York, NY.

121 Rice, D. and D. Eisenberg. 1997. A 3D-1D substitution matrix for protein fold recognition that includes predicted secondary structure of the sequence. J. Mol. Biol. **267**: 1026–38.

122 RYCHLEWSKI, L. AND D. FISCHER. 2005. LiveBench-8: the large-scale, continuous assessment of automated protein structure prediction. Protein Sci. **14**: 240–5.

123 RYCHLEWSKI, L., L. JAROSZEWSKI, W. LI, AND A. GODZIK. 2000. Comparison of sequence profiles. Strategies for structural predictions using sequence information. Protein Sci. **9**: 232–41.

124 RYCHLEWSKI, L., B. ZHANG, AND A. GODZIK. 1998. Fold and function predictions for *Mycoplasma genitalium* proteins. Fold. Design **3**: 229–38.

125 SADREYEV, R. AND N. GRISHIN. 2003. COMPASS: a tool for comparison of multiple protein alignments with assessment of statistical significance. J. Mol. Biol. **326**: 317–36.

126 SAIGO, H., J.-P. VERT, N. UEDA, AND T. AKUTSU. 2004. Protein homology detection using string alignment kernels. Bioinformatics **20**: 1682–9.

127 SAMUDRALA, R. AND M. LEVITT. 2000. Decoys 'R' Us: a database of incorrect conformations to improve protein structure prediction. Protein Sci. **9**: 1399–401.

128 SAMUDRALA, R. AND J. MOULT. 1998. An all-atom distance-dependent conditional probability discriminatory function for protein structure prediction. J. Mol. Biol. **275**: 895–916.

129 SÁNCHEZ, R. AND A. SALI. 1997. Evaluation of comparative protein structure modeling by MODELLER-3. Proteins **29 (Suppl. 1)**: 50–8.

130 SAUDER, J., J. ARTHUR, AND R. DUNBRACK. 2000. Large-scale comparison of protein sequence alignment algorithms with structure alignments. Proteins **40**: 6–22.

131 SCHÄFFER, A., Y. WOLF, C. PONTING, E. KOONIN, L. ARAVIND, AND S. ALTSCHUL. 1999. IMPALA: matching a protein sequence against a collection of PSI-BLAST-constructed position-specific score matrices. Bioinformatics **15**: 1000–11.

132 SCHÖLKOPF, B. AND A. SMOLA. 2002. *Learning with Kernels*. MIT Press, Cambridge, MA.

133 SCHWEDE, T., J. KOPP, N. GUEX, AND M. C. PEITSCH. 2003. SWISS-MODEL: An automated protein homology-modeling server. Nucleic Acids Res. **31**: 3381–5.

134 SIBBALD, P. AND P. ARGOS. 1990. Weighting aligned protein or nucleic acid sequences to correct for unequal representation. J. Mol. Biol. **216**: 813–8.

135 SIEW, N., A. ELOFSSON, L. RYCHLEWSKI, AND D. FISCHER. 2000. MaxSub: an automated measure for the assessment of protein structure prediction quality. Bioinformatics **16**: 776–85.

136 SIMOSSIS, V., J. KLEINJUNG, AND J. HERINGA. 2005. Homology-extended sequence alignment. Nucleic Acids Res. **33**: 816–24.

137 SIPPL, M. 1990. Calculation of conformational ensembles from potentials of mean force. An approach to the knowledge-based prediction of local structures in globular proteins. J. Mol. Biol. **213**: 859–83.

138 SIPPL, M. (1993a). Boltzmann's principle, knowledge-based mean fields and protein folding. An approach to the computational determination of protein structures. J. Comput. Aided Mol. Des. **7**: 473–501.

139 SIPPL, M. (1993b). Recognition of errors in three-dimensional structures of proteins. Proteins **17**: 355–62.

140 SIPPL, M. J. AND H. FLÖCKNER. 1996. Threading thrills and threats. Structure **4**: 15–9.

141 SKOLNICK, J., L. JAROSZEWSKI, A. KOLINSKI, AND A. GODZIK. 1997. Derivation and testing of pair potentials for protein folding. When is the quasichemical approximation correct? Protein Sci. **6**: 676–88.

142 SKOLNICK, J. AND D. KIHARA. 2001. Defrosting the frozen approximation: PROSPECTOR – a new approach to threading. Proteins **42**: 319–31.

143 SKOLNICK, J., D. KIHARA, AND Y. ZHANG. 2004. Development and large scale benchmark testing of the PROSPECTOR_3 threading algorithm. Proteins **56**: 502–18.

144 SKOLNICK, J., A. KOLINSKI, AND A. ORTIZ. 2000. Derivation of protein-specific pair potentials based on weak

sequence fragment similarity. Proteins **38**: 3–16.

145 SMITH, T. F. AND M. S. WATERMAN. 1981. Identification of common molecular subsequences. J. Mol. Biol. **147**: 195–7.

146 SÖDING, J. 2005. Protein homology detection by HMM-HMM comparison. Bioinformatics **21**: 951–60.

147 SOMMER, I., A. ZIEN, N. VON ÖHSEN, R. ZIMMER, AND T. LENGAUER. 2002. Confidence measures for protein fold recognition. Bioinformatics **18**: 802–12.

148 SPOLAR, R. S. AND M. T. RECORD. 1994. Coupling of local folding to site-specific binding of proteins to DNA. Science **263**: 777–84.

149 STERNBERG, M. J., P. A. BATES, L. A. KELLEY, AND R. M. MACCALLUM. 1999. Progress in protein structure prediction: assessment of CASP3. Curr. Opin. Struct. Biol. **9**: 368–73.

150 SUNYAEV, S., F. EISENHABER, I. RODCHENKOV, B. EISENHABER, V. TUMANYAN, AND E. KUZNETSOV. 1999. PSIC: profile extraction from sequence alignments with position-specific counts of independant observations. Protein Eng. **12**: 387–94.

151 TANAKA, H., M. ISHIKAWA, K. ASAI, AND A. KONAGAYA. 1993. Hidden Markov models and iterative aligners: study of their equivalence and possibilities. Proc. ISMB **1**: 395–401.

152 TANG, C. L., L. XIE, I. Y. Y. KOH, S. POSY, E. ALEXOV, AND B. HONIG. (2003). On the role of structural information in remote homology detection and sequence alignment: new methods using hybrid sequence profiles. J. Mol. Biol. **334**: 1043–62.

153 TATUSOV, R., S. ALTSCHUL, AND E. KOONIN. 1994. Detection of conserved segments in proteins: Iterative acanning of sequence databases with alignment blocks. Proc. Natl Acad. Sci. USA **91**: 12 091–5.

154 THIELE, R., R. ZIMMER, AND T. LENGAUER. 1999. Protein threading by recursive dynamic programming. J. Mol. Biol. **290**: 757–79.

155 Thompson, J. D., D. G. Higgins, and T. J. Gibson (1994a, Nov). CLUSTAL W: improving the sensitivity of progressive multiple sequence alignment through sequence weighting, position-specific gap penalties and weight matrix choice. Nucleic Acids Res. **22**: 4673–80.

156 THOMPSON, J. D., D. G. HIGGINS, AND T. J. GIBSON. (1994b). Improved sensitivity of profile searches through the use of sequence weights and gap excision. Comput. Appl. Biosci. **10**: 19–29.

157 TOMPA, P. (2003a). Intrinsically unstructured proteins. Trends Biochem. Sci. **27**: 527–533.

158 TOMPA, P. (2003b). Intrinsically unstructured proteins evolve by repeat expansion. BioEssays **25**: 847–55.

159 TOSATTO, S. C. E. 2005. The Victor/FRST function for model quality estimation. J. Comput. Biol. **12**: 1316–27.

160 TOSATTO, S. C. E. AND S. TOPPO. 2006. Large-scale prediction of protein structure and function from sequence. Curr. Pharm. Des. **12**: 2067–86.

161 TRESS, M., C.-H. TAI, G. WANG, I. EZKURDIA, G. LÓPEZ, A. VALENCIA, B. LEE, AND R. L. DUNBRACK. 2005. Domain definition and target classification for CASP6. Proteins **61 (Suppl. 7)**: 8–18.

162 VAN HOLDE, K. E., W. C. JOHNSON, AND P. S. HO. 1998. *Principles of Physical Biochemistry*. Prentice Hall, Upper Saddle River, NJ.

163 VON GROTTHUSS, M., J. PAS, L. WYRWICZ, K. GINALSKI, AND L. RYCHLEWSKI. 2003. Application of 3D-Jury, GRDB, and Verify3D in fold recognition. Proteins **53 (Suppl. 6)**: 418–23.

164 VON ÖHSEN, N. 2005. A novel profile-profile alignment method and its application in fully automated protein structure prediction. Ph. D. thesis, Ludwig Maximilians Universität München.

165 VON ÖHSEN, N., I. SOMMER, AND R. ZIMMER. 2003. Profile–profile alignment: a powerful tool for protein structure prediction. Pac. Symp. Biocomput. **8**: 252–63.

166 VON ÖHSEN, N., I. SOMMER, R. ZIMMER, AND T. LENGAUER. 2004. Arby: automatic protein structure prediction using profile–profile alignment and

confidence measures. Bioinformatics **20**: 2228–35.

167 von Öhsen, N. and R. Zimmer. 2001. Improving profile-profile alignment via log average scoring. Proc. WABI **1**: 11–26.

168 Vucetic, S., C. J. Brown, A. K. Dunker, and Z. Obradovic. 2003. Flavors of protein disorder. Proteins **52**: 573–84.

169 Wallner, B. and A. Elofsson. 2003. Can correct protein models be identified? Protein Sci. **12**: 1073–86.

170 Wallner, B., H. Fang, T. Ohlson, J. Frey-Skött, and A. Elofsson. 2004. Using evolutionary information for the query and target improves fold recognition. Proteins **54**: 342–50.

171 Wang, G. and R. L. Dunbrack. 2004. Scoring profile-to-profile sequence alignments. Protein Sci. **13**: 1612–26.

172 Wang, G., Y. Jin, and R. L. Dunbrack. 2005. Assessment of fold recognition predictions in CASP6. Proteins **61 (Suppl. 7)**: 46–66.

173 Wang, Y., S. Bryant, R. Tatusov, and T. Tatusova. 2000. Links from genome proteins to known 3-D structures. Genome Res. **10**: 1643–7.

174 Ward, J. J., J. S. Sodhi, L. J. McGuffin, B. F. Buxton, and D. T. Jones. (2004). Prediction and functional analysis of native disorder in proteins from the three kingdoms of life. J. Mol. Biol. **337**: 635–45.

175 Waterman, M. and M. Vingron. 1994. Rapid and accurate estimates of statistical significance for sequence base searches. Proc. Natl Acad. Sci. USA **91**: 4625–8.

176 Webber, C. and G. J. Barton. 2001. Estimation of *P*-values for global alignments of protein sequences. Bioinformatics **17**: 1158–67.

177 Weston, J., C. Leslie, E. Ie, D. Zhou, A. Elisseeff, and W. S. Noble. (2005). Semi-supervised protein classification using cluster kernels. Bioinformatics **21**: 3241–7.

178 Wolf, Y., S. Brenner, P. Bash, and E. Koonin. 1999. Distribution of protein folds in the three superkingdoms of life. Genome Res. **9**: 17–26.

179 Wood, M. J. and J. D. Hirst. 2005. Protein secondary structure prediction with dihedral angles. Proteins **59**: 476–81.

180 Wu, T., C. Nevill-Manning, and D. Brutlag. 1999. Minimal-risk scoring matrices for sequence analysis. J. Comput. Biol. **6**: 219–35.

181 Xu, J. and M. Li. 2003. Assessment of RAPTOR's linear programming approach in CAFASP3. Proteins **53 (Suppl. 6)**: 579–84.

182 Xu, J., M. Li, D. Kim, and Y. Xu. 2003. Raptor: optimal protein threading by linear programming. J. Bioinform. Comput. Biol. **1**: 95–117.

183 Xu, J., M. Li, G. Lin, D. Kim, and Y. Xu. 2003. Protein threading by linear programming. Pac. Symp. Biocomput. **8**: 264–75.

184 Yang, A. and B. Honig. 2000. An integrated approach to the analysis and modeling of protein sequences and structures. III. A comparative study of sequence conservation in protein structural families using multiple structural alignments. J. Mol. Biol. **301**: 691–711.

185 Yona, G. and M. Levitt. 2002. Within the twilight zone: a sensitive profile–profile comparison tool based on information theory. J. Mol. Biol. **315**: 1257–75.

186 Zemla, A. 2003. LGA: A method for finding 3D similarities in protein structures. Nucleic Acids Res. **31**: 3370–4.

187 Zhang, L. and J. Skolnick. 1998. How do potentials derived from structural databases relate to "true" potentials? Protein Sci. **7**: 112–22.

188 Zhang, Y., A. K. Arakaki, and J. Skolnick. 2005. TASSER: An automated method for the prediction of protein tertiary structures in CASP6. Proteins **61 (Suppl. 7)**: 91–8.

189 Zhang, Y. and J. Skolnick. (2004). Automated structure prediction of weakly homologous proteins on a genomic scale. Proc. Natl Acad. Sci. USA **101**: 7594–9.

190 Zhang, Y. and J. Skolnick. (2004b). Scoring function for automated assessment of protein structure template quality. Proteins **57**: 702–10.

191 Zhou, H. and Y. Zhou. 2005. Fold recognition by combining sequence profiles derived >from evolution and from depth-dependent structural alignment of fragments. Proteins **58**: 321–8.

12
De Novo Structure Prediction: Methods and Applications

Richard Bonneau

1 Introduction

1.1 Scope of this Review and Definition of *De Novo* Structure Prediction

This review will focus on the questions: (i) what are the features common to methods that represent the current state of the art in *de novo* structure prediction and (ii) how can these methods benefit biologists whose primary aim is a systems-wide description of a given organism or system of organisms. The role and capabilities of *de novo* structure prediction as well as the relationship of *de novo* structure prediction to other sequence and structure-based methods is far from simple. The literature on this subject is rapidly evolving; for balance in coverage and opinion the reader is also referred to recent reviews of *de novo* structure prediction methods [11,27,32,39,50].

Many methods that are today referred to as *de novo* have alternately or previously been referred to *ab initio* or "new folds" methods. For the purpose of this review I will classify a method as *de novo* structure prediction if that method does not rely on homology between the query sequence and a sequence in the Protein Data Bank (PDB) to create a template for structure prediction. *De novo* methods, by this definition, are forced to consider much larger conformational landscapes than fold recognition and comparative modeling techniques that limit the exploration of conformational space to those regions close to the initial structural template or templates.

Another common pedagogical distinction between structure prediction methods has been the distinction between methods based on statistical principles, on the one hand, and physical or first principles, on the other hand. I will not discuss this distinction here at great length except for noting that one of the shortcomings of this artificial division is that most effective structure prediction methodologies are in fact a combination of these two camps. For example, several methods that are described as based on physical or first-principles employ energy functions and parameters that are statistical approximations of data (e.g. the Lennard–Jones representation of van der

Bioinformatics - From Genomes to Therapies Vol. 1. Edited by Thomas Lengauer

ISBN: 978-3-527-31278-8

Waals forces is often thought of as a physical potential, but is a heuristic fit to data). Most current successful *de novo* structure prediction methods fall into the statistics camp. A more useful distinction may be the distinction between reduced complexity models and models that use atomic detail. Throughout this chapter I will discuss low-resolution (models containing drastic reductions in complexity such as unified atoms and centroid representations of side-chain atoms) and high-resolution methods (methods that represent protein and sometimes solvent in full atomic detail) focusing on this practical classification/division of methods in favor of distinctions based on a given method's derivation or parameterization.

1.2 The Role of Structure Prediction in Biology

What is the main application of structure prediction to biology? At present this is an open question that will take many years to develop, as the answer relies on the relative rate of progress in several fields. In short, I will argue that the main current application of structure prediction in biology lies in understanding protein function. Structure predictions can offer meaningful biological insights at several functional levels depending on the method used to generate the structure prediction, the expected resolution and the comprehensiveness or scale on which predictions are available for a given system.

At the highest levels of detail/accuracy (comparative modeling) there are several similarities between the uses of experimental and computational/predicted protein structure and the types of functional information that can be extracted from models generated by both methods [4]. For example, experimentally determined structures and structures resulting from comparative modeling can be used to help understand the details of protein function at an atomic scale, map conservation and mutagenesis data onto a structural framework, and explore detailed functional relationships between protein with similar folds or active sites.

At the other end of the prediction resolution spectrum, *de novo* structure prediction and fold recognition methods produce models of lower resolution than comparative models (see Chapter 10). These models can be used to assign putative functions to proteins for which little is known [15]. At the most basic level we can use structural similarities between a predicted structure and known structures to explore possible distant evolutionary relationships between query proteins of unknown function and other well-studied proteins for which structures have been experimentally determined. A query protein is likely to share some functional aspects with proteins in the PDB that show strong structure–structure matches to a high confidence predicted structure for that protein. This is based on the assumption that detectable structure relationships are conserved across a greater evolutionary distance than are detectable sequence similarities. This assumption is well supported by multiple surveys of the distributions of folds and their related functions in the PDB [48, 68, 76, 83]. The relationship between fold and function, however, is by no means a simple subject, and I refer the reader to several works that discuss this relationship in greater detail [56, 70, 84, 107]. Another way of exploring the functional significance of high confidence predicted structures is to use libraries of three-dimensional (3-D) functional motifs to search for conserved active site or functional motifs on the predicted structures [33, 72, 103]. Both basic methods, fold–fold matching and the use of small 3-D functional motif searches, can in principle be combined to form the basis for deriving functional hypothesis from predicted structure, thereby extending the completeness of genome annotations based only on primary sequence. For more details on how to infer protein function from protein structure, see Chapter 34.

1.3 *De novo* Structure Prediction in a Genome Annotation Context, Synergy with Other Methods

To date, the annotation of protein function in newly sequenced genomes relies on a large array of tools based ultimately on primary sequence analysis [3, 9, 19, 100]. These tools have afforded great progress in genome annotation including large improvements in gene detection, sequence alignment and detection of homologous sequences across genomes as well as the creation of databases of common protein families and primary sequence functional motifs. Comparative modeling methods have been highly successful on many fronts, creating large databases of highly accurate structure predictions for many organisms, but are based on primary sequence matches between PDB and query sequences [87] (see Chapter 10). Primary sequence methods also exist for the prediction of basic local structure qualities (some of these patterns being lower complexity patterns) of sequences such as the location of coiled-coil, transmembrane and disordered regions [52, 80, 99, 104]. Efforts to use *de novo* structure prediction (and/or fold recognition) must employ these sequence-based methods, as these methods provide a solid foundation on which all *de novo* methods discussed herein are reliant (see Figure 1). Any organization of these methods into an annotation pipeline must properly account for the fact that the accuracy/reliability is quite different between sequence and structure-based methods. One approach is to use structure prediction as part of a hierarchy where methods yielding high-confidence results are exhausted prior to computationally expensive and less accurate *de novo* structure prediction and fold recognition [12]. I will describe some early results from these approaches/pipelines that include structure prediction, the

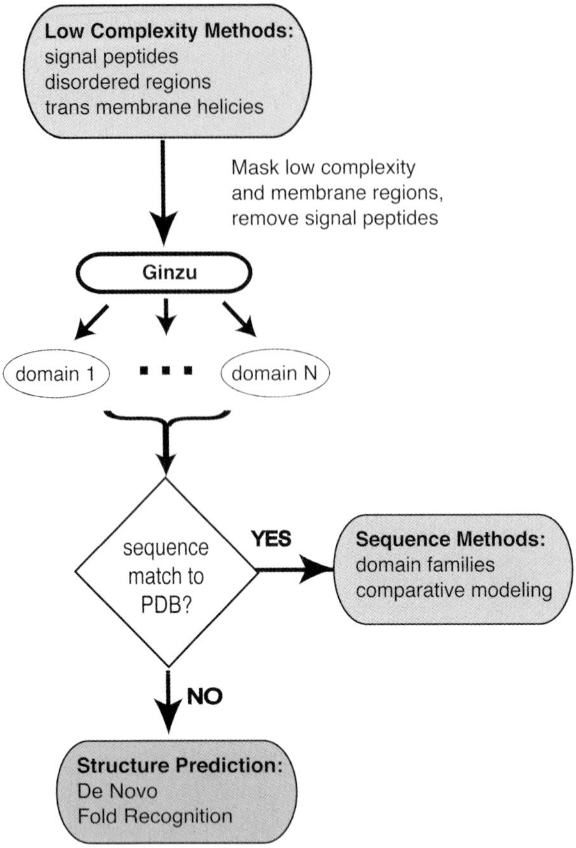

Figure 1 Idealized proteome structure annotation pipeline. Low-complexity regions such as transmembrane helices, signal peptides and disordered regions are masked, and domains dominated by these low-complexity or transmembrane sequence are treated separately. Remaining sequences are parsed to separate regions into structural domains to the degree that such domains are detectable (here, Ginzu is shown as the domain parsing algorithm, see Figure 4). Domains that do not have strong sequence matches to the PDB or other matches to well-annotated domains (Pfam, COG) are forwarded to structure-based methods. The use of structure prediction methods is positioned within this hierarchy of methods to increase comprehension of the resulting annotation without compromising the results obtained by sequence-based methods.

details of these pipelines and the technical and research challenges that remain in applying these pipelines to genome annotation [6,45,86].

The need for methods for predicting transmembrane proteins and understanding membrane–protein interactions is not discussed in this work (see Chapter 9 for this topic), the focus here is instead on soluble domains (including soluble domains excised from proteins containing transmembrane regions). Part of the difficulty in predicting transmembrane protein structure lies in the paucity of membrane protein structures deposited in the PDB

[28, 99]. It is only with access to the PDB, an ideal and comprehensive gold standard, by many criteria, that we can approach the problem of predicting soluble protein structure.

2 Core Features of Current Methods of *De Novo* Structure Prediction

We will now discuss core concepts that are common to multiple successful current *de novo* methods. This review is not intended to be encyclopedic and will invariably fail to mention several methods that are innovative and/or accurate in its attempt to focus on core concepts instead of distinct methodologies. The omission of any specific method should not be interpreted as commentary on the relative accuracy of the omitted method, but is simply due to the scope of this work and the state of rapid development in this field.

2.1 Rosetta *De Novo*

Throughout this work I will use examples of key concepts in *de novo* structure prediction with several examples drawn from the Rosetta *de novo* structure prediction protocol and will thus provide a brief overview of Rosetta before continuing to discuss key elements of the procedure in greater detail [13, 90, 97, 98] (see Figures 2 and 3). Results from the fourth and fifth Critical Assessments of Structure Prediction (CASP4, CASP5 and CASP6; see also Chapter 11) have shown that Rosetta is currently one of the best methods for *de novo* protein structure prediction and distant fold recognition [16, 18, 26, 65]. Rosetta was initially developed as a computer program for *de novo* fold prediction, but has been expanded to include design, docking, experimental determination of structure from partial datasets, protein–protein interaction and protein–DNA interaction prediction [25, 41, 42, 57, 59, 60, 88, 89]. When referring to Rosetta in this work I will be primarily referring to the *de novo* or *ab initio* mode of the Rosetta code base. Early progress in high-resolution structure prediction has been achieved via combinations of low-resolution approaches (for initially searching the conformational landscape) and higher-resolution potentials (where atomic detail and physically derived energy functions are employed). Thus, Rosetta structure prediction is carried out in two phases: (i) a low-resolution phase where overall topology is searched using a statistical scoring function and fragment assembly, and (ii) an atomic-detail refinement phase using rotamers and small backbone angle moves, and a more physically relevant (detailed) scoring function. The algorithms for searching the landscape are Monte-Carlo-type in both phases.

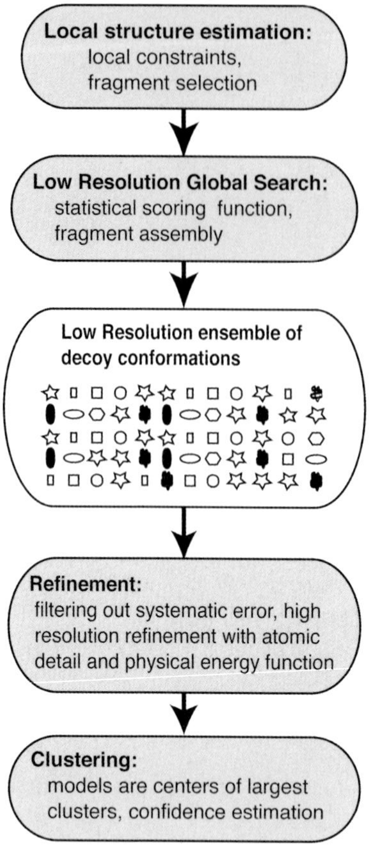

Figure 2 Schematic outline of Rosetta structure prediction protocol. Single sequences enter at the top of this schematic and confidence-ranked structure predictions are produced by the last/bottom step.

In the first phase, Rosetta *de novo* (Rosetta) uses information from the PDB to estimate the possible conformations for local sequence segments. The procedure first generates libraries of local sequence fragments excised from the PDB on the basis of local sequence similarity (three- and nine-residue matches between the query sequence and a given structure in the PDB). See Figure 1 for a schematic overview of the low-resolution (or fold prediction) phase of the Rosetta method, and see Tables 1 and 2 for a complete description of the Rosetta score. Rosetta fragment generation works well even for sequences that have no homologs in the known sequence databases; the structures in the PDB cover possible local sequence well at the three- and nine-residue length according to the current method. Rosetta then assembles these pre-computed local structure fragments by minimizing a global scoring function that favors hydrophobic burial/packing, strand pairing, compactness and highly proba-

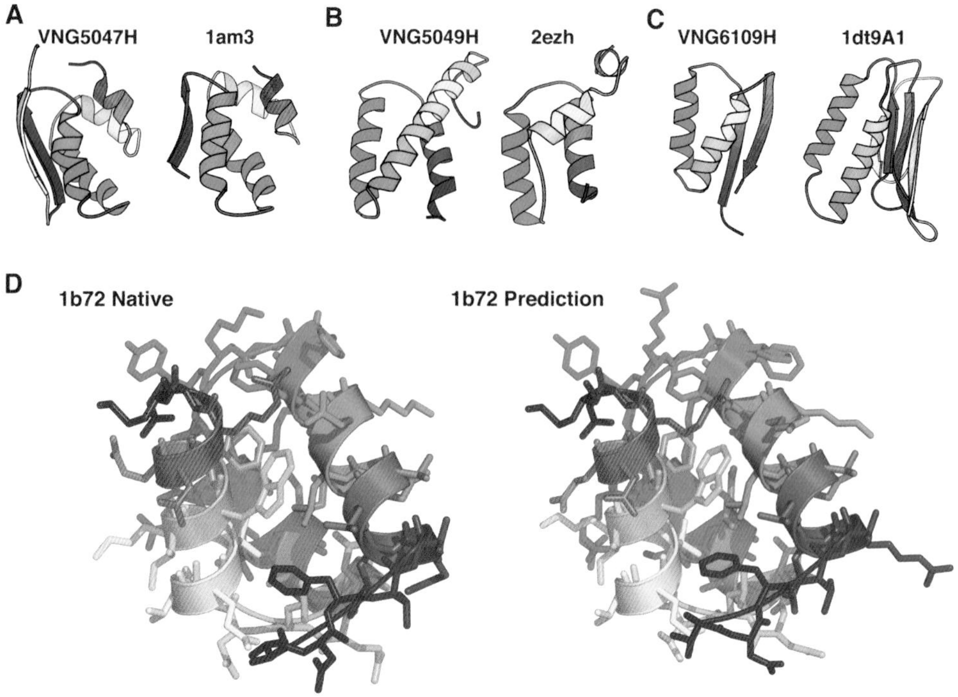

Figure 3 Examples of *de novo* structure predictions generated using Rosetta. (A–C) Examples from our genome-wide prediction of domains of unknown function in *Halobacterium* NRC-1 [12]. In each case the predicted structure is shown next to the correct native. For (A–C) only the backbone ribbons are shown, as these predictions were not refined using the all-atom potential and are examples of the utility of low-resolution prediction in determining function. (D) A recent prediction where high-resolution refinement subsequent to the low-resolution search produced the lowest energy conformation, a prediction of unprecedented accuracy (provided by Phil Bradley) [17].

ble residue pairings. The Rosetta score for this initial low-resolution stage is described in its entirety in Table 1. For the second, refinement, stage centroid representations of amino acid side-chains are replaced with atomic detail (rotamer representations). The scoring function used during this refinement phase includes solvation terms, hydrogen bond terms and other terms with direct physical interpretation. See Table 2 for a full description of the all-atom Rosetta score. Features of the high- and low-resolution phases of the Rosetta method are described below as I discuss key components of *de novo* structure prediction universal to all successful methods.

Using Rosetta generated structure predictions we were able to recapitulate many functional insights not evident from sequence based methods alone [14, 15]. We have reported success in annotating proteins and protein families without links to known structure with Rosetta [8, 14]. Various aspects of this

Table 1 Low-resolution, centroid-based Rosetta scoring function[a]

Name	Description (physical origin)	Functional form	Parameters (values)
env[b]	residue environment (solvation)	$\sum_i -\ln[P(aa_i \mid nb_i)]$	i = residue index aa = amino acid type nb = number of neighboring residues[c] (0, 1, 2, ..., 30. >30)
pair[b]	residue pair interactions (electrostatics, disulfides)	$\sum_i \sum_{j>i} -\ln\left[\frac{P(aa_i, aa_j \mid s_{ij} d_{ij})}{P(aa_i \mid s_{ij} d_{ij}) P(aa_j \mid s_{ij} d_{ij})}\right]$	i, j = residue indices aa = amino acid type d = centroid–centroid distance (10–12, 7.5–10, 5–7.5, <5 Å) s = sequence separation (>8 residues)
vdw[g]	steric repulsion	$\sum_i \sum_{j>i} \frac{(r_{ij}^2 - d_{ij}^2)^2}{r_{ij}^2}; \quad d_{ij} < r_{ij}$	i, j = residue (or centroid) indices d = interatomic distance r = summed van der Waals radii[h]
rg	radius of gyration (van der Waals attraction; solvation)	$\sqrt{\langle d_{ij}^2 \rangle}$	i, j = residue indices d = distance between residue centroids
cbeta	C_β density (solvation; correction for excluded volume effect introduced by simulation)	$\sum_i \sum_{sh} -\ln\left[\frac{P_{\text{compact}}(nb_{i,sh})}{P_{\text{random}}(nb_{i,sh})}\right]$	i = residue index sh = shell radius (6, 12 Å) nb = number of neighboring residues within shell[f] $P_{compact}$ = probability in compact structures assembled from fragments P_{random} = probability in structures assembled randomly from fragments

overall protocol will be reviewed in greater detail below. We also encourage the reader to refer to several prior works where the Rosetta method is described in its entirety.

2.2 Evaluation of Structure Predictions

In general the most effective methods for predicting structure *de novo* depend on parameters ultimately derived from the PDB. Several methods use the PDB directly to estimate local sequence and even explicitly use fragments of local sequence from the PDB to build global conformations. These uses of the PDB require that methods be tested using structures not present in the sets

Table 1 continued

Name	Description (physical origin)	Functional form	Parameters (values)
SS[d]	strand pairing (hydrogen bonding)	Scheme A: $SS_{\phi,\theta} + SS_{hb} + SS_d$ Scheme B: $SS_{-\phi,\theta} + SS_{hb} + SS_{d\sigma}$ where $SS_{\phi,\theta} = \sum_m \sum_{n>m} - \ln[P(\phi_{mn}, \theta_{mn} \mid d_{mn}, sp_{mn}, s_{mn})]$ $SS_{hb} = \sum_m \sum_{n>m} - \ln[P(hb_{mn}, \mid d_{mn}, s_{mn})]$ $SS_d = \sum_m \sum_{n>m} - \ln[P(d_{mn}, \mid s_{mn})]$ $SS_{d\sigma} = \sum_m \sum_{n>m} - \ln[P(d_{mn}, \sigma_{mn} \mid \rho_m, \rho_n)]$	m, n = strand dimer indices; dimer is two consecutive strand residues $\hat{V}$ = vector between first N and last C atom of dimer $\hat{m}$ = unit vector between $\hat{V}m$ and $\hat{V}n$ midpoints $\hat{x}$ = unit vector along carbon-oxygen bond of first dimer residue $\hat{y}$ = unit vector along oxygen-carbon bond of second dimer residue ϕ, θ = polar angles between $\hat{V}m$ and $\hat{V}n$ (10, 36° bins) hb = dimertwist, $\sum_{k=m,n} 0.5(\lvert\hat{m} \cdot \hat{x}_k\rvert + \lvert\hat{m} \cdot \hat{y}_k\rvert)$ (<0.33, 0.33–0.66, 0.66–1.0, 1.0–1.33, 1.33–1.6, 1.6–1.8, 1.8–2.0) d = distance between $\hat{V}m$ and $\hat{V}n$ midpoints (<6.5 Å) σ = angle between $\hat{V}m$ and $\hat{M}$ (18° bins) sp = sequence separation between dimer-containing strands (<2, 2–10, >10 residues) s = sequence separation between dimers (>5 or >10) ρ = mean angle between vectors $\hat{m}, \hat{x}$ and $\hat{m}, \hat{y}$ (180° bins)
sheet[e]	strand pair arrangement into sheets	$- \ln[P(n_{\text{sheets}} n_{\text{lone\_strands}} \mid n_{\text{strands}})]$	n_{sheets} = number of sheets $n_{\text{lone\_strands}}$ = number of unpaired strands n_{strands} = total number of strands
HS	helix-strand packing	$\sum_n \sum_n - \ln[P(\phi_{mn}, \psi_{mn} \mid sp_{mn} d_{mn})]$	m = strand dimer index; dimer is two consecutive strand residues n = helix dimer index; dimer is central two residues of four consecutive helical residues $\hat{V}$ = vector between first N and last C atom of dimer ϕ, θ = polar angles between $\hat{V}m$ and $\hat{V}n$ (36° bins) sp = sequence separation between dimer-containing helix and strand (binned <2, 2–10, >10 residues) d = distance between $\hat{V}m$ and $\hat{V}n$ midpoints (<12 Å)

of protein structures used to train these methods (or present in the sets of structures used to predict local structure fragments). The first such evaluation of structure prediction, CASP (see Chapter 11 for a more detailed description), showed that published estimates of prediction error were smaller than prediction error measured on a set of novel proteins outside the training set (this is not surprising given the difficulties of avoiding overfitting in as complex a data space as protein structure) [64]. Indeed, early experiments showed that no methods for *de novo* structure prediction were effective outside of carefully chosen benchmarks containing only the smallest proteins. Spurred on by these early evaluations the field returned to the drawing board and two years later produced multiple methods with much higher accuracies in the new folds or *de novo* category (CASP3) [73,75,82]. Thus, the CASP experiments proved to be invaluable to the field at that point in the development of the field, provoking a renewed interest in the *de novo* structure prediction and properly realigned interest in techniques according to effectiveness.

Arguably, CASP has the flaw that predictors are allowed to intervene and manually curate their predictions prior to submission to the CASP evaluators. Thus, the results of CASP are a convolution of: (i) the art of prediction (each group's intuition and skill using their tools) and (ii) the relative performance

Footnotes to Table 1:

[a] The individual components in the Rosetta score (the score used by Rosetta during low-resolution/centroid mode *de novo* structure prediction) are given as described originally in Simons [96–98].

[b] Binned function values are linearly interpolated, yielding analytic derivatives.

[c] Neighbors within a 10 Å radius. Residue position defined by C_β coordinates (C_α for glycine).

[d] Interactions between dimers within the same strand are neglected. Favorable interactions are limited to preserve pairwise strand interactions, i.e. dimer m can interact favorably with dimers from at most one strand on each side, with the most favorable dimer interaction ($SS_{\phi s\theta}$ + SS_{hb} + SS_d) determining the identity of the interacting strand. $SS_{d\sigma}$ is exempt from the requirement of pairwise strand interactions. SS_{hb} is evaluated only for m, n pairs for which $SS_{\phi,\theta}$ is favorable. $SS_{d\sigma}$ is evaluated only for m, n pairs for which $SS_{\phi,\theta}$g and SS_{hb} are favorable. A bonus is awarded for each favorable dimer interaction for which $|m - - n| > 11$ and strand separation is more than eight residues

[e] A sheet is comprised of all strands with dimer pairs less than 5.5 Å apart, allowing each strand having at most one neighboring strand on each side. Discrimination between alternate strand pairings is determined according the most favorable dimer interaction. Probability distributions fitted to $c(n_{\text{strands}}) - 0.9n_{\text{sheets}} - 2.7n_{\text{lone\_strands}}$ where $c(n_{\text{strands}})$ = (0.07, 0.41, 0.43, 0.60, 0.61, 0.85, 0.86, 1.12).

[f] Residue position defined by C_β coordinates (C_α for glycine).

[g] Not evaluated for atom (centroid) pairs whose interatomic distance depends on the torsion angles of a single residue.

[h] Radii determined from (i) 25th closest distance seen for atom pair in pdbselect25 structures, (ii) the fifth closest distance observed in X-ray structures with better than 1.3-Å resolution and less than 40% sequence identity or (iii) X-ray structures of less than 2 Å resolution, excluding i, $i + 1$ contacts (centroid radii only).

of the core methods (the performance of each method in an automatic setting). Although this convolution reflects the reality when workers aim to predict proteins of high interest, such as proteins involved in a specific function or proteins critical to a given disease or process being experimentally studied, it does not reflect the demands placed on a method when trying to predict whole genomes, where the shear number of predictions does not allow for much manual intervention. Several additional tests similar to CASP (in that they are blind tests of structure prediction) have been organized in response to the concerns of many that it is important to remove the human aspects of CASP. The Critical Assessment of Fully Automatic Structure Prediction (CAFASP) is an experiment running parallel with CASP that aims to test fully automated methods' performance on CASP targets, mainly testing servers instead of groups [35, 36]. Several groups have also raised concerns that there are problems associated with the small numbers of proteins tested in each CASP experiment, and thus EVA and LiveBench were organized to test methods using larger numbers of proteins [20, 92, 94]. Both use proteins that have structures that are unknown to the participating prediction groups, but that have been recently submitted to the PDB and are not open to the public at the time their sequences are released to those participating in LiveBench or EVA. The participating groups then have the time it takes for the new PDB entries to be validated to predict the structures. Although groups with amazing computer-hacking skills could in principle access this information, these efforts effectively create a CAFASP equivalent for a larger number of proteins.

All four of these tests of prediction methods, as well as benchmarks carried out by authors of any methods in question, are valuable ways of judging the performance of *de novo* methods. The methods, and elements of methods, I describe herein are generally accepted to be the best performers by the five above measures (four blind tests and author benchmarks). I will not focus on the details of the CASP, CAFASP, EVA and LiveBench methods, as they are described in detail elsewhere and instead attempt to focus on common elements of top performing methods.

2.3 Domain Prediction is Key

As the size of a protein increases, so to does the size of the conformational space associated with that protein. Thus, *de novo* methods, which must sample this space, have run times that increase dramatically with sequence length. Current *de novo* methods are limited to proteins and protein domains less than 150 amino acids in length (with Rosetta the limit is around 150 residues for α/β proteins, 80 for β-folds and more than 150 residues for α-only-folds). This limit means that roughly half of the protein domains seen so far in the

Table 2 All-atom Rosetta scoring function: the components of the all-atom score (centroids are expanded using a rotamer description of side-chains) [31, 44, 58, 62, 77, 105]

Name	Description	Functional form	Parameters, variables	References
rama	Ramachandran torsion preferences	$\sum_i - \ln[P(\phi_i, \psi_i \mid aa_i ss_i)]$	i = residue index ϕ, ψ = backbone torsion angles (10°, 36° bins) aa = amino acid type ss = secondary structure type[a]	Bowers, 2000 [16a]
LJ[c]	Lennard–Jones interactions	$\sum_i \sum_{j>i} \begin{cases} \left[\left(\frac{r_{ij}}{d_{ij}}\right)^{12} - 2\left(\frac{r_{ij}}{d_{ij}}\right)^{6}\right] e_{ij} & \text{if } \frac{d_{ij}}{r_{ij}} > 0.6 \\ \left[-8759.2\left(\frac{d_{ij}}{r_{ij}}\right) + 5672.0\right] e_{ij}, & \text{else} \end{cases}$	i,j = residue indices d = interatomic distance e = geometric mean of atom well depths[d] r = summed van der Waals radii[e]	Kuhlman, 2000 [59]
hb[f]	hydrogen bonding	$\sum_i \sum_j (-\ln[p(d_{ij} \mid h_j ss_{ij})]$ $-\ln[P(\cos\theta_{ij} \mid d_{ij} h_j ss_{ij})]$ $-\ln[P(\cos\theta_{ij} \mid d_{ij} h_j ss_{ij})]$ $-\ln[P(\cos\psi_{ij} \mid d_{ij} h_j ss_{ij})]$	i = donor residue index j = acceptor residue index d = acceptor-proton interatomic distance h = hybridization (sp^2, sp^3) ss = secondary structure type[g] θ = proton–acceptor–acceptor base bond angle ψ = donor–proton–acceptor bond angle	Kortemme, 2003 [58]
solv	solvation	$\sum_i \left[\Delta G_i^{\text{ref}} - \sum_j \left(\frac{2\Delta G_i^{\text{free}}}{4\pi^{3/2}\lambda_i r_{ij}^2} e^{-d_{ij}^2} V_j + \frac{2\Delta G_i^{\text{freee}}}{4\pi^{3/2}\lambda_j r_{ij}^2} e^{-d_{ij}^2} V_i\right)\right]$	i, j = atom indices d = distance between atoms r = summed van der Waals radii[e] λ = correlation length[h] V = atomic volume[h] ΔG^{ref}, ΔG^{free} = energy of a fully solvated atom[h]	Lazaridis, 1999 [62]
pair	residue pair interactions (electrostatics, disulfides)	$\sum_i \sum_{j>i} - \ln\left[\frac{P(aa_i, aa_j) \mid d_{ij})}{P(aa_i \mid d_{ij}) P(aa_i \mid d_{ij})}\right]$	i,j = residue indices aa = amino acid type d = distance between residues[i]	Kuhlman, 2000 [59]
dun	rotamer self energy	$\sum_i - \ln\left[\frac{P(\text{rot}_i \mid \phi_i \psi_i) P(aa_i \mid \phi_i, \psi_i)}{P(aa_i)}\right]$	i,j = residue indices rot = Dunbrack backbone-dependent rotamer aa = amino acid type ϕ, ψ = backbone torsion angles	Dunbrack, 1997 [31]
ref	unfolded state reference energy	$\sum_{aa} n_{aa}$	aa = amino acid type n = number of residues	Kuhlman, 2000 [59]

PDB are within the size limit of *de novo* structure prediction. Two approaches to circumventing this size limitation are: (i) increasing the size range of *de novo* structure prediction and (ii) dividing proteins into domains prior to attempting to predict structure. Dividing query sequences into their smallest component domains prior to folding is one straightforward way to dramatically increase the reach of *de novo* structure prediction. For many proteins domain divisions can be easily found (as would be the case for a protein where one domain was unknown and one domain was a member of a well-known protein family) while several domains remain beyond our ability to correctly detect them. The determination of domain family membership and domain boundaries for multi-domain proteins is a vital first step in annotating proteins on the basis of primary sequence and has ramifications for several aspects of protein sequence annotation; multiple works describe methods for detecting such boundaries. In short, most protein domain parsing methods rely on hierarchically searching for domains in a query sequence with a collection of primary sequence methods, domain library searches and matches to structural domains in the PDB [26, 55, 66].

Some notable works use coarse-grained structural simulations/predictions coupled with methods for assigning structural domain boundaries to 3-D structures to detect protein domains from sequence. The guiding principle behind this approach is that very low-resolution predictions will pick up overall patterns of the polypeptide packing into distinct structural domains. Another recent work attempted to use local sequence signals to detect structure domain boundaries under the assumption that there would be detectable differences in local sequence propensities at domain boundaries [37]. As of yet these

Footnotes to Table 2:

[a] All binned function values are linearly interpolated, yielding analytic derivatives, except as noted.
[b] Three-state secondary structure type as assigned by DSSP.
[c] Not evaluated for atom pairs whose interatomic distance depends on the torsion angles of a single residue.
[d] Well depths taken from CHARMm19 parameter set (Neria 1996 [77]).
[e] Radii determined from fitting atom distances in protein X-ray structures to the 6–12 Lennard–Jones potential using CHARMm19 well depths.
[f] Evaluated only for donor acceptor pairs for which $1.4 \leq d \leq 3.0$ and $90^\circ \leq \psi, \theta \leq 180^\circ$. Side-chain hydrogen bonds in involving atoms forming main-chain hydrogen bonds are not evaluated. Individual probability distributions are fitted to eighth-order probability distributions and analytically differentiated.
[g] Secondary structure types for hydrogen bonds are assigned as helical ($j - - i = 4$, main-chain), strand: ($|j - - i| > 4$, main-chain) or other.
[h] Values taken from Lazaridis and Karplus [62].
[i] Residue position defined by C_β coordinates (C_α of glycine).
* Also described in Rohl 2005 [90].

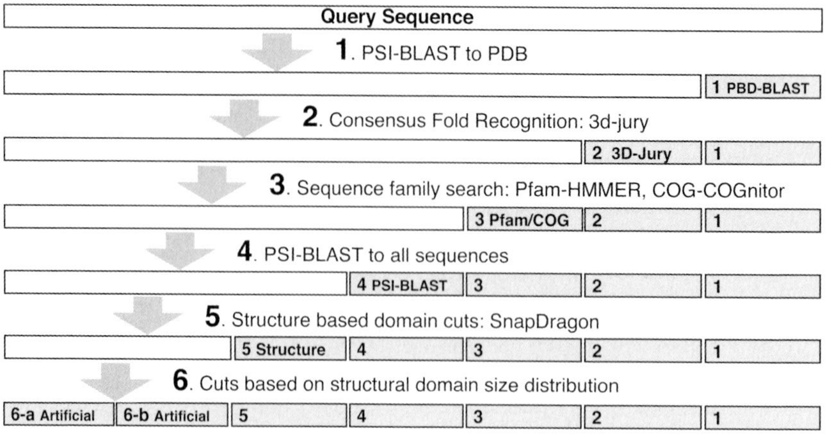

Figure 4 Schematic outline of an ideal hierarchical approach to domain parsing. Methods with higher reliability are used first, with sequence matches to the PDB being the highest-quality information. As higher-reliability/interpretability methods are exhausted, noisier methods are used (such as parsing multiple sequence alignments, step 4, and guessing domain boundaries based on the distribution of domain sizes in the PDB). Sequence regions hit by higher confidence methods (represented as gray rectangles) are masked and the remaining sequence (represented by white rectangles) are forwarded onto the remaining methods. Steps 1–4 and 6 are currently implemented in the Ginzu program; step 5 (adding sequence homolog independent methods such as structure-based domain parsing from sequence to the procedure) represent future work. Although we recognize domains in this schematic from left to right this direction is merely schematic, and Ginzu recognizes and parses domains in a fully general (discontinuous, depending on where the strong hits are at any given level) manner.

methods have unacceptably high error rates and are far too computationally demanding for use in genome wide predictions (David Kim, personal communication) [38]. In spite of the limitations mentioned above, these methods (that are not dependent on detecting sequence homologs for a given query sequence) are attractive for proteins that have no detectable homologs or matches to protein domain families and future work on this front could increase the number of proteins within reach of *de novo* methods considerably. It is likely that a method which successfully combines these coarse structure-based methods with existing sequence-based methods into a hierarchically organized domain detection program (e.g. Ginzu) will eventually outperform any existing method at domain parsing and greatly increase the accuracy of downstream structure prediction. Figure 4 shows a schematic domain detection program (this schematic is implemented as the program Ginzu).

2.4 Local Structure Prediction and Reduced Complexity Models are Central to Current *De Novo* Methods

Several methods for reducing the combinatorial complexity of the protein folding problem have been employed including lattice models (confining possible special coordinates to a predefined 3-D grid) and several discrete-state off-lattice models (e.g. reducing degrees of freedom along the backbone to a set of discrete angles). For a more exhaustive description of these methods and their reduced-complexity move sets I refer the reader to earlier reviews of *de novo* structure prediction methods [11, 27].

Instead, I will focus on the use of local structure information to constrain global structure prediction simulations to only conformations consistent with local structure prediction. Local sequences excised from protein structures often have stable structures in the absence of their global contacts, demonstrating that local sequences can have a strong, sequence-dependent, structural bias towards one or more well defined structures [10, 24, 69, 74, 106]. This experimental observation is a result of the fact that the polypeptide chain is heavily constrained by local structure bias in a sequence dependent manner. The strength of this local, sequence-dependent, structure bias can vary from strong (a local sequence that exhibits a single well defined local structure) to weak (local sequences that are disordered or completely determined by their global environment) with most protein sequences falling into some intermediate regime (local sequences that fold into multiple well-defined local structures depending on their global environment) [21, 46]. Prediction methods that accurately predict the type, strength and possible multiplicity of local structure bias for any given query sequence segment drastically reduce the size of the available conformational landscape. Using either fragment substitution (assembling fragments of local structure) as a move set or local structure constraints derived from predicted local structure also has the advantage that the subsequent global search is limited to protein-like regions of the conformational landscape (helices, correct chirality of secondary strand packing, strands and sheets with correct twist, etc.).

There are two main ways to use local structure prediction as an overriding/hard constraint on the global search: (i) using fragments to build up global structures (local structure defining the moveset) and (ii) using local structure as a hard constraint (local structure heavily modifying the objective function).

Rosetta explicitly uses fragments of three and nine residues of local structure to build global structures via fragment assembly. Prior to a Rosetta simulation a library of local structure fragments is generated such that several fragments (25–200) of different local structure are pre-computed for every possible three- and nine-residue window along the query. The simulation (the search for low-energy conformations given the Rosetta scoring function) consists primarily of randomly selecting three- and nine-residue windows along the query and replacing torsion angles at that three- or nine-residue window with torsion angles taken from a different fragment for that position. These fragments are pulled from a nonredundant version of the PDB on the basis of local sequence similarity to the query sequence [97]. This work was inspired by careful studies of the relationship between local sequence and local structure [46], that demonstrated that this relationship was highly variable on a sequence-specific basis and that there is a great deal of sequence-specific local structure that could be recognized even in the absence of global homology. The selection of fragments of local structure on the basis of local sequence matches dramatically reduces the size of the accessible conformational landscape. In practice we see that, as desired, for some local sequence segments there is a strong bias towards a single local structure in the computed local structure fragments, while other local sequences exhibit a wide range of local conformations in the fragment library. Using fragment substitution as a moveset to optimize Rosetta's objective function has one major drawback: as the structure collapses (forms many contacts favorable according to the energy function) late in the simulation the acceptance rate of fragment moves becomes unworkably small. This is due to the fact that the substitution of six or 18 backbone dihedral angles creates large perturbations to the Cartesian coordinates of parts of the protein distant along the polypeptide chain. The likelihood that such large perturbations cause steric clashes and break energetically favorable contacts late in a given simulation is exceedingly large. To recover effective minimization of the Rosetta score after initial collapse several additional move types have been added to the Rosetta moveset. The simplest move type consists of small angle moves (within populated regions of the Ramachandran map). Additional moves, descriptively named "chuck", "wobble" and "gunn" moves, aim to perform fragment insertions that have small effects far from the insertion. These additional move types are also critical to the modeling of loops in homology modeling and are described in detail elsewhere [89].

The TASSER method smoothly combines fragments of aligned protein structure (from threading runs) with regions of unaligned proteins (represented on a lattice for computational efficiency) to effectively scale between the fold recognition and *de novo* regime [108]. Other notable uses of local structure fragments include the use of I-sites to select fragments that are then fed to Rosetta as described by Bystroff and Shao [22]. I-sites is a hidden Markov model (HMM) method designed to detect strong relationships between sequence and structure as defined by a library of local structure–sequence relationships. One potential advantage of this method is that the I-sites method is not constrained to fragments of a fixed length (Rosetta is constrained to three- and nine-length fragments) [23]. Thus larger patterns of local structure bias are expected to be detected better by this method. Karplus and coworkers also use a similar approach to detecting fragments of local structure (a two-stage HMM) as part of their *de novo* method [53]. These methods have the primary advantage of better performance when local sequence–structure bias is high (e.g. when local structure is strongly and/or uniquely determined by sequence).

2.5 Clustering as a Heuristic Approach to Approximating Entropic Determinants of Protein Folding

Several protein structure prediction methods are effectively two-step procedures involving the generation of large ensembles of conformations (each being the result of a minimization or simulation) followed by the clustering of the generated ensemble to produce one or more cluster centers that are taken to be the predicted models. Regardless of how one justifies the use of clustering as a means of selecting small numbers of predictions or models from ensembles of decoys conformations, the justification is indirectly supported by the efficacy of the procedure and the resultant observation that clustering has become a central, seemingly required, feature of successful *de novo* prediction methods. Starting with CASP3 the field has witnessed a proliferation of clustering methods as post-simulation processing steps in protein structure prediction methods [14,51,96,108].

Prediction of protein structure *de novo* using Rosetta relies heavily on a final clustering stage. In the first step a large ensemble of potential protein structures is generated, each conformation being the result of an extensive Monte Carlo search designed to minimize the Rosetta scoring function (see Figure 1). We then apply clustering to find the centers of the largest clusters. These cluster centers are ranked by the size of their originating cluster in the ensemble. The tightness of clustering in the ensemble is also used as a measure of method success (larger tighter clusters indicate a higher probability that the method produced correct fold predictions for a given protein). Each Rosetta simulation/Monte Carlo run can be thought of as a fast quench starting from a random point on the conformational landscape (defined by the local structure estimation/fragments). Many of these fast quenches (individual simulations) results in incorrect conformations that score nearly as well as any correct conformations generated in the full ensemble of decoy conformations, as judged by the Rosetta score (a number of other potentials tested also lack discriminative power at this stage). This lack of discrimination by *de novo* scoring functions is partially the result of inaccuracies in the scoring function, limitations in our ability to search the landscape and the fact that entropic terms are a major contributor to the free energy of folding. In any case, this lack of discrimination is mitigated by a final clustering step and it has been shown that the centers of the largest clusters in a clustered Rosetta decoy ensemble are in most cases the conformations closest to native. The ubiquitous use of clustering can be justified in several ways: clustering can be thought of as (i) a heuristic way to approximate the entropy of a given conformation given the full ensemble of decoy conformations generated for a given protein, (ii) a signal averaging procedure, averaging out errors in the low-resolution scoring function, or (ii) taking advantage of foldable-protein specific energy landscape features such as broad energy wells that are the result of proteins evolving to be robust to sequence and conformational changes from the native sequence or structure (a mix of sequence and configurational entropy) [95].

An interesting alternative to the strategy of clustering ensembles of results from independent minimizations is the use of replica exchange methods. Replica exchange methods employ large numbers of simulations spanning a range of temperatures (defined physically if one uses a physical potential or simply as a constant in the exponent of the Boltzmann equation (see Chapter 11) for probabilistic scoring functions). These independent simulations are carried out in parallel and are allowed to exchange temperatures throughout the run. This simulation strategy ideally allows for a random walk in energy space (and thus better sampling) and can be used to calculate entropic term *post facto*. Replica exchange Monte Carlo has been used successfully in the simulation and prediction of protein structure, and is interesting due to its explicit connection to a physical description of the system and its ability to search low energy states without getting trapped [81].

2.6 Balancing Resolution with Sampling, Prospects for Improved Accuracy and Atomic Detail

Every *de novo* structure prediction procedure must strike a delicate balance between the computational efficiency of the procedure and the level of physical detail used to model protein structure within the procedure. Low-resolution models can be used to predict protein topology/folds and sometimes suggest function [15]. Low-resolution models have also been remarkably successful at predicting features of the folding process such as folding rates and phi values [1,2]. It is clear, however, that modeling proteins (and possibly bound water and other cofactors) at atomic detail and scoring these higher resolution models with physically derived, detailed potentials is a needed development if higher-resolution structure prediction is to be achieved.

Early progress has focused on the use of low-resolution approaches for initially searching the conformational landscape followed by a refinement step where atomic detail and physical scoring functions are used to select and/or generate higher-resolution structures. For example, several studies have illustrated the usefulness of using *de novo* structure prediction methods as part of a two-stage process in which low-resolution methods are used for fragment assembly and the resulting models are refined using a more physical potential and atomic detail (e.g. rotamers) [31] to represent side-chains [18, 71, 102]. In the first step, Rosetta is used to search the space of possible backbone conformations with all side-chains represented as centroids. This process is well described, and has well-characterized error rates and behavior. High-confidence or low-scoring models are then refined using potentials that account for atomic detail such as hydrogen bonding, van der Waals forces and electrostatics.

One major challenge that faces methods attempting to refine *de novo* methods is that the addition of side-chain degrees of freedom combined with the reduced length scale (reduced radius of convergence; one must get much closer to the correct answer before the scoring function recognizes the conformation as correct) of the potentials employed require the sampling of a much larger space of possible conformations. Thus, one has to correctly determine roughly twice the number of bond angles to a higher tolerance if one hopes to succeed. An illustrative example of the difference in length scale (radius of convergence) between low-resolution methods and high-resolution methods is the scoring of hydrogen bonds. In the low-resolution Rosetta procedure backbone hydrogen bonding is scored indirectly by a term designed to pack strands into sheets under the assumption that correct alignment of strands satisfies hydrogen bonds between backbone atoms along the strand and that intra-helix backbone hydrogen bonds are already well accounted for by the local structure fragments. This low-resolution method first reduces strands to vectors, and then scores strand arrangement (and the correct hydrogen bonding implicit in the relative positions/arrangement of all strand vector pairs) via functions dependent on the angular and distance relationships between the two vectors. Thus, the scoring function is robust to a rather large amount of error in the coordinates of individual electron donors and acceptors participating in backbone hydrogen bonds (as large numbers of residues are reduced to the angle and distance between the two vectors representing a given pair of strands). In the high-resolution, refinement mode of Rosetta an empirical hydrogen bond terms with angle and distance dependence between individual electron donors and acceptors is used [88]. This more-detailed hydrogen bond term has a higher fidelity and a more straightforward connection to the calculation of physically realistic energies (meaningful units), but requires more sampling, as smaller changes in the orientation of the backbone can cause large fluctuations in computed energy.

Another major challenge with high-resolution methods is the difficulty of computing accurate potentials for atomic-detail protein modeling in solvent; with electrostatic and solvation terms being among the most difficult terms to accurately model. Full treatment of the free energy of a protein conformation (with correct treatment of dielectric screening) is complicated by the fact that some waters are detectably bound to the surface of proteins and mediate interactions between residues [34]. Another challenge is the computational cost of full treatment of electrostatic free energy by solving the Poisson–Boltzmann or linearized Poisson–Boltzmann equations for large numbers of conformations. In spite of these difficulties several studies have shown that refinement of *de novo* structures with atomic-detail potentials can increase our ability to select and or generate near native structures [78]. These methods can correctly select near native conformations from these ensembles and improve near native structures, but still rely heavily on the initial low-resolution search to produce an ensemble containing good starting structures [63,71,102]. Some recent examples of high-resolution predictions are quite encouraging and an emerging consensus in the field is that higher resolution *de novo* structure prediction (structure predictions with atomic-detail representations of side-chains) will begin to work if sampling is dramatically increased.

Progress in high-resolution structure prediction will invariably be carried out in parallel with methods including, but not limited to, predicting protein–protein interactions, designing proteins and distilling structures from partially assigned experimental data sets. Indeed, many of the scoring and search strategies that high-resolution *de novo* structure refinement methods employ were initially developed in the context of homology modeling and protein design [61,90].

3 Applying Structure Prediction: *De Novo* Structure Prediction in a Systems Biology Context

Sequence databases are growing rapidly, with new genomes being deposited at a phenomenal pace. A large portion of each of these newly sequenced genomes can be expected to contain proteins that have no detectable homologs or only homologs of unknown function. It can be expected that even with the continued progress of large experimental structural biology efforts there will remain a large number of proteins for which *de novo* structure prediction and distant fold recognition methods are the only options.

3.1 Structure Prediction as a Road to Function

The relationship between protein structure and protein function is discussed in detail in Chapter 33, but will be reviewed briefly here in the context of *de novo* structure prediction. One paradigm for predicting the function of proteins of unknown function in the absence of homologs, sometimes referred to as the "sequence-to-structure-to-structure-to-function" paradigm, is based on the assumption that 3-D structure patterns are conserved across a much greater evolutionary distance than recognizable primary sequence patterns [33]. This assumption is based on the results of several structure–function surveys which show that structure similarities (fold matches between different proteins in the PDB) in the absence of sequence similarities imply some shared function in the majority of cases [48,67,70,84,101]. One protocol for predicting protein function based on this observation is to predict the structure of a query sequence of interest and then use the predicted structure to search for fold or structural similarities between the predicted protein structure and experimentally determined protein structures in the PDB or a nonredundant subset of the PDB [49,76,83,85]. There are several problems associated with deriving functional annotation from fold similarity, e.g. old similarities can occur through convergent evolution and thus have no functional implications. Also, aspects of function can change throughout evolution leaving only general function intact across a given fold superfamily [43, 56, 91]. Fold matches between the predicted structures and the PDB are thus treated as sources of putative general functional information, and are functionally interpreted primarily in combination with other methods such as global expression analysis and the predicted protein association network. To circumvent these ambiguities one can (i) use *de novo* structure prediction and/or fold recognition to generate a confidence ranked list of possible structures for proteins or protein domains of unknown function, (ii) search each of the ranked structure predictions against the PDB for fold similarities and possible 3-D motifs, (iii) calculate confidences for the fold predictions and 3-D motif matches, and, finally, (iv) evaluate possible functional roles in the context of the other systems biology data, such as expression analysis, protein interactions, metabolic networks and comparative genomics.

3.2 Initial Application of *De Novo* Structure Prediction

To date there have been few studies using *de novo* structure prediction as a method for genome annotation, due primarily to the computational expense of the calculations and the relative novelty of the methods. These studies have been carried out in combination with a variety of fold recognition and sequence-based methods for gene annotation, and have provided preliminary results that highlight several successes. It is based on these studies that we argue that *de novo* structure prediction is a viable option for exploring genes of unknown function.

The first emergence of *de novo* structure prediction methods for large-scale structure prediction was heavily limited by available computer resources. These studies were essentially pilot studies to evaluate the potential worth of genome-wide *de novo* structure predictions. In one early study workers were limited to generating predictions for 85 proteins in *Mycoplasma genitalium*, producing around 24 correct fold predictions [54]. Another study approached the computational limitation by folding representatives of Pfam protein families of unknown structure and function [14]. Using this method we were able to generate high confidence fold/structure predictions for around 60% of the 510 protein families for which Rosetta predictions were attempted, covering an additional roughly 12% of the sequences available at that time. Subsequent experimental determination for several of these protein families has shown our computed confidence values to be good estimates of our predictive performance, with success rates (rates of correct fold identification) on internal benchmarks and success rates from blind tests (CASP results and recently solved structures) nearly indistinguishable. Alas, the results of this study were not widely used by biologists due partially to the fact that at the time methods for integrating the resultant low-resolution structure predictions with other data types were not in place. The results of these early studies suggested, however, that whole-genome application of *de novo* structure prediction would result in usable annotations if presented to biologists properly, i.e. integrated with other available data types.

3.3 Application on Genome-wide Scale and Examples of Data Integration

Genome-wide measurements of mRNA transcripts, protein concentrations, protein–protein interactions and protein–DNA interactions generate rich sources of data on proteins, both those with known and those with unknown functions [5,7]. These systems-level measurements seldom suggest a unique function for a given protein of interest, but often suggest their association with or perhaps their direct participation in a previously known cellular process. Investigators using genome-wide experimental techniques are thus routinely generating data for proteins of hitherto unknown function that appear to play pivotal roles in their studies.

The first full-genome application of *de novo* structure prediction was to the genome of *Halobacterium* NRC-1 [12]. This archaeon is an extreme halophile that thrives in saturated brine environments such as the Dead Sea and solar salterns. It offers a versatile and easily assayed system with several well-coordinated physiologies that are necessary for survival in its harsh environment. The completely sequenced genome of *Halobacterium* NRC-1 (containing around 2600 genes) has provided insights into many of its physiological capabilities; however, nearly half of all genes encoded in the halobacterial genome had no known function prior to our re-annotation [29, 30, 79, 93]. A multi-institutional effort is currently underway to study the genome-wide response of *Halobacterium* NRC-1 to its environment, elevating the need for applying improved methods for annotating proteins of unknown function found in the *Halobacterium* NRC-1 genome. Rosetta *de novo* structure prediction was used to predict 3-D structures for 1185 proteins and protein domains (less than 150 residues in length) found in *Halobacterium* NRC-1. Predicted structures were searched against the PDB to identify fold matches [85] and were analyzed in the context of a predicted association network composed of several sources of functional associations, such as predicted protein interactions, predicted operons, phylogenetic profile similarity and domain fusion. This annotation pipeline was also applied to the recently sequenced genome of *Haloarcula marismortui* with similar rates of correct fold identification.

An application of *de novo* structure prediction to yeast has also been described. This study focused on the application and integration of several methods (ranging from experimental methods to *de novo* structure prediction) to 100 essential open reading frames (ORFs) in yeast [47]. For these 100 proteins the group applied affinity purification followed by mass spectrometry (to detect protein binding partners), two-hybrid analysis, florescence microscopy (to localize proteins) and *de novo* structure prediction (Ginzu to separate domains [26, 55] and Rosetta to build structures for domains of unknown function). Due to the cost of experiments and the computational cost of Rosetta *de novo* structure prediction, the group was initially able to prototype the method on just these 100 proteins. Function was assigned to 48 of the proteins (as defined by assignment to Gene Ontology categories). In total, 77 of the 100 proteins were annotated (had confident hits) by on of the methods employed. Given that the starting set represented a difficult set of ORFs of no known function this represents a significant milestone. Scaling this sort of approach up to whole genomes (including large eukaryotic genomes) is still a significant challenge. A grid computing solution (Section 3.4) is currently being employed to complete this study (fold the remaining ORFs in the genome) and, due to the wide use of yeast as a model organism, we can expect this complete resource to be a major step in crossing the social and technical barrier that has so far prevented the wide application of *de novo* structure prediction to biology. A similar approach has also been applied to the Y chromosome of *Homo sapiens* [40]. By integrating fold recognition with *de novo* structure prediction folds were assigned to around 42 of the 60 recognized domains examined (these 60 domains originated from the 27 proteins thought to be encoded on this chromosome at the time of the study). In both of these application, yeast and human, careful thought was put into reducing the set of proteins examined and scaling-up *de novo* structure prediction remains a critical bottleneck (the introduction of all-atom or high-resolution refinement of these predicted structures will only exacerbate this critical need for computing).

3.4 Scaling-up *De Novo* Structure Prediction: Rosetta on the World Community Grid

There are several strategies one can use to limit the number of protein domains for which computationally expensive *de novo* structure prediction needs to be carried out, allowing for the calculation of useful *de novo* structure predictions for only the most relevant subsets of larger genomes, as discussed above. In spite of these strategies, finding the required compute resources has been a constant challenge for the application of *de novo* structure prediction to functional annotation and has limited the application of the method. To circumvent this problem we are currently applying a grid, distributed computing, solution to folding over 100 000 domains with the full Rosetta *de novo* structure prediction protocol (www.worldcommunitygrid.org). These domains were chosen by applying Ginzu [26,55] to over 60 complete genomes as well as several other appropriate sequences in public sequence databases. The results will be integrated with data types that are appropriate/available for a given organism in collaboration with several other groups [12,47]. This work is ongoing in collaboration with David Baker, Lars Malmstroem (University of Washington) Rick Alther, Bill Boverman and Viktors Berstis (IBM), and United Devices (Austin, TX). Currently (11:10 AM, pacific coast time, 14 April 2005), there are over 1 million volunteers (people who have downloaded the client to run grid-Rosetta) comprising a virtual grid of over 3 million devices. Interested parties wishing to participate (donate idle CPU time on your desktop computer to this project) can download the grid-enabled Rosetta client at www.worldcommunitygrid.org. This amount of computational power will enable us to remove the barrier represented by the computational cost of *de novo* methods.

4 Future Directions

4.1 Structure Prediction and Systems Biology: Data Integration

Even with dramatically improved accuracy we still face challenges due to the ambiguities of the relationship between fold and function seen for many fold families (indeed, even close sequence homology is not always trivial to interpret as functional similarity, see also Chapter 30). Thus, the full potential of *de novo* structure prediction in a systems biology context can only be realized if structure predictions are integrated into larger analysis, and subsequently made accessible to biologists through better data integration, analysis and visualization tools. One clear example of this is provided by the bacterial transcription factors, for which even strong sequence similarity can imply several possible functions and system-wide information is required to determine a meaningful function (the target of a given transcription factor).

4.2 Need for Improved Accuracy and Extending the Reach of *De Novo* Methods

Although I have argued that data integration is as critical a bottleneck as any other and that there are current applications of *de novo* structure prediction, it is clear that improved accuracy is also essential for progress in the field and for the acceptance of *de novo* structure methods by the end users of whole-genome annotations. There is still a significant amount of error in predictions generated using current structure prediction and domain parsing methods. Extending the size limit of protein folding methods is a promising area of active research as is the development of higher-resolution refinement methods. *De novo* structure prediction requires large amounts of CPU time compared to sequence-based and fold recognition methods (although the use of distributed computing and Moore's law continue to make this less of a bottleneck). Integrating *de novo* predictions with orthogonal sources of general and putative functional information, both experimental and computational, will likely facilitate the annotation of significant portions of the protein sequences resulting from ongoing sequencing efforts, as well as proteins in currently sequenced genomes.

Acknowledgments

I would like to thank Viktors Berstis, Rick Alther and Bill Boverman of IBM for their critical work on the World Community Grid, and Phil Bradley, Lars Malmstroem, David Kim, Dylan Chivian and David Baker for myriad comments, opinions, structure prediction, preprints etc.

References

1 ALM, E. AND D. BAKER. 1999. Matching theory and experiment in protein folding. Curr. Opin. Struct. Biol. **9**: 189–96.

2 ALM, E. AND D. BAKER. 1999. Prediction of protein-folding mechanisms from free-energy landscapes derived from native structures. Proc. Natl Acad. Sci. USA **96**: 11305–10.

3 ALTSCHUL, S. F., T. L. MADDEN, A. A. SCHAFFER, J. ZHANG, Z. ZHANG, W. MILLER AND D. J. LIPMAN. 1997. Gapped BLAST and PSI-BLAST: a new generation of protein database search programs. Nucleic Acids Res. **25**: 3389–402.

4 BAKER, D. AND A. SALI. 2001. Protein structure prediction and structural genomics. Science **294**: 93–6.

5 BALIGA N. S., S. J. BJORK, R. BONNEAU, M. PAN, C. ILOANUSI , M. C. H. KOTTEMANN, L. HOOD AND J. DIRUGGIERO. 2004. Systems level insights into the stress response to UV

radiation in the halophilic archaeon *Halobacterium* NRC-1. Genome Res. **14**: 1025–35.

6 Baliga, N. S., R. Bonneau, M. T. Facciotti, et al. 2004. Genome sequence of *Haloarcula marismortui*: a halophilic archaeon from the Dead Sea. Genome Res. **14**: 2221–34.

7 Baliga, N. S., M. Pan, Y. A. Goo, et al. 2002. Coordinate regulation of energy transduction modules in *Halobacterium* sp. analyzed by a global systems approach. Proc. Natl Acad. Sci. USA **99**: 14913–8.

8 Bateman, A., E. Birney, R. Durbin, S. R. Eddy, K. L. Howe and E. L. Sonnhammer. 2000. The Pfam protein families database. Nucleic Acids Res. **28**: 263–6.

9 Bateman, A., L. Coin, R. Durbin, et al. 2004. The Pfam protein families database. Nucleic Acids Res. **32**: D138–41.

10 Blanco, F. J., G. Rivas and L. Serrano. 1994. A short linear peptide that folds into a native stable beta-hairpin in aqueous solution. Nat. Struct. Biol. **1**: 584–90.

11 Bonneau, R. and D. Baker. 2001. *Ab initio* protein structure prediction: progress and prospects. Annu. Rev. Biophys. Biomol. Struct. **30**: 173–89.

12 Bonneau, R., N. S. Baliga, E. W. Deutsch, P. Shannon and L. Hood. 2004. Comprehensive *de novo* structure prediction in a systems-biology context for the archaea *Halobacterium* sp. NRC-1. Genome Biol. **5**: R52.

13 Bonneau, R., C. E. Strauss and D. Baker. 2001. Improving the performance of Rosetta using multiple sequence alignment information and global measures of hydrophobic core formation. Proteins **43**: 1–11.

14 Bonneau, R., C. E. Strauss, C. A. Rohl, D. Chivian, P. Bradley, L. Malmstrom, T. Robertson and D. Baker. 2002. *De novo* prediction of three-dimensional structures for major protein families. J. Mol. Biol. **322**: 65–78.

15 Bonneau, R., J. Tsai, I. Ruczinski and D. Baker. 2001. Functional inferences from blind *ab initio* protein structure predictions. J. Struct. Biol. **134**: 186–90.

16 Bonneau, R., J. Tsai, I. Ruczinski, D. Chivian, C. Rohl, C. E. Strauss and D. Baker. 2001. Rosetta in CASP4: progress in *ab initio* protein structure prediction. Proteins **Suppl. 5**: 119–26.

16a Bowers, P. M., C. E. Strauss, and D. Baker. 2000. *De novo* protein using sparse NMR data. J. Biomol. NMR **18 (4)**: 311–8.

17 Bradley, P., K. M. S. Misura and D. Baker. 2006. Toward high-resolution *de novo* structure prediction for small proteins. Science **309**: 1868–71.

18 Bradley, P., D. Chivian, J. Meiler, et al. 2003. Rosetta predictions in CASP5: successes, failures, and prospects for complete automation. Proteins **53** (**Suppl. 6**): 457–68.

19 Brenner, S. E., C. Chothia and T. J. Hubbard. 1998. Assessing sequence comparison methods with reliable structurally identified distant evolutionary relationships. Proc. Natl Acad. Sci. USA **95**: 6073–8.

20 Bujnicki, J. M., A. Elofsson, D. Fischer and L. Rychlewski. 2001. LiveBench-1: continuous benchmarking of protein structure prediction servers. Protein Sci. **10**: 352–61.

21 Bystroff, C. and D. Baker. 1998. Prediction of local structure in proteins using a library of sequence–structure motifs. J. Mol. Biol. **281**: 565–77.

22 Bystroff, C. and Y. Shao. 2002. Fully automated *ab initio* protein structure prediction using I-SITES, HMMSTR and ROSETTA. Bioinformatics **18** (**Suppl. 1**): S54–61.

23 Bystroff, C., V. Thorsson and D. Baker. 2000. HMMSTR: a hidden Markov model for local sequence–structure correlations in proteins. J. Mol. Biol. **301**: 173–90.

24 Callihan, D. E. and T. M. Logan. 1999. Conformations of peptide fragments from the FK506 binding protein: comparison with the native and urea-unfolded states. J. Mol. Biol. **285**: 2161–75.

25 Chevalier, B. S., T. Kortemme, M. S. Chadsey, D. Baker, R. J. Monnat and B. L. Stoddard. 2002. Design, activity, and structure of a highly specific artificial endonuclease. Mol. Cells **10**: 895–905.

26 CHIVIAN, D., D. E. KIM, L. MALMSTROM, et al. 2003. Automated prediction of CASP-5 structures using the Robetta server. Proteins **53** (**Suppl. 6**): 524–33.

27 CHIVIAN, D., T. ROBERTSON, R. BONNEAU AND D. BAKER. 2003. *Ab initio* methods. Methods Biochem. Anal. **44**: 547–57.

28 DESHPANDE, N., K. J. ADDESS, W. F. BLUHM, et al. 2005. The RCSB Protein Data Bank: a redesigned query system and relational database based on the mmCIF schema. Nucleic Acids Res. **33**: D233–7.

29 DEVOS, D. AND A. VALENCIA. 2001. Intrinsic errors in genome annotation. Trends Genet. **17**: 429–31.

30 DEVOS, D. AND A. VALENCIA. 2000. Practical limits of function prediction. Proteins **41**: 98–107.

31 DUNBRACK, R. L., JR. AND F. E. COHEN. 1997. Bayesian statistical analysis of protein side-chain rotamer preferences. Protein Sci. **6**: 1661–81.

32 FETROW, J. S., A. GIAMMONA, A. KOLINSKI AND J. SKOLNICK. 2002. The protein folding problem: a biophysical enigma. Curr. Pharm. Biotechnol. **3**: 329–47.

33 FETROW, J. S. AND J. SKOLNICK. 1998. Method for prediction of protein function from sequence using the sequence-to-structure-to-function paradigm with application to glutaredoxins/thioredoxins and T1 ribonucleases. J. Mol. Biol. **281**: 949–68.

34 FINNEY, J. L. 1977. The organization and function of water in protein crystals. Philos. Trans. R. Soc. Lond. B **278**: 3–32.

35 FISCHER, D., C. BARRET, K. BRYSON, et al. 1999. CAFASP-1: critical assessment of fully automated structure prediction methods. Proteins **Suppl. 3**: 209–17.

36 FISCHER, D., L. RYCHLEWSKI, R. L. DUNBRACK, JR., A. R. ORTIZ AND A. ELOFSSON. 2003. CAFASP3: the third critical assessment of fully automated structure prediction methods. Proteins **53** (**Suppl. 6**): 503–16.

37 GALZITSKAYA, O. V. AND B. S. MELNIK. 2003. Prediction of protein domain boundaries from sequence alone. Protein Sci. **12**: 696–701.

38 GEORGE, R. A. AND J. HERINGA. 2002. SnapDRAGON: a method to delineate protein structural domains from sequence data. J. Mol. Biol. **316**: 839–51.

39 GINALSKI, K., N. V. GRISHIN, A. GODZIK AND L. RYCHLEWSKI. 2005. Practical lessons from protein structure prediction. Nucleic Acids Res. **33**: 1874–91.

40 GINALSKI, K., L. RYCHLEWSKI, D. BAKER AND N. V. GRISHIN. 2004. Protein structure prediction for the male-specific region of the human Y chromosome. Proc. Natl Acad. Sci. USA **101**: 2305–10.

41 GRAY, J. J., S. MOUGHON, C. WANG, O. SCHUELER-FURMAN, B. KUHLMAN, C. A. ROHL AND D. BAKER. 2003. Protein–protein docking with simultaneous optimization of rigid-body displacement and side-chain conformations. J. Mol. Biol. **331**: 281–99.

42 GRAY, J. J., S. E. MOUGHON, T. KORTEMME, O. SCHUELER-FURMAN, K. M. MISURA, A. V. MOROZOV AND D. BAKER. 2003. Protein–protein docking predictions for the CAPRI experiment. Proteins **52**: 118–22.

43 GRISHIN, N. V. 2001. Fold change in evolution of protein structures. J. Struct. Biol. **134**: 167–85.

44 GUNN, J. R. 1997. Sampling protein conformations using segment libraries and a genetic algorithm. J. Chem. Phys. **106**: 4270.

45 HAAS, B. J., J. R. WORTMAN, C. M. RONNING, et al. 2005. Complete reannotation of the *Arabidopsis* genome: methods, tools, protocols and the final release. BMC Biol. **3**: 7.

46 HAN, K. F., C. BYSTROFF AND D. BAKER. 1997. Three-dimensional structures and contexts associated with recurrent amino acid sequence patterns. Protein Sci. **6**: 1587–90.

47 HAZBUN, T. R., L. MALMSTROM, S. ANDERSON, et al. 2003. Assigning function to yeast proteins by integration of technologies. Mol. Cells **12**: 1353–65.

48 HOLM, L. AND C. SANDER. 1997. Dali/FSSP classification of three-dimensional protein folds. Nucleic Acids Res. **25**: 231–4.

49 HOLM, L. AND C. SANDER. 1993. Protein structure comparison by alignment of distance matrices. J. Mol. Biol. **233**: 123–38.

50 HUANG, E. S., R. SAMUDRALA AND B. H. PARK. 2000. Scoring functions for *ab initio* protein structure prediction. Methods Mol. Biol. **143**: 223–45.

51 HUNG, L. H. AND R. SAMUDRALA. 2003. PROTINFO: secondary and tertiary protein structure prediction. Nucleic Acids Res. **31**: 3296–9.

52 JONES, D. T. 1999. Protein secondary structure prediction based on position-specific scoring matrices. J. Mol. Biol. **292**: 195–202.

53 KARPLUS, K., R. KARCHIN, J. DRAPER, J. CASPER, Y. MANDEL-GUTFREUND, M. DIEKHANS AND R. HUGHEY. 2003. Combining local-structure, fold-recognition, and new fold methods for protein structure prediction. Proteins **53 (Suppl. 6)**: 491–6.

54 KIHARA, D., Y. ZHANG, H. LU, A. KOLINSKI AND J. SKOLNICK. 2002. *Ab initio* protein structure prediction on a genomic scale: application to the *Mycoplasma genitalium* genome. Proc. Natl Acad. Sci. USA **99**: 5993–8.

55 KIM, D. E., D. CHIVIAN AND D. BAKER. 2004. Protein structure prediction and analysis using the Robetta server. Nucleic Acids Res. **32**: W526–31.

56 KINCH, L. N. AND N. V. GRISHIN. 2002. Evolution of protein structures and functions. Curr. Opin. Struct. Biol. **12**: 400–8.

57 KORTEMME, T., L. A. JOACHIMIAK, A. N. BULLOCK, A. D. SCHULER, B. L. STODDARD AND D. BAKER. 2004. Computational redesign of protein–protein interaction specificity. Nat. Struct. Mol. Biol. **11**: 371–9.

58 KORTEMME, T., A. V. MOROZOV AND D. BAKER. 2003. An orientation-dependent hydrogen bonding potential improves prediction of specificity and structure for proteins and protein–protein complexes. J. Mol. Biol. **326**: 1239–59.

59 KUHLMAN, B. AND D. BAKER. 2000. Native protein sequences are close to optimal for their structures. Proc. Natl Acad. Sci. USA **97**: 10383–8.

60 KUHLMAN, B., G. DANTAS, G. C. IRETON, G. VARANI, B. L. STODDARD AND D. BAKER. 2003. Design of a novel globular protein fold with atomic-level accuracy. Science **302**: 1364–8.

61 KUHLMAN, B., J. W. O'NEILL, D. E. KIM, K. Y. ZHANG AND D. BAKER. 2002. Accurate computer-based design of a new backbone conformation in the second turn of protein L. J. Mol. Biol. **315**: 471–7.

62 LAZARIDIS, T. AND M. KARPLUS. 1999. Discrimination of the native from misfolded protein models with an energy function including implicit solvation. J. Mol. Biol. **288**: 477–87.

63 LEE, M. R., J. TSAI, D. BAKER AND P. A. KOLLMAN. 2001. Molecular dynamics in the endgame of protein structure prediction. J. Mol. Biol. **313**: 417–30.

64 LESK, A. M. 1997. CASP2: report on *ab initio* predictions. Proteins **Suppl. 1**: 151–66.

65 LESK, A. M., L. LO CONTE AND T. J. HUBBARD. 2001. Assessment of novel fold targets in CASP4: predictions of three-dimensional structures, secondary structures, and interresidue contacts. Proteins **Suppl. 5**: 98–118.

66 LIU, J. AND B. ROST. 2004. CHOP: parsing proteins into structural domains. Nucleic Acids Res. **32**: W569–71.

67 LO CONTE, L., B. AILEY, T. J. HUBBARD, S. E. BRENNER, A. G. MURZIN AND C. CHOTHIA. 2000. SCOP: a structural classification of proteins database. Nucleic Acids Res. **28**: 257–9.

68 LO CONTE, L., S. E. BRENNER, T. J. HUBBARD, C. CHOTHIA AND A. G. MURZIN. 2002. SCOP database in 2002: refinements accommodate structural genomics. Nucleic Acids Res. **30**: 264–7.

69 MARQUSEE, S., V. H. ROBBINS AND R. L. BALDWIN. 1989. Unusually stable helix formation in short alanine-based peptides. Proc. Natl Acad. Sci. USA **86**: 5286–90.

70 MARTIN, A. C., C. A. ORENGO, E. G. HUTCHINSON, et al. 1998. Protein folds and functions. Structure **6**: 875–84.

71 MISURA, K. M. AND D. BAKER. 2005. Progress and challenges in high-resolution

refinement of protein structure models. Proteins **59**: 15–29.

72 Moodie, S. L., J. B. Mitchell and J. M. Thornton. 1996. Protein recognition of adenylate: an example of a fuzzy recognition template. J. Mol. Biol. **263**: 486–500.

73 Moult, J., T. Hubbard, K. Fidelis and J. T. Pedersen. 1999. Critical assessment of methods of protein structure prediction (CASP): round III. Proteins **Suppl. 3**: 2–6.

74 Munoz, V. and L. Serrano. 1996. Local versus nonlocal interactions in protein folding and stability – an experimentalist's point of view. Fold. Des. **1**: R71–7.

75 Murzin, A. G. 1999. Structure classification-based assessment of CASP3 predictions for the fold recognition targets. Proteins **Suppl. 3**: 88–103.

76 Murzin, A. G., S. E. Brenner, T. Hubbard and C. Chothia. 1995. SCOP: a structural classification of proteins database for the investigation of sequences and structures. J. Mol. Biol. **247**: 536–40.

77 Neria, E., S. Fischer and M. Karplus. 1996. Simulation of activation free energies in molecular systems. J. Chem. Phys. **105**: 1902.

78 Neves-Petersen, M. T. and S. B. Petersen. 2003. Protein electrostatics: a review of the equations and methods used to model electrostatic equations in biomolecules – applications in biotechnology. Biotechnol. Annu. Rev. **9**: 315–95.

79 Ng, W. V., S. P. Kennedy, G. G. Mahairas, et al. 2000. Genome sequence of *Halobacterium* species NRC-1. Proc. Natl Acad. Sci. USA **97**: 12176–81.

80 Nielsen, H., S. Brunak and G. von Heijne. 1999. Machine learning approaches for the prediction of signal peptides and other protein sorting signals. Protein Eng. **12**: 3–9.

81 Okamoto, Y. 2004. Generalized-ensemble algorithms: enhanced sampling techniques for Monte Carlo and molecular dynamics simulations. J. Mol. Graph. Model. **22**: 425–39.

82 Orengo, C. A., J. E. Bray, T. Hubbard, L. LoConte and I. Sillitoe. 1999. Analysis and assessment of *ab initio* three-dimensional prediction, secondary structure, and contacts prediction. Proteins **Suppl. 3**: 149–70.

83 Orengo, C. A., F. M. Pearl and J. M. Thornton. 2003. The CATH domain structure database. Methods Biochem. Anal. **44**: 249–71.

84 Orengo, C. A., A. E. Todd and J. M. Thornton. 1999. From protein structure to function. Curr. Opin. Struct. Biol. **9**: 374–82.

85 Ortiz, A. R., C. E. Strauss and O. Olmea. 2002. MAMMOTH (matching molecular models obtained from theory): an automated method for model comparison. Protein Sci. **11**: 2606–21.

86 Ouzounis, C. A. and P. D. Karp. 2002. The past, present and future of genome-wide re-annotation. Genome Biol. **3**: COMMENT2001.

87 Pieper, U., N. Eswar, A. C. Stuart, V. A. Ilyin and A. Sali. 2002. MODBASE, a database of annotated comparative protein structure models. Nucleic Acids Res. **30**: 255–9.

88 Rohl, C. A. 2005. Protein structure estimation from minimal restraints using Rosetta. Methods Enzymol. **394**: 244–60.

89 Rohl, C. A., C. E. Strauss, D. Chivian and D. Baker. 2004. Modeling structurally variable regions in homologous proteins with Rosetta. Proteins **55**: 656–77.

90 Rohl, C. A., C. E. Strauss, K. M. Misura and D. Baker. 2004. Protein structure prediction using Rosetta. Methods Enzymol. **383**: 66–93.

91 Rost, B. 1997. Protein structures sustain evolutionary drift. Fold. Des. **2**: S19–24.

92 Rost, B. and V. A. Eyrich. 2001. EVA: large-scale analysis of secondary structure prediction. Proteins **Suppl. 5**: 192–9.

93 Rost, B. and A. Valencia. 1996. Pitfalls of protein sequence analysis. Curr. Opin. Biotechnol. **7**: 457–61.

94 Rychlewski, L. and D. Fischer. 2005. LiveBench-8: the large-scale, continuous assessment of automated protein structure prediction. Protein Sci. **14**: 240–5.

95 Shortle, D., K. T. Simons and D. Baker. 1998. Clustering of low-energy conformations near the native structures of small proteins. Proc. Natl Acad. Sci. USA **95**: 11158–62.

96 Simons, K. T., R. Bonneau, I. Ruczinski and D. Baker. 1999. *Ab initio* protein structure prediction of CASP III targets using ROSETTA. Proteins **Suppl. 3**: 171–6.

97 Simons, K. T., C. Kooperberg, E. Huang and D. Baker. 1997. Assembly of protein tertiary structures from fragments with similar local sequences using simulated annealing and Bayesian scoring functions. J. Mol. Biol. **268**: 209–25.

98 Simons, K. T., I. Ruczinski, C. Kooperberg, B. A. Fox, C. Bystroff and D. Baker. 1999. Improved recognition of native-like protein structures using a combination of sequence-dependent and sequence-independent features of proteins. Proteins **34**: 82–95.

99 Sonnhammer, E. L., G. von Heijne and A. Krogh. 1998. A hidden Markov model for predicting transmembrane helices in protein sequences. Proc. ISMB **6**: 175–82.

100 Tatusov, R. L., N. D. Fedorova, J. J. Jackson, et al. 2003. The COG database: an updated version includes eukaryotes. BMC Bioinformatics **4**: 41.

101 Todd, A. E., C. A. Orengo and J. M. Thornton. 2001. Evolution of function in protein superfamilies, from a structural perspective. J. Mol. Biol. **307**: 1113–43.

102 Tsai, J., R. Bonneau, A. V. Morozov, B. Kuhlman, C. A. Rohl and D. Baker. 2003. An improved protein decoy set for testing energy functions for protein structure prediction. Proteins **53**: 76–87.

103 Wallace, A. C., R. A. Laskowski and J. M. Thornton. 1996. Derivation of 3D coordinate templates for searching structural databases: application to Ser–His–Asp catalytic triads in the serine proteinases and lipases. Protein Sci. **5**: 1001–13.

104 Ward, J. J., L. J. McGuffin, K. Bryson, B. F. Buxton and D. T. Jones. 2004. The DISOPRED server for the prediction of protein disorder. Bioinformatics **20**: 2138–9.

105 Wedemeyer, W. J. and D. Baker. 2003. Efficient minimization of angle-dependent potentials for polypeptides in internal coordinates. Proteins **53**: 262–72.

106 Yi, Q., C. Bystroff, P. Rajagopal, R. E. Klevit and D. Baker. 1998. Prediction and structural characterization of an independently folding substructure in the src SH3 domain. J. Mol. Biol. **283**: 293–300.

107 Zhang, B., L. Rychlewski, K. Pawlowski, J. S. Fetrow, J. Skolnick and A. Godzik. 1999. From fold predictions to function predictions: automation of functional site conservation analysis for functional genome predictions. Protein Sci. **8**: 1104–15.

108 Zhang, Y. and J. Skolnick. 2004. Tertiary structure predictions on a comprehensive benchmark of medium to large size proteins. Biophys. J. **87**: 2647–55.

13
Structural Genomics

Philip E. Bourne and Adam Godzik

1 Overview

1.1 What is Structural Genomics?

Inspired by the success of the genome-sequencing projects, particularly the Human Genome Project [41], research-funding bodies in the US, Japan and Europe decided to embark on an equally ambitious project of large-scale macromolecular structure determination. Looking at biology in terms of increasing complexity and scale, this made sense – from the sequence of genomes comes structure from which molecular function can be derived. Individual functions define processes that occur in different parts of the cell, different cell types make up an organism and so on. Thus, the next logical step in understanding living systems was large-scale macromolecular structure determination. These efforts, weakly correlated and distributed over several institutions on several continents, became collectively known as structural genomics. The hope was that structural genomics would continue an explosive growth in raw data, knowledge and technology [32,35]. This chapter describes structural genomics on its fifth anniversary. Where representative examples of the work being undertaken are needed, they are taken from the Joint Center for Structural Genomics (JCSG), one of the US structural genomics centers that is close to the authors both in space (it is located in San Diego) and in spirit (one of us, A. G., leads the bioinformatics core at JCSG).

1.2 What are the Motivators?

Whereas the goal of the human genome project was straightforward, i.e. determine the 3 billion nucleotides that comprise the specific genome (human), the goal of structural genomics is less so. The most often stated goal was to provide "structural coverage" of protein space, by solving enough structures that all known proteins could be accurately modeled [7,42]. Other goals, such as targeting disease-relevant genes, were also listed [8]. Last, but

Bioinformatics – From Genomes to Therapies Vol. 1. Edited by Thomas Lengauer

ISBN: 978-3-527-31278-8

not least, structural genomics aimed to develop new structure determination technologies that would lower the costs and time needed to solve protein structures. Without a formal definition of its goals, structural genomics was adopted as a broad research goal by a loose coalition of researchers from around the world. A list of the current projects is given in Table 1 and up-to-date information is available from the Research Collaboratory for Structural Bioinformatics (RCSB) Protein Data Bank (PDB) [4] which tracks all projects (http://sg.pdb.org/target_centers.html) [10, 22].

In the US, structural genomics efforts resulted in the launch of the National Institutes of Health (NIH) Protein Structure Initiative (PSI), which in time developed its own goals and milestones [30] that partly overlapped the original overall goals of structural genomics. The same could be said of developments in other parts of the world. Recently, the PSI initiatives in the US have received their second round of 5-year funding and goals and milestones are being further refined. Table 1 indicates major US centers from round 1 as PSI-1 and the subset of those with major funding in round 2 as PSI-2. Regardless of the stated objectives, structural genomics already accomplished (or perhaps coincided with) a major paradigm shift in structural biology – moving from a strictly functionally driven endeavor to a genomically driven endeavor. As we discuss subsequently, this both requires and is driven by contributions from a variety of communities. From the perspective of bioinformatics research, this includes problems in defining the universe of protein structures, recognition of natural units of protein evolution (domains), understanding the complex relationship between protein function(s) and sequence and structure, development of protein structure prediction in general, and homology modeling in particular, and many others.

1.2.1 Fold Coverage as a Motivator

The often repeated goal of structural genomics is "coverage of protein structural space". However, there are at least two levels on which this goal can be achieved. On one level, solving at least one example of all possible folds would provide some information about how large and complex is protein fold space. On another level, many structures from each fold would have to be solved to provide structural templates for possible comparative modeling for every existing protein. The first level, coarse-grained coverage of fold space, seems to be tractable in terms of number of structures that need to be solved, with a total number of folds estimated to be in the 5000–10 000 range. However, choosing structures that would have to be solved would be a formidable task, as it is difficult to predict from sequence which proteins would have a new fold – once the sequence identity drops below 25% the relationship to an existing structure may not be detected, yet folds are often the same below 10% sequence identity [31] (see Chapters 10 and 12). This

Table 1 Structural genomics projects worldwide

Project	Major objectives (as described on their website)
Protein structure initiative centers US	
Berkeley Structural Genomics Center (BSGC) PSI-1	two minimal genomes, *Mycoplasma genitalium* and *Mycoplasma pneumoniae*
Center for Eukaryotic Structural Genomics (CESG) PSI-2	variety of eukaryotic targets
Joint Center for Structural Genomics (JCSG) PSI-2	targets from all superkingdoms with emphasis on *T. maritima*, mouse and yeast
Midwest Center for Structural Genomics (MCSG) PSI-2	various
NorthEast Structural Genomics Consortium (NESG) PSI-2	coverage of fold space
New York Structural Genomics Research Consortium (NYSGRC) PSI-2	technology development
Southeast Collaboratory for Structural Genomics (SECSG) PSI-1	*Caenorhabditis elegans*, *Homo sapiens* and homologs from *Pyrococcus furiosus*
Structural Genomics of Pathogenic Protozoa Consortium (SGPP) PSI-1	various diseases
Mycobacterium tuberculosis Structural Genomics Consortium (TB) PSI-1	study of tuberculosis
Structural genomics projects in Europe and environs	
Bacterial Targets at IGS-CNRS, France (BIGS)	rickettsia, ORFan targets from *Escherichia. coli*, antibacterial gene targets
deCode – decode Genetics, Iceland	various diseases
Israel Structural Proteomics Center (ISPC)	various
Marseilles Structural Genomics Program, France (MSGP)	unknown
NWSGC – North West Structural Genomics Centre, UK	tuberculosis
Oxford Protein Production Facility, England (OPPF)	technology development
Protein Structure Factory, Germany (PSF)	various
Structural Proteomics in Europe, England (SPINE)	structures related to human health and disease
Mycobacterium Tuberculosis Structural Proteomics Project, Germany (XMTB)	tuberculosis
Paris-Sud Yeast Structural Genomics, France (YSG)	relevant proteins with homologs in *Schizosaccharomyces pombe*
Structural genomics projects in North America	
Montreal-Kingston Bacterial Structural Genomics Initiative, Canada (BSGI)	various
OCSP – Ontario Centre for Structural Proteomics, Canada	various
Project CyberCell, Canada	various

Table 1 continued

Project	Major objectives (as described on their website)
SGC – Structural Genomics Consortium, Canada, UK	various proteins of medical relevance
Structure 2 Function Project, US (S2F)	*Haemophilus influenzae*
Structural genomics projects in Asia	
KSPRO – Korean Structural Proteomics Research Organization, Korea	*Helicobacter pylori* and human cancer genes
RIKEN Structural Genomics Initiative, Japan (RSGI)	various
SGCGES – Structural Genomics Consortium for Research on Gene Expression System, Japan	proteins associated with protein synthesis

difficulty is illustrated in Figure 1, where the growth in the number of new folds is shown not to have increased significantly since the advent of the structural genomics projects in 2000. The picture is complicated by how one defines a new fold. Figure 1 is based upon the Structure Classification of Proteins (SCOP) [1], but other definitions exist as will be described subsequently. Further, proteins not homologous to already crystallized proteins are significantly more difficult to handle and have lower success rates at almost every stage of the structure determination process. On the second level, covering protein space at a fine-grained level requires a number of potential targets counted in hundreds of thousands, thus making the goals of structural genomics essentially unattainable. It is clear that the practical strategy must steer clear from both these extremes and, as we will discuss below, this is indeed what most structural genomics centers have been doing in practice.

Despite improvements in structure determination technology, it is clear that we have not achieved a breakthrough that could be compared to, for example, what shotgun sequencing did for accelerating genome-sequencing projects. Automation and streamlining have been applied at every step, but overall improvements have been incremental rather than dramatic, and as of today even the best centers do not produce more than 10–15 structures a month at a cost of US$50 000–$60 000 per structure. While this is impressive by the standards of structural biology from even a few years ago, at this rate and cost it is clear that fine-grained coverage of protein structural space is impossible to attain. Therefore, the ultimate goal of structural genomics could only be achieved by combining experimental and theoretical approaches, and improvements in comparative modeling are necessary to improve the convergence radius of successful model building.

Another means of defining a goal for structural genomics is to consider that goal from a biological perspective.

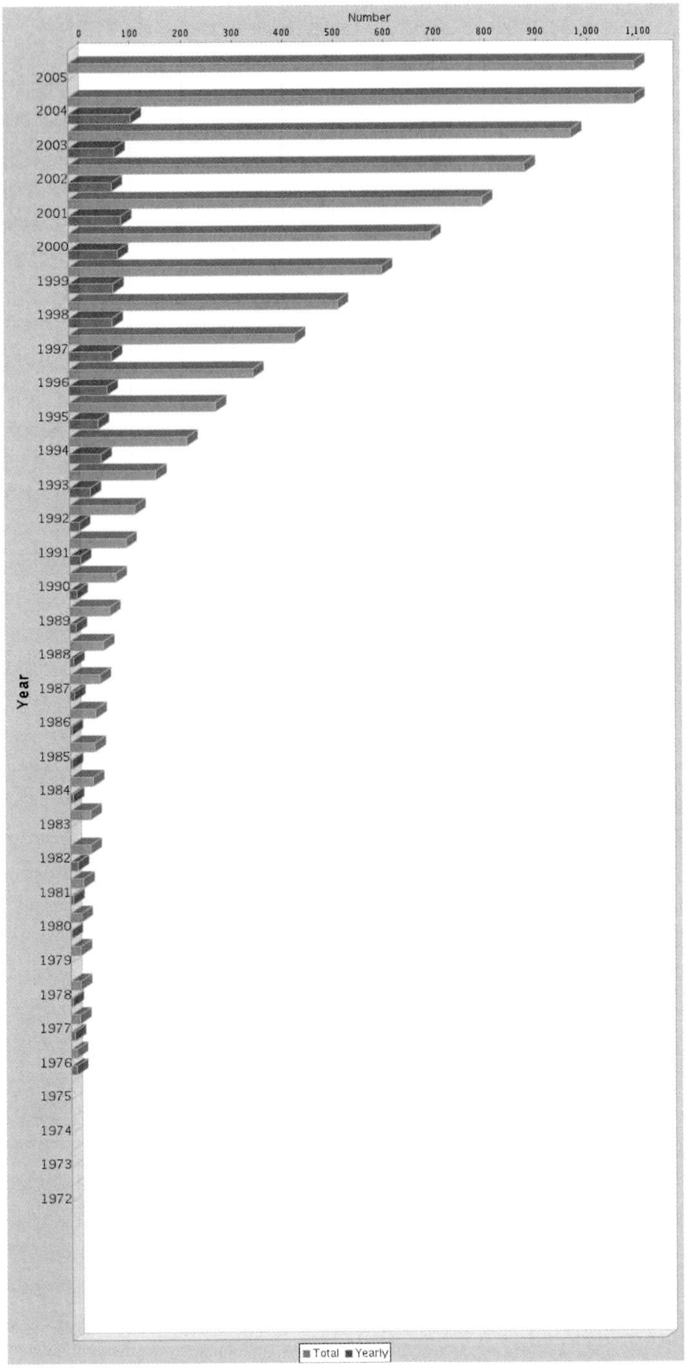

Figure 1 Growth in the number of folds per year as defined by SCOP.

1.2.2 Structural Coverage of an Organism as a Motivator

One stated and ambitious goal of structural genomics is to provide an understanding of an entire organism at the molecular level. Some structural genomics centers focused on single organisms, e.g. the JCSG cloned and attempted the expression of 1777 of the predicted 1877 open reading frames from *Thermotoga maritima* leaving aside some of the putative genes with obvious problems in predicted boundaries, etc. From a predicted 1377 soluble proteins, 705 were expressed and 581 made it to crystallization trials. To date, JCSG have solved 155 *T. maritima* structures, which, when combined with structures of *T. maritima* proteins solved at other centers, gives direct structural coverage of 25% of the expressed soluble proteins and around 12% of this organism's proteome. After taking into account structures that can be modeled through homology and fold recognition, this percentage rises to over 70% (over 90% of predicted crystallizable proteins). Thus, we are only a few dozens structures away from having complete structural coverage of an entire organism. However, as we shall see subsequently, based on a detailed discussion of the coverage of the human genome, what constitutes genome coverage by models is open to interpretation. Beyond that, Brenner and colleagues have pointed out [9] that even the determination of several complete archaeal or bacterial proteomes would still leave many protein families structurally uncovered.

1.2.3 Structure Coverage of Central Metabolism Pathways as a Motivator

Here we consider a specific example. Structures solved at JCSG and other PSI centers allow us to take a structure-based view of the metabolic pathways in *T. maritima*. In collaboration with the SEED project (http://theseed.uchicago.edu/FIG/index.cgi), an integrated *T. maritima* annotation project was initiated combining structural, functional and genomic annotations (Figure 2). Results of this effort, which will soon be available on the *T. maritima* annotation website, will provide a unique genome annotation resource. At this point, all *T. maritima* metabolic pathways can be covered by experimental or modeled structures, providing a first of its kind structural view of an organism's metabolome. Structure determination has resolved many outstanding issues in what seemed to be incomplete or redundant pathways and identified novel aspects of *T. maritima* metabolism. For instance, *T. maritima* is one of only four known organisms that do not depend on biotin decarboxylase for fatty acid metabolism. About 40% of all JCSG-solved *T. maritima* structures had their functional annotations changed or significantly updated after their structures were determined.

1.2.4 Disease as a Motivator

According to a recent study [45] the PDB currently covers approximately 70% of the human disease categories described by the Online Mendelian Inheritance in Man (OMIM) resource [15], but that coverage is not even. For example, diseases of the central nervous system have a disproportionately large number of solved structures and structural genomics targets relative to the number of proteins associated with this class of disease in the human genome. Blood- and lymph-based diseases have a disproportionately large number of solved structures in the PDB, yet an appropriate underrepresentation of structural genomics targets being attempted. Diseases of the ear, nose and throat, which are currently structurally underrepresented in the PDB, have few targets being attempted and yet there are a significant number of proteins identified as being responsible for this class of disease in the human genome [45]. Some structural genomics projects (Table 1) are targeting specific diseases, e.g. *Mycobacterium tuberculosis*, Chagas' disease and malaria, and a more balanced coverage of proteins associated with human diseases is expected in the next 5 years.

1.3 How Does Structural Genomics Relate to Conventional Structural Biology?

What should be apparent from the above discussion of motivators is that structural genomics changes the conventional paradigm of "I know something about the function of this protein from biochemical evidence, let me determine the single structure of this protein to better elucidate the mechanism" to an almost reverse approach at a different scale of biology, "I see a protein that seems to be important – it is conserved, essential, sits on a virulence island, but its function is unknown, lets solve a structure to start the functional characterization process". One outcome of this paradigm shift is that, for the first time, we are seeing a number of structures which have yet to be functionally characterized offering new challenges for computational biologists to determine function from sequence and structure, not necessarily a trivial undertaking, but success is possible.

Taking a specific example, for 42 of the *T. maritima* structures solved by the JCSG, structural analysis provided strong indications for the possible functions of proteins which were previously listed as "hypothetical proteins". Further, by incorporating structural information into the annotation process, functional annotations of 90 out of 122 structures have been modified, usually by making the function annotation more specific and occasionally by correcting it. Importantly, for some proteins even knowledge of their three-dimensional (3-D) structure did not help to elucidate their function. For instance, TM0875 (a specific JCSG target), with a unique fold and no known homologs, remains uncharacterized a few years after structure determination.

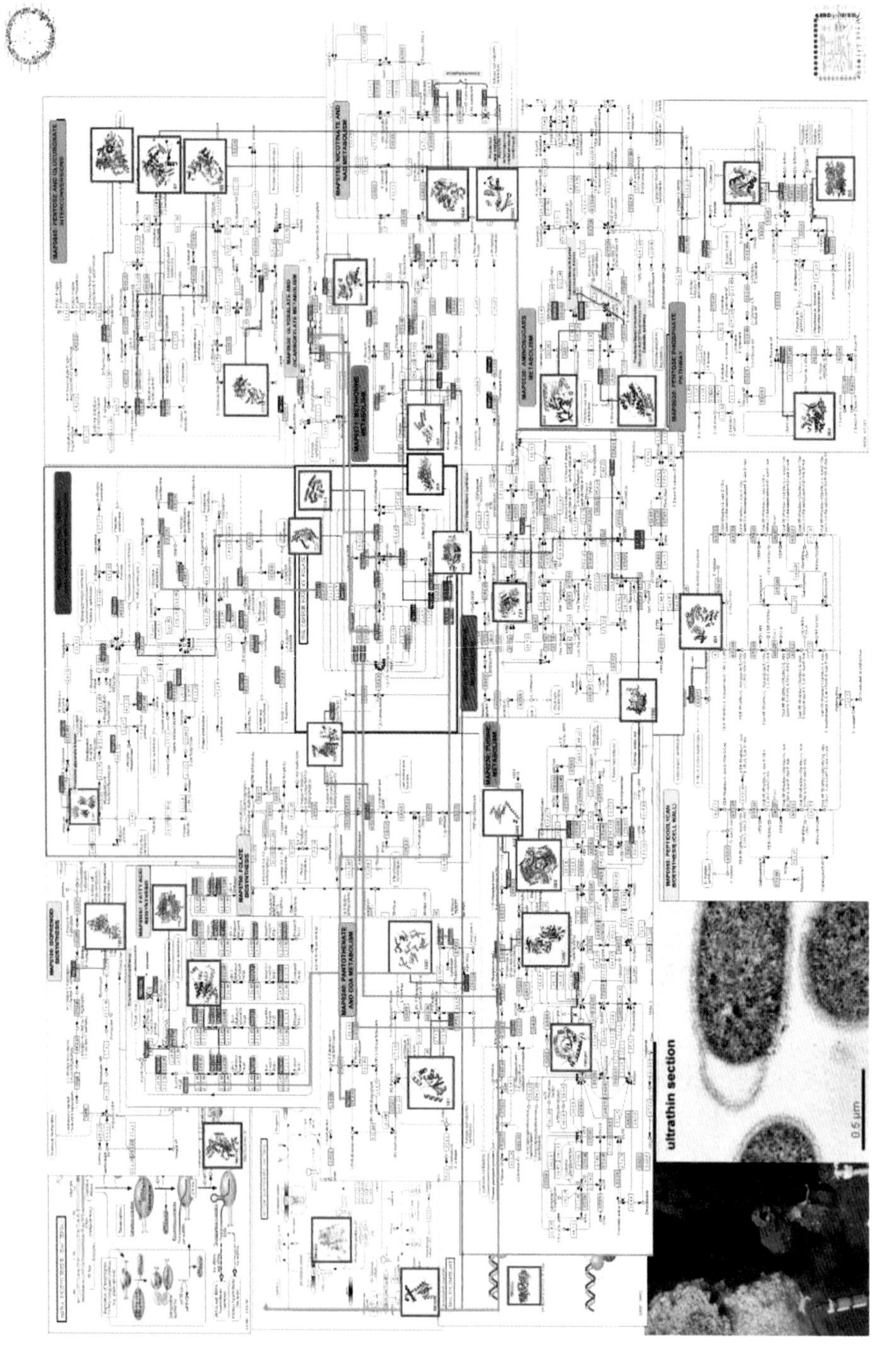
ultrathin section
0.5 µm

Currently, over 900 proteins in PDB are classified as "structural genomics unknown function".

Interestingly, while the value of structural genomics was questioned by some structural biologists at the outset, the consensus opinion now seems to be that advances made in structure determination through structural genomics have fed back to impact conventional structural biology laboratories through, for example, improved software, streamlined procedures at the synchrotron beamlines, and improved techniques for expression, purification and crystallization. We now consider some of these advances.

2 Methodology

Structural genomics employ a variety of methods – X-ray crystallography, nuclear magnetic resonance (NMR), electron microscopy, neutron diffraction, mass spectrometry, etc. It is beyond the scope of this chapter to describe all of these. Rather, we consider X-ray crystallography as the most prevalent of the methods (85% of structures in the PDB to date) to illustrate the impact that structural genomics is having on the methods employed to solve structures by single crystal X-ray diffraction. Figure 3 is a schematic of the basic process. We consider each step and demonstrate what has been achieved by one project, the JCSG, as an example of progress that is being made.

2.1 Target Selection

The motivation for selecting targets was introduced above. The goal of target selection is to ascertain that the sequence of the protein meets the criteria defined for the anticipated structural outcome. That could be biological, i.e. it has a particular function usually determined by identification of homology to another protein known to have this function (see Chapters 30–35), or methodological, i.e. a specific globular domain can be identified which is likely to be amenable to crystallization. Neither recognizing distant homologs nor domains from sequence is a solved problem, although these are active areas of endeavor. See, for example, Ref. [20] for the latest on domain recognition from the Sixth Critical Assessment of Structure Prediction (CASP) [39].

Figure 2 Fragment of the metabolic map of *T. maritima* with experimental structures identified by ribbon diagrams and models identified by a green highlight of the enzyme name. Most of the pathways have complete structural coverage and the remaining proteins are being targeted in the fifth year of JCSG.

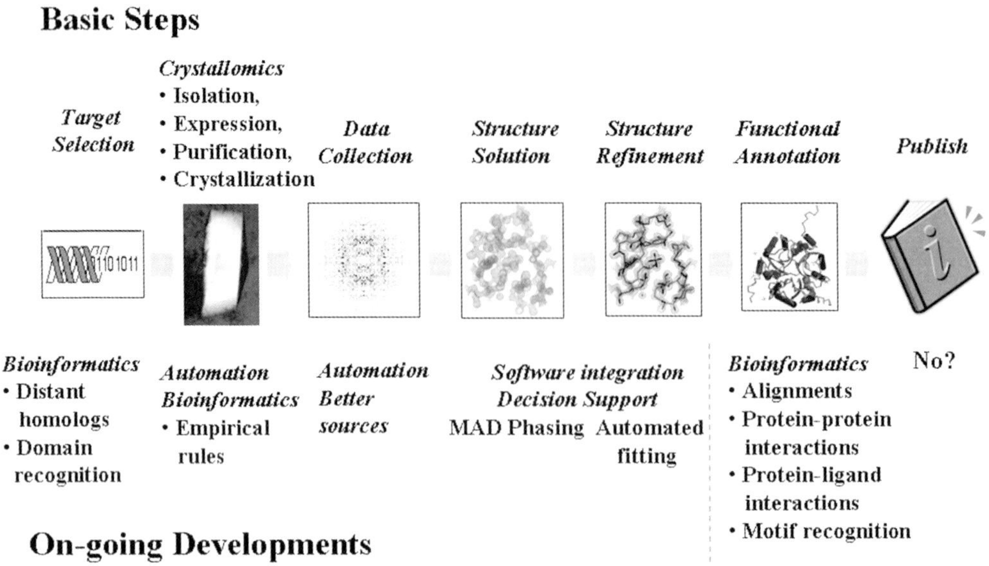

Figure 3 The basic steps in a single-crystal X-ray diffraction experiment (top) and associated on-going developments (bottom) being catalyzed by structural genomics efforts worldwide.

2.2 Crystallomics

Crystallomics is a term to collectively define the steps of protein isolation, expression, purification and crystallization. The initial phase of structural genomics has yielded great progress here. Large-scale, fully automated facilities for protein production have been developed by all structural genomics centers and, increasingly, many commercial solutions are available. Structural genomics centers cloned over 56 000 targets, expressed over 30 000 targets, purified over 10 000 targets and crystallized about 4000 targets in the period since program inception in 2000 to October 2005 (Figure 4). Additional constantly updated statistics are available at http://targetdb.pdb.org/. Consider the approach of JCSG. Many options for creating expression systems were evaluated to maximize flexibility and minimize cost. Ultimately the JCSG team chose to automate a conventional cloning approach. They developed a robotic platform and were able to provide up to 384 validated clones per week. To date, over 2500 clones have been generated and expressed with this system from over 30 000 attempts. The system is efficient, needing only a single person to operate. *Escherichia coli* has been used as the expressions system. To purify the expressed protein, a two-process system has been adopted, both processes being controlled by robotics. Together, they can produce 48–96

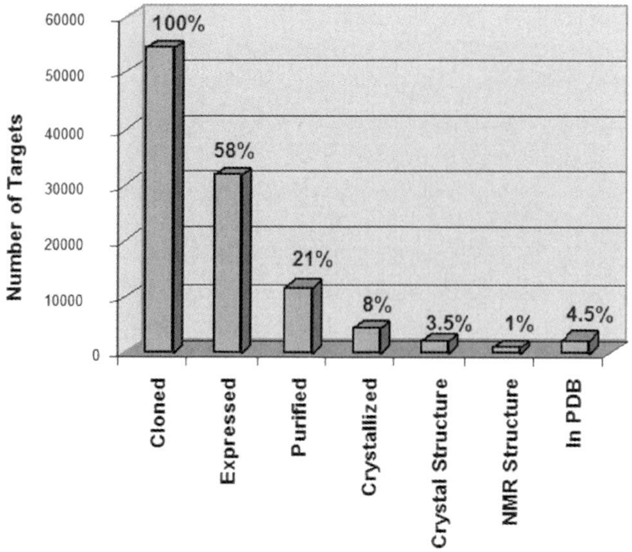

Figure 4 Progress within the structural genomics initiative worldwide as of 17 November 2005.

proteins a week on a 10–50 mg scale. Further details on this process can be found at http://www.jcsg.org/scripts/prod/technologies1.html.

Crystallization strategies vary but are high throughput involving multi-well robotic systems. The JCSG system includes an automated system for shooting digital images of each well and automatic recognition of crystalline samples. To date over 3 million images have been taken and analyzed. Once a sample is identified a prescreening is undertaken to determine if the crystal is of diffraction quality before being sent to the synchrotron for data collection. In excess of 500 crystals are prescreened by JCSG on a monthly basis.

2.3 Data Collection

Crystal samples are transported to the synchrotron in specially designed compact, cylindrical, aluminum crystal cassettes, which holds 96 crystals. Upon reaching the synchrotron samples are automatically mounted, entered into the high-energy X-ray beam and aligned automatically, a process taking approximately 30 s. Using a video system it is intended that data collection will eventually be done remotely without the need for the researcher to travel to the beamline itself. The online report of each crystalline sample is automatically augmented with information from the data collection process.

2.4 Structure Solution

Data collection provides X-ray diffraction patterns from the ordered crystal lattice which appear as a set of diffraction spots. The positions of the spots is defined by the size and type of crystal lattice; however, the intensity of the spots is a function of their amplitude based on the types of atoms (known) and their phases based on the positions of the atoms (unknown at the outset). This is referred to as the phase problem in crystallography. The majority of structures determined by the structure genomics centers worldwide solve the phase problem by establishing a starting set of phases using multi-wavelength anomalous dispersion [17] which requires collecting data at slightly different wavelengths, a process that is possible using synchrotron beamlines which can be tuned for this purpose. The discreet scattering of the X-ray beam by the electron cloud from specific atom types (anomalous scatterers), usually selenium introduced into the structure in the form of selenomethionine, when collected at different wavelengths, provides a starting set of phases, based on the atomic positions of the anomalous scatterers which can now be resolved. A disadvantage of this approach is that data must be collected at each wavelength, lengthening the data collection time. An alternative approach to establishing a starting set of phases is to have a model which approximates the final atomic positions. From this model a starting set of phases can be derived, hence solving the phase problem assuming, first, that the model is accurate enough and, second, that it can be positioned correctly within the crystal lattice. This technique, known as molecular replacement (MR), works well for similar structures, e.g. taking an existing solved native protein structure and using it to phase a mutant structure containing post-translational modifications. An exciting prospect is to routinely use a theoretically derived model structure to determine a staring set of phases. Likelihood for the routine success of this approach relies on having a wide variety of experimentally determined structures from which to model. While this represents somewhat of a chicken-and-egg situation, the number of available structures is increasing rapidly.

In general, MR solutions are seldom attempted (and are even less often successful) against templates with less than 35% sequence identity. Using a fold recognition approach [13] as opposed to an approach based solely on homology, to date, the JCSG MR pipeline was successfully applied to over 20 cases with less than 35% sequence identity, 10 cases with less than 30% and several cases where sequence identity was close to 15%. The analysis shows that fold recognition models, derived from work done in the discipline of protein structure prediction [44], have a significantly higher success rate than homology modeling, especially when the unknown structure and the search model share less than 35% sequence identity. Using the software programs

MOLREP [40] and EPMR [21], three out of 26 MR targets under 35% sequence identity could only be solved with models derived from fold recognition methods, and six showed significantly better statistics and behavior in subsequent refinement [34] than those defined by homology.

2.5 Structure Refinement

Structure refinement implies maximizing the agreement between the intensities observed on the diffraction pattern spots and those calculated from the atomic model. This can be roughly divided into two tasks: (i) getting the main chain and side chains into an optimal position and (ii) performing a final minimization. Ideally both steps must be completely automated to maximize throughput. Improvements to algorithms and usability of software are key factors in this process. A final check must be made to be sure the stereochemical quality of the model is reasonable and this is done primary by checking the relatively low-resolution structure of the complete macromolecule against what is known about macromolecular structure *en masse* (e.g. allowable dihedral angles) and what is known from high-resolution structures of small molecules, such as single amino acids and nucleotides. Despite significant progress in recent years, large parts of the refinement process are still done by hand and high-quality refinement is one of the most time-consuming tasks in high-throughput structure determination. Despite these difficulties, the quality of structures determined by structural genomics centers matches and often exceeds the quality of structures coming from standard structural biology laboratories [38].

2.6 PDB Deposition

Deposition within the PDB (http://www.pdb.org) [4] is a requirement for all structural genomics centers, thereby facilitating the maintenance of a single worldwide archive of macromolecular structures. An ideal goal is to fully automate the deposition process whereby a full structure submission contains not just the final atomic coordinates and experimental data (NMR restraints or X-ray structure factors), but all experimental information including experimental protocols for all the above steps. This would then be validated by the PDB prior to submission. This process is not yet in place, but significant progress has been made. A key element of the process is an ontology which defines in a formal way the items to be collected and their interrelationships. While details of the structure itself have been defined in this way using an ontology referred to as the macromolecular crystallographic information file (mmCIF) [14], additional ontological terms for the various detailed experimental protocols have yet to be fully defined. The progress that has been made thus far is reflected in the Protein Expression Purification and Crystallization Database (PEPCdb; http://pepcdb.pdb.org) which collects experimental protocols according to the beginnings of a full ontological description.

2.7 Functional Annotation

One of the underling goals of structural genomics is to study the relationship between gene sequence, protein structure and protein function, thereby expanding the knowledge of the underlying biology. However, at present most structural genomics centers, by design, stop after structure determination. As a result, a large number of proteins solved by structural genomics groups are listed in the PDB as "hypothetical proteins". This growing list provides raw data from which bioinformatics groups worldwide can apply a variety of methods to assigning putative function(s) to these uncharacterized structures (see Chapters 30–35, especially Chapter 33) [46]. Here, we outline a few of the approaches that have been adopted. In the results section some success stories are introduced. Popular approaches are as follows.

2.7.1 Biological Multimeric State

The structure solved in many cases does not comprise the biologically active molecule. Rather, it represents a unique component. That component may be one domain in a multi-domain protein, a situation found in the PDB in general. For example, multiple SH2 and SH3 domains have been solved and are known to be part of a larger macromolecular complex. Alternatively, the application of crystallographic symmetry can be used to construct the biologically active molecule. Identifying what components comprise the biologically active molecule often requires expert input, although efforts have been made to automate this determination. The Protein Quaternary Server (PQS) uses the notion of proximity of components to define the multimeric state [18].

2.7.2 Active-site Determination

This is an active area of research (see [37] for a review and Chapter 33). Active sites include a small number of residues involved in catalysis, substrate and cofactor binding sites, sites of protein–protein interaction, phosphorylation sites, glycosylation sites, fatty acylation sites, prosthetic group binding sites, hinge regions, domain–domain contacts, sites of membrane association and more. The complexity and importance of the problem is well illustrated by subtilisin and chrymotrypsin. Both are endopeptidases, yet share no sequence identity and their folds are unrelated. However, they share an identical 3-D motif comprising a Ser–His–Asp catalytic triad. The challenge becomes one of identifying an identical motif in two different structures undoubtedly resulting from convergent evolution. Methods are varied, but all comprise basic steps of protein structure representation, application of a search algorithm and assessment of the reliability of the result. Early work used a graph theoretic approach [2], progressing to the use of fuzzy functional forms [11], template modeling [43] and, most recently, an elegantly simple approach based on the proximity of the active site of an enzyme to the centroid of the molecule [3] has emerged. Each method builds upon empirical observations made as more structures are determined and functions classified by biochemical analysis. Thus, we have a rich repertoire from which functional prediction can only improve.

The need for improved methods for function prediction from structure leads to increased research into automated function prediction. The first two meetings of the Automated Functional Prediction (AFP) special interest group were held in 2005 and 2006 (http://biofunctionprediction.org) and the proceedings will be published in 2006. The first metaserver, collecting and analyzing prediction for several servers is now in beta testing (http://jafa.burnham.org).

2.8 Publishing

The original macromolecular structures represented a scientist's life's work. We are now faced with a situation where the rate-determining steps may well be writing the publication. Hence, a number of structures, while deposited in the PDB, remain unpublished. At the time of writing only a small percentage of structures determined by structural genomics are described by peer-reviewed publications, placing additional emphasis on the individual centers websites and the PDB to disseminate as much information about these structures as possible.

JCSG structures are shared with the scientific community not only through deposition in the PDB, but also through publication of a "structure note". Structure notes are short papers describing the annotation, biology, structure and functional implications of each protein. The process of collecting all relevant data from all stages of the JCSG pipeline has been streamlined through the central JCSG database, which includes information on the sequence, annotation, cloning, purification, crystallization, data collection, structure solution, tracing, refinement and structural evaluation. The structure note automatically captures any functional information in the JCSG annotation system (functional annotation is described above). The paper introduction, for example, includes annotation information, with a brief biological background taken and curated from the Pfam [12], Interpro [28], SwissProt [6], BRENDA [33] and SEED databases (http://theseed.uchicago.edu/FIG/index.cgi). Methodological and experimental data, as well as all crystallographic statistics, are automatically harvested from the JCSG database, and assembled into purification, crystallization, structure solution and refinement paragraphs. The structure description and the preparation of figures are done manually using PyMOL (http://pymol.sourceforge.net/). Structures are analyzed, compared and evaluated for biological significance using a plethora of structure analysis tools including structural homology searches (DALI [19], CE [36], FATCAT [25]) and extensive literature searches. In this way the preparation of a structure note is a semiautomated process.

3 Results – Number and Characteristics of Structures Determined

As of 17 November 2005 there were 90 613 targets under consideration by structural genomics projects. Of these 57 019 had been cloned and could be considered under investigation. Figure 4 from http://targetdb.pdb.org/ [10] shows the success rate for the steps described in Section 2. A total of 2540 structures have appeared in the PDB, which is 4.5% of the targets under investigation; 1% has come from NMR structure determination and 3.5% from X-ray crystallography. At that time 7.5% of all structures in the PDB could be considered from structural genomics, with an overall contribution of between 15 and 20% per year. An earlier study from Todd and coworkers [38], when only 316 structures had been deposited, indicated that the quality and size of structures determined by structural genomics versus functionally driven structure determination were comparable. Further, 29% of the domains solved revealed evolutionary relationships not apparent from sequence alone. Similarly, 19 and 11% contributed new superfamilies and folds, respectively. While this number of folds is significantly higher that the contribution from all structures (2% based on the SCOP definition of fold as indicated earlier), it reflects on the difficulty of finding new folds and surely indicates that protein fold space is indeed quite limited.

To get a sense of what structural genomics is contributing, it is first necessary to get some measure of what structure is contributing overall to our understanding of living systems. Clearly, this contribution is somewhat intangible and can be defined in different ways. One recent approach was to review what both structures and targets contribute to the functional and disease coverage of the human genome [45]. In some sense this measure cuts across the various criteria for choosing structural genomics targets that was outlined above. This contribution was measured by looking at the functional coverage of the human genome using either EC numbers (http://www.chem.qmul.ac.uk/iubmb/enzyme/) or Gene Ontology (GO) classifiers [16] and disease via the OMIM [15] classifiers and comparing what the solved structures and targets contributed. Human genome annotation was taken from Ensembl [5] and structure data from the PDB and targets from targetdb [10], the repository maintained by the RCSB PDB of protein sequences from all the structural genomics centers that are being considered for structure determination (http://targetdb.pdb.org). Comparisons were made for both single domains and whole structures. In addition, the ability to homology model was ascertained based on results from SUPERFAMILY [26]. SUPERFAMILY identifies proteins within complete proteomes based on their structural characterization. As such it represents the percentage of a given proteome that can be modeled by existing structures. The results can be summarized as follows:

- Single domains cover 37% of the GO molecular function classes identified in the human genome
- Whole structures cover 25% of the human genome.
- The 37% domain coverage extends to 56% using homology modeling.
- The 25% whole structure coverage extends to 31% using homology models.
- If all current structural genomics targets were solved ($\sim$3 times the current PDB):

  37% goes to 69%
  25% goes to 44%

4 Discussion

4.1 Follow-up Studies

One of the ultimate measures of the impact of a new structure is the number of follow-up studies and publications. This impact may not be apparent for several years and is difficult to assess in this comparatively short time frame. However, taking as an example JCSG, their structures have evoked numerous individual collaboration agreements (over 50), associations with larger consortiums detailed in Section 3.7, and numerous requests for clones and proteins for biochemical studies. As an example, TM0449 [23, 24] (PDB code: 1kg4) represents a novel fold which has inspired studies from three different laboratories [27, 29] and has led to the elucidation of a novel biochemical pathway of thymidylate synthesis present, among others, in several important human pathogens. TM0449 and its homologs present an attractive antibacterial drug target, since humans and most eukaryotes depend upon the conventional thymidylate synthase.

4.2 Examples of Functional Discoveries

Again we use JCSG to illustrate the power of long range function and structure projections. In these cases, a bacterial structure from a relatively obscure organism such as *T. maritima* proves to have significant biomedical relevance. For example, the *T. maritima* protein TM1620 provided a template to model a human protein PA26, which belongs to GADD (genes active in DNA defense) family that is highly upregulated in several cancers. It also highlights an interesting conservation of DNA antioxidant defense from bacteria to humans. TM1620 is the second protein solved in this family (the first was AhpD from *Mycobacterium tuberculosis*), which is closely related to 1300 proteins from all kingdoms of life. The mechanism of action of proteins from this family is not clear.

As a second example, TM0813 was shown to be unexpectedly similar to a domain from a known protein involved in antibiotic resistance. Interestingly, a homologous domain is also present in sacsin – a protein whose mutations are responsible for a human neurogenerative disease resulting in autosomal recessive spastic ataxia, often found in Quebec, but also in Turkey and several other areas of the world. In these and other similar cases there is a chance that bacterial proteins, which are easier to characterize and study, would provide hints as to specific mechanism of action of their (very distant) human homologs.

5 The Future

Remaining PSI centers in the US (labeled PSI-2 in Table 1) have just received a second round of 5-year funding and are working together to define the most valuable target list of proteins to be structurally determined. While the objectives of structural genomics remain relatively nebulous relative to the completion of the human genome, solving a significant number of protein structures on that final target list will have a significant outcome. That outcome will be measured in an improved understanding of structure-function relationships, improved coverage of protein fold space and improved technologies for all concerned within the science of structural biology.

References

1 Andreeva, A., D. Howorth, S. E. Brenner, T. J. Hubbard, C. Chothia and A. G. Murzin. 2004. SCOP database in 2004: refinements integrate structure and sequence family data. Nucleic Acids Res. **32**: D226–9.

2 Artymiuk, P. J., A. R. Poirrette, H. M. Grindley, D. W. Rice and P. Willett.

1994. A graph-theoretic approach to the identification of three-dimensional patterns of amino acid side-chains in protein structures. J. Mol. Biol. **243**: 327–44.

3 Ben-Shimon, A. and M. Eisenstein. 2005. Looking at enzymes from the inside out: the proximity of catalytic residues to the molecular centroid can be used for detection of active sites and enzyme–ligand interfaces. J. Mol. Biol. **351**: 309–26.

4 Berman, H. M., J. Westbrook, Z. Feng, G. Gilliland, T. N. Bhat, H. Weissig, I. N. Shindyalov and P. E. Bourne. 2000. The Protein Data Bank. Nucleic Acids Res. **28**: 235–42.

5 Birney, E., T. D. Andrews, P. Bevan, et al. 2004. An overview of Ensembl. Genome Res. **14**: 925–8.

6 Boeckmann, B., A. Bairoch, R. Apweiler, et al. 2003. The SWISS-PROT protein knowledgebase and its supplement TrEMBL in 2003. Nucleic Acids Res. **31**: 365–70.

7 Brenner, S. E. 2001. A tour of structural genomics. Nat Rev Genet **2**: 801–9.

8 Burley, J. 1999. The ethics of therapeutic and reproductive human cloning. Semin. Cell Dev. Biol. **10**: 287–94.

9 Chandonia, J. M. and S. E. Brenner. 2005. Implications of structural genomics target selection strategies: Pfam5000, whole genome, and random approaches. Proteins **58**: 166–79.

10 Chen, L., R. Oughtred, H. M. Berman and J. Westbrook. 2004. TargetDB: a target registration database for structural genomics projects. Bioinformatics **20**: 2860–2.

11 Fetrow, J. S. and J. Skolnick. 1998. Method for prediction of protein function from sequence using the sequence-to-structure-to-function paradigm with application to glutaredoxins/thioredoxins and T1 ribonucleases. J. Mol. Biol. **281**: 949–68.

12 Finn, R. D., J. Mistry, B. Schuster-Bocklerv 2006. Pfam: clans, web tools and services. Nucleic Acids Res. **34**: D247–51.

13 Friedberg, I., L. Jaroszewski, Y. Ye and A. Godzik. 2004. The interplay of fold recognition and experimental structure determination in structural genomics. Curr. Opin. Struct. Biol. **14**: 307–12.

14 Greer, D. S., J. D. Westbrook and P. E. Bourne. 2002. An ontology driven architecture for derived representations of macromolecular structure. Bioinformatics **18**: 1280–1.

15 Hamosh, A., A. F. Scott, J. S. Amberger, C. A. Bocchini and V. A. McKusick. 2005. Online Mendelian Inheritance in Man (OMIM), a knowledgebase of human genes and genetic disorders. Nucleic Acids Res. **33**: D514–7.

16 Harris, M. A., J. Clark, A. Ireland, et al. 2004. The Gene Ontology (GO) database and informatics resource. Nucleic Acids Res. **32**: D258–61.

17 Hendrickson, W. A. 1991. Determination of macromolecular structures from anomalous diffraction of synchrotron radiation. Science **254**: 51–8.

18 Henrick, K. and J. M. Thornton. 1998. PQS: a protein quaternary structure file server. Trends Biochem. Sci. **23**: 358–61.

19 Holm, L. and C. Sander. 1997. Dali/FSSP classification of three-dimensional protein folds. Nucleic Acids Res. **25**: 231–4.

20 Kim, D. E., D. Chivian, L. Malmstrom and D. Baker. 2005. Automated prediction of domain boundaries in CASP6 targets using Ginzu and RosettaDOM. Proteins **61 (Suppl. 7)**: 193–200.

21 Kissinger, C. R., D. K. Gehlhaar and D. B. Fogel. 1999. Rapid automated molecular replacement by evolutionary search. Acta Crystallogr. D **55**: 484–91.

22 Kouranov, A., L. Xie, J. de la Cruz, L. Chen, J. Westbrook, P. E. Bourne and H. M. Berman. 2006. The RCSB PDB information portal for structural genomics. Nucleic Acids Res. **34**: D302–5.

23 Kuhn, P., S. A. Lesley, I. I. Mathews, et al. 2002. Crystal structure of thy1, a thymidylate synthase complementing protein from *Thermotoga maritima* at 2.25 Å resolution. Proteins **49**: 142–45.

24 Lesley, S. A., P. Kuhn, A. Godzik, et al. 2002. Structural genomics of the *Thermotoga maritima* proteome implemented in a high-throughput structure determination pipeline. Proc. Natl Acad. Sci. USA **99**: 11664–11669.

25 Li, Z., Y. Ye and A. Godzik. 2006. Flexible Structural Neighborhood – a database of protein structural similarities and alignments. Nucleic Acids Res. **34**: D277–80.

26 Madera, M., C. Vogel, S. K. Kummerfeld, C. Chothia and J. Gough. 2004. The SUPERFAMILY database in 2004: additions and improvements. Nucleic Acids Res. **32**: D235–9.

27 Mathews, I. I., A. M. Deacon, J. M. Canaves, D. McMullan, S. A. Lesley, S. Agarwalla and P. Kuhn. 2003. Functional analysis of substrate and cofactor complex structures of a thymidylate synthase-complementing protein. Structure **11**: 677–90.

28 Mulder, N. J., R. Apweiler, T. K. Attwood, et al. 2005. InterPro, progress and status in 2005. Nucleic Acids Res. **33**: D201–5.

29 Murzin, A. G. 2002. Biochemistry. DNA building block reinvented. Science **297**: 61–2.

30 NIGMS. 2005. *Protein Structure Initiative Mission Statement*. National Institute of General Medical Sciences, Bethesda, MD.

31 Rost, B. 1999. Twilight zone of protein sequence alignments. Protein Eng. **12**: 85–94.

32 Sali, A. 1998. 100,000 protein structures for the biologist. Nat. Struct. Biol. **5**: 1029–32.

33 Schomburg, I., A. Chang, C. Ebeling, M. Gremse, C. Heldt, G. Huhn and D. Schomburg. 2004. BRENDA, the enzyme database: updates and major new developments. Nucleic Acids Res. **32**: D431–3.

34 Schwarzenbacher R., A. Godzik, S. K. Grzechnik and L. Jaroszewski. 2004. The importance of alignment accuracy for molecular replacement. Acta Crystallgr. D **60**: 1229–36.

35 Shapiro, L. and C. D. Lima. 1998. The Argonne Structural Genomics Workshop: Lamaze class for the birth of a new science. Structure **6**: 265–7.

36 Shindyalov, I. N. and P. E. Bourne. 1998. Protein structure alignment by incremental combinatorial extension (CE) of the optimal path. Protein Eng. **11**: 739–47.

37 Thornton, J. M., A. E. Todd, D. Milburn, N. Borkakoti and C. A. Orengo. 2000. From structure to function: approaches and limitations. Nat. Struct. Biol. **7 (Suppl.)**: 991–4.

38 Todd, A. E., R. L. Marsden, J. M. Thornton and C. A. Orengo. 2005. Progress of structural genomics initiatives: an analysis of solved target structures. J. Mol. Biol. **348**: 1235–60.

39 Tress, M., I. Ezkurdia, O. Grana, G. Lopez and A. Valencia. 2005. Assessment of predictions submitted for the CASP6 comparative modelling category. Proteins **61 (Suppl. 7)**: 27–45.

40 Vagin, A. and A. Teplyakov. 1997. MOLREP: an automated program for molecular replacement. J. Appl. Crystallogr. **30**: 1022–1025.

41 Venter, J. C., M. D. Adams, E. W. Myers, et al. 2001. The sequence of the human genome. Science **291**: 1304–51.

42 Vitkup, D., E. Melamud, J. Moult and C. Sander. 2001. Completeness in structural genomics. Nat. Struct. Biol. **8**: 559–66.

43 Wallace, A. C., N. Borkakoti and J. M. Thornton. 1997. TESS: a geometric hashing algorithm for deriving 3D coordinate templates for searching structural databases. Application to enzyme active sites. Protein Sci. **6**: 2308–23.

44 Wang, G., Y. Jin and R. L. Dunbrack, Jr. 2005. Assessment of fold recognition predictions in CASP6. Proteins **61 (Suppl. 7)**: 46–66.

45 Xie, L. and P. E. Bourne. 2005. Functional coverage of the human genome by existing structures, structural genomics targets, and homology models. PLoS Comput. Biol. **1**: e31.

46 Yakunin, A. F., A. A. Yee, A. Savchenko, A. M. Edwards and C. H. Arrowsmith. 2004. Structural proteomics: a tool for genome annotation. Curr. Opin. Chem. Biol. **8**: 42–8.

14
RNA Secondary Structures

Ivo L. Hofacker and Peter F. Stadler

1 Secondary Structure Graphs

1.1 Introduction

The tendency of complementary strands of DNA to form double helices is well known since the work of Watson and Crick. Single-stranded nucleic acid sequences will in general contain many complementary regions that have the potential to form double helices when the molecule folds back onto itself. The resulting pattern of double-helical stretches interspersed with loops is what is called the *secondary structure* of an RNA or DNA. Secondary structure elements may in turn be arranged in space to form three-dimensional (3-D) tertiary structure, leading to additional noncovalent interactions (an example is shown in Figure 1). Energetically, however, these tertiary interactions are weaker than secondary structure. As a consequence RNA folding can be regarded as a hierarchical process in which secondary structure forms before tertiary structure [129, 130]. Since formation of tertiary structure usually does not induce changes in secondary structure, the two processes can be described independently. Functional RNA molecules (tRNAs, rRNAs, etc., as opposed to pure coding sequences), usually have characteristic spatial structures – and therefore also characteristic secondary structures – that are prerequisites for their function. As a consequence, secondary structures are highly conserved in evolution for many classes of RNA molecules.

Both the experimental determination of full spatial RNA structures and computational predictions of RNA 3-D structures are very hard tasks – arguably even harder than the corresponding problems for proteins [62, 82]. Computational approaches to RNA tertiary structure thus have been successful only for selected cases (see Chapter 15). RNA secondary structures, on the other hand, not only have a definite physical meaning as folding intermediates and are useful tools in the interpretation RNA molecules, but they give rise to efficient computational techniques. Secondary structure

Bioinformatics - From Genomes to Therapies Vol. 1. Edited by Thomas Lengauer

ISBN: 978-3-527-31278-8

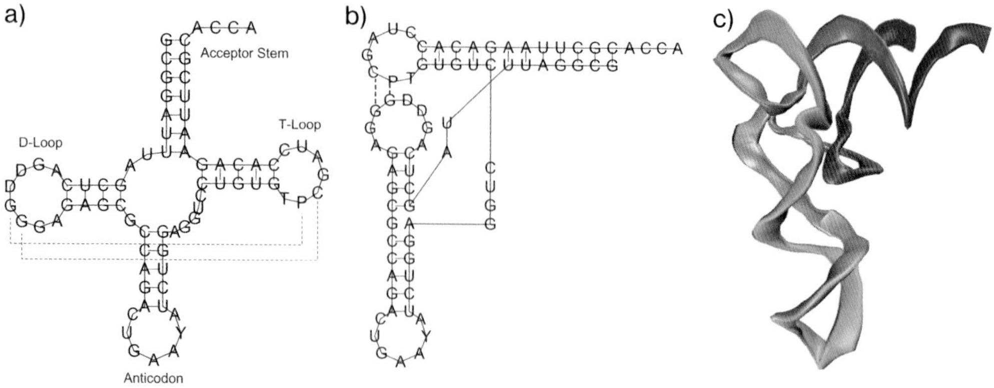

Figure 1 Secondary and tertiary structure of yeast phenylalanine tRNA. (a) The cloverleaf shaped secondary structure consisting of four helices. The dotted blue lines mark evolutionary conserved tertiary contacts. (b) Coaxial stacking results in two extended helices that form the L-shaped tertiary structure. (c) Tertiary structure taken from Protein Data Bank entry 4TRA. The color code (from red to blue) indicates the position along the chain.

prediction and comparison, the focal topics of this chapter, have therefore become a routine tool in the analysis of RNA function.

RNA secondary structures consists of two distinct classes of residues: those that are incorporated in double-helical regions (so called stems) and those that are not part of helices. For RNA, the double-helical regions consist almost exclusively of Watson–Crick C–G and A–U pairs as well as the slightly weaker G–U wobble pairs. All other combinations of pairing nucleotides, called *noncanonical* pairs, are neglected in secondary structure prediction, although they do occur, especially in tertiary structure motifs.

1.2 Secondary Structure Graphs

A secondary structure is primarily a list of base pairs Ω. To ensure that the structure is feasible, a valid secondary structure should fulfill the following constraints:

(i) A base cannot participate in more than one base pair, i.e. Ω is a match on the set of sequence positions.

(ii) Bases that are paired with each other must be separated by at least three (unpaired) bases.

(iii) No two base pairs (i, j) and $(k, l) \in \Omega$ "cross" in the sense that $i < k < j < l$. Matchings that contain no crossing edges are known as loop matchings or circular matchings.

Condition (i) excludes tertiary structure motifs such as base triplets and G-quartets; condition (ii) takes into account that the RNA backbone cannot bend too sharply.

Base pairs that violate condition (iii) are said to form a pseudoknot. While pseudoknots do occur in RNA structures, our definition (somewhat arbitrarily) classifies them as tertiary structure motifs. This is done in part because most dynamic programming algorithms cannot deal with pseudoknots. However, including pseudoknots entails other complications, since most hypothetical structures that violate condition (iii) would also be sterically impossible. Furthermore, little is known about the energetics of pseudoknots, except for some data on H-type pseudoknots [43], the simplest and most common type of pseudoknot (Figure 2). Pseudoknots should therefore be regarded as a first step toward prediction of RNA tertiary structure.

Secondary structures can be represented by "secondary structure graphs" (first two panels in Figure 3). In this representation one creates a graph whose nodes represent nucleotides. There are two kinds of edges: one kind representing the adjacency of nucleotides along the RNA sequence and the other kind representing base pairings. Condition (iii) above assures that this graph is planar, more precisely an *outer-planar graph*, in which all nodes can be arranged along a single face of the planar embedding made up by the edges forming the RNA sequence. We can therefore draw the secondary structure by placing the backbone on a circle and drawing a chord for every base pair such that no two chords intersect.

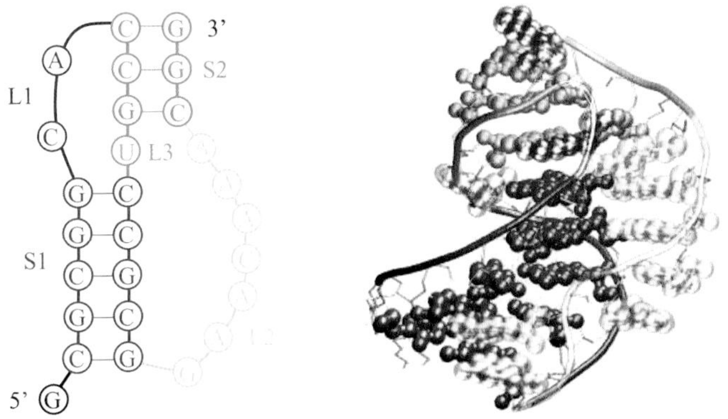

Figure 2 Example of an H-type pseudoknot from beet western yellow virus. The crystal structure (right) shows that the two helices S1 and S2 are coaxially stacked to form a single 3-D helix.

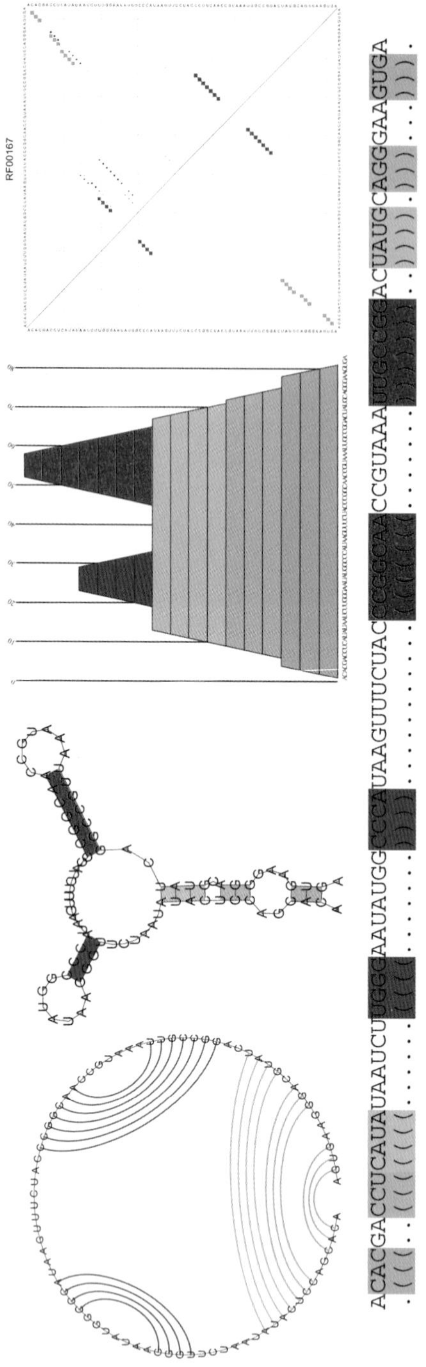

Figure 3 Representations of secondary structures. From left to right: Circle plot, conventional secondary structure graph, mountain plot and dot plot. Removing the backbone edges from the first two representations leaves the matching Ω. Below, the structure is shown in "bracket notation", where each base pair corresponds to a pair of matching parentheses. The structure shown is the purine riboswitch (Rfam RF00167).

1.3 Mountain Plots and Dot Plots

A representation that works well for large structures and is well suited for comparing structures is the so-called mountain representation. In the mountain representation a single secondary structure is represented in a 2-D graph, in which the x-coordinate is the position k of a nucleotide in the sequence and the y-coordinate the number $m(k)$ of base pairs that enclose nucleotide k (third panel in Figure 3).

Another possible representation is the dot plot, where each base pair (i, j) is represented by a dot or box in row i and column j of a rectangular grid, representing the contact matrix of the structure. Dot plots are well suited to represent structure *ensembles* by superimposing structural possibilities. In particular, they are used to represent thermodynamic ensembles by plotting for each pair a box with area proportional to the equilibrium probability of the pair p_{ij}. Similarly, mfold uses colors to indicate the best possible energy for structures containing a particular pair (right-most panel in Figure 3).

1.4 Trees and Forests

Secondary structures can also be stored compactly in strings consisting of dots and matching brackets: For any pair between positions i and j ($i < j$) we place an open bracket "(" at position i and a closed bracket ")" at j, while unpaired positions in the molecule are represented by a dot (".") (bottom of Figure 3). Since base pairs may not cross, the representation is unambiguous.

An ordered forest F is a sequence of rooted ordered trees $T_1, T_2, \dots, T_m$ such that within each tree T_i the left-to-right order of siblings (children of the same parent) is given. In order to represent RNA secondary structures as ordered forests, we will need to associate a label from a suitable alphabet $\mathcal{A}$ with each node.

This representation of secondary structures in terms of matched parentheses suggests to interpret the structure as a tree [117, 119]. In the *full-tree* representation [32] each base pair corresponds to an interior node and each unpaired base is represented by a leaf (Figure 4). A virtual root vertex is added mostly for graphical reasons.

Leaves may be labeled with the corresponding unpaired base, while interior nodes are labeled with the corresponding base pair. In an extended representation, two leaves, one labeled with the $5'$ and one labeled with the $3'$ nucleotide of the base pair, are attached as the left-most and right-most children to each interior vertex. In this representation the sequence of the molecule can be read of the leaves of the Bielefeld tree in pre-order.

Various coarse-grained representations have been considered. Homeomorphically irreducible trees represent entire helices as interior nodes, while

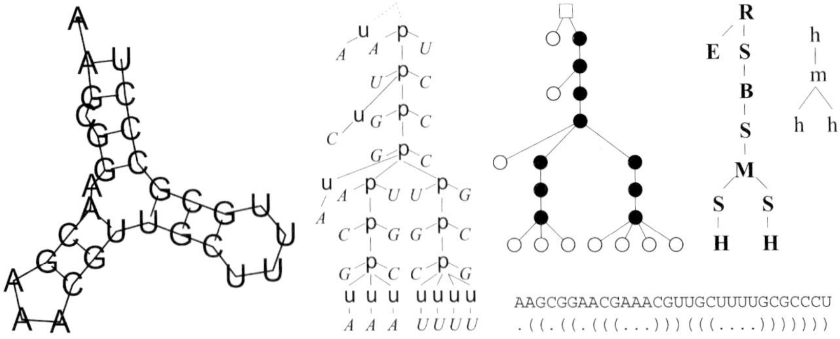

Figure 4 A variety of tree and forest representations of RNA secondary structures have been described in the literature. From left to right: conventional drawing, sequence annotated trees (as e.g. used in RNAforester [51]), "full tree" [32], Shapiro-style tree [117] and branching structure. For comparison, we also show the "bracket notation"

leaves correspond to runs of unpaired bases. Optionally, the length of such a structural element can be used as a weight. Shapiro–Zhang trees [117, 119] explicitly represent the different loop types (hairpin loop, interior loops, bulges, multiloops) as well as stacked regions with special labels. Figure 4 summarizes a few examples.

1.5 Notes

Since RNA secondary structures are planar graphs, they can always be drawn on paper without intersections. Nevertheless, finding a visually pleasing layout is difficult especially for large structures. Layout algorithms for RNA typically make use of the tree-like topology of secondary structure (e.g. Refs. [18, 47, 98, 118]). The problem becomes more complicated when pseudoknots are allowed [46].

2 Loop-based Energy Model

2.1 Loop Decomposition

Secondary structures can be uniquely decomposed into loops, i.e. the faces of the planar drawing of the structure. More formally, we call a position k *immediately interior* of the pair (i, j) if $i < k < j$ and there exists no other base pair (p, q) such that $i < p < k < q < j$. A loop then consists of the closing pair (i, j) and all positions immediately interior of (i, j). As a special case the exterior loop contains all positions not interior of any pair. The loops form a

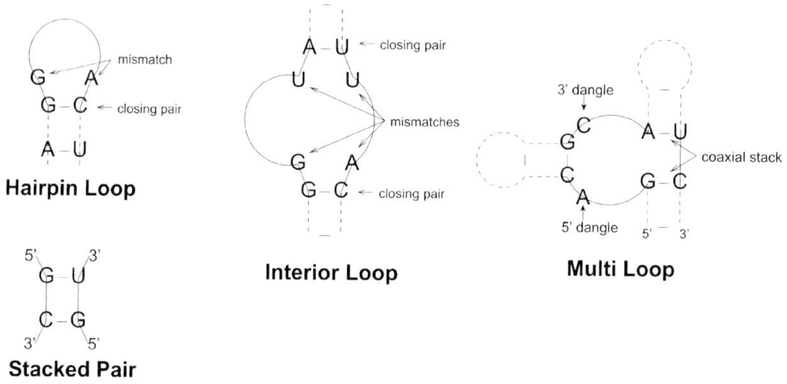

Figure 5 The major types of loops in RNA secondary structures.

minimal cycle basis of the secondary structure graph and this basis is unique for pseudoknot-free structures [80].

A loop is characterized by its length, i.e. the number of unpaired nucleotides in the loop, and its degree, given by the number of base pairs delimiting the loop (including the closing pair). Loops of degree 1 are called hairpin loops, interior loops have degree 2 and loops with degree above 2 are called multiloops (Figure 5). Bulge loops are a special cases of interior loops in which there are unpaired bases only on one side, while stacked pairs correspond to an interior loop of size zero.

The loop decomposition forms the basis of the standard energy model for RNA secondary structures, where the total free energy of a structure is assumed to be a sum over the energies of its constituent loops. As the energy contribution of a base pair in a helix now depends on the preceding and following pair, the model is often called the *nearest-neighbor* model.

2.2 Energy Parameters

Note that a secondary structure corresponds to an ensemble of conformations of the molecule at atomic resolution, i.e. the set of all conformations compatible with a certain base-pairing (hydrogen-bonding) pattern. For example, no information is assumed about the spatial conformation of unpaired regions. The entropic contributions of these restricted conformations have to be taken into account; hence, we are dealing with free energies which will be dependent on external parameters such as temperature and ionic conditions.

Qualitatively, the major energy contributions are base stacking, hydrogen bonds and loop entropies. While hydrogen bond and stacking energies *in vacuo* can be computed using quantum chemistry, the secondary structure model considers free energy differences between folded and unfolded states in an aqueous solution with rather high salt concentrations. As a consequence one has to rely on empirical energy parameters.

Energy parameters are typically derived by following the unfolding of RNA oligomers using a collection of energy parameters is maintained by the group of David Turner [90, 91, 145]. These standard parameters are measured in a buffer of 1 M NaCl at 37°C. As examples we list the free energies for stacked pairs in Table 1. Stacked pairs confer most of the stabilizing energy to a secondary structure, a single additional base pair can stabilize a structure by up to -3.4 kcal mol^{-1}. For comparison, the thermal energy at room temperature is about $RT = 0.6$ kcal mol^{-1}, i.e. the stabilizing energy contribution of a single base pair is typically of the same order of magnitude as the thermal energy. (RNA energy parameters are still published in kcal mol^{-1} to facilitate comparison with previous parameter sets; 1 kcal mol^{-1} $\approx$ 4.2 kJ mol^{-1} in SI units.)

In general, loop energies depend on the loop type and its size. Except for small loops (which are tabulated exhaustively [90]), sequence dependence is conferred only through the base pairs closing the loop and the unpaired bases directly adjacent to the pair. Thus, the loop energy takes the form:

$$E_{\text{loop}} = E_{\text{mismatch}} + E_{\text{size}} + E_{\text{special}}, \tag{1}$$

where E_{mismatch} is the contribution from unpaired bases inside the closing pair and the base pairs immediately interior to the closing pair. The last term is used, for example, to assign bonus energies to unusually stable tetra loops, such as hairpin loops with the sequence motif GNRA. Polymer theory predicts that for large loops E_{size} should grow logarithmically. For multiloops, however, energies that are linear in loop size and loop degree have to be used in order to allow efficient dynamic programming algorithms for structure prediction. While the model allows only Watson–Crick (AU, UA, CG and GC) and wobble pairs (GU, UG), nonstandard base pairs in helices are treated as special types of interior loops in the most recent parameter sets.

The energy model above contains inaccuracies, on the one hand because it assumes that loop energies are strictly additive, on the other hand, because

Table 1 Free energies for stacked pairs (in kcal mol^{-1}).

	CG	GC	GU	UG	AU	UA
CG	−2.4	−3.3	−2.1	−1.4	−2.1	−2.1
GC	−3.3	−3.4	−2.5	−1.5	−2.2	−2.4
GU	−2.1	−2.5	1.3	−0.5	−1.4	−1.3
UG	−1.4	−1.5	−0.5	0.3	−0.6	−1.0
AU	−2.1	−2.2	−1.4	−0.6	−1.1	−0.9
UA	−2.1	−2.4	−1.3	−1.0	−0.9	−1.3

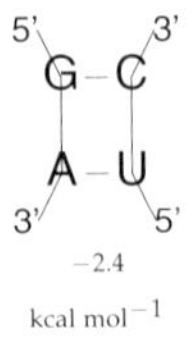

Note that both base pairs have to be read in 5′–3′ direction.

energy parameters carry experimental errors (typically about 0.1 kcal mol^{-1}). Most seriously, the sequence dependence of loop energies has to be kept relatively simple in order to deduce the parameters from a limited number of experiments.

2.3 Notes

Adjacent helices in multiloops may stack coaxially to form a single extended helix. tRNA structures are prominent examples of this. In the four-armed multiloop the acceptor stem coaxially stacks on the T-stem and the anticodon stem stacks on the D-arm. This results in two extended helices which then form the L-shaped tertiary structure characteristic for tRNAs (Figure 1). Strictly speaking, coaxial stacking goes beyond the secondary structure model, since one has to know *which* helices in the loop will stack in order to include the energetic effect; the list of base pairs is no longer sufficient information to compute the energy. Coaxial stacking is also cumbersome to include in structure prediction algorithms. It has, however, been shown to improve prediction quality [135]. Useful energy parameters for structures with pseudoknots have so far only been collected for simple H-type pseudoknots [43].

3 The Problem of RNA Folding

3.1 Counting Structures and Maximizing Base Pairs

In order to understand the basic ideas behind the dynamic programming algorithms for RNA folding, it is instructive to first consider the underlying combinatorial problem: given an RNA sequence x of length n, enumerate all secondary structures on x. Let x_i denote the i-th position of x. We will simply write "(i, j) pairs" to mean that the nucleotides x_i and x_j *can* form a Watson–Crick or a wobble pair, i.e. $x_i x_j$ is one of GC, CG, AU, UA, GU or UG. A subsequence (substring) will be denoted by $x[i, \ldots, j]$. For notational convenience we interpret $x[j+1, \ldots, j]$ as the empty sequence and associate a single empty structure with it.

The basic idea is that a structure on n nucleotides can be formed in only two distinct ways from shorter structures: either the first nucleotide is unpaired, in which case it is followed by an arbitrary structure on the shorter sequence $x[i+1, \ldots, j]$, or the first nucleotide is paired with some partner base, say k. In the latter case the rule that base pairs must not cross implies that we have independent secondary structures on the subintervals $x[i+1, \ldots, k-1]$ and $x[k+1, \ldots, j]$. Graphically, we can write this decomposition of the set of structures like this:

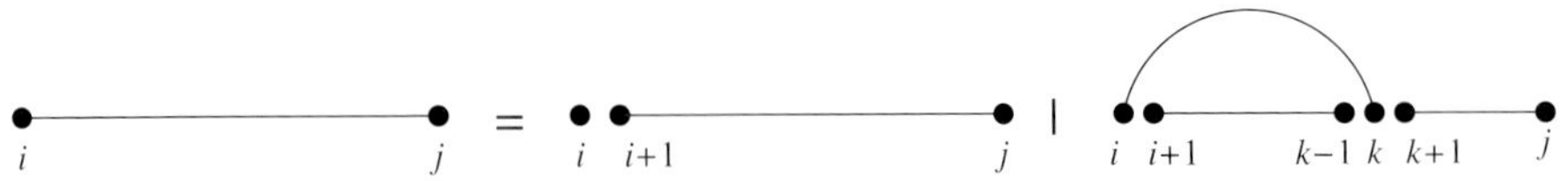

It is now easy to compute the number N_{ij} of secondary structures on the subsequence $x[i,\ldots,j]$ from positions i to j [139, 140]:

$$N_{ij} = N_{i+1,j} + \sum_{k,\,(i,k)\text{ pairs}} N_{i+1,k-1} N_{k+1,j}, \tag{2}$$

with $N_{ii} = 1$. The independence of the structures on $x[i+1,\ldots,k-1]$ and $x[k+1,\ldots,j]$ implies that we can simply multiply their numbers. This simple combinatorial structure of secondary structures was realized by Waterman in the late 1970s [139, 140].

Historically, the first attempts at secondary structure prediction tried to maximize the number of base pairs in the structure. The solution to this problem by the Nussinov algorithm [101] is very similar to the combinatorial recursion above. Denote by E_{ij} the maximal number of base pairs in a secondary structure on $x[i,\ldots,j]$. Using the decomposition of the structure set, we see that E_{ij} is the optimal choice among each of the alternatives. In this context, independence of two substructures in the paired cases implies that we have to optimize these substructures independently. If we like, we can associate each base pair with a weight (negative energy) β_{ij} which depends on x_i and x_j; we arrive immediately at the recursion:

$$E_{ij} = \max\left\{ E_{i+1,j},\ \max_{k,\,(i,k)\text{ pairs}} \left\{ E_{i+1,k-1} + E_{k+1,j} + \beta_{ik} \right\} \right\}. \tag{3}$$

Replacing the weights by binding energies (which are negative for stabilizing interactions) we simply have to replace max by min in the above recursions. In practice, this simplified energy model does not lead to reasonable predictions in most cases. We use it here for didactic purposes and relegate a more detailed description of the complete RNA folding problem to Section 3.3.

The energy contributions of individual base pairs are of the same order of magnitude as the thermal energy at room temperature. Thus, RNA molecules exist in a distribution of structures rather than in a single ground-state structure. Thermodynamics dictates that, in equilibrium, the probability of a particular structure Ψ is proportional to its Boltzmann factor $\exp[-E(\Psi)/RT]$. Here $E(\Psi)$ is the energy of the sequence in conformation (secondary structure) Ψ, R is the molar gas constant (Boltzmann's constant in molar units) and T is the absolute ambient temperature in Kelvin. This ensemble of structures is determined by its *partition function*:

$$Z = \sum_{\Psi} \exp(-E(\Psi)/RT), \tag{4}$$

or, equivalently, by the free energy $\Delta G = -RT \ln Z$. The partition function Z can be computed in analogy to Eq. (3). Using Z_{ij} as the partition function over all structures on subsequence $x[i, \ldots, j]$ we obtain [93]:

$$Z_{ij} = Z_{i+1,j} + \sum_{k,\,(i,k)\text{ pairs}} Z_{i+1,k-1} Z_{k+1,j} \exp(-\beta_{ik}/RT). \tag{5}$$

Note that we can transform the recursion for E_{ij} in Eq. (3) into the equation for Z_{ij} simply by exchanging maximum operations with sums, sums with multiplications and energies by their corresponding Boltzmann factors.

The partition function allows us to compute the equilibrium probability of a structure Ψ as $p(\Psi) = \exp[-E(\Psi)/RT]/Z$. The formalism is also used to efficiently compute the equilibrium probability of a base pair $p_{ij} = \sum_{(i,j)\in\Psi} p(\Psi)$. To this end one needs to compute the partition function $\widehat{Z}_{ij}$ of structures *outside* the subsequence $x[i, \ldots, j]$ using a recursion similar to the one above for Z. We can now compute the partition function over all structures containing the pair (i, j) and thus its probability:

$$p_{ij} = \widehat{Z}_{ij} Z_{i+1,j-1} \exp(-\beta_{ij}/RT)/Z. \tag{6}$$

Further variants of this scheme can be employed to compute, for example, the number of states with a given energy, to explicitly list all possible structures or to determine structures that optimize other properties. In Section 5.5 we will briefly mention how such variants can be constructed in a systematic way within the framework of *algebraic dynamic programming* (ADP) [36].

3.2 Backtracing

Recursion (3) computes only the optimal energy, not an optimal structure which realizes this energy. This is typical for most dynamic programming algorithms: one first computes the value of the optimum, then uses *backtracing* (sometimes called *backtracking*) to generate one (or more) structures in a stepwise fashion based on the information collected in the forward recursions. This section closely follows an exposition of the topic in Ref. [29]. The basic object is a partial structure π consisting of a collection Ω_π of base pairs and a collection Υ_π of sequence intervals in which the structure is not (yet) known. Positions that are known to be unpaired can easily be inferred from this information. The completely unknown structure on the sequence interval $[1, n]$ is therefore $\varnothing = (\emptyset, \{[1, n]\})$ while a structure is complete if it is of the form $\pi = (\Omega, \emptyset)$.

Suppose $I = [i, j] \in \Upsilon$ are positions for which the partial structure $\pi = (\Omega, \Upsilon)$ is still unknown. If we know that i is unpaired, then $\pi' = (\Omega', \Upsilon')$ with $\Omega' = \Omega\ \Upsilon' = \Upsilon \setminus \{I\} \cup \{[i+1, j]\}$. If $(i, k), i < k \leq j$, is a base pair, then $\Omega' = \Omega \cup \{(i, k)\}$ and $\Upsilon' = \Upsilon \setminus \{I\} \cup \{[i+1, k-1], [k+1, j]\}$. Here we use the convention that empty intervals are ignored. Furthermore, base pairs can only be inserted within a single interval of the list Υ. We write $\pi' = \pi \blacktriangleleft (i)$ and $\pi' = \pi \blacktriangleleft (i, k)$ for these two cases.

The energy of a partial structure π is defined as:

$$E(\pi) = \sum_{(k,l) \in \Omega} \beta_{kl} + \sum_{I \in \Upsilon} E_{\text{opt}}(I), \tag{7}$$

where $E_{\text{opt}}(I) = E_{ij}$ is the optimal energy for the substructure on the interval $I = [i, j]$.

The standard backtracing for the minimal energy folding starts with the unknown structure. Instead of a recursive version we describe here a variant where incomplete structures are kept on a stack $\mathfrak{S}$. We write $\pi \leftarrow \mathfrak{S}$ to mean that π is popped from the stack and $\pi \rightarrow \mathfrak{S}$ to mean that π is pushed onto the stack.

If we want all optimal energy structures instead of a single representative we simply test all alternatives, i.e. we omit the **next** in Algorithm B1, Table 2. It is now almost trivial to modify the backtracing to produce all structures within an energy band $E_{\text{opt}} \leq E \leq E_{\text{max}}$ above the ground state.

Stochastic backtracing procedures for dynamic programming algorithm such as pairwise sequence alignment are well known [97]. Replacing Z_{ij} by N_{ij} in Algorithm B3 we recover recursions for producing a uniform ensemble of structures similar to the procedure for producing random structures without sequence constraint used in Ref. [127].

Note that the probabilities of $\pi \blacktriangleleft (i+1)$ and $\pi \blacktriangleleft (i, k)$ for all k add to 1 so that in each iteration we take exactly one step. Hence, we simply fill one structure which we output as soon as it is complete. See Table 2.

3.3 Energy Minimization in the Loop-based Energy Model

Using the loop-based energy model is essential in order to achieve reasonable prediction accuracies. As we shall see, the more complicated energy model results in somewhat more complicated recursions and requires additional tables. However, memory and CPU requirements are still $\mathcal{O}(n^3)$ and $\mathcal{O}(n^2)$. The main difference from the simple model discussed in the previous sections is that we now have to distinguish between different types of loops. Thus, we have to further decompose the set of substructures enclosed by the base pair (i, k) according to the loop types: hairpin loop, interior loop

Table 2 Comparison of backtracing recursions for different algorithms

Algorithm B1 [101, 150]: Backtracing a single structure	**Algorithm B2** [144]: Backtracing multiple structures	**Algorithm B3** [21]: Stochastic backtracing

$\varnothing \to \mathfrak{S}$.
while $\mathfrak{S} \neq \varnothing$
$\pi \leftarrow \mathfrak{S}$;
if π is complete **then** output π
$[i,j] = I \in Y_\pi$.
$\pi' = \pi \blacktriangleleft (i+1)$
if $E(\pi') = E_{\text{opt}}$ **then** $\pi' \to \mathfrak{S}$; **next**;
for all $k \in [i,j]$ **do**
$\pi' = \pi \blacktriangleleft (i,k)$
if $E(\pi') \leq E_{\text{opt}}$
then $\pi' \to \mathfrak{S}$; **next**;
	$\varnothing \to \mathfrak{S}$.
	while $\mathfrak{S} \neq \varnothing$
	$\pi \leftarrow \mathfrak{S}$;
	if π is complete **then** output π
	for all $[i,j] = I \in Y_\pi$ **do**
	$\pi' = \pi \blacktriangleleft (i+1)$
	if $E(\pi') \leq E_{\text{opt}} + \Delta E$
	then $\pi' \to \mathfrak{S}$;
	for all $k \in [i,j]$ **do**
	$\pi' = \pi \blacktriangleleft (i,k)$
	if $E(\pi') \leq E_{\text{opt}} + \Delta E$
	then $\pi' \to \mathfrak{S}$;
		$\varnothing \to \mathfrak{S}$.
		while $\mathfrak{S} \neq \varnothing$
		$\pi \leftarrow \mathfrak{S}$;
		if π is complete **then** output π
		for all $[i,j] = I \in Y_\pi$ **do**
		$\pi' = \pi \blacktriangleleft (i+1)$
		$\pi' \to \mathfrak{S}$ with probability $Z(\pi')/Z(\pi)$
		for all $k \in [i,j]$ **do**
		$\pi' = \pi \blacktriangleleft (i,k)$
		$\pi' \to \mathfrak{S}$
		with prob. $Z(\pi')/Z(\pi)$

and multi(branched) loops (Figure 6). The hairpin and interior loop cases are simple since they reduce again to the same decomposition step.

The multiloop case is more complicated, however, since the multiloop energy depends explicitly on the number of substructures ("components") that emanate from the loop. We therefore need to decompose the structures within the multiloop in such a way that we can at least implicitly keep track of the number of components. To this end we represent a substructure within a multiloop as a concatenation of two components: an arbitrary 5′ part that contains *at least* one component, and a 3′ part that starts with a base pair and contains only a single component. These two types of multiloop substructures are now decomposed further into parts that we already know: unpaired inter-

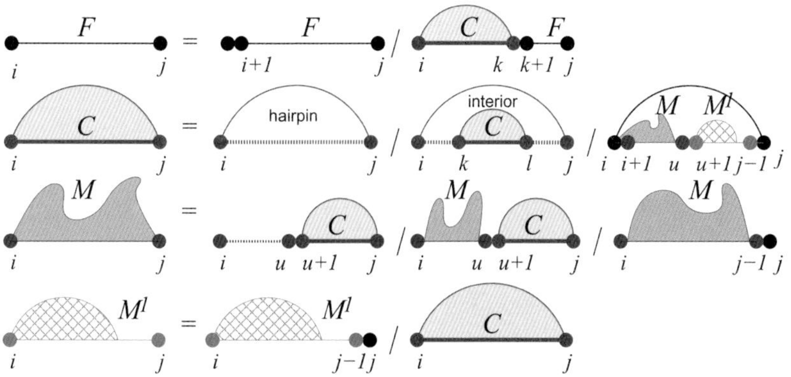

Figure 6 Decomposition of RNA secondary structure. Dotted lines indicate unpaired substructures, while full lines denote arbitrary structures; base pairs are indicated as arcs. Multiloop contributions with an arbitrary number of components are shown as irregular "mountains". See text for further details.

vals, structures enclosed by a base pair and (shorter) multiloop substructures (Figure 6). It is not too hard to check that this decomposition really accounts for all possible structures and that each secondary structure has a unique decomposition.

Given the recursive decomposition of the structures, we can now rather easily derive the associated energy minimization algorithm. We will use the abbreviations $\mathcal{H}(i,j)$ for the energy of a hairpin loop closed by the pair (i,j), similarly $\mathcal{I}(i,j;k,l)$ shall denote the energy of an interior loop determined by the two base pairs (i,j) and (k,l). We will also tabulate the following quantities:

F_{ij} free energy of the optimal substructure on the subsequence $x[i,\ldots,j]$.

C_{ij} free energy of the optimal substructure on the subsequence $x[i,\ldots,j]$ subject to the constraint that i and j form a base pair.

M_{ij} free energy of the optimal substructure on the subsequence $x[i,\ldots,j]$ subject to the constraint that that $x[i,\ldots,j]$ is part of a multiloop and has at least one component.

M^1_{ij} free energy of the optimal substructure on the subsequence $x[i,\ldots,j]$ subject to the constraint that that $x[i,\ldots,j]$ is part of a multiloop and has exactly one component, which has the closing pair i,h for some h satisfying $i < h \le j$.

The recursions for computing the minimum free energy of an RNA molecule in the loop based energy model were first formulated by Zuker and Stiegler [150]. They can be summarized as follows:

$$\begin{aligned}
F_{ij} &= \min\left\{F_{i+1,j}, \min_{i<k\le j}\left(C_{ik}+F_{k+1,j}\right)\right\} \\
C_{ij} &= \min\begin{Bmatrix}\mathcal{H}(i,j), \min_{i<k<l<j}\left(C_{kl}+\mathcal{I}(i,j;k,l)\right), \\ \min_{i<u<j}\left(M_{i+1,u}+M^1_{u+1,j-1}+a\right)\end{Bmatrix} \\
M_{ij} &= \min\begin{Bmatrix}\min_{i<u<j}\left((u-i+1)c+C_{u+1,j}+b\right), \\ \min_{i<u<j}\left(M_{i,u}+C_{u+1,j}+b\right),\ M_{i,j-1}+c\end{Bmatrix} \\
M^1_{ij} &= \min\left\{M^1_{i,j-1}+c,\ C_{ij}+b\right\},
\end{aligned} \tag{8}$$

where we assume linear multiloop energies of the form $E_{\mathrm{ML}} = a + b\cdot\text{degree} + c\cdot\text{size}$. In contrast to most implementations the version shown here decomposes structures in such a way that each substructure occurs exactly once.

While this is not strictly necessary for energy minimization, it allows us to use essentially the same recursions for all variants of the problem, including the computation of the partition function, or the backtracing of all or a sample of suboptimal structures.

3.4 RNA Hybridization

Intermolecular base pairing between two RNA molecules can be treated in the same way as intramolecular interactions. The most straightforward approach is to concatenate the two molecules. One can then apply the folding algorithms for single molecules. There is only a single necessary modification to the folding algorithms: the energy contribution of the loop that contains the cut point is different. Implementations of this approach are RNAcofold [56] and pairfold [5].

From a physics point of view, however, additional effects need to be taken into account: the interaction of two distinct molecules is concentration dependent. Furthermore, there is an additional (entropic) contribution for the initiation of an intermolecular interaction. The extension of the folding algorithms of course compute both inter- and intra-molecular contributions. It is therefore necessary to correct for the initialization energy E^i:

$$\begin{aligned} Z_{AA} &= (Z^\circ_{AA} - Z^2_A)\exp(-E^i/RT) \\ Z_{BB} &= (Z^\circ_{BB} - Z^2_B)\exp(-E^i/RT) \quad \text{and} \\ Z_{AB} &= (Z^\circ_{AB} - Z_A Z_B)\exp(-E^i/RT)\,. \end{aligned} \tag{9}$$

where Z° is the partition function as calculated from the folding algorithm for the concatenated sequences, and Z_A and Z_B are the partition functions of the isolated molecules A and B. Standard statistical thermodynamics can then be used to compute the concentration dependencies of the complex formation (e.g. Ref. [20] for a discussion in the context of RNA hybridization).

Various simplified approaches have been discussed in recent years. In particular, the most common approximation is to neglect the secondary structures of the two interacting molecules. This amounts to a model in which the concatenated structure can only have base pairs and interior loops, and the cut point is located in the single hairpin loop. It does, however, result in a much faster algorithm with time complexity $\mathcal{O}(n \cdot m)$ instead of $\mathcal{O}((n+m)^3)$ for two sequences of length n and m. Algorithms for this case have been described in Refs. [20, 105]; the Vienna RNA Package also provides an implementation. The RNAhybrid program was in particular used to detect microRNA/target interactions.

3.5 Pseudoknotted Structures

Many functionally important RNA structures contain pseudoknots, including rRNAs [12], RNase P RNAs [10, 49] and tmRNA [151]. Recently, algorithms have been described that are able to deal with certain classes of pseudoknotted structures. However, as we shall see below, these are plagued by considerable computational costs. In addition, a common problem of all these approaches is the still very limited information about the energetics of pseudoknots [43, 66].

In the general case of unrestricted pseudoknots the problem is NP-complete when a loop-based energy model is used [3, 86]. Arbitrarily complex pseudoknots, however, are also biologically unrealistic. While every secondary structure has a plausible 3-D realization (this follows directly from the tree structure of secondary structures), this is not true for more general structures: it may well be impossible to embed a given arbitrary set of base pairs in 3-D space such that chemically reasonable distances are maintained. By construction, such constraints cannot be incorporated into our graph-based model of RNA structure. One remedy is to restrict oneself to certain (simple) classes of pseudoknots.

Figure 7 shows the algorithmic problem with pseudoknots. In principle, one could include (a certain type of) pseudoknots as additional structural elements into the dynamic programming recursion. The further decomposition of the structure, however, requires at least two coupled cut points, here k and l, even if we assume that the structures of crossing arc sets are only single stems. This increases the CPU requirements to at least $\mathcal{O}(n^4)$. More realistic models, such as the H-type model, require additional memory as well.

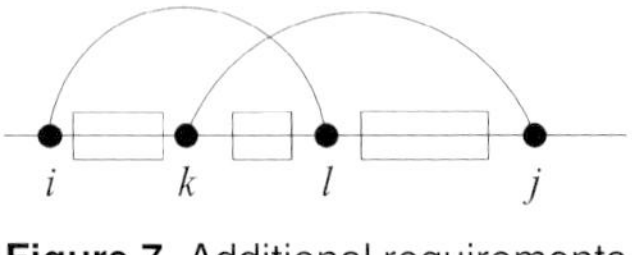

Figure 7 Additional requirements for computing pseudoknotted structures. In order to evaluate the contribution for the pseudoknot on $[i, j]$ we need to iterate over all combinations of cutpoints $k < l$.

Algorithms for a number of different classes of pseudoknots have been published in recent years, (e.g. Refs. [3, 22, 86, 104, 109]). Figure 8 summarizes the relationships between the algorithmic complexities of predicting secondary structures from some of these structure classes [16].

3.6 Notes

The basic counting recursion can be readily modified to enumerate other quantities of interest such as the structures with particular properties and distributions of structural elements (e.g. Ref. [58]). The combinatorics of RNA

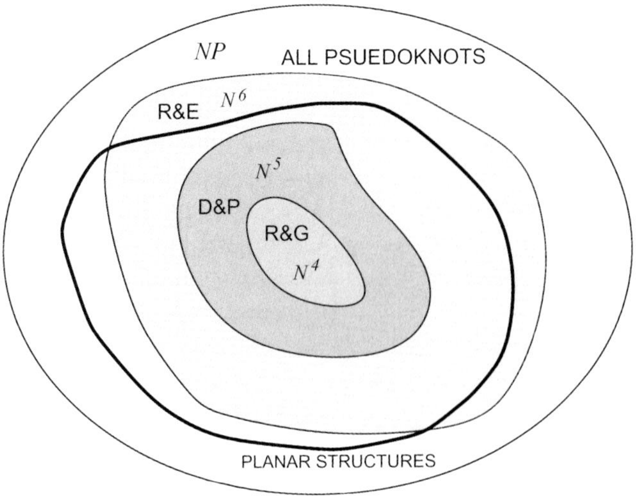

Figure 8 The most prominent classes of pseudoknotted structure are those investigated by Reeder and Giegerich R&G [104], Dirks and Pierce D&P [22], and Rivas and Eddy R&E [109].

secondary structures and related mathematical objects such as ordered trees, Motzkin paths and noncrossing partitions is still an active area of research (e.g. Refs. [13,19,81] and the references therein).

The recursions for the loop-based energy model as displayed above, in fact, give rise to $\mathcal{O}(n^4)$ CPU requirements due to the interior loop contribution. However, very long interior loops are extremely unlikely (and unstable), so that the length of interior can be bounded by a constant, e.g. $M = 30$. The interior loop contribution thus remains quadratic. Under certain plausible assumptions on the interior loop energies, a cubic time algorithm can be designed [87] that takes interior loops of all sizes into account.

A restriction of the folding algorithm to local structure is described in Ref. [57]. Here, the maximum span $|j - i + 1|$ of a base pair (i, j) is bounded by a constant L. The resulting "scanning" algorithms are linear in time and space, and hence can be used to screen entire genomes for locally stable structures.

Circular RNA molecules are rare, but their secondary structures are of considerable interest because structural features are important, e.g. in viroids [108,123]. A straightforward way of dealing with circular RNA molecules is to compute C_{ij} and M_{ij} also for the subsequences of the form $x[j, \dots, n]x[1, \dots, i]$ [149]. The disadvantage of this approach is, however, that it doubles the memory requirements. An alternative is described in Ref. [60].

A secondary structure Ω is *saturated* if none of its stems can be elongated, i.e. if any single base pair that is inserted into Ω does not stabilize the structure by stacking to any other base pairs. The recursions (Figure 6) can be modified to produce only saturated structures [26]. Similarly, we may call Ω *locally base pair optimal* if Ω cannot be expanded by any additional base pair. In Ref. [15] a dynamic programming algorithm is described that computes such locally optimal structures in quartic time with cubic memory requirements.

Prediction of pseudoknotted structures based on maximum matching can be done using algorithms for Maximum Weighted Matching [125]. While this approach requires only $O(n^3)$ time, it cannot take the loop-based energy model into account. "Iterated loop matching", i.e. the repeated (greedy) application of the Nussinov algorithm, is another approximate way of computing pseudoknotted structures [113]. Finally, heuristics such as genetic algorithms can be used to compute pseudoknotted structures [78].

4 Conserved Structures, Consensus Structures and RNA Gene Finding

4.1 The Phylogenetic Method

Most functional RNA molecules have characteristic secondary structures that are highly conserved in evolution. Well-known examples include rRNAs, tRNAs, RNase P and MRP RNAs, the RNA component of signal recognition particles, tmRNA, group I and group II introns, and small nucleolar RNAs. It is therefore of considerable practical interest to efficiently compute the consensus structure of a collection of such RNA molecules.

Given a sufficiently large database of aligned RNA sequences, one can directly infer a consensus secondary structure from the data. The basic idea is that substitutions in the sequence will respect the common structural constraints. Therefore, substitutions in helical regions have to be correlated, since in general only six (GC, CG, AU, UA, UG and GU) out of the 16 combinations of two bases can be incorporated in the helix. Two columns in the alignment thus will covary if they form a base pair.

For concreteness, assume that we are given a multiple sequence alignment $\mathbb{A}$ of N sequences. By $\mathbb{A}_i$ we denote the i-th column of the alignment, while a_i^α is the entry in the α-th row of the i-th column. The length of $\mathbb{A}$, i.e. the number of columns, is n. Furthermore, let $f_i(\mathrm{X})$ be the frequency of base X at aligned position i and let $f_{ij}(\mathrm{XY})$ be the frequency of finding simultaneously X at position i and Y at j.

The most common way of quantifying sequence covariation for the purpose of RNA secondary determination is the *mutual information* score [14,44,45]:

$$MI_{ij} = \sum_{\mathrm{X,Y}} f_{ij}(\mathrm{XY}) \log \frac{f_{ij}(\mathrm{XY})}{f_i(\mathrm{X}) f_j(\mathrm{Y})}. \tag{10}$$

Usually, the mutual information score makes no use of RNA base-pairing rules. For large datasets this is desirable, since it allows us to identify non-canonical base pairs and tertiary interaction. For the small datasets considered in the following subsections, however, neglecting base pairing rules does more harm (by increasing noise) than good. In particular, mutual information does not account at all for consistent noncompensatory mutations, i.e. if we have, say, only GC and GU pairs at positions i and j then $M_{ij} = 0$. Thus, sites with two different types of base pairs are treated just like a pair of conserved positions.

A straightforward measure of covariation takes the form:

$$C_{ij} = \sum_{\mathrm{XY},\mathrm{X'Y'}} f_{ij}(\mathrm{XY}) \mathbf{D}_{\mathrm{XY},\mathrm{X'Y'}} f_{ij}(\mathrm{X'Y'}). \tag{11}$$

where a suitable choice for the 16×16 matrix $\mathbf{D}$ has entries $\mathbf{D}_{\mathrm{XY},\mathrm{X'Y'}} = d_H(\mathrm{XY}, \mathrm{X'Y'})$ if both $\mathrm{XY} \in \mathcal{B}$ and $\mathrm{X'Y'} \in \mathcal{B}$ and $\mathbf{D}_{\mathrm{XY},\mathrm{X'Y'}} = 0$ otherwise. Here $\mathcal{B} = \mathrm{GC}, \mathrm{CG}, \mathrm{AU}, \mathrm{UA}, \mathrm{GU}$ or UG, and $d_H(\mathrm{XY}, \mathrm{X'Y'})$ is the Hamming distance of XY and X′Y′. The idea here is that consistent mutations such as GC $\to$ GU should count less (here half) of a compensatory mutation such as GC $\to$ AU. Note that Eq. (11) is a scalar product, $C_{ij} = \langle f_{ij} \mathbf{D} f_{ij} \rangle$ and hence can be evaluated efficiently. If desired, $\mathbf{D}$ could be replaced by a different kernel that, for example, could incorporate measured substitution rates [35].

The purely phylogenetic approach suffers from two limitations. (i) It requires a very large set of sequences in order to obtain a reliable estimate of covariance or mutual information for each pair of sequences. With the exception of rRNAs and tRNAs, such large datasets are usually not (yet) available. (ii) It is sensitive to alignment errors and hence not applicable to very diverse sets of sequences. A possibly remedy is provided by approaches towards solving the folding and alignment problems simultaneously or iteratively. These are discussed in the following section.

4.2 Conserved Structures

The amount of data that is required for inferring structures can be reduced dramatically by taking thermodynamics of folding into account. Indeed, Ref. [48] suggested to resolve ambiguities in the phylogenetic analysis based on thermodynamic considerations.

However, the converse approach, i.e. to use the information which base pairs are thermodynamically plausible, appears to be more efficient. Most of the alignment-based methods therefore start from thermodynamics-based folding and use the analysis of sequence covariations or mutual information for postprocessing (see, e.g. Refs. [54, 59, 71, 77, 84, 85]). We describe here the alidot algorithm (Figure 9) [54, 59].

For each of the aligned sequences secondary structures are computed separately. The resulting lists of base pairs from either minimum free energy calculations or from a partition function calculation are then superimposed by using the multiple sequence alignment to determine which pairs in different sequences are equivalent. For each pair we now have both thermodynamic and sequence covariation information, which is used to hierarchically rank order base pairs depending on their support across the entire data set. A greedy procedure then extracts contiguous stems from the rank ordered list and combines them to a partial secondary structure which contains only those sequence/structure elements that are significantly conserved throughout the aligned input sequences.

An alternative to the ranking/greedy approach of alidot is to compute a score or weight w_{ij} for each possible base pair. The program ConStruct [84] uses a simple scoring function that exclusively combines the base-pairing probabilities of the individual sequences. Covariance or mutual information score as well as contributions that consider the potential to extend the pair to a longer helix [141] could easily be included. A secondary structure can then be computed using the Nussinov algorithm with weights w_{ij} for the base pairs. The downside of this approach is that it returns a global secondary structure rather than a collection of well-supported local features.

Comparative approaches are based on the fact that RNA secondary structure is quite fragile against randomly placed point mutations. Our earlier computational studies suggest that even with 85% sequence identity we should expect no significant structural similarity [33, 116]. While this result

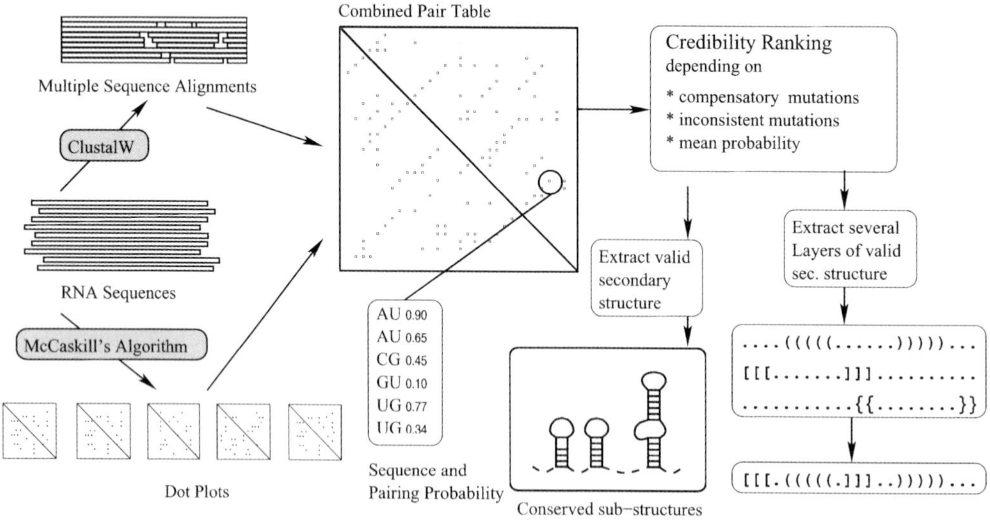

Figure 9 Flow chart of the alidot algorithm.

may seem surprising, there has been convincing experimental evidence (see e.g. Ref. [115]). Methods such as alidot thus can discriminate very well between conserved and nonconserved RNAs. Both alidot and ConStruct require interactive work and are therefore best suited for small genomes as found in RNA viruses [61,131,142].

4.3 Consensus Structures

Sometimes it is known *a priori* that the aligned sequences should fold into a common secondary structure. This is the case, for example, for rRNAs, tRNAs and many other small noncoding RNA (ncRNA) molecules. In this case it makes sense to ask, what is the most stable structure that can be formed simultaneously by all (or almost all) input sequences? This problem is solved in a rather straightforward way by RNAalifold [55]. It treats the entire alignment like a single sequence and solves the secondary structure problem for this "generalized sequence". To this end, of course, an extension of the standard energy model to alignments is required. RNAalifold simply averages the energy contribution over all sequences. In the simple case of base-pair-dependent energies this means:

$$\beta_{ij}^{\mathbb{A}} = \frac{1}{N} \sum_{\alpha} \beta_{x_i^\alpha, x_j^\alpha}. \tag{12}$$

For the realistic energy model, energies for the different loop types are averaged individually.

Both the mutual information score and the covariance score assign a bonus to compensatory mutation. Neither score deals with inconsistent sequences, i.e. with sequences that cannot form a base pair between positions i, j. The simplest ansatz for this purpose is to simply count the number of sequences q_{ij} that cannot form a canonical base pair between columns i and j. Here, combinations of a nucleotide and a gap are counted as inconsistent while gap–gap combinations (i.e. deletions of an entire base pair) are ignored.

In a multiple alignment of a larger number of sequences we have to expect occasional sequencing errors and of course there will be alignment errors. Thus, we cannot simply mark a pair of positions as nonpairing if a single sequence is inconsistent. Furthermore, there is the possibility of a nonstandard base pair [44]. Thus, we define a threshold value for the combined score $B_{ij} = C_{ij} - \phi_1 q_{ij}$ and declare a pair of positions i, j as nonpairing if B_{ij} is too small.

Figure 10 shows the consensus structure of the mir-105 microRNA family as an example. Such consensus structures are needed for the derivation of pattern descriptions that can be used to search for structurally similar RNAs in genomic DNA, as briefly described in the following section.

4.4 RNA Gene Finding

It is, of course, possible to identify genomic sequences that are homologous to known RNA genes, using either BLASTN or, as in the case of tRNAs, more specialized methods. For most functional ncRNA molecules the secondary

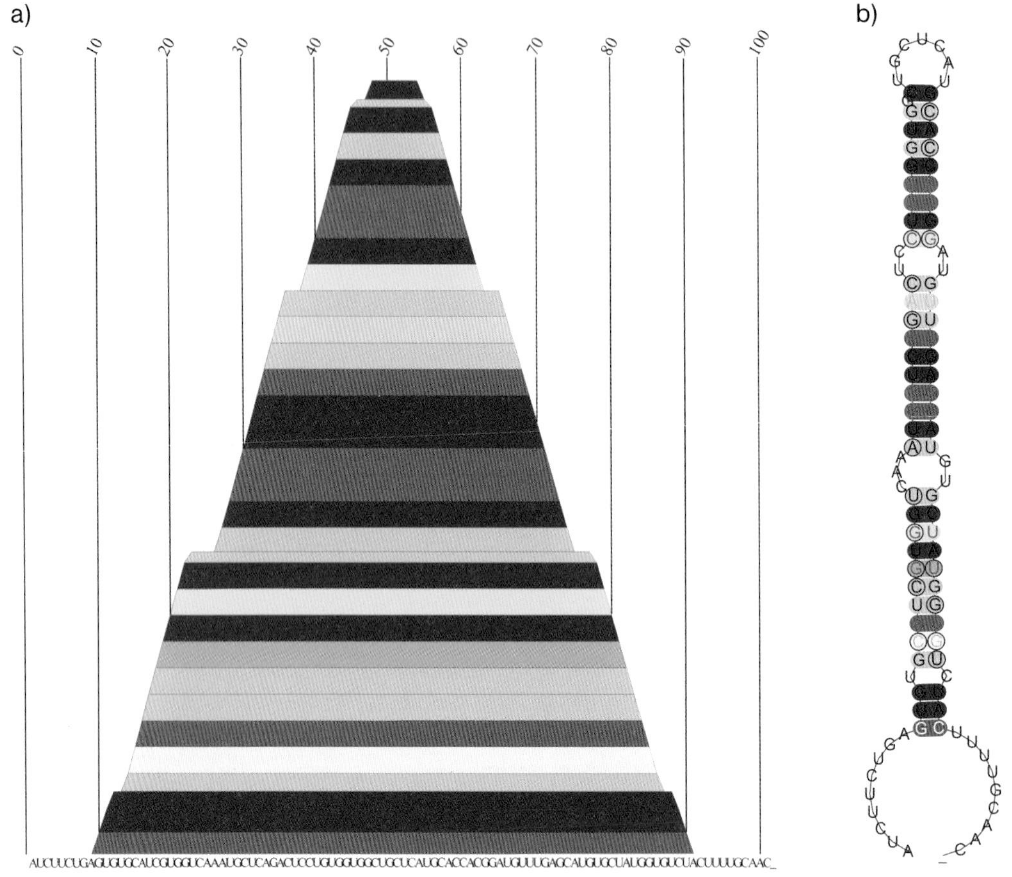

Figure 10 Consensus secondary structure of the 11 sequences from mammalian microRNA mir-105. Sequences are taken from the microRNA Registry (version 6.0) and from BLAST searches in vertebrate genomes. (a) *Mountain plot:* a base pair (i, j) is represented by a slab ranging from i to j. The $5'$ and $3'$ sides of stems thus appear as up-hill and down-hill slopes, respectively, while plateaus indicate unpaired regions. Colors indicate sequence variation by encoding the number of different types of base pairs (GC, CG, AU, UA,GU, UG) that occur in the two paired columns of the alignment. Pairs with conserved sequence are shown in red; ocher, green, cyan, blue and violet indicate two to six types of base pairs. Pairs with one or two inconsistent mutations are shown in (two degrees of) pale colors. (b) In the *conventional secondary structure graph* paired positions are color coded as in the mountain plot. Consistent mutations are indicated by circles around the varying position, compensatory mutations thus are marked by circles around both pairing partners.

structure is much more conserved than their sequence. This can be used to identify putative ncRNA sequences using programs such as RNAmot [34], tRNAscan [83] or HyPa [40]. Nevertheless, all these approaches are restricted to searching for new members of the few well-established families such as tRNAs, small nucleolar RNAs, microRNAs and certain spliceosomal RNAs.

A different approach is taken in the program QRNA [110]. This method for comparative analysis of two aligned homologous sequences can detect novel structural RNA genes by deciding whether the substitution pattern fits better with (i) synonymous substitutions, which are expected in protein-coding regions, (ii) the compensatory mutations consistent with some base-paired secondary structure or (iii) uncorrelated mutations.

The alidot approach has never been used for large-scale gene finding since it has turned out to be nontrivial to assign statistical significance values to its results. Most recently, however, a conceptually related technique has been developed that is efficient and sensitive enough to allow genome-wide screens for RNAs.

The program RNAz [138] combines a comparative approach (scoring conservation of secondary structure) with the observation [8,76,136] that ncRNAs are thermodynamically more stable than expected by chance. This excess stability is conveniently measured in terms of the z-score:

$$z = \frac{E - \overline{E}}{\sigma}, \tag{13}$$

where $\overline{E}$ and σ are mean and standard deviation of the distribution of shuffled sequences. Instead of dealing with individual sequences, RNAz uses multiple sequence alignments of potential RNAs from different species as input. The computation of z by direct sampling is extremely time-consuming. In RNAz it is therefore replaced by a support vector machine that has been trained to solve the regression problems of estimating $\overline{E}$ and σ from properties of the input sequences.

Structural conservation is also quantified in thermodynamical terms. The structure conservation index S is defined as the ratio of the average energy of the consensus structure (as computed by RNAalifold) and the average of the unconstrained folding energies of the individual sequences. An alignment of identical sequences thus has $S = 1$. On the other hand, completely unrelated sequences will not be able to form a consensus structure since there are always some sequences that contradict any particular pairing, thus $S = 0$. Sequences that form a well-conserved consensus sequence in the presence of sequence covariations, finally, will have the same energy contributions in the consensus and in the individual folds. In addition, however, the consensus energy contains the bonus contributions for sequence covariations, so that we obtain $S > 1$. See Figure 11.

RNAz uses a support vector machine (SVM) [17] to determine from the z-score and the structure conservation index whether a given multiple sequence alignment is a structurally conserved RNA. Surveys of animal genomes [96,

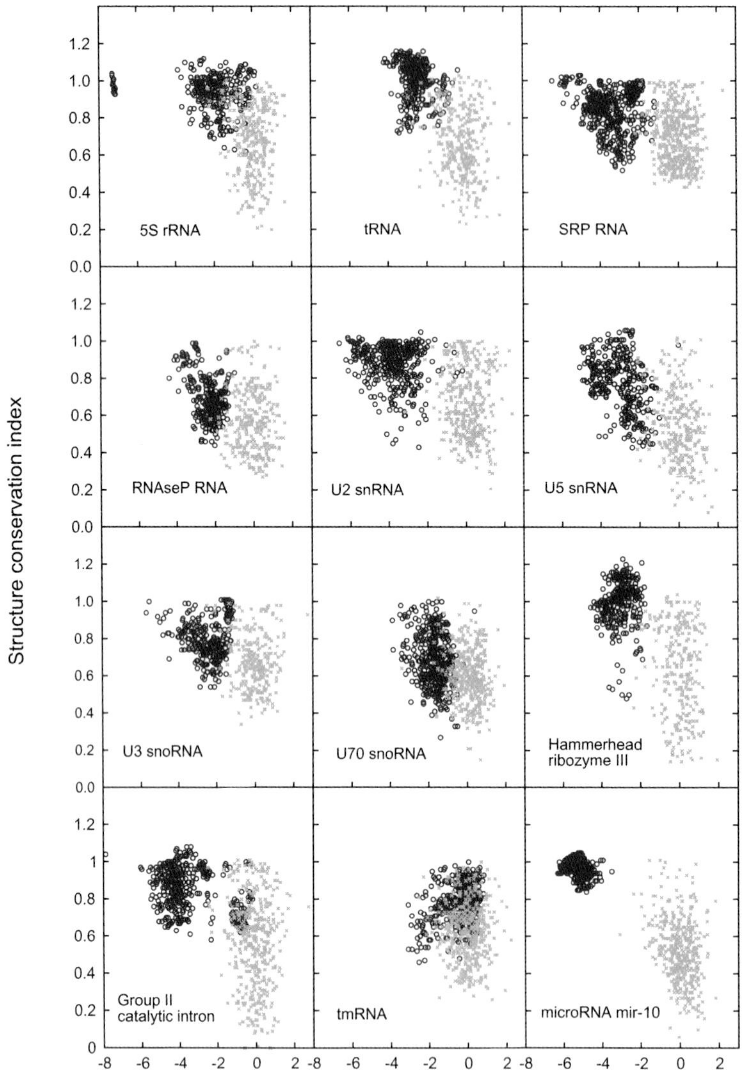

Figure 11 Scatter plot of structure conservation index S (x-axis) and energy z-score (y-axis) for different families of structured ncRNAs. In each panel, the properties of the true sequences (dark) are compared with controls obtained by shuffling the sequence. Data are taken from Ref. [138].

137] reveal a very large number of previously unknown candidates for both independent ncRNAs and structured *cis*-acting elements in mRNAs.

4.5 Notes

Including covariation information is also a good way to improve the accuracy of structure predictions including pseudoknots. One approach is to forgo a loop-based energy model and use base pair scores instead, in which case the resulting Maximum Weighted Matching problem can be solved efficiently [125]. Good accuracies can be achieved by using a combination of covariance and thermodynamic criteria for scoring potential base pairs [141]. The ILM program of Ruan and coworkers [113], uses the Nussinov algorithm iteratively in order to build pseudoknotted structures.

5 Grammars for RNA Structures

5.1 Context-free Grammars (CFGs) and RNA Secondary Structures

The recursions for RNA folding in Figure 6 suggest a close connection with certain grammars. More precisely, we may interpret Figure 6 as the production rules of an "RNA language". The tree representations in Figure 4, on the other hand, are suggestive of a connection between RNA structures and parse trees of a grammar that generates RNA sequence. As we shall see in this section, these connections can be made precise and open the door to the application of learning techniques in RNA bioinformatics.

Recall that a formal language $\mathcal{L}$ is a set of strings over a given alphabet $\mathcal{A}$. A *grammar* G for the language $\mathcal{L}$ consists of:

- A set T of *terminals* which are the letters of the alphabet $\mathcal{A}$ possibly augmented by the null-character ε.
- A set N of *nonterminals* which represent the syntactic categories of $\mathcal{L}$
- A set P of *production* or *derivation rules* which are used to derive the strings in $\mathcal{L}$. Each production consists of a nonterminal "head" that is produced and a string of zero or more nonterminal and terminals (the "body" of the production)
- A single nonterminal $S \in N$ that is designated as the start symbol.

The "dot-parenthesis" grammar for RNA, in the simplest case, can be written as $G_0 = (T, N, P, s)$ with $T = \{(,),.,\varnothing\}$, $N = \{S\}$, $s = S$ and:

$$P = \{S \to S., S \to (S)S, S \to \varnothing\}, \tag{14}$$

where ∅ denotes the empty string. The grammar above is *context-free* since all productions are of the form $V \rightarrow w$, where V is a nonterminal and w is a string consisting of terminals and/or nonterminals. The grammar generates strings of dots and balanced parenthesis; the parse trees of this grammar correspond to the secondary structures. More elaborate grammars can be designed that explicitly encode different types of loops or other substructures. In particular, the decompositions of the structure sets in Section 3.3 can be recast in terms of a grammar:

$$
\begin{aligned}
F &\rightarrow uF|CF|\varnothing \\
C &\rightarrow pL'\bar{p} \mid pLCL\bar{p} \mid pMN\bar{p} \\
M &\rightarrow LC \mid MC \mid Mu \\
N &\rightarrow Nu \mid C \\
L' &\rightarrow uuuL \\
L &\rightarrow uL \mid \varnothing .
\end{aligned}
\tag{15}
$$

This grammar generates RNA sequences, while again the parse trees correspond to secondary structures (Figure 12). The terminal u denotes an unpaired base, while p and $\bar{p}$ is a shorthand for one of the six pairing combinations of bases. The start symbol F represents any structure, L stands for an unpaired sequence within a loop, M and N represent the left and right half of a multi-loop. The production for L' enforces the minimum length of a hairpin loop.

Chomsky normal forms have only productions of the form $V \rightarrow XY$ and $V \rightarrow a$ with $V, X, Y \in N$ and $a \in T$. One can show that every context-free grammar can be converted to normal form, i.e. there is a CFG in normal form that produces the same language $\mathcal{L}$.

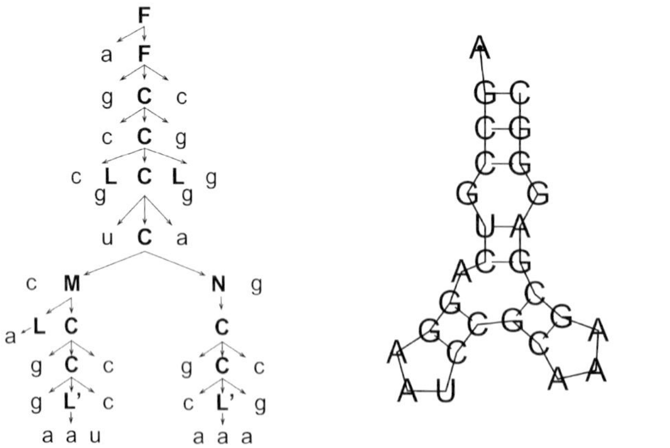

Figure 12 Parse tree and secondary structure drawing for a small example structure, using the grammar from Eq. (15). Productions of the form $L \rightarrow \varnothing$ are left out for simplicity.

Given a CFG $G = (T, N, P, s)$ we obtain a *stochastic CFG (SCFG)* by assigning probabilities $\mathbb{P}(\alpha)$ to all productions $\alpha \in P$ such that $\sum_{\alpha \in P} \mathbb{P}(\alpha) = 1$ is satisfied.

The probabilities associated with the individual productions take on the role of the energy parameters in the previous sections. While the energy parameters must be measured directly, the values of $\mathbb{P}(\alpha)$ can be inferred from training sets of known sequence/structure pairs in a generic machine learning setting. Thus, they can, at least in principle, readily combine different sources of information that can be expressed probabilistically, such as an evolutionary model (derived from a comparative analysis of RNA sequences) and a biophysically motivated model of structure plausibility. In the following three subsection we briefly outline the basic techniques: finding the most likely parse-tree, computing the probability of a given word and the estimation of production probabilities from a given dataset. None of these algorithms is RNA specific; rather, they apply to any SCFG in Chomsky normal form.

5.2 Cocke–Younger–Kasami (CYK) Algorithm

The analog of the minimum free energy folding problem in the SCFG setting can be phrased in the following way: given a string $x \in \mathcal{L}$, find the most likely parse tree for x in a grammar G.

Under the assumption that G is in Chomsky normal form, there is an efficient (polynomial-time) solution to this question, the CYK algorithm [146].

Let $w(i, j, V)$ denote the likelihood of the most likely parse tree on the substring $x[i, \ldots, j]$ rooted at the nonterminal V. Clearly, we have $w(i, i, V) = \log \mathbb{P}(V \to x_i)$ for all i and V. For all larger substrings, $j > i$, we try all productions of the form $V \to XY$ and select the one that maximizes the likelihood. This immediately leads to the recursion

$$w(i, j, V) = \max_X \max_Y \max_{i \le k < j} \left[\log \mathbb{P}(V \to XY) + w(i, k, X) + w(k+1, j, Y) \right] \quad (16)$$

with the initialization $w(i, i, V) = \log \mathbb{P}(V \to x_i)$. The same type of backtracing approach as in the Nussinov algorithm can be used to explicitly recover the parse tree, which corresponds to the secondary structure of the RNA molecule.

5.3 Inside and Outside Algorithms

Instead of retrieving the most likely parse tree one may instead be interested in the probabilities of generating substrings in a particular way. In particular, let $p(i, j, V)$ be the probability that the "inside" substring $x[i, \ldots, j]$ is generated by the nonterminal V. Furthermore, let $q(i, j, V)$ be the probability that the "outside" substrings $x[1, \ldots, i-1] \cup x[j+1, \ldots, n]$ are generated from the start symbol S under the condition that (the parse subtree of) the subsequence $x[i,\ldots,j]$ is rooted at V. Conceptually, these quantities correspond to the partition functions inside and outside of a subsequence $x[i,\ldots,j]$. It is straightforward to derive the corresponding *inside recursion*:

$$p(i,j,V) = \sum_X \sum_Y \sum_{k=i}^{j} \mathbb{P}(V \to XY) p(i,k,X) p(k+1,j,Y), \tag{17}$$

which is initialized with $p(i,i,V) = \mathbb{P}(V \to x_i)$. The *outside recursion* consists of two parts, depending on whether the root V of the interior parse tree is the right or the left nonterminal in the previous production. This yields:

$$\begin{aligned} q(i,j,V) = & \sum_X \sum_Y \sum_{k<i} \mathbb{P}(Y \to XV) p(k,i-1,X) q(k,j,Y) \\ & + \sum_X \sum_Y \sum_{k>j} \mathbb{P}(Y \to VX) q(i,k,Y) p(j+1,k,X), \end{aligned} \tag{18}$$

with the initial conditions $q(1,n,S) = 1$ and $q(1,n,X) = 0$ for all $X \in N \setminus \{S\}$. The probability to produce the sequence x is:

$$\mathbb{P}(x) = p(1,n,S) = \sum_X q(i,i,X) \mathbb{P}(X \to x_i). \tag{19}$$

5.4 Parameter Estimation

One problem with SCFG approaches is that the production probabilities have to be estimated from data. To this end, we compute the expected number $c(V)$ that V is used to parse x and the expected numbers $c(\alpha)$ that production α is used in the derivation of x. It is straightforward to derive:

$$\begin{aligned} c(V) &= \frac{1}{\mathbb{P}(x)} \sum_{i,j=1}^{n} p(i,j,V) q(i,j,V) \\ c(V \to a) &= \frac{1}{\mathbb{P}(x)} \sum_{i:x_i=a} q(i,i,V) \mathbb{P}(V \to a) \\ c(V \to XY) &= \frac{1}{\mathbb{P}(x)} \sum_{i,j=1}^{n} \sum_{k=i}^{j} q(i,j,V) p(i,k,X) p(k+1,j,Y) \mathbb{P}(V \to XY). \end{aligned} \tag{20}$$

Updated estimates for the production probabilities can thus be obtained as $\mathbb{P}'(\alpha) = c(\alpha)/c(V)$ for all $\alpha \in P$. The procedure is then repeated until $\sum_\alpha |\mathbb{P}'(\alpha) - \mathbb{P}(\alpha)| < \varepsilon$, where ε is a user-defined accuracy.

5.5 Algebraic Dynamic Programming

Algebraic Dynamic Programming (ADP) [36] was introduced to facilitate and systematize the development of dynamic programming algorithms. Conceptually, a dynamic programming algorithm consists of three components: a search space of candidate solutions (in our case RNA secondary structures), a scoring scheme (free energies, partition functions, etc.) and an objective function [minimize (energy), sum up (Boltzmann factors)]. The idea behind ADP is to separate these three aspects. For a comprehensive discussion of ADP in the context of bioinformatics we refer to Ref. [36]. We can give here only a very brief, qualitative sketch of the topic.

The search space is defined by a *yield grammar*, i.e. a tree grammar that generates a string language by mapping its terminal symbols at the leaves of the tree into sequences of symbols. A tree grammar is similar to a CFG, with terminal and nonterminal symbols, and productions where the right-hand sides are trees (formulas) from some underlying term algebra. Intuitively, first the search space is "constructed" by enumerating all candidate solutions. This is a parsing problem for which standard solutions, so-called tabulating yield parsers, exist. Scoring and choice are described in terms of an *evaluation algebra*, which is independent of the details of the search space.

The main advantage is that complex variants of folding problems can be implemented very easily. It suffices to modify the grammar to restrict the dynamic programming recursions to all canonical secondary structures, i.e. those that have no isolated base pairs. Conversely, the evaluation algebra can be changed easily. Once the energy model is implemented, one can change the choice function from minimizing energies to adding up Boltzmann factors or listing all structures within an energy range.

The restrictions of the search space can be quite dramatic. One can, for example, restrict oneself to saturated secondary structures, which consist solely of maximally extended stacking regions, i.e. no adjacent single-stranded nucleotides exist that could form a base pair and stack on top of a helix [26]. A particularly interesting application of the ADP framework is RNAshapes [37] which can be used to systematically generate (sub)optimal RNA structures belonging to distinct course-grained structural classes. For example, one can search for the most stable clover-leaf shaped secondary structure that can be formed by the input sequence.

5.6 Notes

Due to space restrictions we only gave a brief sketch of the SCFG approach to RNA secondary structures. A variety of implementations of SCFG-based algorithms are available for different purposes: pfold [74, 75] as an SCFG-approach to "folding an alignment" similar in spirit to the thermodynamics-based RNAalifold.

A general approach to computing suboptimal parse trees, similar in spirit to the backtracing of RNA secondary structures with suboptimal energies, is described in Ref. [70]. A systematic comparison of several alternative grammar models for RNA secondary structures showed that the actual performance of SCFGs can depend considerably on the details of the grammar being used [23].

A practical problem for the application of SCFGs is that one needs a grammar that is both unambiguous and in Chomsky normal form. The decomposition of Figure 6, for example, does not satisfy this requirement, because the last case in the second line, for example, requires nonterminals for the closing base pair as well as for the two enclosed multiloop components. Without discussing the details here, this creates problems in particular with the multiloop decomposition.

Sean Eddy's Infernal [25] creates a covariance model from local alignments and can be used to search a sequence database for sequences that are likely to be produced from this SCFG. Rsearch [73] aligns an RNA query to target sequences, using SCFG algorithms to score both secondary structure and primary sequence alignment simultaneously.

So-called pair SCFGs can be used to solve the combined folding and alignment problem in analogy to Sankoff's algorithm described in the next section, (e.g. Ref. [63]). The QRNA program [111] uses a pair SCFG to compute the probability that the substitution pattern in a pairwise alignment is derived from RNA secondary structure conservation. It has been used successfully to predict ncRNA candidates in *Escherichia coli* and *Saccharomyces cerevisiae* [94, 112]. Most recently, Pedersen and coworkers [102, 103] devised an SCFG-based algorithm for detecting conserved secondary structure motifs specifically within coding sequences. An SCFG-like approach to pseudoknotted structures can be found in Ref. [11].

6 Comparison of Secondary Structures

Many classes of functional RNA molecules, including tRNAs, rRNAs and many other "classical" ncRNAs, are characterized by highly conserved secondary structures, but little detectable sequence similarity. Reliable multiple alignments can therefore be constructed only when the shared structural features are taken into account. Since multiple alignments are used as input for many subsequent methods of data analysis, structure-based alignments are an indispensable necessity in RNA bioinformatics. This problem is far from being solved in a satisfactory way, both because the available approaches are computationally expensive and because little is known about the evolution of RNA at the structural level, and hence on the appropriate edit cost parameters.

6.1 String-based Alignments

The problem of comparing two structures Ψ_1 and Ψ_2 of the *same* RNA molecule is trivial. Since a secondary structures is simply a set of base pairs one may use, for example, the size of the symmetric difference between the two sets $|\Psi_1 \triangle \Psi_2|$ as a distance measure that is obviously a metric. In other words, we simply count the number of base pairs that occur in one of the structures, but not in both,

The question immediately becomes nontrivial, however, if we do not assume that the two structures have the same underlying sequence length, i.e. if we do not know *a priori* which sequence positions in the two molecules correspond to each other.

As we have seen, RNA secondary structures can be faithfully represented as strings over the alphabet $\{(,),.\}$. Clearly, we can use this string representation to compute a metric on secondary structures by means of standard sequence alignment methods, e.g. using the Needleman–Wunsch algorithm [99].

This approach can be generalized to a comparison of base pair probability matrices [7]. From the pairing probabilities of base i we construct a vector containing the probabilities of being paired upstream $p^{<}(i) = \sum_{j>i} P_{ij}$, downstream $p^{>}(i) = \sum_{j<i} P_{ji}$ or unpaired $p^{\circ}(i) = 1 - p^{<}(i) - p^{>}(i)$. The resulting profiles can be aligned by means of a standard string/profile alignment algorithm in $\mathcal{O}(n^2)$ time using:

$$\rho = \sqrt{p_A^{>} p_B^{>}} + \sqrt{p_A^{<} p_B^{<}} + \sqrt{p_A^{\circ} p_B^{\circ}}, \tag{21}$$

as the match score (or $1 - \rho$ as an edit cost). While this approach of "string-like alignments" is fast, it often produces misaligned pairs (Figure 13).

Sequence alignment	Structure alignment
CAGUCUCAGGUGGUUGGGCU-	CAGUCUCAGGUGGUUG-GGCU
.((((.(((....))))))))-	.((((.(((....)))-))))
UAG-CUGAGGUG-UCGUGCUA	-UAGC-UGAGGUGUCGUGCUA
(((-((((....-))).))))	-((((-(((....))).))))

Figure 13 Sequence versus structure alignment. Compared to the structural alignment (right), the sequence alignment (from ClustalW) misaligns five of the seven base pairs.

6.2 Tree Editing

The string-based alignments above essentially use only the information whether a nucleotide is paired or unpaired, but neglect the connectivity information who pairs with whom. This limitation can be overcome by methods based on the tree representation of secondary structures. Of particular interest are tree editing and the related tree alignment, since they are still fast enough to be applicable to genome wide surveys. We present these approaches in detail here since there does not appear be a good textbook exposition of this topic.

The three most natural operations ("moves") that can be used to convert ordered trees (and, more generally, ordered forests) into each other are depicted in Figure 14:

(i) *Substitution* $(x \to y)$ consists of replacing a single vertex label x by another vertex label y.

(ii) *Insertion* $(\varnothing \to z)$ consists of adding a vertex z as a child of x, thereby making z the parent of a consecutive subsequence of children of x. A node z can also be inserted at the "top level", thereby becoming the root of a tree.

(iii) *Deletion* $(z \to \varnothing)$ consists of removing a vertex z, its children thereby become children of the parent x of z. Removing the root of a tree produces a forest in which the children of z become roots of trees.

Naturally, we associate a *cost* with each edit operation, which we will denote by $\gamma(x \to y)$, $\gamma(\varnothing \to z)$ and $\gamma(z \to \varnothing)$ for substitutions, insertions and deletions, respectively. We assume that γ is a metric on the extended alphabet $\mathcal{A} \cup \{\varnothing\}$. By using an appropriate alphabet of vertex labels, one can easily include sequence information in the cost function.

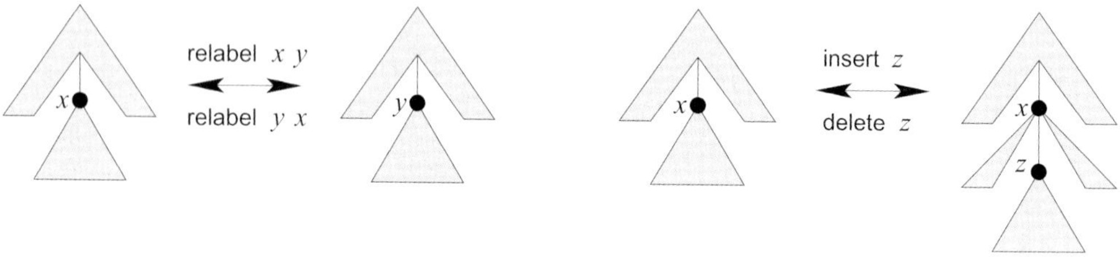

Figure 14 Elementary operations in tree editing

A sequence of moves that transforms a forest F_1 into a forest F_2 is known as an *edit script*. Its cost is the sum of the costs of edit operations in the script.

A *mapping* from F_1 to F_2 is a binary relation $M \in V(F_1) \times V(F_2)$ between the vertex sets of the two forests such that for pairs $(x, y), (x', y') \in M$ holds

(i) $x = x'$ if and only if $y = y'$ (one-to-one condition).

(ii) x is an ancestor of x' if and only if y is an ancestor of y' (ancestor condition).

(iii) x is to the left of x' if and only if y is to the left of y' (sibling condition).

By definition, for each $x \in F_1$ there is a unique "partner" in $y \in F_2$ such that $(x, y) \in M$ or there is no partner at all. In the latter case we write $x \in M_1'$. Analogously, we write $y \in M_2'$ if $y \in F_2$ does not have a partner in F_1. With each mapping we can associate the cost:

$$\gamma(M) = \sum_{(x,y)\in M} \gamma(x \to y) + \sum_{y\in M_2'} \gamma(\varnothing \to y) + \sum_{x\in M_1'} \gamma(x \to \varnothing). \qquad (22)$$

Clearly, each edit operation gives rise to a corresponding mapping between the initial and the final tree. In the case of a substitution, all vertices have partners; in the case of insertion and deletion, there is exactly one vertex without partner.

Mappings are relations and hence they can be composed in a natural way. Consider three forests F_1, F_2 and F_3 and mappings M_1 from F_1 to F_2 and M_2 from F_2 to F_3. Then:

$$M_1 \circ M_2 = \left\{(x, z) \,\middle|\, \exists y \in V(F_2) \text{ such that } (x, y) \in M_1 \text{ and } (y, z) \in M_2\right\}, \qquad (23)$$

is a mapping from F_1 to F_3. It is easy convince oneself that the cost function defined in Eq. (22) is subadditive under composition, $\gamma(M_1 \circ M_2) \leq \gamma(M_1) + \gamma(M_2)$. Using this result and the fact that every mapping can be obtained as a composition of edit operations one can show that the minimum cost mapping is equivalent to the minimum cost edit script [128].

For a given forest F we note by $F - x$ the forest obtained by deleting x and $F \setminus T(x)$ is the forest obtained from F by deleting with x all descendants of x. Note that $T(x) - x$ is the forest consisting of all trees whose roots are the children of x.

Now consider two forests F_1 and F_2, and let v_i be the root of the right-most tree in F_i, $i = 1, 2$ and an optimal mapping M. Apart from the trivial cases, in which one of the two forests is empty, we have to distinguish three cases. (i) v_2 has no partner in the optimal mapping. In this case, v_2 is inserted and the optimal mapping consists of an optimal mapping from F_1 to $F_2 - v_2$ composed with the insertion of v_2. (ii) v_1 has no partner. This corresponds to the deletion of v_1. (iii) both v_1 and v_2 have partners. In this case $(v_1, v_2) \in M$.

To see this, one can argue as follows. Suppose $(v_1, h) \in M$, $h \neq v_2$ and $(k, v_2) \in M$. By the one-to-one condition, $k \neq v_1$. By the sibling condition, if v_1 is to the right of k, then h must be to the right of v_2. If v_1 is a proper ancestor of k, then h must be a proper ancestor of v_2 by the ancestor condition. Both cases are impossible, however, since both v_1 and v_2 are by construction right-most roots.

For each of the three cases it is now straightforward to recursively compute the optimal cost of M. We arrive directly at the dynamic programming recursion:

$$D(F_1, F_2) = \min \begin{cases} D(F_1 - v_1, F_2) + \gamma(v_1 \to \varnothing), \\ D(F_1, F_2 - v_2) + \gamma(\varnothing \to v_2), \\ D(T(v_1) - v_1, T(v_2) - v_2) + \\ \qquad\qquad D(F_1 \setminus T(v_1), F_2 \setminus T(v_2)) + \gamma(v_1 \to v_2). \end{cases} \tag{24}$$

which allows us to compute the tree edit distance $D(F_1, F_2)$ from smaller subproblems. The initialization is the distance between $D(\emptyset, \emptyset) = 0$ of two empty forests. In the cases where one of the two forests is empty, Eq. (24) reduces to $D(\emptyset, F_2) = D(\emptyset, F_2 - v_2) + \gamma(\varnothing \to v_2)$ and $D(F_1, \emptyset) = D(F_1 - v_1, \emptyset) + \gamma(v_1 \to \varnothing)$.

One can show that the time complexity of this algorithm is bounded by $\mathcal{O}(|F_1|^2|F_2|^2)$. Various more efficient implementations exist (see in particular Refs. [72, 148]). A detailed performance analysis of the algorithm by Zhang and Shasha [148] is given in Ref. [24].

A common feature of all tree representations discussed above is that each subtree $T(x)$ rooted at a vertex x corresponds to an interval I_x of the underlying RNA sequence. We can thus regard every pair (v_1, v_2) as a prescription to match up the intervals I_{v_1} with J_{v_2} between the two input sequences. In particular, if v_1 and v_2 are leaves in the forests F_1 and F_2, then they correspond to individual bases. Interior nodes serve as delimiters of intervals in Giegerich's encoding, while they correspond to base pairs in the encoding used in the Vienna RNA Package. In either case, one can derive all (mis)matches directly from M. The sibling and ancestor properties of M guarantee that (mis)matches preserve the order in which they appear on the RNA sequence. All other nucleotides, i.e. those that correspond to vertices $v_1 \in M_1'$ and $v_2 \in M_2'$, are deleted or inserted, respectively, in the appropriate positions. Every mapping M therefore implies a (canonical) pairwise alignment $\mathbb{A}(M)$ of the underlying sequences.

6.3 Tree Alignments

An alternative way of defining the difference of two forests is using *tree alignments* [69]. Consider a forest G with vertex labels taken from $(\mathcal{A} \cup \{-\}) \times (\mathcal{A} \cup \{-\})$. Then we obtain restrictions $\pi_1(G)$ and $\pi_2(G)$ by considering only the first or the second coordinate of the labels, respectively, and by then deleting all nodes that are labeled with the gap character "–" (Figure 15). We say that G is an alignment of the two forests F_1 and F_2 if $F_1 = \pi_1(G)$ and $F_2 = \pi_2(G)$. Naturally, we score the alignment G by adding up the costs of the

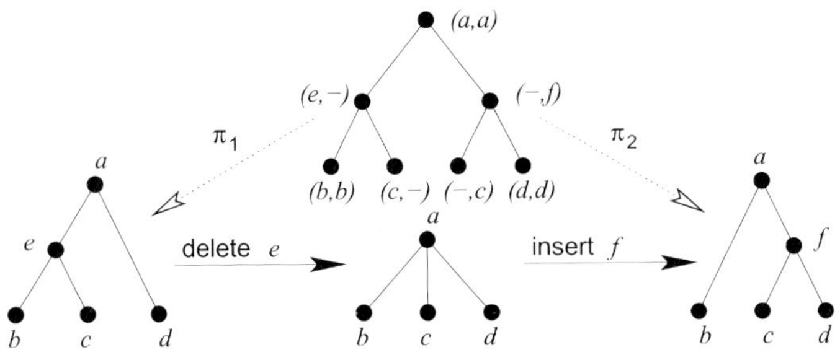

Figure 15 Alignment of two forests F_1 and F_2 and a mapping from F_1 to F_2 that cannot be derived from an alignment.

label pairs:

$$\gamma(G) = \sum_{(v_1,v_2)\in G} \gamma(v_1 \to v_2), \tag{25}$$

where a pair $(v_1, -)$ corresponds to the edit operation $(v_1 \to \varnothing)$. On the other hand, an alignment G defines a mapping M_G from F_1 to F_2 by setting $(v_1, v_2) \in M_G$ iff $(v_1, v_2) \in G$ and neither v_1 nor v_2 is a gap character. One easily verifies that the three defining properties of mapping are satisfied. Furthermore, it follows that $\gamma(M_G) = \gamma(G)$ as pair of the form $(v_1, -)$ and $(-, v_2)$ corresponds to deletion and insertion operations, respectively. Note that, in the special case of two a totally disconnected forests, the problem reduces to ordinary sequence alignment with additive gap costs.

However, as the example in Figure 15 shows, not all mappings derive from alignments. It follows, therefore, that the minimum cost alignment is more costly than the minimum cost edit script, in general.

In order to compute the optimal alignment, let us first investigate the decomposition of an alignment at a particular (mis)match (v_1, v_2) or in/del $(-, v_2)$ or $(v_1, -)$. We will need a bit of notation (Figure 16). Let F be an ordered forest. By $i : F$ we denote the subforest consisting of the first i trees, while $F : j$ denotes the subforest starting with the $j + 1$-th tree. By $F^{\downarrow}$ we denote forest consisting of the children trees of the root $v = r_F$ of the first tree in F. $F^{\rightarrow} = F : 1$ is the forest of the right siblings trees of F.

Now consider an alignment A of two forests F_1 and F_2. Let $a = r_A$ be the root of its first tree. We have either:

(i) $a = (v_1, v_2)$. Then $v_1 = r_{F_1}$ and $v_r = r_{F_2}$; $A^{\downarrow}$ is an alignment of $F_1^{\downarrow}$ and $F_2^{\downarrow}$; $A^{\rightarrow}$ is an alignment of $F_1^{\rightarrow}$ and $F_2^{\rightarrow}$.

(ii) $a = (-, v_2)$. Then $v_2 = r_{F_2}$; for some k, $A^{\downarrow}$ is an alignment of $k : F_1$ and $F_2^{\downarrow}$ and $A^{\rightarrow}$ is an alignment of $F_1 : k$ with $F_2^{\rightarrow}$.

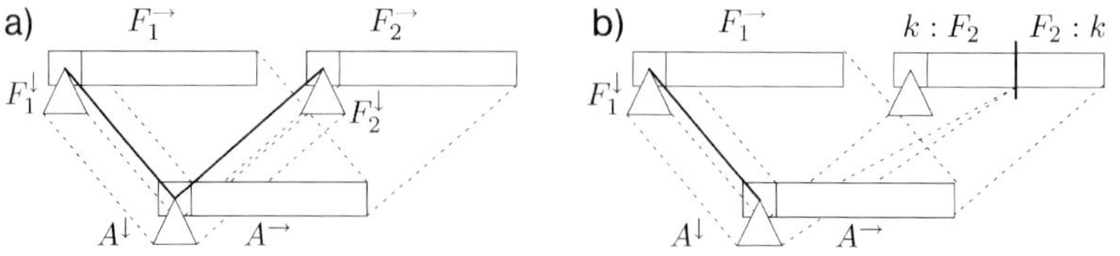

Figure 16 Decomposition of tree alignments. (a) In the match case the subtrees $F_1^{\downarrow}$ and $F_2^{\downarrow}$ are aligned to form $A^{\downarrow}$ and, correspondingly, the sibling subforests $F_1^{\rightarrow}$ and $F_2^{\rightarrow}$ must be aligned to yield $A^{\rightarrow}$. (b) In the deletion case the subforest $F_1^{\downarrow} - v_1$ must be aligned with a part $k : F_2$ of the second forest. $F_1^{\rightarrow}$ then must be aligned with remainder of $F_2 : k$ of the top-level trees of F_2. The insertion case is analogous to the deletion case, with the roles of F_1 and F_2 exchanged.

(iii) $a = (v_1, -)$. Then $v_1 = r_{F_1}$; for some k, $A^{\downarrow}$ is an alignment of $F_1^{\downarrow}$ and $k : F_2$ and $A^{\rightarrow}$ is an alignment of $F_1^{\rightarrow}$ with $F_2 : k$.

See Figure 16 for a graphical representation.

Let $S(F_1, F_2)$ be the optimal score of an alignment of the forests F_1 and F_2. For easier comparison with the tree-editing algorithm in the previous section we formulate the problem here as a minimization problem. One can, however, just as well maximize appropriate similarity scores. The three cases discussed above and in Figure 16 imply the following dynamic programming recursion:

$$S(F_1, F_2) = \min \begin{cases} S(F_1^{\downarrow}, F_2^{\downarrow}) + S(F_1^{\rightarrow}, F_2^{\rightarrow}) + \gamma(v_1 \rightarrow v_2) \\ \min_k S(k : F_1, F_2^{\downarrow}) + S(F_1 : k, F_2^{\rightarrow}) + \gamma(\varnothing \rightarrow r_{F_2}) \\ \min_k S(F_1^{\downarrow}, k : F_2) + S(F_1^{\rightarrow}, F_2 : k) + \gamma(r_{F_1} \rightarrow \varnothing). \end{cases} \quad (26)$$

In the special cases where one of the forests is empty this reduces to $S(\emptyset, F_2) = S(\emptyset, F_2^{\downarrow}) + S(\emptyset, F_2^{\rightarrow}) + \gamma(\varnothing \rightarrow r_{F_2})$ for the insertion case, and $S(F_1, \emptyset) = S(F_1^{\downarrow}, \emptyset) + S(F_1^{\rightarrow}, \emptyset) + \gamma(r_{F_1} \rightarrow \varnothing)$ for the deletion case. The initial condition is again $S(\emptyset, \emptyset) = 0$.

In order to estimate the resource requirements for this algorithm, we observe that we have to consider only those subforests of F_1 and F_2 that consist of trees rooted at an uninterrupted interval of sibling nodes. These forests have been termed *closed subforests* in Ref. [51]. If d_i is the maximum of the number of trees and the numbers of children of the nodes in F_i, we see that there are at most $\mathcal{O}(d_i^2)$ closed subforests at each node and hence at most $\mathcal{O}(|F_1|\,|F_2|d_1^2 d_2^2)$ entries $S(F_1, F_2)$ need to be computed, each of which requires $\mathcal{O}(d_1 + d_2)$ operations, i.e. tree alignments can be computed in polynomial time. A compact, memory-efficient encoding of the subforests is described in detail in Ref. [51], where a careful analysis shows that pairwise tree alignments can be computed in $\mathcal{O}(|F_1|d_1\,|F_2|d_2)$ space and $\mathcal{O}(|F_1|\,|F_2|\,d_1\,d_2(d_1 + d_2))$ time.

6.4 The Sankoff Algorithm and Variants

David Sankoff described an algorithm that simultaneously allows the solution of the structure prediction and the sequence alignment problem [114]. The basic idea is to search for a maximal secondary structure that is common to two RNA sequences. Given a score $\sigma_{ij,kl}$ for the alignment of the base pairs (i,j) and (k,l) from the two sequences (as well as gap penalties γ and scores α_{ik} for matches of unpaired positions) we compute the optimal alignment recursively from alignments of the subsequences $x[i,\ldots,j]$ and $y[k,\ldots,l]$. Let $S_{ij,kl}$ be the score of the optimal alignment of these fragments. We have:

$$\begin{aligned} S_{ij;kl} = \max\Big\{ & S_{i+1,j;kl} + \gamma, S_{ij;k+1,l} + \gamma, S_{i+1,j;k+1,l} + \alpha_{ik}, \\ & \max_{(p,q)\text{ paired}} \left\{ S_{i+1,p-1;k+1,q-1} + \sigma_{ij,pq} + S_{p+1,j;q+1,l} \right\} \Big\} . \end{aligned} \tag{27}$$

Backtracing is just as easy as in the RNA folding case. Only now π is a partial alignment of two structures and we insert aligned positions instead of positions in individual structures. More precisely we have to insert individual columns or pairs of columns of the form:

$$\pi \blacktriangleleft \begin{pmatrix} i. \\ - \end{pmatrix} \quad \pi \blacktriangleleft \begin{pmatrix} - \\ j. \end{pmatrix} \quad \pi \blacktriangleleft \begin{pmatrix} i. \\ j. \end{pmatrix} \qquad \pi \blacktriangleleft \begin{pmatrix} i(& , & j) \\ p(& , & q) \end{pmatrix}, \tag{28}$$

into a growing partial alignment π, just as we insert unpaired bases or base pairs in the backtracing of the folding algorithm in Section 3.2.

This algorithm is computationally very expensive, however. It requires $\mathcal{O}(n^4)$ memory and $\mathcal{O}(n^6)$ CPU time. Currently available software packages such as foldalign [39,65] and dynalign [92] therefore implement only restricted versions. The simple, maximum matching style version is used in pmcomp [52] as an approach to comparing base pairing probability matrices.

6.5 Multiple Alignments

Pairwise alignment methods, be they for sequences or structures, can be readily generalized to alignments of many objects. Usually, it is too costly to compute optimal multiple alignments exactly and one therefore resorts to heuristics such as progressive multiple alignments. pmmulti [52], for example, produces multiple structural alignments in the context of the Sankoff algorithm by calling pmcomp for pairwise alignments. For tree alignments, the RNAforester programs can be used to compute both pairwise and progressive multiple alignments.

As we have seen above, the mappings produced by tree editing do not correspond to tree alignments in general. These methods can therefore not be used for comparing multiple structures. The edit scripts can, however, be interpreted in terms of a sequence alignment. One may therefore still use these methods as the starting point for multiple sequence alignments. This is the central idea of the MARNA program [121] which uses pairwise structural alignments as input to the multiple alignment program T-Coffee [100].

6.6 Notes

Various variants, specializations and generalizations of the tree-editing approach have been described in recent years. Examples include efficient algorithms for similar trees [67] and with simplified edit cost models [120]. Tree-editing with restricted mappings M satisfying stronger requirements on structural conservation are described in Ref. [147]. Let $\text{lca}(a,b)$ denote the "last common ancestor" of a and b. For all $(x',x''),(y',y''),(z',z'') \in M$ holds: $\text{lca}(x',y')$ is a proper ancestor of z' if and only if $\text{lca}(x'',y'')$ is a proper ancestor of z''. Other variants of tree edit distances have also been discussed (e.g. Ref. [132]). A tree-edit model for RNA that allows additional "node-fusion" and "edge-fusion" events is described in Ref. [4].

More general edit models with application to RNA structures are described in Refs. [68, 88], an alignment distance for pseudoknotted structures can be found in Ref. [9].

A very different approach to the pairwise comparison of RNA structures, with or without pseudoknots, converts the RNA alignment problem into an integer programming problem [79]. Recently, efficient algorithms based on Lagrangian relaxation have been developed [6], that have helped to make the performance of this approach comparable to other methods.

A partition function version of the Sankoff algorithm, which can be used to compute the probabilities of all possible (mis)matches in a structural alignment of two RNA base pairing probability matrices is described in Ref. [53]. RNA structure comparison can also be recast in the SCFG framework [64]. The corresponding pair-SCFG algorithms correspond to the Sankoff algorithm.

7 Kinetic Folding

7.1 Folding Energy Landscapes

The folding dynamics of a particular RNA molecule can also be studied successfully within the framework of secondary structures. The folding process is determined by the energy landscape [or potential energy surface (PES) in the terminology of theoretical chemistry]. Instead of considering all possible spatial conformations, it is meaningful to partition the conformation space into sets of conformations that belong to a given secondary structure. Instead of a smooth surface defined on a space of real-valued coordinate vectors we are therefore dealing with a landscape on a complex graph [107]. The vertices of this graph are the secondary structures that can be formed by the given RNA sequence, the edges are determined by a rule specifying which structures can be interconverted and the height of the landscape at a structure x is its free energy $E(x)$. Typically, one considers a "move set" that allows the insertion and deletion of single base pairs. In addition, a shift-move that changes (i, j) to (i, k) or (h, j) is sometimes included [30]. Further coarse-grainings of this landscape can be achieved, e.g. by considering secondary structures as composed of stacks instead of individual base pairs.

7.2 Kinetic Folding Algorithms

Several groups have designed kinetic folding algorithms for RNA secondary structures, mostly in an attempt to obtain more accurate predictions or in order to include pseudoknots (see e.g. Refs. [2, 41, 89, 95, 126]). Only a few papers have attempted to reconstruct folding pathways [42, 50, 124]. These algorithms generally operate on a list of all possible helices and consequently use move sets that destroy or form entire helices in a single move. Such a move set can introduce large structural changes in a single move and, furthermore, *ad hoc* assumptions have to be made about the rates of helix formation and disruption. A more local move set is, therefore, preferable if one hopes to observe realistic folding trajectories.

The process of kinetic folding itself can be modeled as homogeneous Markov chain. The probability p_x that a given RNA molecule will have the secondary structure x at time t is given by the master equation:

$$\frac{\mathrm{d}p_x}{\mathrm{d}t} = \sum_{y \in X} r_{xy} p_y(t), \tag{29}$$

where r_{xy} is the rate constant for the transition from secondary structure y to secondary structure x in the deterministic description [38]. The transition state model dictates an expression of the form:

$$r_{yx} = r_0 \mathrm{e}^{-\frac{E^{\neq}_{yx} - E(x)}{RT}} \quad \text{for } x \neq y \quad \text{and} \quad r_{xx} = -\sum_{y \neq x} r_{yx}, \tag{30}$$

where the transition state energies $E^{\neq}_{yx}$ must be symmetric, $E^{\neq}_{yx} = E^{\neq}_{xy}$, and r_0 is a scaling constant. In the simplest case one can use:

$$E^{\neq}_{yx} = \max\{E(x), E(y)\}. \tag{31}$$

For short sequences or very restricted subsets of conformations Eq. (29) can be solved exactly or integrated numerically [126]. Solving the master equation for larger conformation spaces is out of the question. In such cases the dynamics can be obtained by simulating the Markov chain directly by a rejection-less Monte Carlo algorithm [27] and sampling a large number of trajectories.

7.3 Approximate Folding Trajectories and Barrier Trees

An alternative approach to the direct simulation of the master equation (29) starts with a more detailed analysis of folding energy landscape. Let us start with a few definitions.

A conformation x is a global minimum if $E(x) \leq E(y)$ for all $y \in X$ and a local minimum if $E(x) \leq E(y)$ for all neighbors y of x. The energy $\hat{E}$ of the lowest saddle point separating two local minima x and y is:

$$\hat{E}[x,y] = \min_{\mathbf{p} \in \mathbb{P}_{xy}} \max_{z \in \mathbf{p}} E(z), \tag{32}$$

where $\mathbb{P}_{xy}$ is the set of all paths $\mathbf{p}$ connecting x and y by a series of consecutive transformations taken from the move set. If the energy function is nondegenerate then there is a unique saddle point $s = s(x,y)$ connecting x and y characterized by $E(s) = \hat{E}[x,y]$. To each saddle point s there is a unique collection of conformations $B(s)$ that can be reached from s by a path along which the energy never exceeds $E(s)$. In other words, the conformations in $B(s)$ are mutually connected by paths that never go higher than $E(s)$. This property warrants to call $B(s)$ the *basin of attraction* below the saddle s.

Two situations can arise for any two saddle points s and s' with energies $E(s) < E(s')$. Either the basin of s is a "subbasin" of $B(s')$ or the two basins are disjoint. This property arranges the local minima and the saddle points in a unique hierarchical structure which is conveniently represented as a tree, termed a *barrier tree* (Figure 17).

An efficient flooding algorithm [31] can be used to identify local minima and saddle points starting, for example, from the complete list of suboptimal secondary structures produced by the RNAsubopt program [144]. Consider a stack Σ which initially contains all secondary structures in the order of ascending energy. We pop the element z from the top of Σ and check which of its neighbors we have already seen before, i.e. which of its neighbors have a lower energy. There are three cases:

(i) z has no neighbor with lower energy, then it is a local minimum, i.e. a new leaf of the barrier tree.

(ii) z has only lower energy neighbors that all belong the same basin, say $B(x)$. Then z itself also belongs to $B(x)$.

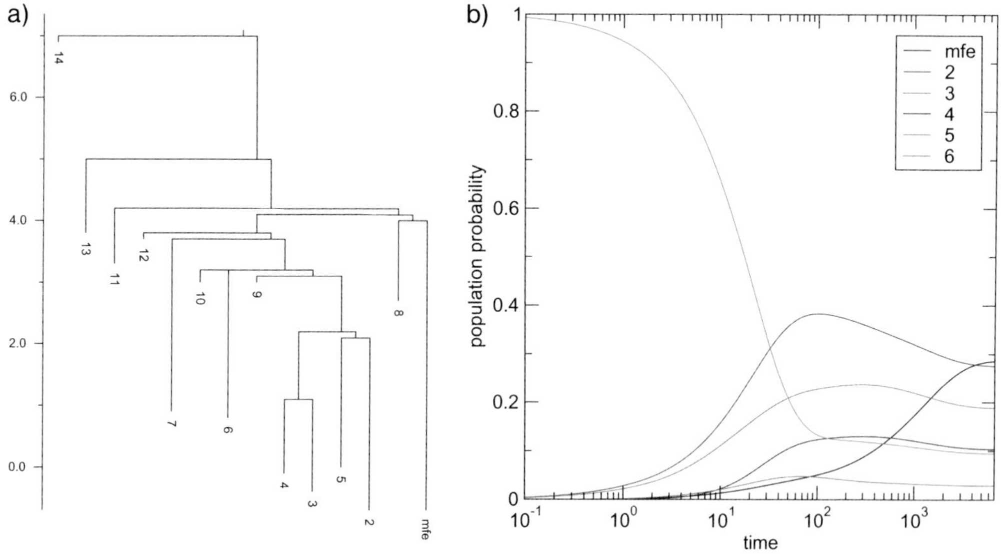

Figure 17 (a) Barrier tree of short artificial sequence UAUGCUGCGGCCUAGGC. The leaves of the tree are the local minima of the energy landscape. (b) Folding kinetics from the open structure. Population density p_α for the basin containing the local minimum α is shown for the six largest basins as a function of time.

(iii) z has lower-energy neighbors in two or more different basins. In this case z is the saddle point separating these basins, i.e. an interior vertex of the barrier tree. For the subsequent computation we now unify all basins connected by z into a new basin $B(z)$ and remove its subbasins from the list of "active" basins.

At the end, we are left with the barrier tree of the landscape. As a byproduct we also obtain the assignment of each secondary structure to its basin $B(x)$. Instead of searching through the list of all previously encountered structures it is more efficient to generate all neighbors of z and to check whether they have already been seen before by means of a hash-table lookup. The procedure thus runs in $\mathcal{O}(LD)$ time, where L is the length of the list of structures and D is the maximal number of moves that can be applied to a secondary structure. The barriers program implements this algorithm [31].

A description of the energy landscape or the dynamics of an RNA molecule based on all secondary structures is feasible only for very small sequences. We therefore need to coarse-grain the representation of the energy landscape. Let $\Pi = \{\alpha, \beta, \dots\}$ be a partition of the state space. The classes of such a partition are *macrostates*. As a concrete example consider the partition of X defined by the gradient basins $\mathcal{B}(z)$ of the local energy minima. To each macrostate α we can assign the partition function:

$$Z_\alpha = \sum_{x\in\alpha} e^{-E(x)/RT}, \tag{33}$$

and the corresponding free energy:

$$G(\alpha) = -RT \ln Z_\alpha. \tag{34}$$

The transition rates between macrostates can be obtained at least approximately from the elementary rate constants using the assumption that the random process is equilibrated within each macrostate [143]. Then:

$$r_{\beta\alpha} = \sum_{y\in\beta}\sum_{x\in\alpha} r_{yx} \frac{e^{-E(x)/RT}}{Z_\alpha} \quad \text{for } \alpha \neq \beta. \tag{35}$$

We can use the transition state model to define the free energies of the transition state $G^{\neq}$ by setting:

$$r_{\beta\alpha} = r_0 e^{-\frac{G^{\neq}_{\beta\alpha} - G(\alpha)}{RT}}. \tag{36}$$

A short computation then yields:

$$G^{\neq}_{\beta\alpha} = -RT \ln \sum_{y\in\beta}\sum_{x\in\alpha} e^{-\frac{E^{\neq}_{yx}}{RT}}, \tag{37}$$

as one would expect.

In practice one can compute $r_{\beta\alpha}$ "on the fly" while executing the barriers program if two conditions are satisfied: (i) for each x we can efficiently determine to which macrostate it belongs and (ii) the double sum in Eq. (35) needs to be evaluated only for pairs of neighboring conformations (x, y). Condition (i) is easily satisfied for each of the gradient basins: in each step of the barriers algorithm all neighbors y of the newly added structures x that have a smaller energy have already been processed. Condition (ii) is satisfied by construction of the microscopic transition rates r_{xy}, which vanish unless x is a neighbor of y. In the case of short sequence, both the microscopic model and the macro-state model can be solved exactly. In many cases (e.g. Figure 17) the macro-state model provides a very good approximation of the dynamics.

7.4 RNA Switches

Some RNA molecules exhibit two meta-stable conformations, whose equilibrium can be shifted easily by external events, such as binding of another molecule. This can be used to regulate gene expression, when the two mutually exclusive alternatives correspond to an active and in-active conformation of the transcript. The best known example of such behavior are the riboswitches [133] found in the 5′ untranslated regions of bacterial mRNAs, where the conformational change is triggered by binding of a small organic molecule.

Molecules that may be RNA switches can be recognized by inspection of the barrier tree, but this is feasible only for rather short sequences. The paRNAss program [134] instead uses a sample of suboptimal structures, and computes for every pair of structures "morphological" distance (e.g. tree edit distance) and a simple estimate of the energy barrier. The structures are then clustered according to these two measures, RNA switches are expected to exhibit two well separated clusters.

Interestingly, for any two secondary structures there exist sequences that are compatible with both structures, i.e. that can form both structures in principle [106]. If both structures are reasonably stable, it is not hard to design switching sequences with these two structures as stable conformations [28].

7.5 Notes

The analysis of landscapes becomes technically more complicated when structures, in particular adjacent structures, may have the same energy. In this case there is no unique definition of gradient basins and a variety of concepts, all related to saddle points, have to be distinguished (see Ref. [31] for further details).

The notion of barrier trees can be generalized to multivalued landscapes, which arise, for example, in the context of multiobjective optimization problems with conflicting constraints [122].

A computationally simpler alternative to the macrostate approach for transition rate is to assume an Arrhenius law $r_{\beta\alpha} \sim \exp(E^{\neq}/RT)$ and to approximate the transition state energy $E^{\neq}$ by the energy of the saddle point between the local minima α and β [143]

A generalization of the "intersection theorem" characterizes sets of more than two secondary structures that can be realized simultaneously by a common RNA sequence [28]. This observation can be used as a starting point for computational designs of switches with multiple states [1].

8 Concluding Remarks

Secondary structure drives the RNA-folding process, arguably even more so than it is the case for proteins. This renders the prediction of RNA secondary structure highly relevant for the prediction of RNA structure and analysis, in general. As this chapter shows, the field of RNA structure prediction is comparatively well developed. As a matter of fact, it is one of the fields in bioinformatics that benefits most comprehensively from algorithmic methods derived from computer science. The comparatively technical makeup of this chapter is a mirror of this phenomenon. Notably, very different questions, which in the protein world require different mathematical models, can be described and analyzed in the RNA case at the level of secondary structures: the thermodynamics of folding as well as the thermodynamics of RNA–RNA interactions are accessible via the same parameters and the same algorithms that can also be used to compute consensus structures in an evolutionary context or to investigate the dynamics of the folding process itself.

With the increased importance of RNA in biology, in general (consider, for instance, the recent surge in work on RNA interference (see also Chapter 45) and in the analysis of structural aspects of mRNA in the context of gene regulation), RNA secondary structure prediction is rapdily becoming an obligatory tool in the arsenal of bioinformatics analysis methods.

References

1 ABFALTER, I., C. FLAMM, AND P. F. STADLER. 2003. Design of multi-stable nucleic acid sequences. In Proc. Proc. German Conf. Bioinformatics, München, 1–7.

2 ABRAHAMS, J. P., M. VAN DEN BERG, E. VAN BATENBURG, AND C. PLEIJ. 1990. Prediction of RNA secondary structure, including pseudoknotting, by computer simulation. Nucleic Acids Res. **18**: 3035–44.

3 AKUTSU, T.. 2001. Dynamic programming algorithms for RNA secondary structure prediction with pseudoknots. Discr. Appl. Math. **104**: 45–62.

4 ALLALI, J. AND M.-F. SAGOT. 2005. A new distance for high level RNA secondary structure comparison. IEEE/ACM Trans. Comp. Biol. Bioinf. **2**: 3–14.

5 ANDRONESCU, M., Z. ZHANG, AND A. CONDON. 2005. Secondary structure prediction of interacting RNA molecules. J. Mol. Biol. **345**: 987–1001.

6 BAUER, M. AND G. KLAU. 2004. Structural alignment of two RNA sequences with Lagrangian relaxation. Int. Symp. on Algorithms and Computation. Hong Kong: 113–25.

7 BONHOEFFER, S., J. S. MCCASKILL, P. F. STADLER, AND P. SCHUSTER. 1993. RNA multi-structure landscapes. a study based on temperature dependent partition functions. Eur. Biophys. J. **22**: 13–24.

8 BONNET, E., J. WUYTS, P. ROUZÉ, AND Y. VAN DE PEER. 2004. Evidence that microRNA precursors, unlike other non-coding RNAs, have lower folding free energies than random sequences. Bioinformatics **20**: 2911–17.

9 BRINKMEIER, M.. 2005. Structural alignments of pseudo-knotted RNA-molecules in polynomial time. Technical Report, TU Ilmenau.

10 BROWN, J. W., J. M. NOLAN, E. S. HAAS, M. A. T. RUBIO, F. MAJOR, AND N. R. PACE. 1996. Comparative analysis of ribonuclease P RNA using gene sequences from natural microbial populations reveals tertiary structural elements. Proc. Natl Acad. Sci. USA **93**: 3001–6.

11 CAI, L., R. L. MALMBERG, AND Y. WU. 2003. Stochastic modeling of RNA

pseudoknotted structures: a grammatical approach. Bioinformatics **19**(Suppl. 1): i66–73.

12 CANNONE, J. J., S. SUBRAMANIAN, M. N. SCHNARE, ET AL. 2002. The comparative RNA web (CRW) site: an online database of comparative sequence and structure information for ribosomal, intron, and other RNAs. BMC Bioinformatics **3**: 2.

13 CHEN, W. Y. C., E. Y. P. DENG, AND R. R. X. DU. 2005. Reduction of *m*-regular noncrossing partitions. Eur. J. Comb. **26**: 237–43.

14 CHIU, D. K. AND T. KOLODZIEJCZAK. 1991. Inferring consensus structure from nucleic acid sequences. CABIOS **7**: 347–52.

15 CLOTE, P.. 2005. An efficient algorithm to compute the landscape of locally optimal RNA secondary structures with respect to the Nussinov–Jacobson energy model. J. Comput. Biol. **12**: 83–101.

16 CONDON, A., B. DAVY, B. RASTEGARI, S. ZHAO, AND F. TARRANT. 2004. Classifying RNA pseudoknotted structures. Theor. Comput. Sci. **320**: 35–50.

17 CRISTIANINI, N. AND J. SHAWE-TAYLOR. 2000. *An Introduction to Support Vector Machines*. Cambridge University Press, Cambridge, UK.

18 DE RIJK, P. AND R. DE WACHTER. 1997. RnaViz, a program for the visualisation of RNA secondary structure. Nucleic Acids Res. **25**: 4679–84.

19 DEUTSCH, E. AND L. W. SHAPIRO. 2002. A bijection between ordered trees and 2-Motzkin paths and its many consequences. Discr. Math. **256**: 655–70.

20 DIMITROV, R. A. AND M. ZUKER. 2004. Prediction of hybridization and melting for double-stranded nucleic acids. Biophys. J. **87**: 215–26.

21 DING, Y., C. CHAN, AND C. LAWRENCE. 2004. Sfold web server for statistical folding and rational design of nucleic acids. Nucleic Acids Res. **32**: W135–41.

22 DIRKS, R. AND N. PIERCE. 2003. A parition function algorithm for nucleic acid secondary structure including pseudoknots. J. Comput. Chem. **24**: 1664–77.

23 DOWELL, R. D. AND S. R. EDDY. 2004. Evaluation of several lightweight stochastic context-free grammars for RNA secondary structure prediction. BMC Bioinformatics **5**: 71.

24 DULUCQ, S. AND L. TICHIT. 2003. RNA secondary structure comparison: exact analysis of the Zhang–Shasha tree-edit algorithm. Theor. Comput. Sci. **306**: 471–84.

25 EDDY, S.. 2002. A memory-efficient dynamic programming algorithm for optimal alignment of a sequence to an RNA secondary structure. BMC Bioinformatics **3**: 18.

26 EVERS, D. J. AND R. GIEGERICH. 2001. Reducing the conformation space in RNA structure prediction. Proc. German Conf. on Bioinformatics, Braunschweig. 118–24.

27 FLAMM, C., W. FONTANA, I. HOFACKER, AND P. SCHUSTER. 2000. RNA folding kinetics at elementary step resolution. RNA **6**: 325–38.

28 FLAMM, C., I. L. HOFACKER, S. MAURER-STROH, P. F. STADLER, AND M. ZEHL. 2001. Design of multi-stable RNA molecules. RNA **7**: 254–65.

29 FLAMM, C., I. L. HOFACKER, AND . P. F. STADLER. 2004. Computational chemistry with RNA secondary structures. Kemija u industriji **53**: 315–22.

30 FLAMM, C., I. L. HOFACKER, AND P. F. STADLER. 1999. RNA *in silico*: the computational biology of RNA secondary structures. Adv. Complex Syst. **2**: 65–90.

31 FLAMM, C., I. L. HOFACKER, P. F. STADLER, AND M. T. WOLFINGER. 2002. Barrier trees of degenerate landscapes. Z. Phys. Chem. **216**: 155–73.

32 FONTANA, W., D. A. M. KONINGS, P. F. STADLER, AND P. SCHUSTER. 1993. Statistics of RNA secondary structures. Biopolymers **33**: 1389–404.

33 FONTANA, W., P. F. STADLER, E. G. BORNBERG-BAUER, ET AL. 1993. RNA folding landscapes and combinatory landscapes. Phys. Rev. E **47**: 2083–99.

34 GAUTHERET, D., F. MAJOR, AND R. CEDERGREN. 1990. Pattern searching/alignment with RNA primary and secondary structures: an effective descriptor for tRNA. Comput. Appl. Biosci. **6**: 325–31.

35 GIBSON, A., V. GOWRI-SHANKAR, P. G. HIGGS, AND M. RATTRAY. 2005. A comprehensive analysis of mammalian mitochondrial genome base composition and improved phylogenetic methods. Mol. Biol. Evol. **22**(2): 251–64.

36 GIEGERICH, R.. 2000. A systematic approach to dynamic programming in bioinformatics. Bioinformatics **16**: 665–77.

37 GIEGERICH, R., B. VOSS, AND M. REHMSMEIER. 2004. Abstract shapes of RNA. Nucleic Acids Res. **32**: 4843–51.

38 GILLESPIE, D. T.. 1976. A general method for numerically simulating the stochastic time evolution of coupled chemical reactions. J. Comput. Phys. **22**: 403.

39 GORODKIN, J., L. J. HEYER, AND G. D. STORMO. 1997. Finding the most significant common sequence and structure motifs in a set of RNA sequences. Nucleic Acids Res. **25**: 3724–32.

40 GRÄF, S., D. STROTHMANN, S. KURTZ, AND G. STEGER. 2001. HyPaLib: a database of RNAs and RNA structural elements defined by hybrid patterns. Nucleic Acids. Res. **29**: 196–98.

41 GULTYAEV, A. P.. 1991. The computer simulation of RNA folding involving pseudoknot formation. Nucleic Acids Res. **19**: 2489–93.

42 GULTYAEV, A. P., VAN BATENBURG, AND C. W. A. PLEIJ. 1995. The computer simulation of RNA folding pathways using an genetic algorithm. J. Mol. Biol. **250**: 37–51.

43 GULTYAEV, A. P., F. H. D. VAN BATENBURG, AND C. W. A. PLEIJ. 1999. An approximation of loop free energy values of RNA H-pseudoknots. RNA **5**: 609–17.

44 GUTELL, R. R., A. POWER, G. Z. HERTZ, E. J. PUTZ, AND G. D. STORMO. 1992. Identifying constraints on the higher-order structure of RNA: continued development and application of comparative sequence analysis methods. Nucleic Acids Res. **20**: 5785–95.

45 GUTELL, R. R. AND C. R. WOESE. 1990. Higher order structural elements in ribosomal RNAs: Pseudo-knots and the use of noncanonical pairs. Proc. Natl Acad. Sci. USA **87**: 663–67.

46 HAN, K. AND Y. BYUN. 2003. PSEUDOVIEWER2: Visualization of RNA pseudoknots of any type. Nucleic Acids Res. **31**: 3432–40.

47 HAN, K., D. KIM, AND H. J. KIM. 1999. A vector-based method for drawing RNA secondary structure. Bioinformatics **15**: 286–97.

48 HAN, K. AND H.-J. KIM. 1993. Prediction of common folding structures of homologous RNAs. Nucleic Acids Res. **21**: 1251–57.

49 HARRIS, J. K., E. S. HAAS, D. WILLIAMS, AND D. N. FRANK. 2001. New insight into RNase P RNA structure from comparative analysis of the archaeal RNA. RNA **7**: 220–32.

50 HIGGS, P. G.. 1995. Thermodynamic properties of transfer RNA: a computational study. J. Chem. Soc. Faraday Trans. **91**, 2531–40.

51 HÖCHSMANN, M., T. TÖLLER, R. GIEGERICH, AND S. KURTZ. 2003. Local similarity in RNA secondary structures. Proc. Computational Systems Bioinformatics Conf., Stanford, CA, 159–68.

52 HOFACKER, I. L., S. H. F. BERNHART, AND P. F. STADLER. 2004. Alignment of RNA base pairing probability matrices. Bioinformatics **20**: 2222–27.

53 HOFACKER, I. L. AND P. F. STADLER. 2004. The partition function variant of Sankoff's algorithm. In Proc. Int. Conf on Computational Science, Krakow, 728–35.

54 HOFACKER, I. L., M. FEKETE, C. FLAMM, M. A. HUYNEN, S. RAUSCHER, P. E. STOLORZ, AND P. F. STADLER. 1998. Automatic detection of conserved RNA structure elements in complete RNA virus genomes. Nucleic Acids Res. **26**: 3825–36.

55 HOFACKER, I. L., M. FEKETE, AND P. F. STADLER. 2002. Secondary structure prediction for aligned RNA sequences. J. Mol. Biol. **319**: 1059–66.

56 HOFACKER, I. L., W. FONTANA, P. F. STADLER, L. S. BONHOEFFER, M. TACKER, AND P. SCHUSTER. 1994. Fast folding and comparison of RNA secondary structures. Monatsh. Chem. **125**: 167–88.

57 HOFACKER, I. L., B. PRIWITZER, AND P. F. STADLER. 2004. Prediction of locally stable RNA secondary structures for genome-wide surveys. Bioinformatics **20**: 191–98.

58 HOFACKER, I. L., P. SCHUSTER, AND P. F. STADLER. 1998. Combinatorics of RNA secondary structures. Discr. Appl. Math. **89**: 177–207.

59 HOFACKER, I. L. AND P. F. STADLER. 1999. Automatic detection of conserved base pairing patterns in RNA virus genomes. Comp. & Chem. **23**: 401–14.

60 HOFACKER, I. L. AND P. F. STADLER. 2006. Memory efficient folding algorithms for circular RNA secondary structures. Bioinformatics **22**: 1172–6.

61 HOFACKER, I. L., R. STOCSITS, AND P. F. STADLER. 2004. Conserved RNA secondary structures in viral genomes: a survey. Bioinformatics **20**: 1495–99.

62 HOLBROOK, S. R.. 2005. RNA structure: the long and the short of it. Curr. Opin. Struct. Biol. **15**: 302–8.

63 HOLMES, I.. 2005. Accelerated probabilistic inference of RNA structure evolution. BMC Bioinformatics **6**: 73.

64 HOLMES, I. AND G. M. RUBIN. 2002. Pairwise RNA structure comparison with stochastic context-free grammars. *Pac. Symp. Biocomput.* 2002: 163–74.

65 HULL HAVGAARD, J., R. LYNGSØ, G. STORMO, AND J. GORODKIN. 2005. Pairwise local structural alignment of RNA sequences with sequence similarity less than 40%. Bioinformatics **21**: 1815–24.

66 ISAMBERT, H. AND E. D. SIGGIA. 2000. Modeling RNA folding paths with pseudoknots: application to hepatitis delta virus ribozyme. Proc. Natl Acad. Sci. USA **97**: 6515–20.

67 JANSSON, J. AND A. LINGAS. 2003. A fast algorithm for optimal alignment between similar ordered trees. Fund. Inf. **56**: 105–20.

68 JIANG, T., G. LIN, B. MA, AND K. ZHANG. 2002. A general edit distance between beteen RNA structures. J. Comput. Biol. **9**: 371–88.

69 JIANG, T., J. WANG, AND K. ZHANG. 1995. Alignment of trees – an alternative to tree edit. Theor. Comput. Sci. **143**: 137–48.

70 JIMÉNEZ, V. M. AND A. MARZAL. 2000. Computation of the n best parse trees for weighted and stochastic context-free grammars. Proc. Joint Int. Workshops on Advances in Pattern Recognition, Spain: 183–92.

71 JUAN, V. AND C. WILSON. 1999. RNA secondary structure prediction based on free energy and phylogenetic analysis. J. Mol. Biol. **289**: 935–47.

72 KLEIN, P.. 1998. Computing the edit distance between unrooted ordered trees. In Proc. Annu. Eur. Symp. on Algorithms, Venice, 91–102.

73 KLEIN, R. J. AND S. R. EDDY. 2003. RSEARCH: finding homologs of single structured RNA sequences. *BMC Bioinformatics 4*(44), 1471–2105.

74 KNUDSEN, B. AND J. HEIN. 2003. Pfold: RNA secondary structure prediction using stochastic context-free grammars. Nucleic Acids Res. **31**: 3423–28.

75 KNUDSEN, B. AND J. J. HEIN. 1999. Using stochastic context free grammars and molecular evolution to predict RNA secondary structure. Bioinformatics **15**: 446–54.

76 LE, S.-Y., J.-H. CHEN, K. CURREY, AND J. MAIZEL. 1988. A program for predicting significant RNA secondary structures. CABIOS **4**: 153–59.

77 LE, S. Y. AND M. ZUKER. 1991. Predicting common foldings of homologous RNAs. J. Biomol. Struct. Dyn. **8**: 1027–44.

78 LEE, D. AND K. HAN. 2002. Prediction of RNA pseudoknots – comparative study of genetic algorithms. Genome Inf. **13**: 414–5.

79 LENHOF, H.-P., K. REINERT, AND M. VINGRON. 1998. A polyhedral approach to RNA sequence structure alignment. J. Comput. Biol. **5**: 517–30.

80 LEYDOLD, J. AND P. F. STADLER. 1998. Minimal cycle basis of outerplanar graphs. Elec. J. Comb. **5**: 209–22.

81 LIAO, B. AND T. WANG. 2002. An enumeration of RNA secondary structure. Math. Appl. **15**: 109–12.

82 LOUISE-MAY, S., P. AUFFINGER, AND E. WESTHOF. 1996. Calculations of

nucleic acid conformations. Curr. Opin. Struct. Biol. **6**: 289–98.

83 Lowe, T. M. and S. Eddy. 1997. tRNAscan-SE: a program for improved detection of transfer RNA genes in genomic sequence. Nucleic Acids Res. **25**: 955–64.

84 Lück, R., S. Gräf, and G. Steger. 1999. ConStruct: a tool for thermodynamic controlled prediction of conserved secondary structure. Nucl. Acids Res. **27**: 4208–17.

85 Lück, R., G. Steger, and D. Riesner. 1996. Thermodynamic prediction of conserved secondary structure: application to the RRE element of HIV, the tRNA-like element of CMV, and the mRNA of prion protein. J. Mol. Biol. **258**: 813–26.

86 Lyngsø, R. B. and C. N. S. Pedersen. 2000. RNA pseudoknot prediction in energy-based models. J. Comput. Biol. **7**: 409–27.

87 Lyngsø, R. B., M. Zuker, and C. N. Pedersen. 1999. Fast evaluation of internal loops in RNA secondary structure prediction. Bioinformatics **15**: 440–45.

88 Ma, B., L. Wang, and K. Zhang. 2002. Computational similarity between RNA structures. Theor. Comput. Sci. **276**: 111–32.

89 Martinez, H. M.. 1984. An RNA folding rule. Nucl. Acid Res. **12**: 323–35.

90 Mathews, D. H., M. D. Disney, J. L. Childs, S. J. Schroeder, M. Zuker, and D. H. Turner. 2004. Incorporating chemical modification constraints into a dynamic programming algorithm for prediction ofRNA secondary structure. Proc. Natl Acad. Sci. USA **101**: 7287–92.

91 Mathews, D. H., J. Sabina, M. Zuker, and H. Turner. 1999. Expanded sequence dependence of thermodynamic parameters provides robust prediction of RNA secondary structure. J. Mol. Biol. **288**: 911–40.

92 Mathews, D. H. and D. H. Turner. 2002. Dynalign: an algorithm for finding secondary structures common to two RNA sequences. J. Mol. Biol. **317**: 191–203.

93 McCaskill, J.. 1990. The equilibrium partition function and base pair binding probabilities for RNA secondary structure. Biopolymers **29**: 1105–19.

94 McCutcheon, J. P. and S. R. Eddy. 2003. Computational identification of non-coding RNAs in *Saccharomyces cerevisiae* by comparative genomics. Nucleic Acids Res. **31**: 4119–28.

95 Mironov, A. A., L. P. Dyakonova, and A. E. Kister. 1985. A kinetic approach to the prediction of RNA secondary structures. J. Biomol. Struct. Dyn. **2**: 953.

96 Missal, K., D. Rose, and P. F. Stadler. 2005. Non-coding RNAs in *Ciona intestinalis*. Bioinformatics **21 (Suppl. 2)**: i77–8.

97 Mückstein, U., I. L. Hofacker, and P. F. Stadler. 2002. Stochastic pairwise alignments. Bioinformatics **18**: 153–60.

98 Muller, G., C. Gaspin, A. Etienne, and E. Westhof. 1993. Automatic display of RNA secondary structures. Comput. Appl. Biosci. **9**: 551–61.

99 Needleman, S. B. and C. D. Wunsch. 1970. A general method applicable to the search for similarities in the aminoacid sequences of two proteins. J. Mol. Biol. **48**: 443–52.

100 Notredame, C., D. Higgins, and J. Heringa. 2000. T-coffee: a novel method for multiple sequence alignments. J. Mol. Biol. **302**: 205–17.

101 Nussinov, R., G. Pieczník, J. R. Griggs, and D. J. Kleitman. 1978. Algorithms for loop matching. SIAM J. Appl. Math. **35**: 68–82.

102 Pedersen, J. S., I. M. Meyer, R. Forsberg, and J. Hein. 2004. An evolutionary model for protein-coding regions with conserved RNA structure. Mol. Biol. Evol. **21**: 1913–22.

103 Pedersen, J. S., I. M. Meyer, R. Forsberg, P. Simmonds, and J. Hein. 2004. A comparative method for finding and folding RNA secondary structures within protein-coding regions. Nucleic Acids Res. **32**: 4925–36.

104 Reeder, J. and R. Giegerich. 2004. Design, implementation and evaluation of a practical pseudoknot folding algorithm based on thermodynamics. BMC Bioinformatics **5**: 104.

105 REHMSMEIER, M., P. STEFFEN, M. HÖCHSMANN, AND R. GIEGERICH. 2004. Fast and effective prediction of microRNA/target duplexes. RNA **10**: 1507–17.

106 REIDYS, C., P. F. STADLER, AND P. SCHUSTER. 1997. Generic properties of combinatory maps: neutral networks of RNA secondary structures. Bull. Math. Biol. **59**: 339–97.

107 REIDYS, C. M. AND P. F. STADLER. 2002. Combinatorial landscapes. SIAM Rev. **44**: 3–54.

108 Repsilber, D., S. Wiese, M. Rachen, A. W. Schroder, D. Riesner, and G. Steger . 1999. Formation of metastable RNA structures by sequential folding during transcription: time-resolved structural analysis of potato spindle tuber viroid (−)-stranded RNA by temperature-gradient gel electrophoresis. RNA **5**: 574–84.

109 RIVAS, E. AND S. R. EDDY. 1999. A dynamic programming algorithm for RNA structure prediction including pseudoknots. J. Mol. Biol. **285**: 2053–68.

110 RIVAS, E. AND S. R. EDDY. 2001a. Noncoding RNA gene detection using comparative sequence analysis. BMC Bioinformatics **2**: 19.

111 RIVAS, E. AND S. R. EDDY. 2001b. Noncoding RNA gene detection using comparative sequence analysis. BMC Bioinformatics **2**: 8.

112 RIVAS, E., R. J. KLEIN, T. A. JONES, AND S. R. EDDY. 2001. Computational identification of non-coding RNAs in *E. coli* by comparative genomics. Curr. Biol. **11**: 1369–73.

113 RUAN, J., G. D. STORMO, AND W. ZHANG. 2004. An iterated loop matching approach to the prediction of RNA secondary structures with pseudoknots. Bioinformatics **20**: 58–66.

114 SANKOFF, D.. 1985. Simultaneous solution of the RNA folding, alignment, and proto-sequence problems. SIAM J. Appl. Math. **45**: 810–25.

115 SCHULTES, E. A. AND D. P. BARTEL. 2000. One sequence, two ribozymes: Implications for the emergence of new ribozyme folds. Science **289**: 448–52.

116 SCHUSTER, P., W. FONTANA, P. F. STADLER, AND I. L. HOFACKER. 1994. From sequences to shapes and back: a case study in RNA secondary structures. Proc. R. Soc. Lond. B **255**: 279–84.

117 SHAPIRO, B. A.. 1988. An algorithm for comparing multiple RNA secondary stuctures. CABIOS **4**: 387–93.

118 SHAPIRO, B. A., J. MAIZEL, L. E. LIPKIN, K. CURREY, AND C. WHITNEY. 1984. Generating non-overlapping displays of nucleic acid secondary structure. Nucleic Acids Res. **12**: 75–88.

119 SHAPIRO, B. A. AND K. ZHANG. 1990. Comparing multiple RNA secondary structures using tree comparisons. CABIOS **6**: 309–18.

120 SHASHA, D. AND K. ZHANG. 1990. Fast algorithm for the unit cost editing distance between trees. J. Algorithms **11**: 581–621.

121 SIEBERT, S. AND R. BACKOFEN. 2003. MARNA: a server for multiple alignment of RNAs. In Proc. German Conf. on Bioinformatics, München, 135–140.

122 STADLER, P. F. AND C. FLAMM. 2003. Barrier trees on poset-valued landscapes. Genet. Prog. Evolv. Mach. **7(20)**: 4.

123 STEGER, G., H. HOFMANN, J. FORTSCH, H. J. GROSS, J. W. RANDLES, H. L. SANGER, AND D. RIESNER. 1984. Conformational transitions in viroids and virusoids: comparison of results from energy minimization algorithm and from experimental data. J. Biomol. Struct. Dyn. **2**: 543–71.

124 SUVERNEV, A. AND P. FRANTSUZOV. 1995. Statistical description of nucleic acid secondary structure folding. J. Biomol. Struct. Dyn. **13**: 135–44.

125 TABASKA, J. E., R. B. CARY, H. N. GABOW, AND G. D. STORMO. 1998. An RNA folding method capable of identifying pseudoknots and base triples. Bioinformatics **14**(8): 691–9.

126 TACKER, M., W. FONTANA, P. F. STADLER, AND P. SCHUSTER. 1994. Statistics of RNA melting kinetics. Eur. Biophys. J. **23**: 29–38.

127 TACKER, M., P. F. STADLER, E. G. BORNBERG-BAUER, I. L. HOFACKER, AND P. SCHUSTER. 1996. Algorithm

independent properties of RNA structure prediction. Eur. Biophys. J. **25**: 115–30.

128 TAI, K.. 1979. The tree-to-tree correction problem. J. ACM **26**: 422–33.

129 THIRUMALAI, D.. 1998. Native secondary structure formation in RNA may be a slave to tertiary folding. Proc. Natl Acad. Sci. USA **95**: 11506–8.

130 THIRUMALAI, D., N. LEE, S. A. WOODSON, AND D. K. KLIMOV. 2001. Early events in RNA folding. Annu. Rev. Phys. Chem. **52**: 751–62.

131 THURNER, C., C. WITWER, I. HOFACKER, AND P. F. STADLER. 2004. Conserved RNA secondary structures in Flaviviridae genomes. J. Gen. Virol. **85**: 1113–24.

132 VALIENTE, G.. 2001. An efficient bottom-up distance between trees. In Proc. Int. Symp. on String Processing and Information Retrieval, Laguna De San Raphael: 212–9.

133 VITRESCHAK, A. G., D. A. RODIONOV, A. A. MIRONOV, AND M. S. GELFAND. 2004. Riboswitches: the oldest mechanism for the regulation of gene expression? Trends Genet. **20**: 44–50.

134 VOSS, B., C. MEYER, AND R. GIEGERICH. 2004. Evaluating the predictability of conformational switching in RNA. Bioinformatics **20**: 1573–82.

135 WALTER, A., D. TURNER, J. KIM, M. LYTTLE, P. MÜLLER, D. MATHEWS, AND M. ZUKER. 1994. Coaxial stacking of helixes enhances binding of oligoribonucleotides and improves predicions of RNA folding. Proc. Natl Acad. Sci. USA **91**: 9218–22.

136 WASHIETL, S. AND I. L. HOFACKER. 2004. Consensus folding of aligned sequences as a new measure for the detection of functional RNAs by comparative genomics. J. Mol. Biol. **342**: 19–30.

137 WASHIETL, S., I. L. HOFACKER, M. LUKASSER, A. HÜTTENHOFER, AND P. F. STADLER. 2005. Mapping of conserved RNA secondary structures predicts thousands of functional non-coding RNAs in the human genome. Nat. Biotechnol. **23**: 1383–90.

138 WASHIETL, S., I. L. HOFACKER, AND P. F. STADLER. 2005. Fast and reliable prediction of noncoding RNAs. Proc. Natl Acad. Sci. USA **102**: 2454–59.

139 WATERMAN, M. S.. 1978. Secondary structure of single-stranded nucleic acids. Studies on foundations and combinatorics. Adv. Math. Supplement. Studies **1**: 167–212.

140 WATERMAN, M. S. AND T. F. SMITH. 1978. RNA secondary structure: a complete mathematical analysis. Math. Biosci. **42**: 257–66.

141 WITWER, C., I. L. HOFACKER, AND P. F. STADLER. 2004. Prediction of consensus RNA secondary structures including pseudoknots. IEEE/ACM Trans. Comput. Biol. Bioinf. **1**: 65–77.

142 WITWER, C., S. RAUSCHER, I. L. HOFACKER, AND P. F. STADLER. 2001. Conserved RNA secondary structures in picornaviridae genomes. Nucleic Acids Res. **29**: 5079–89.

143 WOLFINGER, M. T., W. A. SVRCEK-SEILER, C. FLAMM, I. L. HOFACKER, AND P. F. STADLER. 2004. Exact folding dynamics of RNA secondary structures. J. Phys. A **37**: 4731–41.

144 WUCHTY, S., W. FONTANA, I. L. HOFACKER, AND P. SCHUSTER. 1999. Complete suboptimal folding of RNA and the stability of secondary structures. Biopolymers **49**: 145–165.

145 XIA, T., J. SANATLUCIA JR., M. E. BURKARD, R. KIERZEK, S. J. SCHROEDER, X. JIAO, C. COX, AND D. H. TURNER. 1998. Parameters for an expanded nearest-neighbor model for formation of RNA duplexes with Watson–Crick pairs. Biochemistry **37**: 14719–35.

146 YOUNGER, D. H.. 1967. Recognition and parsing of context-free languages in time n^3. Inf. Control **10**: 189–208.

147 ZHANG, K.. 1995. Algorithms for the constrained editing problem between ordered labeled trees and related problems. Pattern Recogn. **28**: 463–74.

148 ZHANG, K. AND D. SHASHA. 1989. Simple fast algorithms for the editing distance between trees and related problems. SIAM J. Comput. **18**: 1245–62.

149 ZUKER, M.. 1989. On finding all suboptimal foldings of an RNA molecule. Science **244**: 48–52.

150 Zuker, M. and P. Stiegler. 1981. Optimal computer folding of larger RNA sequences using thermodynamics and auxiliary information. Nucleic Acids Res. **9**: 133–48.

151 Zwieb, C., I. Wower, and J. Wower. 1999. Comparative sequence analysis of tmRNA. Nucleic Acids Res. **27**: 2063–71.

15
RNA Tertiary Structure Prediction

François Major and Philippe Thibault

1 Introduction

During the last decade, the number of high-quality X-ray crystallographic RNA three-dimensional (3-D) structures has increased significantly, and the resolution of the large ribosomal subunit crystal structure was considered a major step towards a better understanding of RNA tertiary structure and folding. The recent discovery of the RNA interference (RNAi) pathway (see Chapter 45) has also contributed greatly to the popularity of RNAs, by suggesting their direct implication in genetic expression and regulation. More than ever, determining rapidly and precisely the tertiary structure and function of noncoding RNAs is a crucial step towards our understanding of several cellular metabolic pathways.

This chapter is dedicated to the RNA tertiary structure prediction problem – the determination of the complete set of chemical interactions (and therefore 3-D fold) of an RNA from sequence data. To achieve RNA structure prediction, one needs to discover and apply its structural and architectural principles, which can be learnt from thermodynamics, as well as from structural data gathered from X-ray crystallography and other high-resolution, but also low-resolution, experimental methods. Here, we present a series of nomenclatures and formalisms to describe RNA tertiary structure, as well as computer data structures and algorithms that implement three important research activities with the aim of solving RNA tertiary structure prediction: annotation, motif discovery and modeling.

We present in Section 2 a series of RNA structure components and the terms employed by the RNA specialists to discuss them – their universe of discourse (nowadays referred to as their ontology). First, we present an ontology of nucleotide conformations and binary interactions. Then, visual or automated inspection of RNA 3-D structures is necessary to depict higher-order architectural principles (the next abstraction levels). In Section 3, we introduce a definition of n-ary nucleotide interactions to describe RNA higher-order motifs, which are found repeated in RNA structures, and are often

Bioinformatics - From Genomes to Therapies Vol. 1. Edited by Thomas Lengauer

ISBN: 978-3-527-31278-8

linked to specific structural and biochemical functions, and an approach to search them. Finally, in Section 4, we present how accurate computer models of RNA tertiary structures can be generated and how, by challenging them experimentally, they bring insights about function.

The flowchart in Figure 1 shows the relationships between structural data and hypotheses, and how the research activities that aim at solving the RNA tertiary structure prediction problem are intimately linked. The high-resolution (better than 3 Å) X-ray crystal structures of the Protein Data Bank (PDB) (www.rcsb.org) [1] constitute an excellent structural (learning) data set that aids the research, and from which RNA tertiary structure prediction algorithms can be inspired and tested. The characterization and formalization of RNA structural data (annotation), the discovery of high-order components (motif discovery) and the building of RNA tertiary structure models (modeling) contribute directly to the learning and discovery processes, leading to new knowledge that is fed back to research.

An ultimate solution to the tertiary structure prediction problem will provide us with invaluable structural information, and will allow us to determine the function and the evolutionary relationships of RNAs. Knowledge of RNA tertiary structure impacts on molecular medicine techniques to control genetic expression, and to inhibit and activate specific cellular functions. The cell controls its own genetic expression by processing micro RNAs through the RNAi pathway. As we discover and characterize the elements of RNAi, we learn how to design RNAs that interfere and block the expression of several genes. Knowledge of the structure and of the interplay between the RNAs and the other RNAi elements is fundamental. Alternatively, we could target the natural micro RNAs of the cell using drugs. Again, knowledge of the targeted RNA structure is necessary to design accurate drugs. Targeting the noncoding RNAs of the cell allows us to manipulate its fundamental mechanisms prior to protein translation; like playing with the "source code" of the cell. Antibiotics such as aminoglycosides and macrolides target the site-A of prokaryotic ribosomes, blocking protein translation. The search and discovery of other sites in the ribosome or in other RNAs involved in such fundamental mechanisms require the determination of their tertiary structure if we want to design drugs capable of inhibiting them. Ribozymes are catalytic RNAs that can cleave a substrate efficiently and precisely. For instance, ribozymes can be used to cleave a messenger prior to its translation by the ribosome. Here, again, knowledge of the tertiary structure of ribozymes and of their complex with the substrate is essential for rational design.

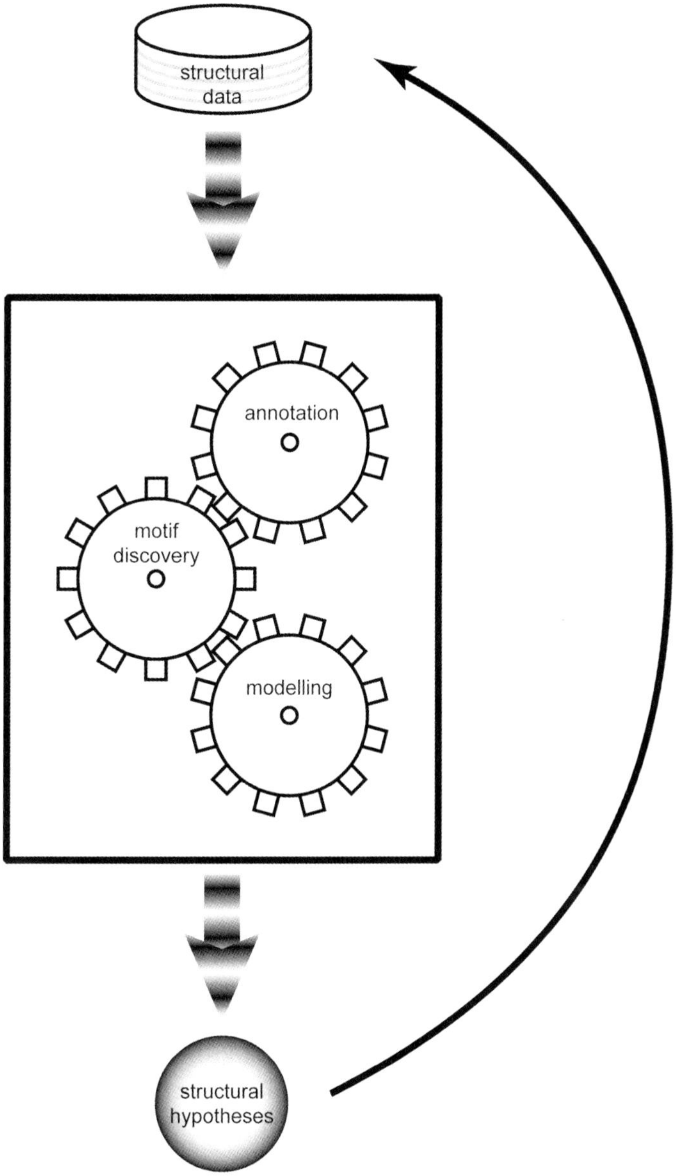

Figure 1 Data flow and research activities of tertiary structure prediction. Structural data are used to develop the (computational) tools employed by the researchers to annotate, inspect and model RNA tertiary structures. Structural hypotheses are generated and, when challenged experimentally, bring new structural data to research.

2 Annotation

The annotation of an RNA tertiary structure is the assignment (manual or automated) of appropriate symbols, taken from the RNA ontology, that apply to a given RNA. One can see annotation as a data refinement process that complements the 3-D atomic coordinates – a different and perhaps higher level of abstraction which can be thought of as an efficient and sound data format to study further tertiary structures.

A human RNA expert recognizes the attributes of tertiary structures by visualization using interactive computer graphics and can therefore annotate a given RNA 3-D structure. An automated procedure loads the RNA 3-D atomic coordinates in memory and then computes the annotations. Gendron and coworkers have developed a computer program, MC-Annotate, which annotates a fraction of the current RNA ontology, in particular the terms related to the nucleotide conformations, as well as base stacking and base pairing types [2] (see Section 2.2). MC-Annotate can be run over the web (www-lbit.iro.umontreal.ca; under Research and MC-Annotate). Westhof and coworkers, in collaboration with the PDB, developed RNAView, a computer program that draws the secondary structure of an RNA while using the *LW* nomenclature (see Sections 2.2.2) to display the base pair types [3]. RNAView is accessible on the web (ndbserver.rutgers.edu/services).

In this section, we present a series of RNA tertiary structure attributes and how they can be computed from 3-D atomic coordinates from X-ray crystallographic structures of the PDB. We present the nucleotide conformations (Section 2.1) and interactions (Section 2.2) that are needed to define higher-order RNA components (Section 3), and to build and describe RNA tertiary structure (Section 4).

2.1 Nucleotide Conformations

RNAs are polymeric molecules. The monomer unit is a ribonucleotide, or simply nucleotide, which divides in three units: the nucleobase (or simply base), the ribose and the phosphate group (see Figure 2). There are four bases: adenine (A), guanine (G), cytosine (C) and uracil (U). The four bases partition in two families: the pyrimidines (Y) C and U, which are composed of a single pyrimidine ring, and the purines (R) A and G, which are composed of the fusion of the pyrimidine ring ($C_4H_4N_2$) and an imidazole ring ($C_3H_4N_2$). The

Figure 2 RNA chemical structure. The polynucleotide chain (on the left) is made of the bases (hexagons), riboses (pentagons) and phosphate groups (diamonds). The four common types of bases (on the right): the two purines adenine and guanine, and the two pyrimidines cytosine and uracil. The phosphodiester linkage (middle) connects two nucleotides. The ribose (center) links two phosphate groups: one to its 5′ oxygen (above) and the other to its 3′ oxygen (below). The conventional atomic numbering system is used. Small black circles represent the carbon atoms and their complementary hydrogen atoms are not shown.

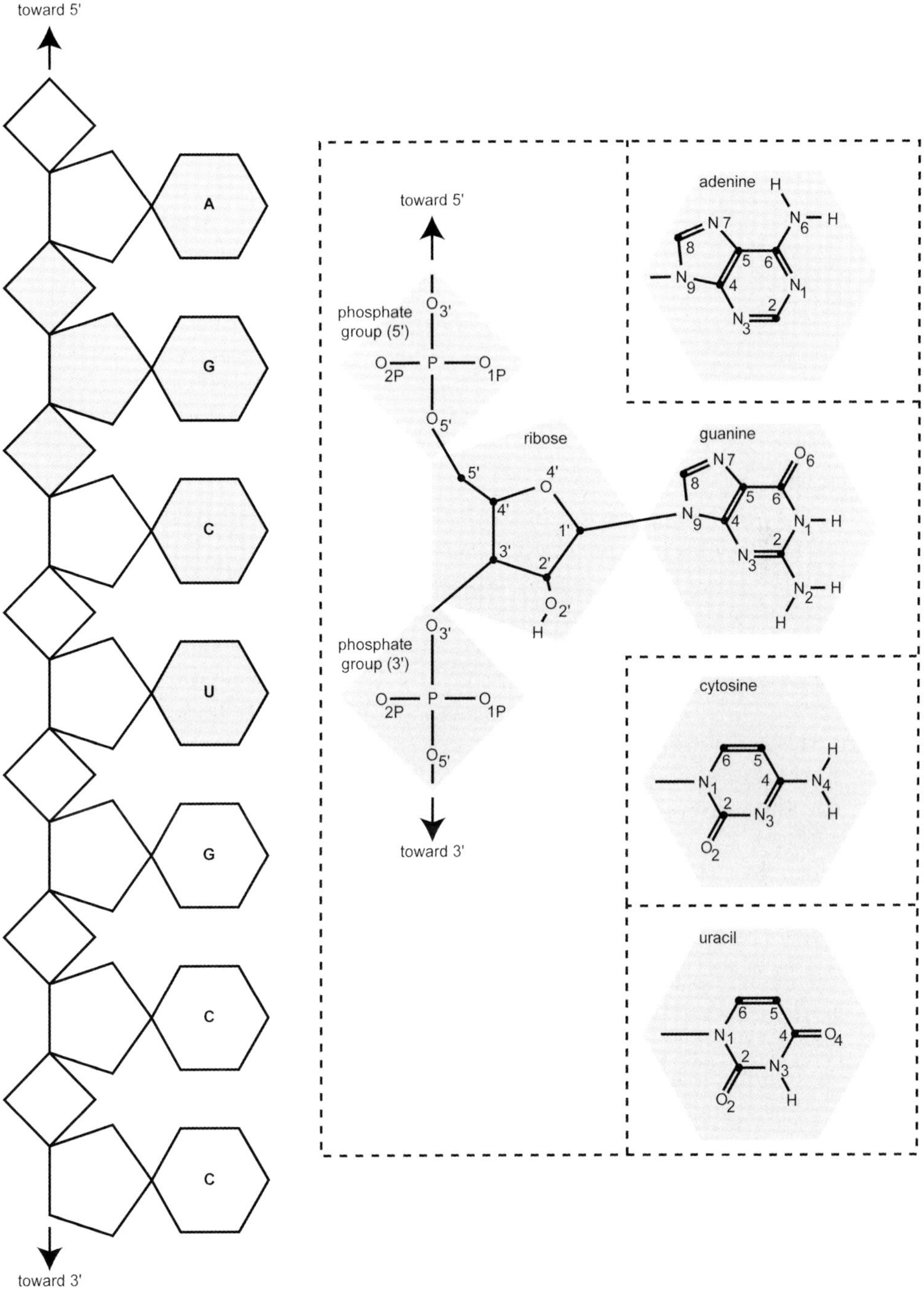
toward 5'
A
G
C
U
G
C
C
toward 3'
toward 5'
phosphate group (5')
ribose
phosphate group (3')
toward 3'
adenine
guanine
cytosine
uracil

International Union of Pure and Applied Chemistry (IUPAC) defined a one-letter code for all possible subsets of {A, C, G, U} (shown in Table 1).

Table 1 IUPAC nucleotide nomenclature

Code	Nucleotide subset
M	{A, C}
R	{A, G}
W	{A, U}
S	{C, G}
Y	{C, U}
K	{G, U}
V	{A, C, G}
H	{A, C, U}
D	{A, G, U}
B	{C, G, U}
N	{A, C, G, U}

The ribose links the phosphate groups to which the bases are attached by the glycosidic bond: C1′–N9 in purines and C1′–N1 in pyrimidines. The riboses and the phosphate groups constitute the backbone, and are linked through diester bonds: C5′–O5′ and C3′–O3′. The chain C3′–O3′–P–O5′–C5′ from one ribose to another is referred to as the phosphodiester linkage that ties the nucleotides together (see Figure 2).

When 3-D points represent the center of the atoms in the structure, the covalent bond lengths, and the bond and torsion angles can be computed directly. The covalent bond length between atoms A and B is simply defined by the Cartesian distance between points A and B. The covalent bond angle between atoms A, B and C is defined by the angle between vectors $\overrightarrow{BA}$ and $\overrightarrow{BC}$. Finally, the covalent bond torsion angle between atoms A, B, C and D is defined by the angle between the projection of $\overrightarrow{BA}$ and $\overrightarrow{CD}$ in a plane perpendicular to $\overrightarrow{CD}$. In general, bond lengths and angles are considered constant in most computer prediction systems. Consequently, the 3-D conformation of a nucleotide can be described by its torsion angles. The phosphodiester linkage has six torsions (α, β, γ, δ, ε and ζ), the ribose has five torsions (θ_0–θ_4) and there is one torsion around the glycosidic bond (χ) (see Figure 3a).

Note the δ and θ_1 torsions are measured on the same covalent bond, C3′–C4′, but from different end-points, respectively, C2′ and O4′ in the ribose for θ_1, and C5′ and O3′ in the phosphodiester chain for δ. The glycosidic torsion, χ, is measured respectively in purines and pyrimidines from atoms O4′–C1′–N9–C4 and O4′–C1′–N1–C2. The furanose ring stereochemistry imposes interdependent relations on θ_{0-4}, which is expressed by the cosine function:

$$\theta_j = \theta_{\max} \cos(\rho + j\varphi), \tag{1}$$

where $j = 0, \ldots, 4$ and $\varphi = 144°$ (720°/5).

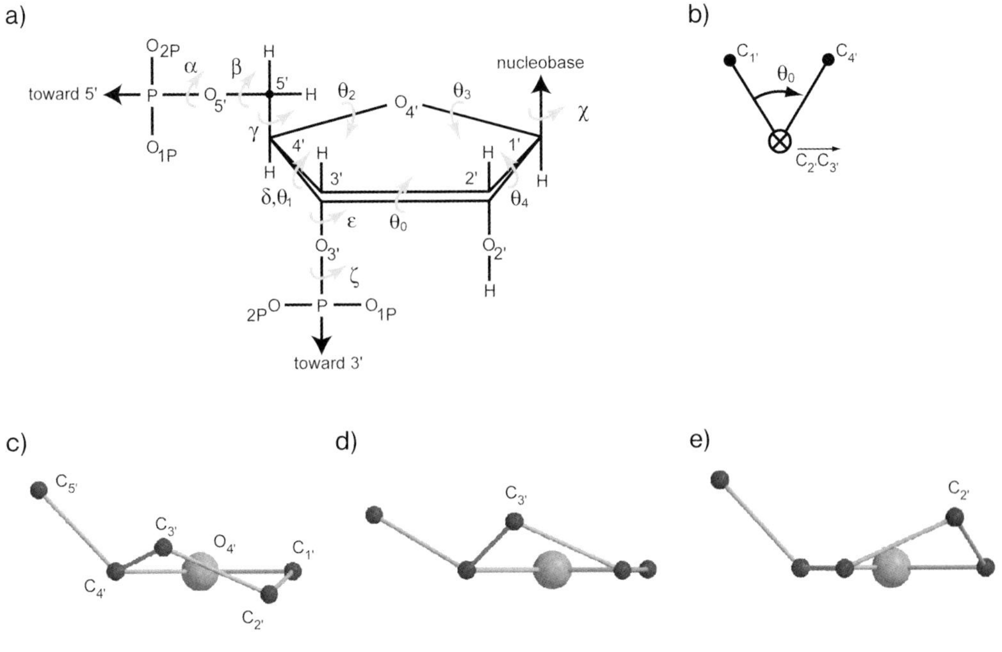

Figure 3 The nucleotide torsion angles. (a) Individual torsions are shown using grey arrows: χ on the glycosidic bond between the base and the ribose, θ_{0-4} around the ribose, and α, β, γ, δ, ε and ζ along the phosphodiester chain (θ_1 and δ are defined on the same covalent bond). (b) θ_0 measurement. The torsion is computed as the angle between the projection of vectors $\overrightarrow{C_{2'}C_{1'}}$ and $\overrightarrow{C_{3'}C_{4'}}$ in the plane perpendicular to vector $\overrightarrow{C_{2'}C_{3'}}$ (crossed circle). (c) The $^2T_3$ twist shape C2′-*exo*–C3′-*endo* ribose pucker mode. (d) The ^3E envelope shape C3′-*endo* ribose pucker mode. (e) The ^2E envelope shape C2′-*endo* ribose pucker mode.

When $j = 0$, we have:

$$\theta_0 = \theta_{\max}\cos(\rho). \tag{2}$$

In Eqs. (1) and (2), ρ is the pseudorotation of the ribose ring [4]. By varying ρ from 0 to 360° by steps of 90°, θ_0 goes from $\theta_{\max}$ to 0, to θ–$\theta_{\max}$, back to 0 and, finally, back to $\theta_{\max}$. The $\theta_{\max}$ value is reached twice – at the initial conformation [$\rho = 0$, as $\cos(0) = 1$] and at $\rho = 360°$. At each step of $\rho + 180°$, the sign of all torsions is inversed, corresponding to the mirror image of the conformation at ρ. A useful equation is derived from Eq. (1), which determines ρ:

$$\tan\rho = \frac{(\theta_2 + \theta_4) - (\theta_1 + \theta_3)}{2\theta_0(\sin 36° + \sin 72°)}. \tag{3}$$

Equation (2) determines $\theta_{\max}$.

Two geometric shapes characterize the stereochemistry of the ribose: envelope and twist (or half-chair). The ribose forms an envelope when only one of the five atoms of the furanose (C1′, C2′, C3′, C4′ or O4′) is out of the plane formed by the four others. The ribose forms a twist when two atoms are out of the plane formed by the remaining three. In a 360° period of the pseudorotation angle, the ribose stereochemistry alternates from the envelope to the twist shapes, successively, on each atom. At $\rho = 0°$, the ribose is in the twist shape with the C3′ atom above and the C2′ atom below the plane, which is conveniently denoted by $^{3}T_{2}$ (numbered atom above the plane in superscript and below the plane in subscript), as illustrated in Figure 3(c). All molecule 3-D rendered images were generated using MolScript [5] and Raster3D [6], as well as PyMOL (www.pymol.org). When the ball-and-stick representation is used, sphere radii are proportional to atomic masses ($C < N < O < P$). Then, the geometry of the ribose shifts at each 18°; for $\rho \in [18°, 36°]$, the ribose forms an envelope with the C3′ atom above the plane, ^{3}E (see Figure 3d); for $\rho \in [36°, 54°]$, the geometry changes to $^{3}T_{4}$; $_{4}E$ for $\rho \in [54°, 72°]$ and so forth for the 20 different geometries [7]. The ribose geometries are referred to as the sugar pucker modes. Another widely used ribose pucker mode nomenclature among RNA structure specialists is one where the atom(s) bulging out of the plane are suffixed with either *endo* or *exo*, respectively, for above and below the plane. Thus, in the example of Figure 3(c), the $^{3}T_{2}$ shape is equally named C2′-*exo*-C3′-*endo*; C3′-*endo* for the example in Figure 3(d) and C2′-*endo* for the example in Figure 3(e).

Single-stranded RNAs fold back on themselves to form double-stranded helices in the A-RNA conformation which is similar to the A-DNA double helix. Among all 20 ribose pucker modes, the C3′-*endo* is the most common, as it is the conformation of the riboses in the Watson–Crick base pairs of the A-RNA double helix (Figure 4). The asymmetry of the Watson–Crick base pair geometry (Section 2.2.2) induces the formation of two grooves in the helix. The major groove of the A-RNA double helix is narrow and deep, whereas the minor groove is broad and shallow (Figure 4b). Theoretically, the C2′-*endo* mode, adopted by the nucleotides in the B-RNA double-helical form, is unstable because of the proximity of the 2′-OH groups to the bases. Nevertheless, a good fraction of RNAs contain nucleotides in the C2′-*endo* conformations, as in loop regions and at the extremities of a double helix.

Figure 4 Type A-RNA double helix. (a) The bases form stacked Watson–Crick base pairs. The 5′-strand is shown in dark; the 3′-strand in light. The thread follows the phosphorus atoms. The hydrogen atoms are not shown. (b) Major and minor grooves. The bases are in red; the backbone in blue. The 3-D structures were generated by MC-Sym (see Section 4.2).

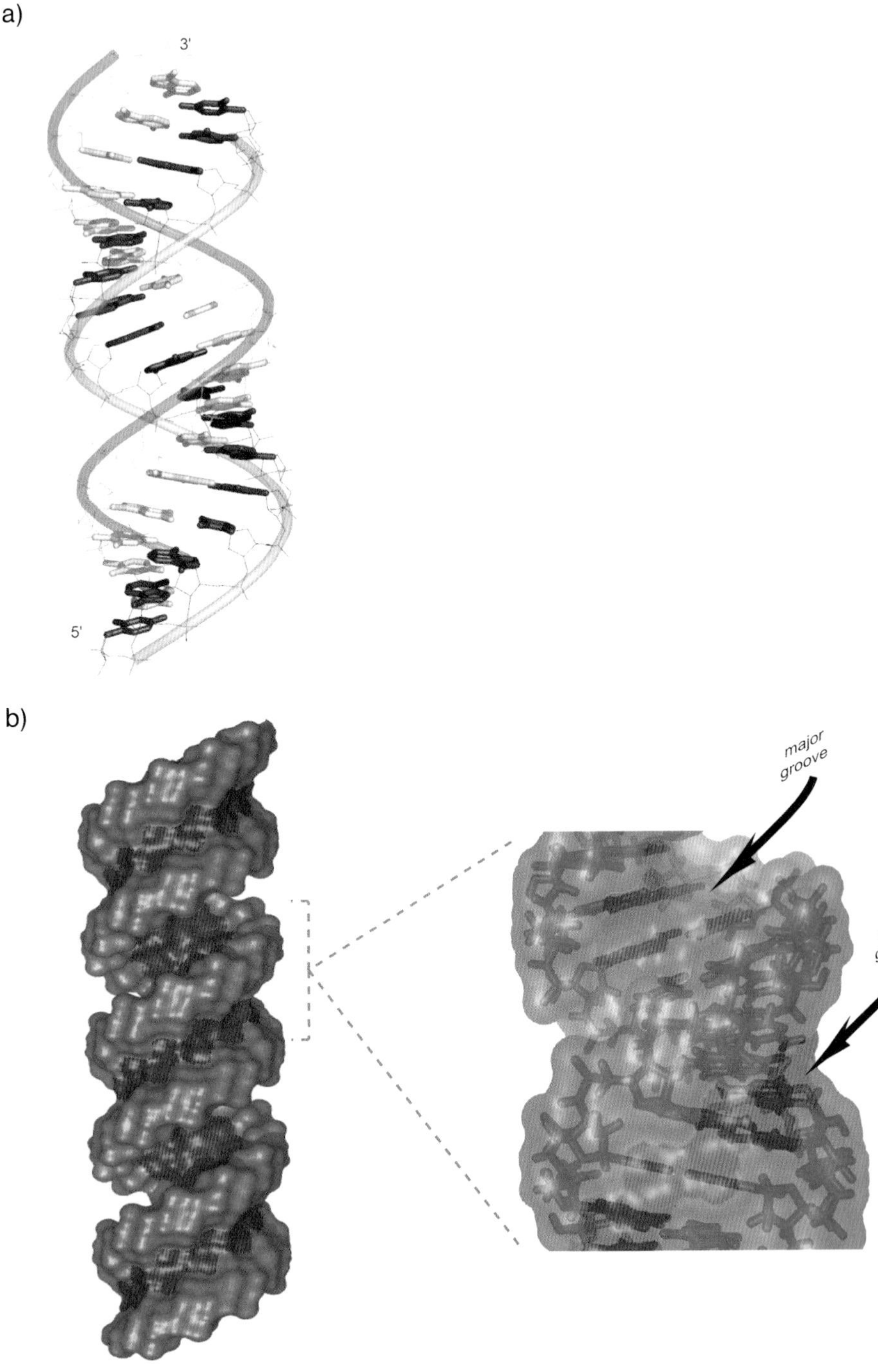
a)
3'
5'
b)
major groove
minor groove

In addition to the pucker mode, the glycosidic torsion χ is also divided in a range of values. The *anti* conformation characterizes a base oriented away from the ribose. From $\chi = 180°$, where the plane of the base is aligned with the O4′–C1′ bond in a direction away from O4′, the *anti* conformation covers a rotation of $\pm 90°$: $\chi \in [-180°, -90°]$ and $\chi \in [90°, 180°]$ (see Figure 5a). At $\chi = 0°$, where the plane of the base is aligned with the O4′–C1′ bond in a direction towards the O4′ atom, the *syn* conformation covers the remaining rotations of $\pm 90°$: $\chi \in [-90°, 90°]$ (see Figure 5b). The nucleotide conformations in both the A-RNA and B-RNA double-helical forms adopt the *anti* conformation.

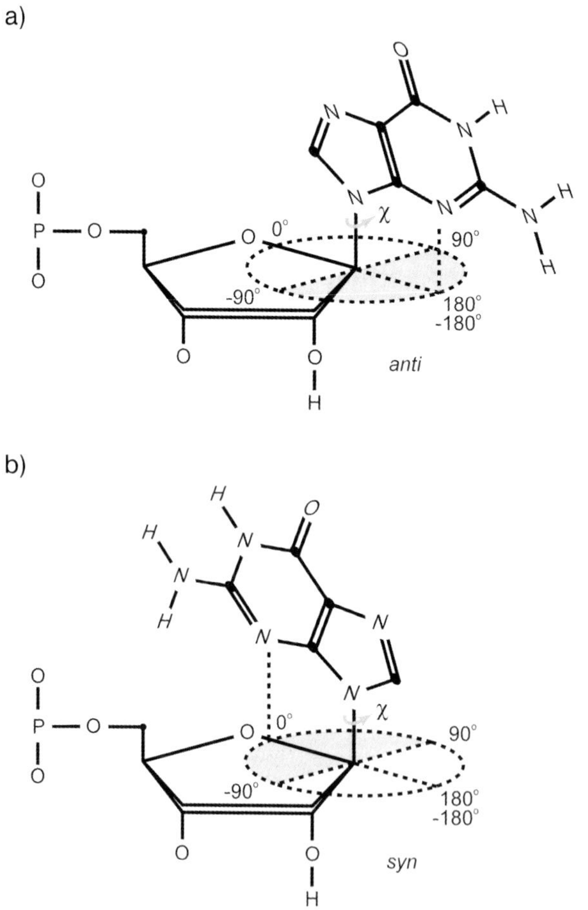

Figure 5 The glycosidic bond torsion. The ranges of χ values are shown in grey. (a) The *anti* conformation aligns the base away from the ribose. (b) The *syn* conformation aligns the base towards the ribose.

The nucleotide conformations can be annotated directly from their torsion angles: θ_{0-4} to determine the ribose puckering modes (C3′-*endo*, C2′-*endo*, etc.) and χ to determine the base to ribose relative orientation (*anti* or *syn*).

As mentioned above, the nucleotide conformation is mainly, if not completely, characterized by its free torsion angles. Consequently, many attempts aim at classifying nucleotide conformations according to torsion angles. In the late 1970s, Olson reduced the six phosphodiester chain torsions to two pseudotorsions of virtual bonds spanning the chain in two C–C–O–P segments (C4′–C5′–O5′–P and C4′–C3′–O3′–P). She reported a statistical correlation between the individual torsions and those spanned by the two pseudotorsions [8]. Gautheret and coworkers proposed a different approach that analyzed the clustering of dinucleotide conformations. They used a RMSD distance to compare pairs of dinucleotides aligned by their P–O3′ bond. The clustering discriminated families of dinucleotides with similar P–P orientations, and was used as the basis of a conformational search space in early versions of the MC-Sym computer program (see Section 4.2) [9, 10]. They observed with this approach the "crankshaft effect" [11], as different torsion patterns lead to similar 3-D conformations. In more recent studies, Duarte and coworkers extended Olson's work by defining two pseudotorsions and identified recurrent torsion patterns as well [12].

Hershkovitz and coworkers analyzed individual Gaussian distribution data fitting of the four backbone torsions α, γ, δ and ζ in the crystal structure of *Haloarcula marismortui* 23S rRNA. They identified 37 different conformers from which they defined nucleotide signatures [13]. Others, such as Murray and her colleagues, have classified three-torsion patterns (α, β, γ) and (δ, ε, ζ) into 42 conformers by applying quality filtering to high-resolution X-ray crystal structures [14]. According to them, each nucleotide conformer represents a high-quality reference nucleotide conformation. Schneider and coworkers have analyzed the torsion angles of dinucleotides by Fourier averaging of six selected 3-D distributions. They found 18 conformers, apart from the overrepresented A-RNA helical conformation [15]. Similarly to Gautheret and coworkers, they concluded the structural conformational space of RNA 3-D structures could be sampled by a small number of dinucleotide conformers.

2.2 Nucleotide Interactions

Inter-nucleotide interactions contribute to the overall stability of RNA tertiary structure. The obvious example is the stacked Watson–Crick base pairs that forms the A-RNA double helix. Interactions outside double helices that are distant in sequence are often referred to as tertiary interactions and play a major role in RNA folding. Here, we define and present a nomenclature to describe base stacking and base pairing information.

2.2.1 Base Stacking

Base stacking involves London dispersion inter-molecular interactions between two bases that induce a 3-D arrangement where one base is stacked on top of the other (see Figure 4). Bases can stack towards each side and therefore there are four different base stacking types. To identify on which side a base is stacked, a vector normal to the plane of the base is defined so that any base in a classical A-form helix have their normal vectors oriented in the same direction; towards the 3′-strand end-point. In pyrimidines, this normal vector is the rotational vector $\vec{n}_Y$ obtained by a right-handed rotation from N1 to N6 around the pyrimidine ring. The pyrimidine ring in purines is reversed with respect to that of pyrimidines, as stacked in the A-form helix, and therefore the pyrimidine ring normal vector for purines must be reversed. We define $\vec{\sigma}$ as the normal vector for any base: $\vec{\sigma} = \vec{n}_Y$ in cytosine and uracil, whereas $\vec{\sigma} = -\vec{n}_Y$ in adenine and guanine (see Figure 6a). When bases A and B stack, $\vec{\sigma}_A$ is in the same or opposite direction to $\vec{\sigma}_B$, and B is either above or below A. Therefore, a base stack is "straight" or "reverse" and the second base is either "above" or "below".

The four cases are shown in Figure 6(b). The "upward" stacking corresponds to "straight" and "above"; "downward" to "straight" and "below"; "inward" to "reverse" and "above"; and "outward" to "reverse" and "below". Consequently, the base stack in A-form helices is "upward". The four cases can be written using the less than ($<$) and greater than ($>$) characters. For instance, if A and B stack inward, then we can simply write "A $><$ B" (see Figure 6b).

Note that base stacking is independent of the backbone direction. Two adjacent bases in a sequence can be stacked in any of the four cases. As an example, consider the A-riboswitch aptamer module adenine-sensing messenger RNA (mRNA) crystal structure from *Vibrio vulnificus* (PDB ID 1Y26), where U22 and A23 are stacked downward (U22 $<<$ A23) (see Figure 7). This particular stacking interaction occurs at a junction that connects two fragments inside the adenine-sensing pocket. Both U22 and A23 participate in base triples (a base simultaneously pairs to two other bases) [16].

MC-Annotate implements base stacking as in Gabb and coauthors [17], by using the distance between the ring centers, and the dihedral angle and horizontal shift between the rings. As purines are made of two rings, the pyrimidine and imidazole, both are verified. The base stacking interactions are labeled according to the nomenclature above. Note that biased cutoffs on each parameter are needed to decide whether two bases stack.

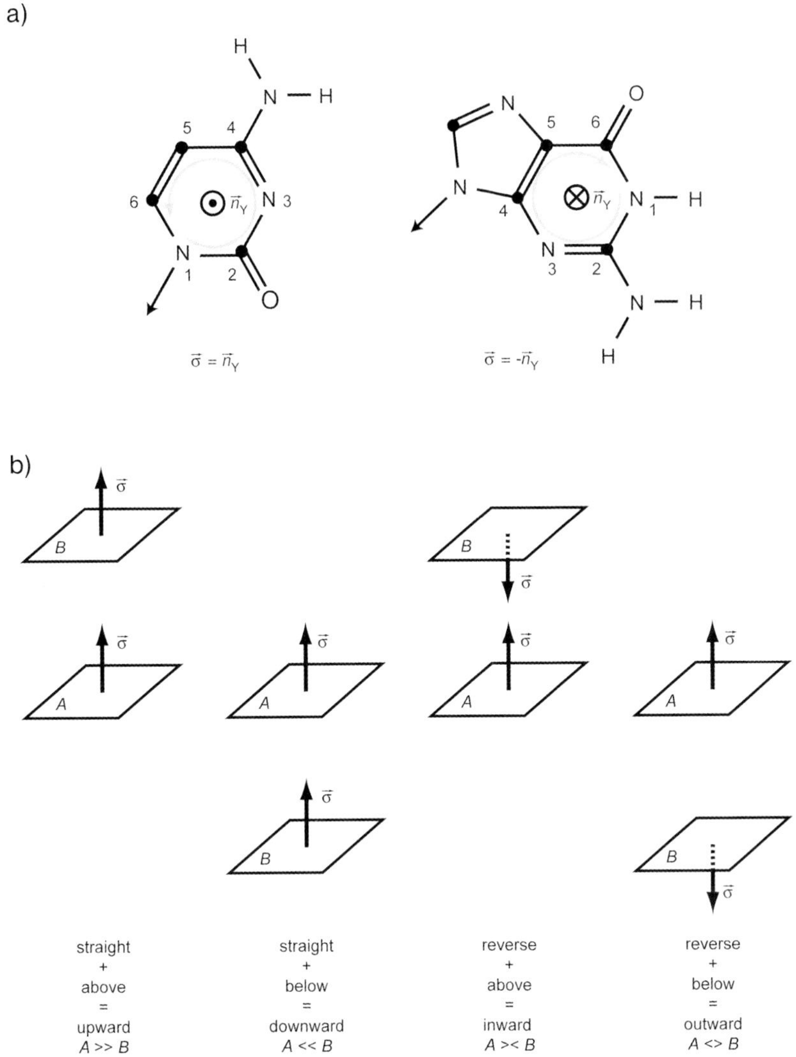

Figure 6 Base stacking. (a) The base normal vector $\vec{\sigma}$ in terms of the pyrimidine ring normal vector $\vec{n}_Y$. In pyrimidines (left, here a cytosine), $\vec{\sigma}$ is defined as $\vec{n}_Y$, the rotational vector obtained from a right-handed rotation around the pyrimidine ring from atoms 1–6. In purines (right, here a guanine), $\vec{\sigma}$ is defined as $-\vec{n}_Y$. (b) Nomenclature of the four stacking cases. Bases A and B are represented by planes.

a)

b)

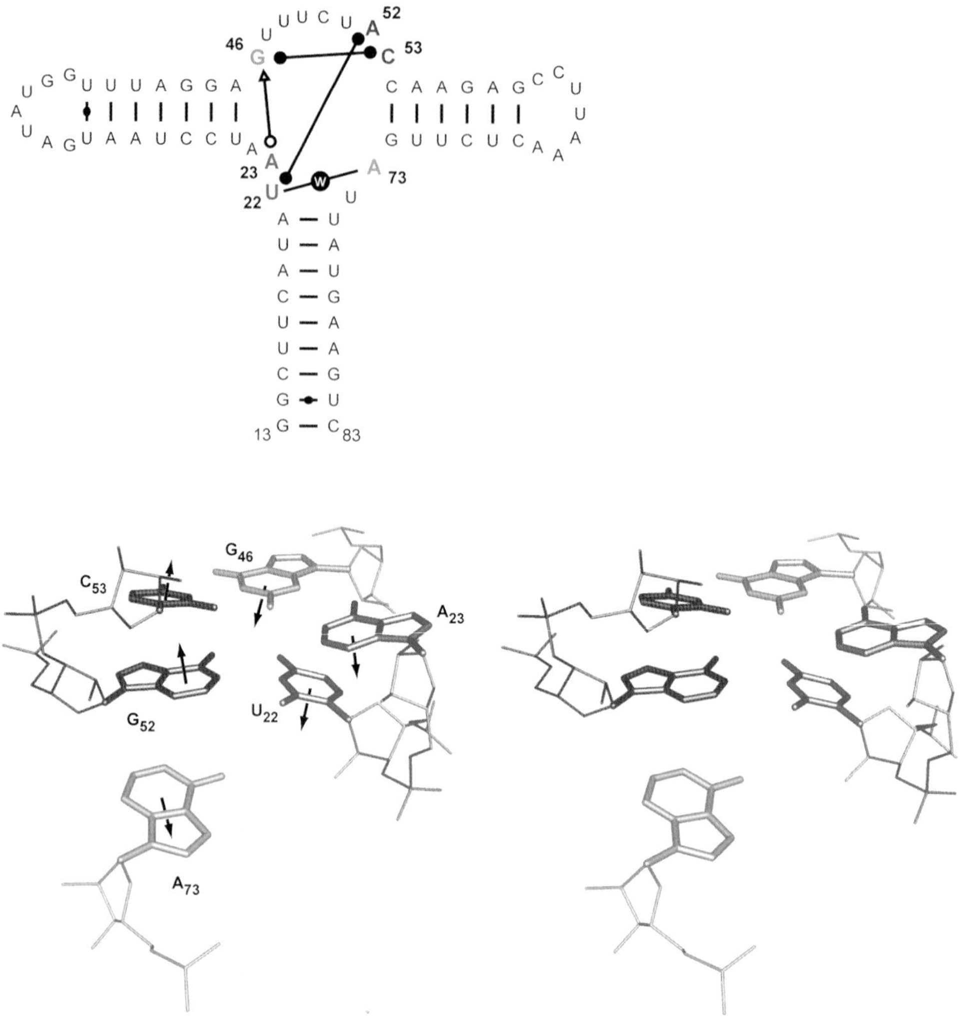

Figure 7 A-riboswitch aptamer. (a) Secondary structure. The bases shown in color are involved in two base triples: U22 (blue) with A52 (red) and A73 (orange), and G46 (green) with U23 (blue) and C53 (red). U22 and A23 are stacked downward. The *LW* nomenclature is used: w for water-mediated; circle for W; square for H; triangle for S. (b) Tertiary structure of the two base triples. The arrows indicate each normal vector, $\vec{\sigma}$

2.2.2 Base Pairing

Base pairing involves the formation of hydrogen bonds between exocyclic hydrogen donor groups (mainly NH and NH_2) and acceptor groups (mainly CO and N). The well-known canonical Watson–Crick G=C and A–U base pairs have three and two hydrogen bonds, respectively. Successive Watson–Crick base pairs that stack upward result in the A-form helix, also called stem (see Figure 4). The determination of the helices of an RNA from sequence data is the goal of secondary structure prediction (see Chapter 14). As stems have a tight and local 3-D structure, they are often manipulated as rigid objects in computer modeling. Other than Watson–Crick base pairs abound in RNAs; near 20% in the yeast phenylalanine transfer RNA (tRNA-Phe) and near 50% in the large ribosomal subunit), and are often qualified as "non-canonical", or non-Watson–Crick.

In his famous 1984 book, Saenger compiled 28 base pairing patterns involving at least two hydrogen bonds [7]. Each base pair was assigned a roman number: for instance XIX for the G=C Watson–Crick base pair, XX for the A–U Watson–Crick base pair, XXIII for the "Hoogsteen" A–U base pair, XXIV for the "reverse Hoogsteen" A–U base pair and XI for the "sheared" G–A base pair; see Ref. [7], p. 120 for the complete list).

More recently, Leontis and Westhof proposed a new nomenclature, *LW* [18]. In *LW*, three hydrogen bond contact edges (*W* = Watson–Crick, *H* = Hoogsteen and *S* = Sugar) were defined in each base (see Figure 8). To describe a base pair, one has simply to name its interacting edges. For the Watson–Crick base pair, since the hydrogen bonds are formed by chemical groups on the *W* edges of each base, we refer to it as *W*/*W*. In addition, the relative orientation of the riboses with respect to the plane of the base pair is annotated as *cis* or *trans*, respectively, if the glycosidic bonds extend towards the ribose are in the same (as in the A-form helix) or opposite orientation (see Figure 9a).

To introduce more precision and distinguished among possible ambiguities of the *LW*, and in particular in one-hydrogen-bond base pairs, Lemieux and Major divided each contact edge in three regions they named faces [19]. In this extension of *LW*, *LW+*, each possible hydrogen bond face is named by its corresponding *LW* edge, to which one of three possible orientations was added: *w*, *h* and *s*. For instance, the *W* edge has the *Ww* face at the center of the edge, the *Wh* face towards the *H* edge and the *Ws* face towards the *S* edge (see Figure 8). The wobble GU base pair is annotated *W*/*W* in *LW*, and more precisely *Ww/Ws* in *LW+*. Bifurcated hydrogen bonds that oscillate between two *LW* edges have their own faces in *LW+*: *Bh* between *W* and *H* edges, and *Bs* between *W* and *S* edges.

Finally, the normal vector $\vec{\sigma}$ used to annotate base stacking can also be used in base pairing to address the relative orientation (parallel or antiparallel) of

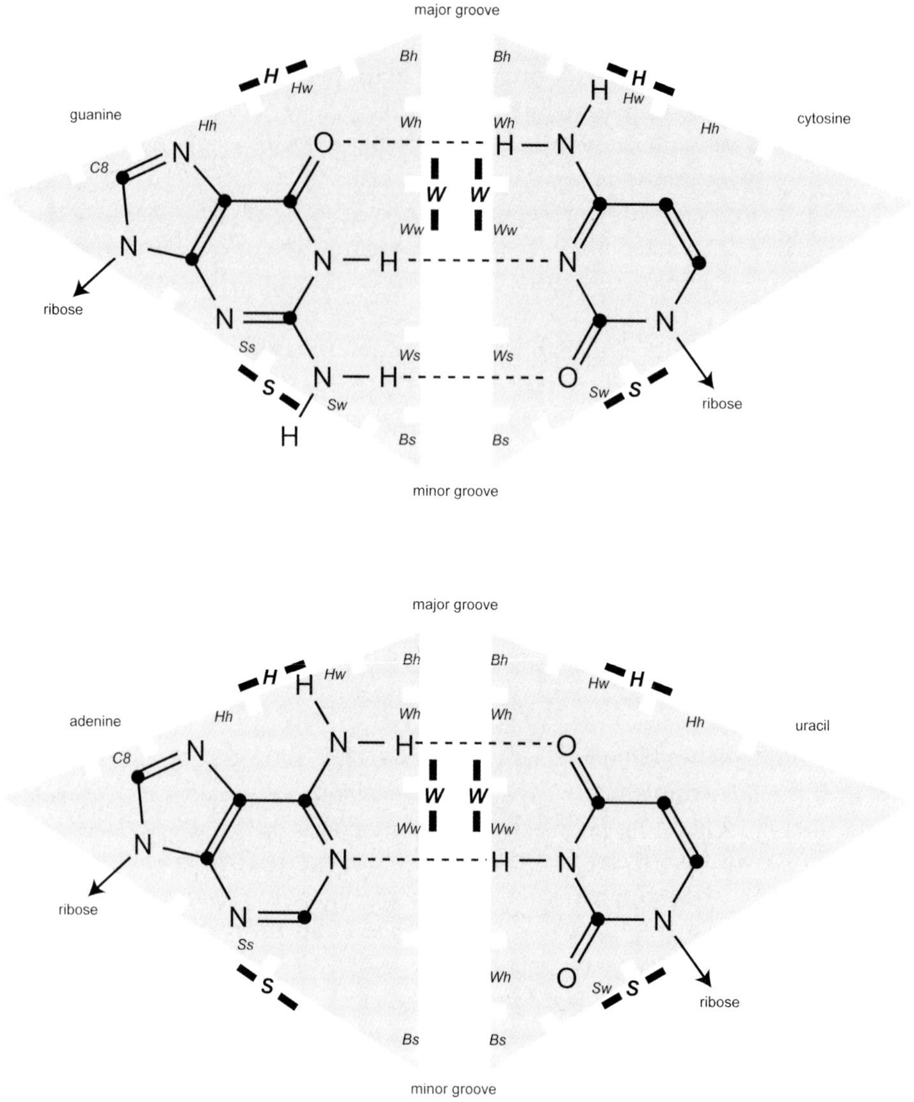

Figure 8 Base pairing patterns. The two canonical Watson–Crick base pairs and their hydrogen bonds (dashed lines) are shown (top: G=C; bottom: A–U). The *LW* nomenclature is shown by engulfing shaded triangles, where the arcs represent the *W*, *H* and *S* edges on the four standard bases. Here, the *W* edges are in contact in both base pairs. Notches along each edge delimit the *LW+* faces for each base. The major groove of the A-RNA double helix is on the *H* side, whereas the minor groove is on the *S* side.

the two bases in a base pair. The *cis W/W* base pairs in the A-form helix are characterized by the antiparallel orientation (see Figure 9b).

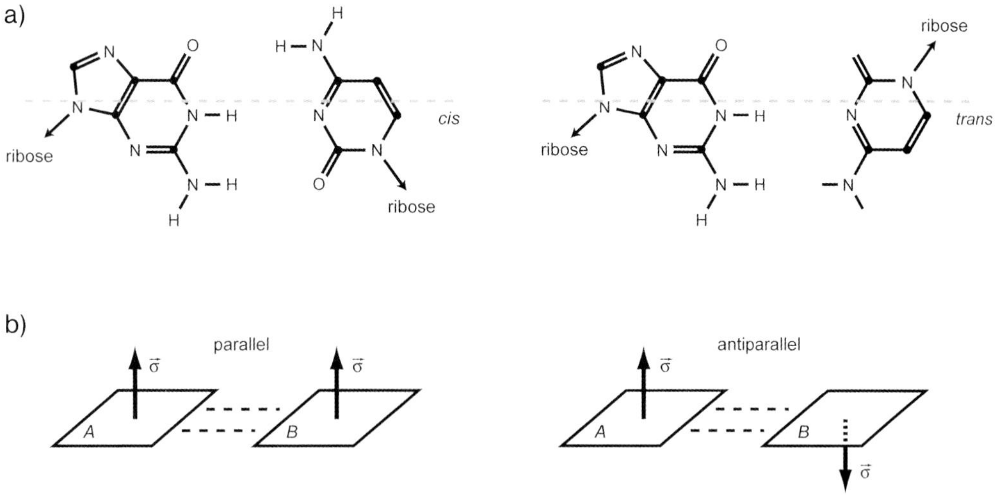

Figure 9 Base pairing orientations. (a) From the plane of the base pair, if both glycosidic bonds are oriented on the same side of the line that splits the plane evenly in two (dashed gray line), the base pair is *cis* (left). Else, if the glycosidic bonds are on each side, the base pair is *trans* (right). (b) Base normal vector $\vec{\sigma}$ relative orientation. Two bases, A and B, represented here by planes in perspective view, are either "parallel" if their $\vec{\sigma}$ are oriented in the same direction (left) or "antiparallel" if their $\vec{\sigma}$ are oriented in opposite directions (right).

In order to limit the bias of using cutoffs, Gendron and coworkers, in MC-Annotate, implemented the detection of the base pair types by using unsupervised learning [19]. Single hydrogen bonds are selected by Gaussian distribution fitting of three geometric parameters calculated between all pairs of bases: the hydrogen, the donor, the acceptor and the lone electron pair (positioned 1 Å away from the acceptor in the direction of the orbital). The subset of hydrogen bonds between two bases is selected by solving the equilibrium state of the maximal flow of the directed bipartite graph formed by all possible hydrogen bonds between the two bases. The base pair is labeled according to the *LW+* nomenclature. The relative orientations of the glycosidic bonds (*cis* or *trans*) and of the normal vectors (parallel or antiparallel) are also computed. The RNAView annotation procedure of base pairs differs considerably from the one implemented in MC-Annotate, as it is based on geometrical cutoffs.

Lee and Gutell proposed an alternative topological nomenclature, *LG*. Starting with the Watson–Crick C=G or A–U, or even with the wobble G–U base pair, they defined 14 families by successively manipulating the base plane and glycosidic bond relative orientations: shearing, flipping, reversing, paralleling or slipping. The resulting 14 families are the Watson–Crick (*WC*), wobble (*Wb*), slipped Watson–Crick (*sWC*), slipped wobble (*sWb*), reversed Watson–Crick (*rWC*), reversed wobble (*rWb*), Hoogsteen (*H*), reversed Hoogsteen (*rH*), sheared (*S*), reversed sheared (*rS*), flipped sheared (*fS*), parallel flipped sheared (*pfS*), parallel sheared (*pS*) and reversed parallel sheared (*rpS*) [20].

2.2.3 **Isosteric Base Pairs**

Base pairs are isosteric if they preserve a local tertiary structure and, thus, function, as observed in evolutionary related RNAs whose sequences may diverge. Leontis and coworkers have superimposed the geometry of all possible base pairs according to their C1′–C1′ distances and *cis*/*trans* base orientations [21]. Then, they mapped all 16 combinations to the 12 families of the *LW* nomenclature, which resulted in isostericity matrices from which it can be shown that all canonical Watson–Crick combinations are isosteric and that the wobble G–U base pair is isosteric to the protonated A–C base pair. Walberer and coworkers defined isostericity from a theoretical analysis [22]. They generated a base pair database for all possible hydrogen bond arrangements and deduced an isostericity measure based on glycosidic bond overlap. They observed high isostericity values in helical base pairs, validating their approach, and more surprisingly in several purine–purine and pyrimidine–pyrimidine combinations.

Accurate RNA sequence comparison requires precise (structural) alignments in order to ensure the positions compared truly correspond in the tertiary structure. Including isosteric information in sequence alignment gives better insights into the sequence requirement of structural motifs (see Section 3) across RNA phylogenies [23]. The presence of isosteric base pairs is another fact that supports a structural rather than sequential RNA evolution.

3 Motif Discovery

In the previous section, we presented a series of nomenclature and formalisms to describe some already acknowledged components of RNA tertiary structure: the nucleotide conformations and interactions. Here, we take one step further and describe higher-order RNA components.

The increase in high-resolution X-ray crystallographic structures, in particular the resolution of the large ribosomal subunit [24–26], has increased the literature describing repeated RNA fragments or motifs [27]. Many occurrences of each of these fragments can be found in one or among several different 3-D structures, they are conserved among evolutionary related RNAs, and they are often related to specific structural and biochemical functions.

One can think of RNA motifs as fundamental RNA building blocks. Therefore, finding and characterizing all of them should provide us with invaluable knowledge about RNA folding and aid substantially in tertiary structure prediction. Here, we present some classical RNA motifs, and introduce a formal definition allowing us to computationally represent, search for and discover them.

3.1 RNA Motifs

The most obvious RNA motif is the double helix, which is composed of a succession of stacked Watson–Crick base pairs. Similarly to the double helix, RNA motifs are thermodynamically stable and fold into similar tertiary structures that can be found in various structural contexts.

Let us define an RNA motif as a graph of nucleotide conformations and interactions, where the nucleotides are the vertices of the graph. An arc between two nucleotides is present if the two nucleotides are adjacent in the sequence or if their bases interact. Note that if we use the nomenclature introduced in the previous section, then this definition is equivalent to our formal representation of an annotated RNA tertiary structure and is, in fact, the output of the MC-Annotate computer program.

While RNA graphs are easily represented in computer programs by classical data structures, there is currently no consensus in the RNA ontology nor is there a data file format to represent them. RNA graphs in computer programs such as MC-Annotate are serialized into opaque binary files using the C++ MC-Core library developed in our laboratory (also freely available at sourceforge.net/projects/mccore). The RNAML format, derived from XML (extensible markup language), can handle RNA graphs and is portable among many different RNA applications [28,29].

Since RNA motifs can be represented by characteristic RNA graphs, they can be searched within hosting RNA tertiary structures via graph isomorphism; the occurrences of an RNA motif are simply the isomorphic subgraphs in the hosting graphs. Our laboratory implemented the classical graph isomorphism algorithm [30] in a computer program called MC-Search. The input to MC-Search is a description of the target RNA structure, or pattern, from which MC-Search returns all occurrences of the target motif found in a set of pre-selected PDB files.

3.1.1 Classical Examples

Consider the sarcin/ricin motif (Figure 10), which has been predicted to occur in many different locations of the large ribosomal subunit by comparative sequence analysis [31]. The MC-Search descriptor file for the sarcin/ricin motif is shown in Figure 10(b). MC-Search finds seven occurrences of this motif in the crystal structure of the *H. marismortui* 23S rRNA (PDB ID 1JJ2). All occurrences found share a maximum RMSD of 0.93 Å. Interestingly, the annotation of the found occurrences revealed four conserved base stacking interactions (see Figure 10c), including those among nonadjacent nucleotides:

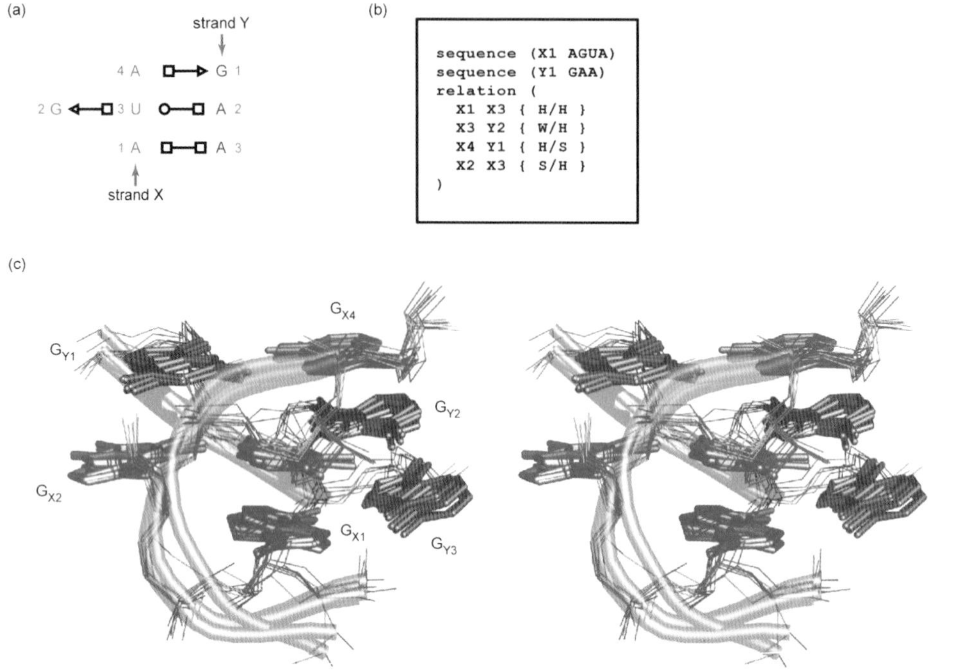

Figure 10 Sarcin/ricin motif. The motif is made of two strands, here named X (shown in blue) and Y (red), and their nucleotides, respectively, numbered 1–4 and 1–3. (a) A schematic representation using *LW* (circle for *W*; square for *H*; triangle for *S*). (b) MC-Search input. (c) The seven occurrences in *H. marismortui* 23S rRNA crystal structure optimally aligned (stereoview). The threads follow the phosphorus atoms in each strand. The hydrogen atoms are not shown.

A_{X1} <> U_{X3} (outward), G_{X2} >< G_{Y1} (inward) and A_{X4} <> A_{Y2} (outward). Here, the two strands involved in the motif were arbitrarily named X and Y, and their nucleotides numbered, respectively, 1–4 and 1–3.

A motif corresponding to the T-loop conserved across tRNAs was matched in the ribosome. In tRNAs, the loop capping the T-stem is characterized by a *trans* W/H U–A base pair stacked on the last W/W G=C base pair of the stem with a two- or three-nucleotide bulge on the A side. Several instances of the motif where found by visual inspection in *H. marismortui* 23S and in *Thermus thermophilus* 16S subunits [32]. These tRNA T-loop motifs in the ribosome were found to interact with other elements of their rRNA through tertiary interactions, similarly to the interactions found between the T- and the D-loop in tRNAs. Two instances of the two-nucleotide bulge version are found by MC-Search in the *H. marismortui* 23S subunit (Figure 11).

The frequently observed A-minor motif [33] is made of an adenine that interacts with the minor groove of a double helix and is of particular interest since it is involved in the selection of tRNAs by the ribosome [34]. Nine

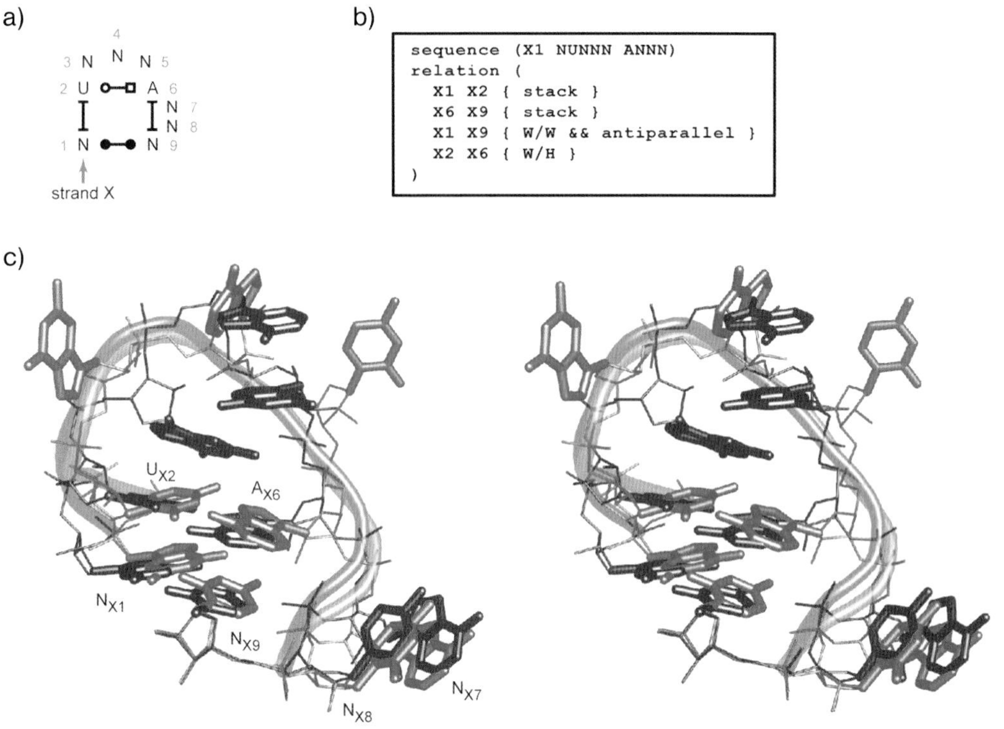

Figure 11 T-loop motif. The motif is made of one strand, here named X, and the nucleotides, respectively, numbered 1–9. (a) A schematic representation using *LW* (circle for *W*; square for *H*; triangle for *S*). (b) MC-Search input. (c) The two occurrences in *H. marismortui* 23S rRNA crystal structure (PDB 1JJ2) optimally aligned (stereoview). The instance found at X1 = 334 is shown in blue; the one at X1 = 1387 in red. The threads follow the phosphorus atoms in each strand. The hydrogen atoms are not shown.

instances of the A-minor motif are found by MC-Search in the *H. marismortui* 23S subunit (Figure 12). Note that the *S* edge in the base pair annotation of the double-helix nucleotide indicates the minor groove interaction. The A-minor motif is a key element of the larger K-turn motif which induces a bend between two double helices [35]. The core of the K-turn motif is constituted of two *S*/*H* G–A base pairs. There is only one occurrence of the K-turn motif in the *H. marismortui* 23S subunit (Figure 13).

The tetraloop/receptor motif is most frequent in RNAs. It stabilizes the conformation of a hairpin loop interacting with the minor groove of a stem. It was discovered in the hammerhead ribozyme [36] and in the group I intron [37]. The tetraloop/receptor participates in protein translation fidelity, and in the association of the rRNA 16S and 30S subunits, as mutations in the motif induce loss of ribosomal activity [38]. In *T. thermophilus* rRNA, a conserved

a)

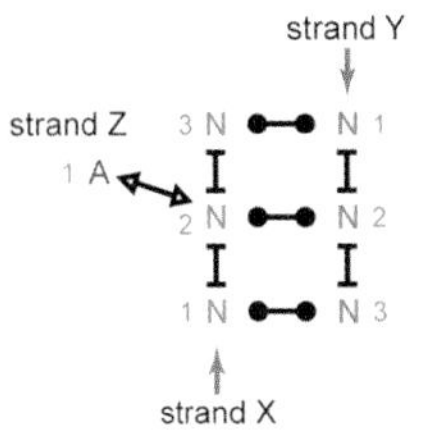

b)

```
sequence  (X1 NNN)
sequence  (Y1 NNN)
sequence  (Z1 A)
relation  (
  X1 X2 { upward }
  X2 X3 { upward }
  Y1 Y2 { upward }
  Y2 Y3 { upward }
  X1 Y3 { W/W && antiparallel }
  X2 Y2 { W/W && antiparallel }
  X3 Y1 { W/W && antiparallel }
  X2 Z1 { S/S }
)
```

c)

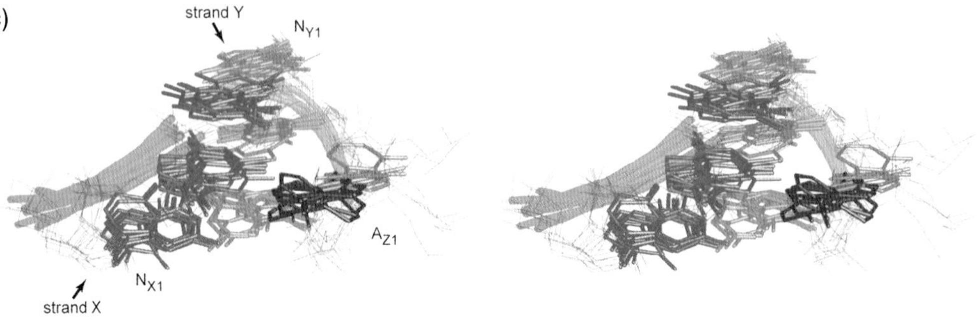

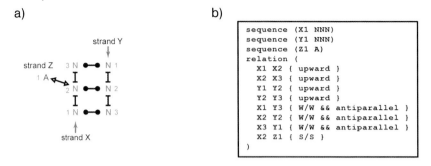

Figure 12 A-minor motif. The motif is made of three strands, here named X (shown in blue), Y (green) and Z (red); the nucleotides are, respectively, numbered 1–3, 1–3 and 1. (a) A schematic representation using *LW* (circle for W; square for H; triangle for S). (b) MC-Search input. (c) The nine occurrences in *H. marismortui* 23S rRNA crystal structure optimally aligned (stereoview). The threads follow the phosphorus atoms in each strand. The hydrogen atoms are not shown.

GCAA tetraloop (referred to as loop 900) caps helix 27 in the 16S subunit and binds to the minor groove of helix 24 in the 30S subunit. Two occurrences of the tetraloop/receptor motif-binding stems of at least three W/W base pairs are found by MC-Search in the *H. marismortui* 23S subunit (Figure 14). Interestingly, the two occurrences found have different base pair orientations.

As found in many pathogenic viruses, the formation of a pseudoknot motif in mRNAs may induce frameshifting in the protein translation [39, 40]. The pseudoknot is made of a hairpin stem–loop whose nucleotides in the loop participate in the formation of a second stem. In fact, a pseudoknot occurs when the four strands, A5′, A3′, B5′ and B3′, involved in the formation of two stems, A and B, interleave in the sequence: A5′–B5′–A3′–B3′. Pseudoknots are reported by the MC-Annotate computer program, and 15 are found in the crystal structure of the *H. marismortui* 23S rRNA (PDB ID 1JJ2). In feline immunodeficiency virus (FIV), a pseudoknot was found by comparative sequence and mutagenesis analyses [41]. The secondary and tertiary structures of the FIV pseudoknot are shown in Figure 15. See Section 4.3.2 for details

a)

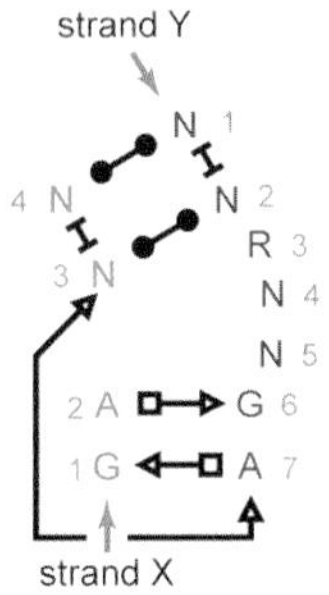

b)

```
sequence (X1 GANN)
sequence (Y1 NNRNN GA)
relation (
  X3 X4 { upward }
  Y1 Y2 { upward }
  X1 Y7 { S/H }
  X2 Y6 { H/S }
  X3 Y7 { S/S }
  X3 Y2 { W/W && antiparallel }
  X4 Y1 { W/W && antiparallel }
)
```

c)

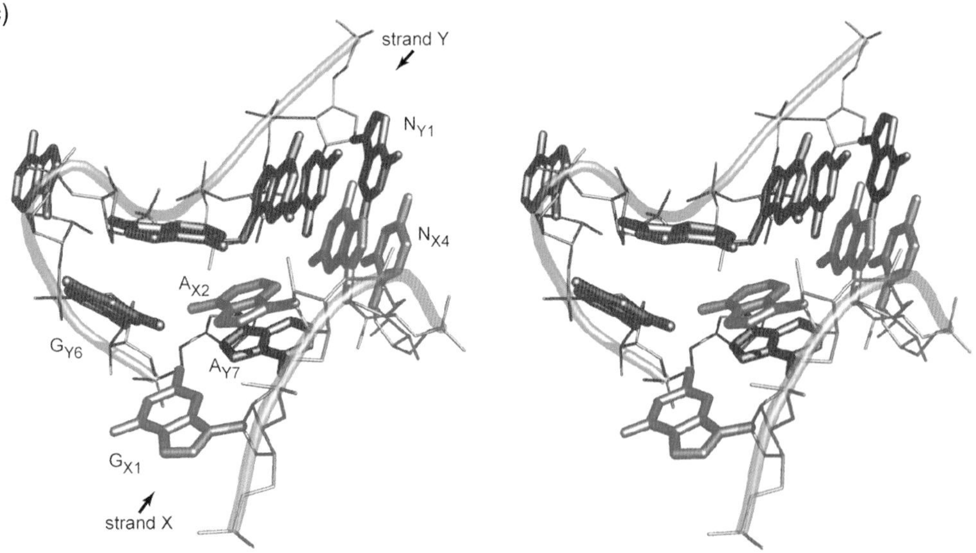

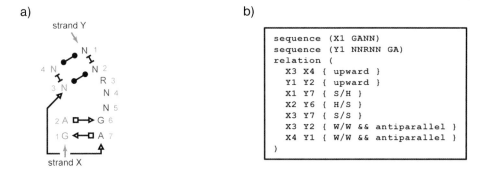

Figure 13 K-turn motif. The motif is made of two strands, here named X (shown in blue) and Y (red), and their nucleotides, respectively, numbered 1–4 and 1–7. (a) A schematic representation using *LW* (circle for W; square for H; triangle for S). (b) MC-Search input. (c) The one occurrence in *H. marismortui* 23S rRNA crystal structure (stereoview). The threads follow the phosphorus atoms in each strand. The hydrogen atoms are not shown.

how Fabris and his collaborators combined mass spectroscopy and computer modeling to determine the FIV pseudoknot tertiary structure [42].

3.2 Catalytic Motifs

The structure and function of catalytic RNAs (ribozymes) have extensively been studied [43–51]. In particular, the crystal structure of the hammerhead ribozyme shows a three-way junction catalytic core [45]. The three-way junc-

a)

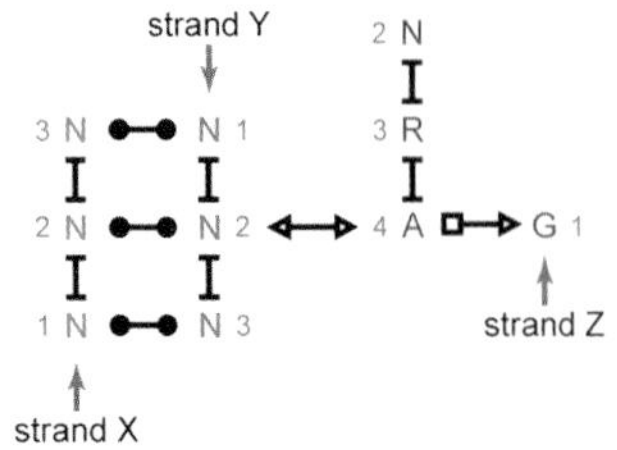

b)

```
sequence (X1 NNN)
sequence (Y1 NNN)
sequence (Z1 GNRA)
relation (
  X1 X2 { upward }
  X2 X3 { upward }
  Y1 Y2 { upward }
  Y2 Y3 { upward }
  X1 Y3 { W/W && antiparallel }
  X2 Y2 { W/W && antiparallel }
  X3 Y1 { W/W && antiparallel }
  Z2 Z3 { stack }
  Z3 Z4 { stack }
  Z1 Z4 { S/H }
  Z4 Y2 { S/S }
)
```

c)

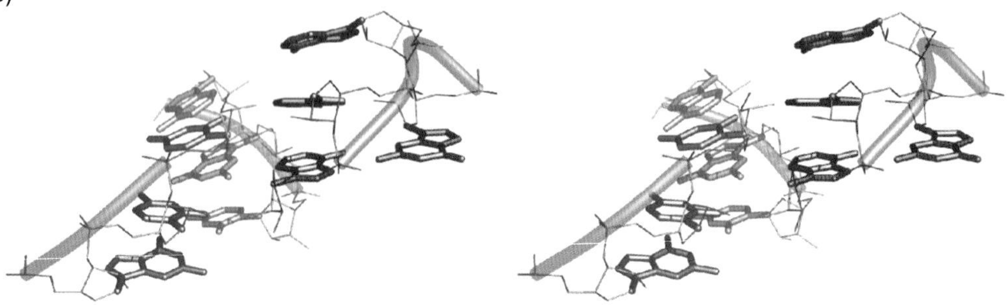

d)

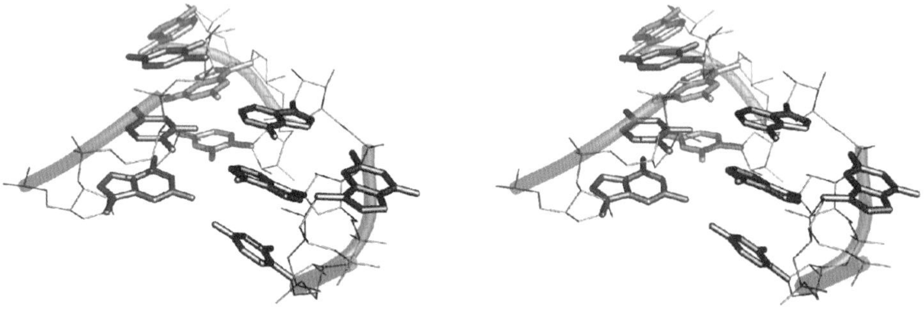

Figure 14 Tetraloop receptor. The motif is made of three strands, here named X (shown in blue), Y (green), and Z (red); the nucleotides, respectively, numbered 1–3, 1–3 and 1–4. (a) A schematic representation using *LW* (circle for *W*; square for *H*; triangle for *S*). (b) MC-Search input. (c) The occurrence in *H. marismortui* 23S rRNA crystal structure at position X1 = 1552, Y1 = 1567 and Z1 = 1629. The Z4–Y2 base pair is parallel. (d) The occurrence in *H. marismortui* 23S rRNA crystal structure at position X1 = 2529, Y1 = 2490 and Z1 = 1055. The Z4–Y2 base pair is antiparallel. The threads follow the phosphorus atoms in each strand. The hydrogen atoms are not shown.

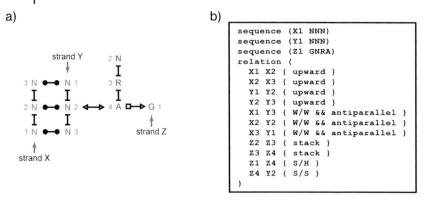

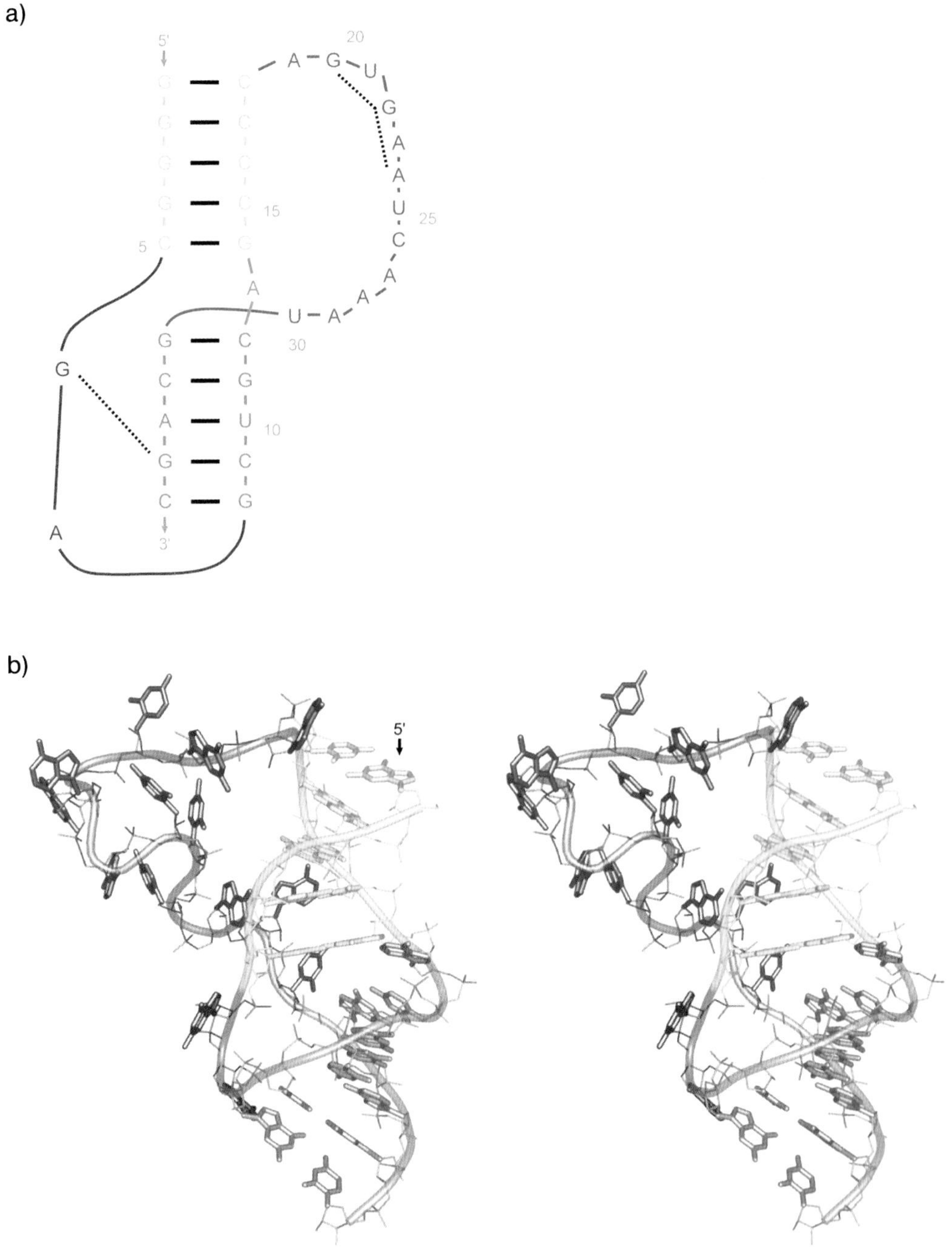

Figure 15 FIV pseudoknot structure. (a) Secondary structure. The phosphodiester linkage is shown by thin lines. The W/W base pairs are shown by thick lines. The tertiary interactions predicted by chemical cross-linking are shown in dashed lines. (b) Tertiary structure (stereoview). The thread shows the phosphodiester backbone. The hydrogen atoms are not shown.

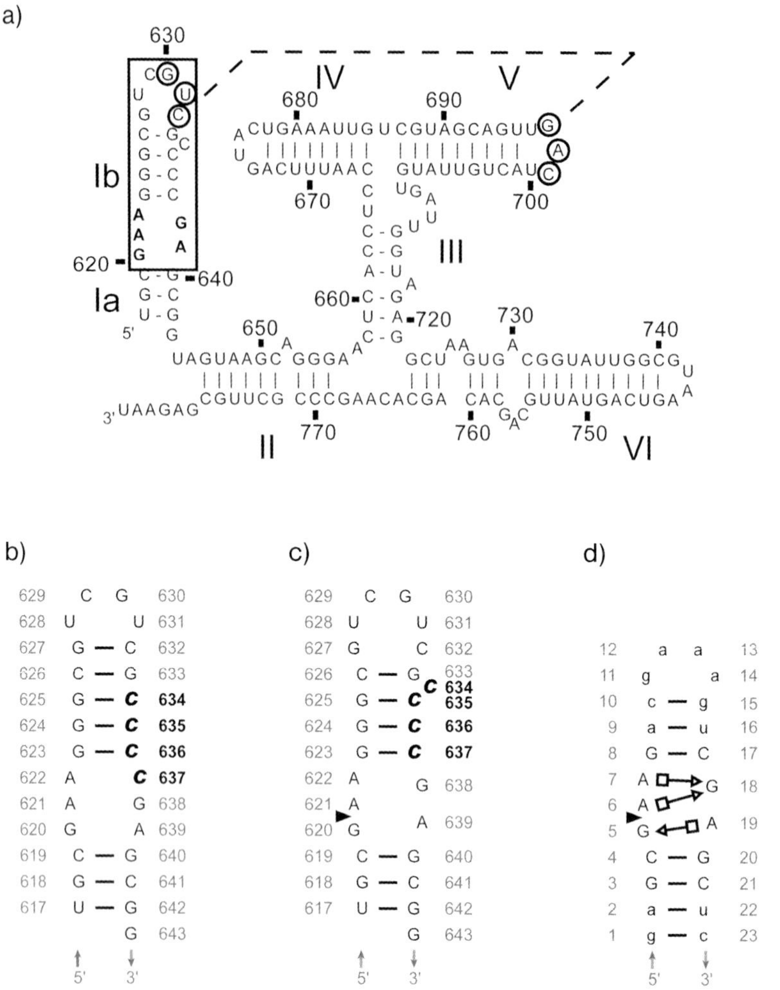

Figure 16 *Neurospora* VS ribozyme. (a) Secondary structure. The substrate domain is shown in the box. The dotted line indicates the tertiary interactions of the circled bases. (b) Inactive state of the substrate domain. C_{637} becomes the base pair partner of G_{623}, which affects the base pair registry of stem I. C_{637} to C_{634} (shown in bold) are shifted by one base pair towards the 5′ strand. As a result, C_{634} looses its base pair partner in the active state. (c) Active state. The arrow points the cleavage site. (d) Mutant stem–loop I (SL1′). Uppercase nucleotides belong to the wild-type; lowercase to mutations. The sheared G–A base pairs identified in the NMR structure are shown using the *LW* nomenclature (circle for *W*; square for *H*; triangle for *S*).

tion motif is made of three double helices, and its topology has been much studied by Lescoute and Westhof [52]. Interestingly, the H/H and S/S loop–loop interactions involved in the catalytic mechanism of the hammerhead have been characterized by 3-D modeling by Massire and Westhof, using their computer program MANIP [53,54].

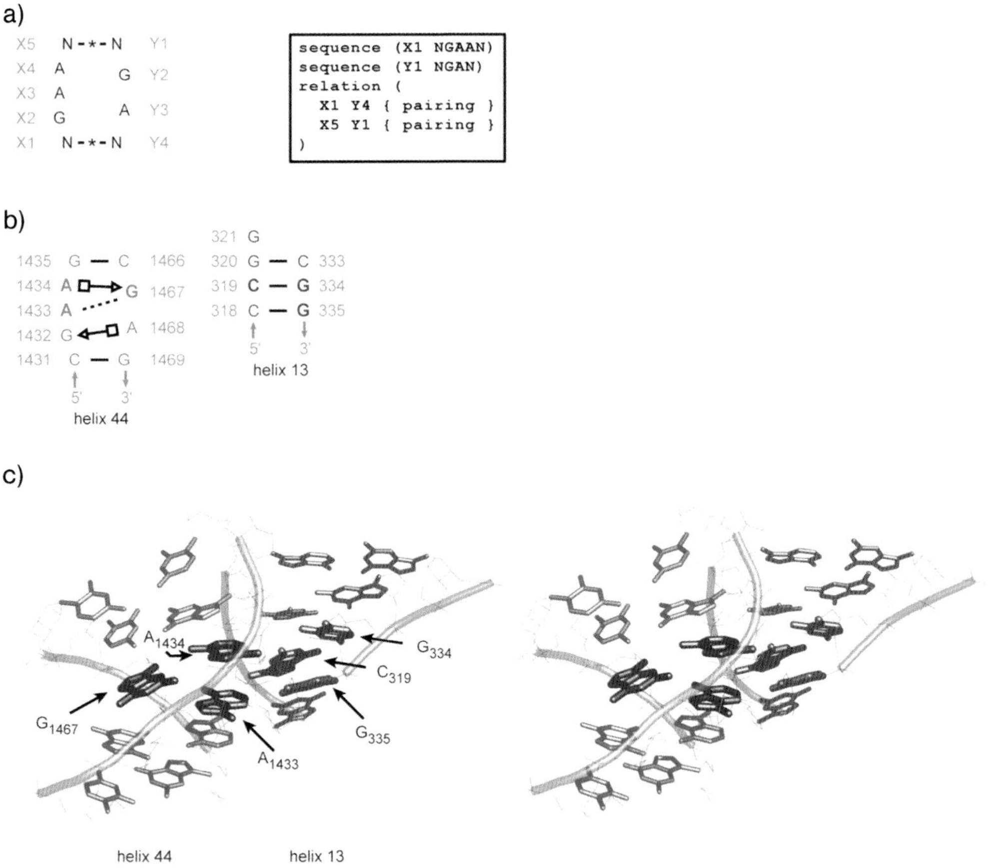

Figure 17 VS ribozyme SL1′ active internal loop motif. (a) MC-Search input. The strands are named "X" and "Y", respectively, for the 5′ and 3′ strands. A star in the schema (left) is used as a wildcard matching any type of base pairs, which is denoted "pairing" in the input file (right). (b) The secondary structure of the occurrence found in *T. thermophilus* 16S rRNA. *LW* is used (circle for *W*; square for *H*; triangle for *S*). This motif matches in helix 44 of the 16S rRNA (left) and forms many ribose–ribose contacts and two A-minor motifs with helix 13 (right). (c) Stereoview of helix 44 (blue) and helix 13 (red). The match is at 1.40 Å of RMSD of the NMR structure of the VS ribozyme SL1′ loop. The shared sheared G–A base pairs are shown in bold. The nucleotides in helix 13 that participate to the A-minor motifs are shown in bold.

Another relationship between the structure and catalytic activity of RNAs was discovered in the *Neurospora* Varkud Satellite (VS) ribozyme. Six helical domains characterize the secondary structure of the self-cleaving VS ribozyme (see Figure 16a). The substrate domain at stem–loop I is recognized by the catalytic core by, so far, unknown loop–loop interactions between stem–loop I and V [55]. The cleavage mechanism of this ribozyme is induced by a modification of the base pair registry in stem I (see Figure 16b and c). C_{637}

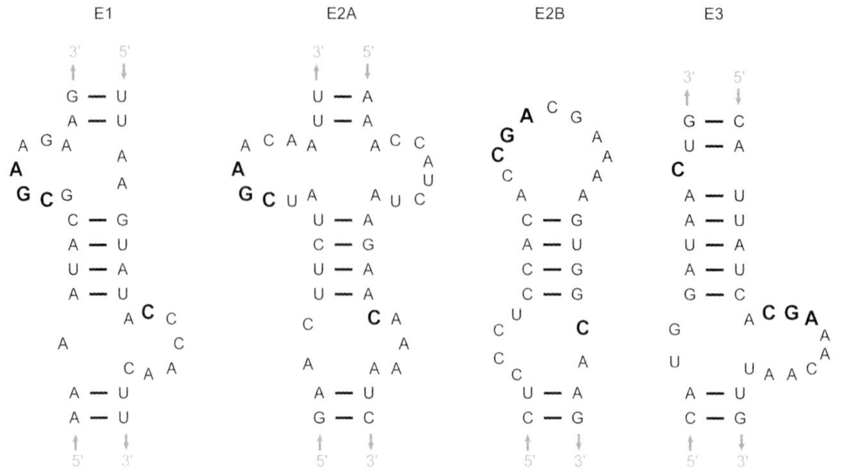

Figure 18 Secondary structures of the four localization elements within *ASH1* mRNA in yeast *S. cerevisiae*: E1, E2A, E2B and E3. Only fragments hosting the loop–stem–loop common motif are shown, with the conserved CGA and C shown in bold.

is paired to G_{623} in the active state, reducing the size of the 3′ strand of the internal loop from three to two.

To get a deeper insight into the 3-D structure of the VS ribozyme active conformation, a nuclear magnetic resonance (NMR) spectroscopy structure of a stem–loop I mutant (SL1′) was built to mimic the active conformation [56]. In previous NMR structures of the inactive stem–loop I, the 3–3 internal loop was composed of three stacked base pairs, two S/H G–A base pairs and a wobble A–C base pair. In the SL1′ active structure, the A–C base pair is broken by the helix shift, resulting in a double S/H base pair shared between both As on the 5′ strand and the G on 3′ strand (see Figure 16c).

This distinctive internal loop motif was searched in other RNAs with the MC-Search computer program without any specific interaction types in the loop, by using the "pairing" keyword. The pattern matched a fragment in helix 44 of the *T. thermophilus* 16S rRNA, which is at 1.4 Å of RMSD to the NMR structure of the interior loop of SL1′ (see Figure 17). Helix 44 in the rRNA interacts with helix 13 to form a ribose zipper motif [57], which is defined by ribose–ribose interactions and the presence of the A-minor motif. Both adenines share one sheared (S/H) G–A base pair. The SL1′ structure became a model for the active substrate element of the VS ribozyme, which can bind to another helix and form a ribose zipper motif. Since the catalytic mechanism is characterized by tertiary interactions between stem–loop I and V (see Figure 16a), the authors suggest that the specific interaction needed for an accurate recognition of the substrate is this ribose zipper.

3.3 Transport and Localization

A joint effort between Chartrand's laboratory in Montreal and our group resulted in a deeper understanding of the molecular basis behind cytoplasmic mRNA transport and localization to the yeast bud [58]. A small RNA motif was found conserved across four localization elements within the mRNA of the *ASH1* protein. The motif was found to interact with She2p, one of the important components of the yeast mRNA localization machinery in *Saccharomyces cerevisiae*.

Three localization elements occur in the coding region (E1, E2A and E2B) [59] and the fourth is located in the 3′ untranslated region (E3) [60]. Conserved nucleotides in each element were identified by *in vivo* selection from a polymerase chain reaction (PCR) library. Similar She2p RNA motif-binding sites were characterized, and their secondary structures predicted. The generalized motif is composed of two loops, separated by a short stem, which contains a conserved C on the 3′ strand and a CGA sequence on the 5′ strand (see Figure 18), but for E3 for which the loops are inversed.

A structural rule of the generalized She2p-binding motif was deduced: *i*:*s*:*j*, where *i* is the number of nucleotides in the loop before the conserved CGA sequence (*cf.* one in the 5′ loop of E1: G<u>CGA</u>AGA), *j* is the number of nucleotides before the conserved C in the 3′ loop (*cf.* one in the 3′ loop of E1: A<u>C</u>CCAAC) and *s* the length of the stem separating the two loops (*cf.* four in E1). The localization element E1 is thus a 1:4:1 motif, E2A and E2B are 2:4:0, and E3 is a 0:5:1. The sum $i + s + j = 6$ is invariant.

To find a structural explanation of the invariant rule, MC-Search was used to scan the high-resolution RNA 3-D structures to seek occurrences of these three types of She2p-binding motifs (see Figure 19). In total, 123 matches were found for type 1:4:1 (E1), 85 for type 2:4:0 (E2A/E2B) and 717 for type 0:5:1 (E3). The distance between the 3′ phosphate groups of the two conserved cytosines in all occurrences was 28.3 ± 0.9 Å for type 1:4:1, 28.0 ± 1.0 Å for type 2:4:0 and 28.2 ± 0.7 Å for type 0:5:1. Increasing the length of the She2p-binding motif stem in each localization elements in the yeast three-hybrid assay resulted in a total loss of interaction with She2p.

The resolution of the X-ray crystal structure of She2p revealed a helical region essential for its interaction with *ASH1* mRNA localization elements. This helical region covers a distance of about 27 Å. It was thus logical to the authors to propose a model of the *ASH1* mRNA localization element motif binding to this region of the protein through interactions between the two conserved Cs. An interesting conclusion from this project is that tertiary structure conservation of the RNA motif is more relevant to the binding function than sequence conservation.

a)

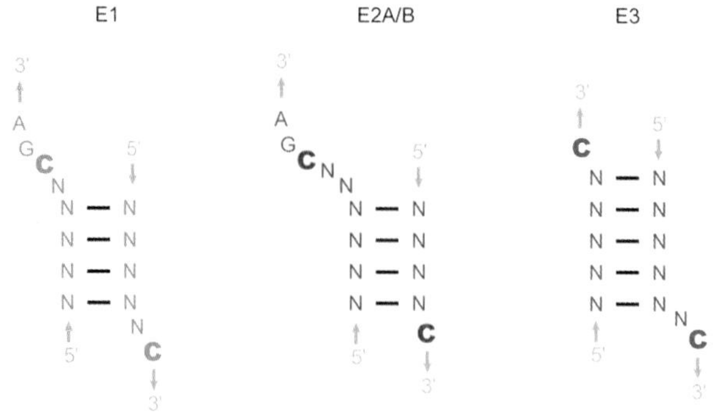

b)

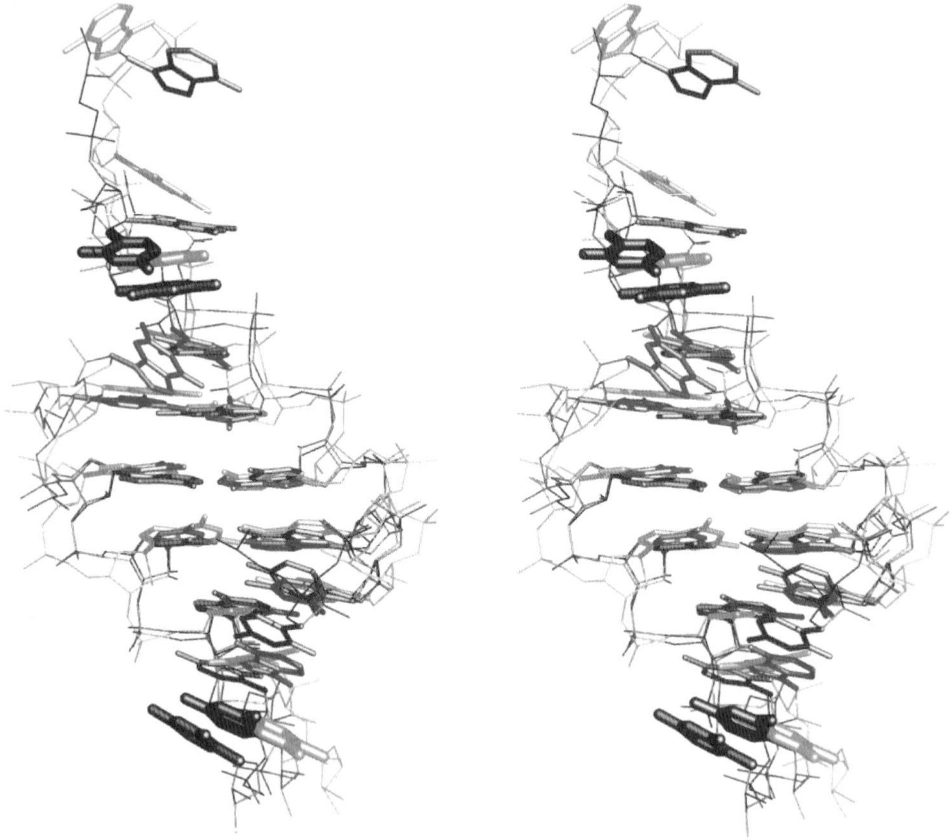

4 Modeling

In the previous sections, we introduced an ontology to model RNA tertiary structure components and motifs. Here, we employ the term "modeling" to describe 3-D model building. Two types of 3-D models exist: physical or handicraft models (wood, plastic, metal or any other artistic materials) and computer models (interactively built or directly generated). Often, scientists start with the former type, mainly to draft ideas and to get a global picture of the RNA tertiary structure, and then, when enough structural data are gathered, they switch to the latter.

The goal of RNA tertiary structure modeling is to summarize and project in 3-D structural data. In the last few decades, many low-resolution techniques, e.g. footprinting, and enzymatic and chemical probing, have improved and produced a large quantity of structural data. Consequently, RNA tertiary structure modeling has become very popular and useful in recent years to translate these structural data in to precise 3-D models. The low-resolution techniques compensate when higher-resolution ones cannot be applied, e.g. in the case of *in vivo* or reactive conformations.

In this section, we focus on computer-assisted model building. There are two major approaches to 3-D modeling. The first approach starts with all nucleotides in an extended or randomized state, which is then successively modified until a folded and satisfactory state is reached. The conformational space of such folding methods is defined by the number of accessible states, e.g. molecular mechanics is a folding method that defines satisfaction as the optimum of an objective function. Harvey's laboratory has developed an RNA objective function composed of penalty terms corresponding to experimentally determined nucleotide interactions and distances. The objective function is penalized if the observed interactions and distances are not present in a state. They simplified the RNA model by using one to five points per nucleotide to speedup the folding operations. Using YAMMP, their computer program, they were able to build models of large ribosomal subunits [61]. In the second approach, one assembles the components of the RNA by using

Figure 19 The She2p-binding motif. (a) MC-Search input for the three motif types, from left to right 1:4:1 (E1), 2:4:0 (E2A/B) and 0:5:1 (E3). The conserved Cs are shown in bold. (b) Stereoview of three occurrences, one for each type, aligned by their stems. Type E2A/B is shown in red and was found in *H. marismortui* 23S rRNA (PDB ID 1M1K, strands 1463–1467 and 1477–1485 in chain 'A'). Type E3 is shown in blue and was found in *Deinococcus radiodurans* 23S rRNA (PDB ID 1JZY, strands 295–300 and 363–369 in chain "A"). Finally, type E1 is shown in green and was found in a NMR structure of a hairpin similar to the P5abc region within group I intron (PDB ID 1EOR, strands 4–11 and 16–21 in chain "A"). The conserved Cs are shown in bold in the three models.

construction operators to position and orient them in 3-D space. MANIP [54], an interactive system, and MC-Sym [62], an automated procedure, are among the most employed computer systems to model RNA tertiary structures.

In this section, we present how RNA tertiary structure modeling can be mapped to the discrete constraint satisfaction problem (CSP) and how the CSP solver was implemented in the MC-Sym computer program.

4.1 The CSP

The CSP can be described by three finite sets: the variables $V = \{v_1, v_2, \dots, v_n\}$, the domains $D = \{d_1, d_2, \dots, d_n\}$ and the constraints $C = \{c_1, c_2, \dots, c_m\}$ [63,64]. A variable v_i is assigned values from domain $d_i = \{d_{i,1}, d_{i,2}, \dots, d_{i,|d^i|}\}$. Each constraint restricts the assignment of a subset of V, called the constraint scope. For example, if the scope of $c_1 \in C$ is $\{v_2, v_4, v_5\} \subset V$, then $c_1 \subset d_2 \times d_4 \times d_5$.

The three sets are defined in the context of the problem application. In RNA tertiary structure prediction, the variables could be the nucleotides and the domains their possible 3-D coordinates. An example of constraint would be two nucleotides that form a base pair. The scope of the base pair constraint is the two nucleotide partners which link their relative positions and orientations in 3-D space.

A solution to the CSP is a complete variable assignment, $v_i \in d_i$ for $i \in \{1 \dots n\}$, so that all constraints in C, $c_j \in C$ for $j \in \{1 \dots m\}$ are satisfied. A constraint c_j is satisfied only if its scope in the solution is assigned according to the relation it defines. Solving the CSP consists in finding one, many or all solutions.

The search space of a CSP is the Cartesian product of all d_i:

$$\prod_{i=1}^{n} |d_i| \,. \tag{4}$$

The size of a CSP search space is exponential in the number of variables domain sizes. The solutions of a CSP are found by exploring the variable assignments of its search space and verifying if they satisfy the constraints. Backtracking is the classical search algorithm to solve a CSP deterministically and exhaustively. In backtracking, the variables are assigned values systematically, one at a time and to the next available value from its associated domain. When all the values of a domain have been tried, the domain is reset and backtracking moves to the previous variable, assigning its next value, before continuing. The search finishes when the domain of the first variable is reset, indicating that all possible assignments have been tried.

Backtracking develops a search tree (see Figure 20), where the nodes are visited in a depth-first manner. A path from the root of the search tree to a leaf represents a solution if all of its nodes are consistent (the black nodes in

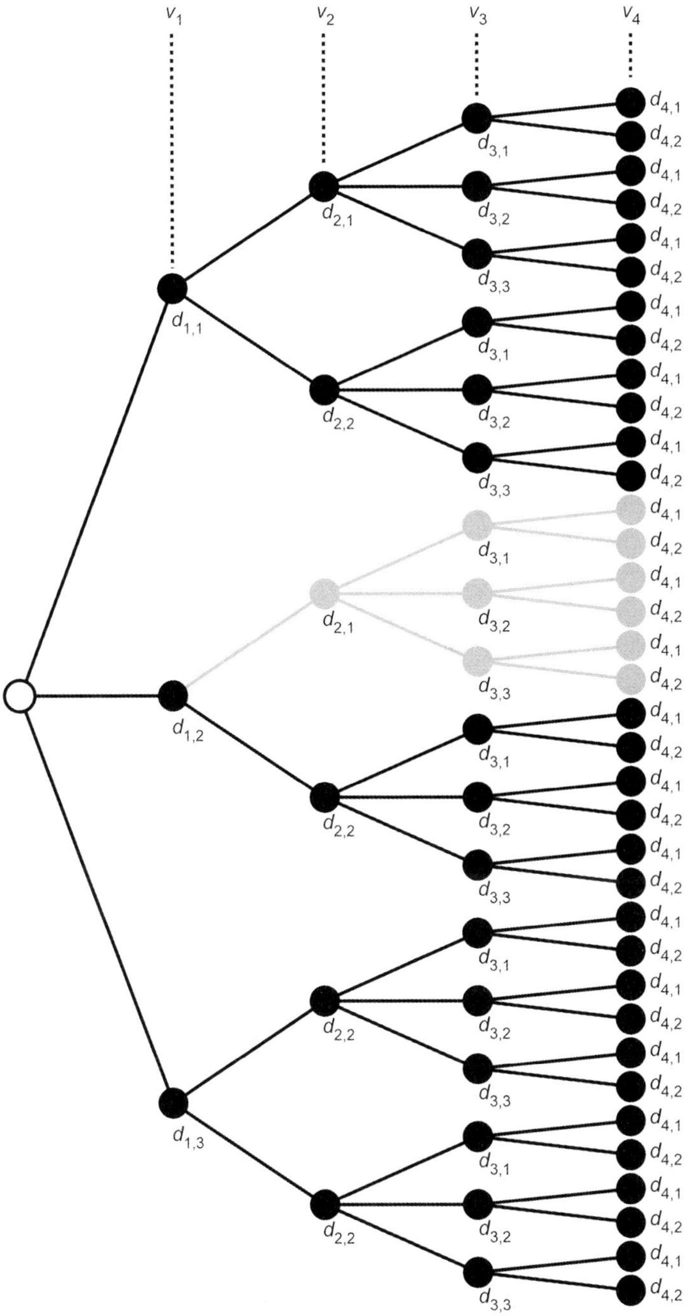

Figure 20 Backtracking search tree. The search space is defined by $V = \{v_1, v_2, v_3, v_4\}$ and $D = \{d_1, d_2, d_3, d_4\}$, where $d_1 = \{d_{1,1}, d_{1,2}, d_{1,3}\}$, $d_2 = \{d_{2,1}, d_{2,2}\}$, $d_3 = \{d_{3,1}, d_{3,2}, d_{3,3}\}$ and $d_4 = \{d_{4,1}, d_{4,2}\}$. Any path from the root (empty circle) to a leaf assigns each variable to a value from its respective domain. A verification of the constraint $\{(v_1, v_2) \in d_1 \times d_2 \mid (v_1, v_2) \neq (d_{1,2}, d_{2,1})\}$ prunes the subtree in grey, as soon as the backtracking assigns v_2 to $d_{2,1}$, and then the search jumps to the next assignment for v_2.

Figure 20). The search tree size is given by Eq. (5):

$$1 + \sum_{i=1}^{n} \sum_{j=1}^{i} |d_i| \,. \tag{5}$$

However, the search can avoid visiting most inconsistent nodes by pruning from the search tree inconsistent branches (those with grey nodes in Figure 20), as soon as one inconsistent node is found. This trick does not change the search time complexity of backtracking, which is exponential in n and $|d_i|$, but can reduce the search time in practice. Using MC-Sym (see next section), the search space of a particular instance of the CSP for the yeast tRNA-Phe tertiary structure was 10^{26}, but the constraints used to explore it defined a consistent search tree of only about 5×10^5, which was explored in seconds and included only about 30 solutions [65]. The best model was at approximately 3.0 Å of RMSD to the yeast tRNA-Phe, as well as with the yeast tRNA-Asp crystal structures (crystal structure resolution). In RNA tertiary structure prediction, it is customary to assess the quality of a prediction method by evaluating its performances at refolding known structures, and RMSD is a very popular quantitative measure.

4.2 MC-Sym

MC-Sym (Macromolecular Conformations by SYMbolic programming) is a computer program for building RNA tertiary structure models from syntactic descriptions of the RNA, similar to the ones used for MC-Search. The computer program implements a CSP solver with backtracking and thus generates models that are consistent with the constraints [6,62,65,66].

In MC-Sym, the variables, V, correspond to the vertices, v, and arcs, a, of the RNA graph (see Section 3.1). The domains of the vertices, D_v, are rigid nucleotides (the relative 3-D atomic coordinates never change) and those of the arcs, D_a, are linear transformation matrices encoding base stacking and base pairing. Consequently, the domains, D, are taken from the Cartesian product, $D = D_v \times D_a$. The goal is to generate consistent 3-D atomic coordinates to each nucleotide in the global frame of the RNA. The rigid sets of nucleotide 3-D atomic coordinates, as well as the linear transformation matrices, are extracted from the PDB [1]. The nucleotide is defined by cutting the phosphodiester chain at the O3′–P bond.

The linear transformation matrices combine nucleotide rotations and translations in 4-D homogeneous coordinates, which represent the spatial relation between two stacked or paired bases. As we saw in Section 2.2, nucleotide interactions involve their bases and thus we represent their spatial relations by coordinate frame relative transformations. The frames are defined at the

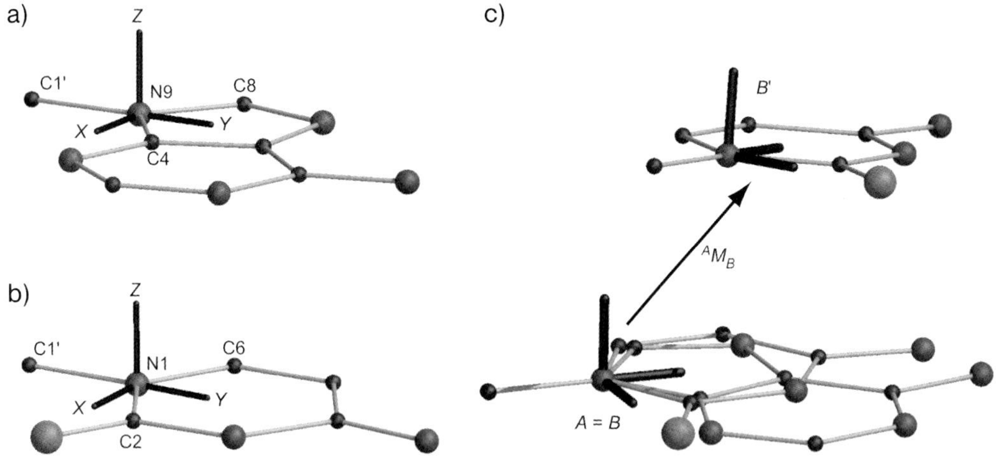

Figure 21 Nucleotide coordinate frame. Hydrogen and backbone atoms are not shown. (a) Frame for a purine (here an adenine). The origin is on the N9 atom, and the *XY* plane is aligned on the plane described by atoms N9, C4 and C8, shared by all purines. (b) Frame for a pyrimidine (here a cytosine). The origin is on the N1 atom, and the *XY* plane is aligned on the plane described by atoms N1, C2 and C6, shared by all pyrimidines. (c) Transformation of cytosine B in adenine A's frame that expresses a stacking interaction. B is moved from a position aligned on A's local frame to B' by applying ^AM_B.

terminal nitrogen atom (N9 in purines and N1 in pyrimidines), with the XY plane aligned with the base plane (see Figure 21a and b).

For any coordinate frames A and B, the relative transformation ^AM_B expresses B's position (right subscript) in A's coordinates (left superscript). For example, let ^AM_B be a relative transformation that expresses a stacking interaction between an adenine (frame A) and a cytosine (frame B). As the relative transformation ^AM_B is defined in A's frame, B can be moved from a position aligned on A's frame to a position that expresses the stacking interaction between the two bases (Figure 21c). This relative transformation places B in the local frame of A, independently of A's position in the global frame. Then, to obtain B's position in the global frame, O, it must be transformed by the relative matrix ^OM_A, which expresses A's position in the global frame.

Using such transformations, all nucleotides of an RNA graph can be positioned relative to another by starting with an initial origin nucleotide, whose frame defines the global frame, O. A construction order is defined by selecting a spanning tree of the RNA graph rooted at the origin nucleotide. The transformations of the arcs of the spanning tree applied to the rigid coordinates of the vertices build the RNA (see Figure 22). Ideally, the chosen spanning tree of the RNA graph would include all arcs. However, RNA graphs are rarely

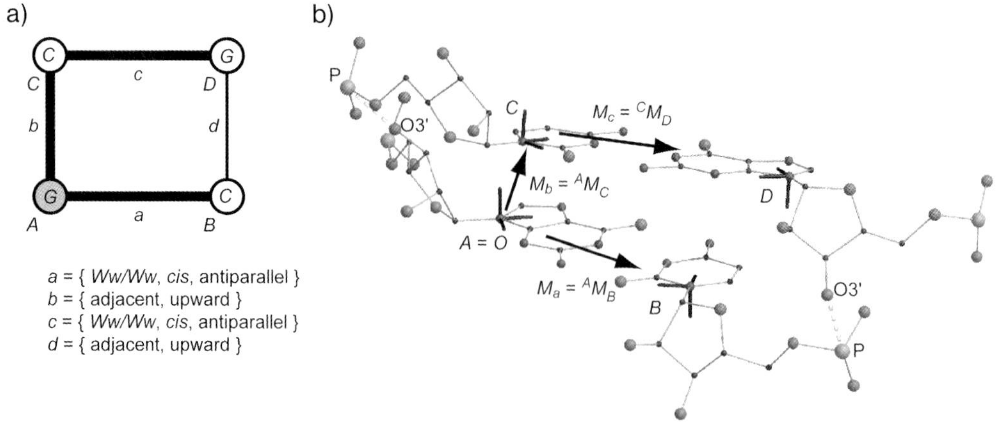

Figure 22 A two-stacked G=C base pairing CSP. (a) RNA graph. The vertices are in uppercase; arcs in lowercase. The arcs of a chosen spanning tree are shown in bold; rooted at vertex A (shaded). (b) One solution: the nucleotide structure for each vertex and a local transformation matrix for each arc (M_i for arc "i"). The root is aligned to the global coordinate frame O. B is positioned by applying $M_a = {}^AM_B$ in the local frame A, which is aligned with O, and thus B is directly placed in the global frame by M_a; vertex C directly placed by $M_b = {}^AM_C$. Finally, to position D, $M_c = {}^CM_D$ is applied in the local frame of C, which is then moved to the global frame by applying $M_b = {}^AM_C$. The O3′–P bond lengths shown in dashed lines are verified by a distance constraint. In this spanning tree, arc d is not considered in the construction.

trees, [1]as they include cycles of interactions. (The only situation where an RNA graph is a tree is when the only information available is the sequence.) Consequently, the missing arcs must be represented by constraints in the CSP, C. As we will see in Section 4.3.3, this problem can be overcome.

The search space for an RNA tertiary structure prediction problem defined with this instance of the CSP is given by the spanning tree n nucleotide variables and $n-1$ interactions. Equation (6) gives the size of the search space, the Cartesian product $D_v \times D_a$, where $x_i \in D_v$, $y_i \in D_a$:

$$\prod_{i=1}^{n} |x_i| \times \prod_{i=1}^{n-1} |y_i| \,. \tag{6}$$

Two types of constraints need to be verified at each variable assignment: the atomic clashes and the O3′–P covalent bond distances. The scope of the atomic clash constraints is all nucleotide pairs and they are needed to insure that no pair of atoms from both nucleotides are overlapping. A threshold inter-atomic distance, typically 1 Å, implements the steric clash constraints.

The adjacency constraint, as we name it, has a scope of two adjacent nucleotides in the sequence and is used to verify the covalent O3′–P bond

1) The only situation where a RNA graph is a tree is when the only information available is the sequence.

distance (dashed lines in Figure 22b), as the nucleotide backbone are positioned independently. The distance threshold is fixed by a user distance range, typically set to [1,2] Å, as the O–P theoretical distance is approximately 1.6 Å (see Section 4.3). Note that choosing a lower bound smaller that the atomic clash threshold would be useless. Increasing the upper bound is equivalent to "relaxing" the CSP. If an overstretched backbone results from choosing a high upper bound, it can easily be fixed by applying numerical refinement [9].

4.2.1 Backbone Optimization

In the above mapping of the RNA tertiary structure CSP, the adjacency constraint is violated more than 50% of the time, even if its range is relaxed to [1,5] Å. This problem comes from the rigid nucleotide conformations that do not reflect well the overall flexibility of the backbone.

To address this problem, we propose a slight modification of the above CSP mapping. We make the assumption that the RNA tertiary structure is driven by rigid base interactions. Consequently, the backbone conformation can be derived from the bases, and we can define a new CSP on the bases alone. The new CSP defines a search space over the transformation matrices, D_a, reducing considerably its size to $\prod_{i=1}^{n-1} |y_i|$ (see Eq. 6).

To reduce the complexity of building a complete backbone from six free torsion angles (11 in total minus the five of the furanose), we define rigid conformations made of the base and the phosphate group. The ribose interconnects a base with two phosphate groups. To position all of the ribose atoms, six free torsions are needed: θ_0, θ_1, θ_2, θ_3, θ_4 and χ (see Figure 3a). However, we know from Eqs. (1) and (3) that θ_0 to θ_4 are all related by the single pseudorotation angle ρ (Section 2.1). Hence, by fixing all covalent bond lengths and angles to their theoretical values, a ribose structure can be built by optimizing a suitable function, $f(\rho, \chi)$. f builds a ribose using torsions ρ and χ and returns the RMSD between the theoretical and the implicit values of the C5′–O5′ and C3′–O3′ bonds (see Figure 23). Equation (7) gives the value of $f(\rho, \chi)$, where l_k are the measured distances and λ_k are the theoretical distances, $k = 5'$ or $k = 3'$, respectively, for the C5′–O5′ or the C3′–O3′ bond

$$f(\rho,\chi) = \sqrt{\frac{(l_{5'} - \lambda_{5'})^2 + (l_{3'} - \lambda_{3'})^2}{2}} \,. \tag{7}$$

Finding the optimal parameters that minimize Eq. (7) builds a ribose attached to its base and interconnecting phosphate groups. The minimization can be solved using classical optimization methods, such as the cyclic coordinate method that does not require derivatives [67]. In this context, each evaluation of the function builds a different ribose conformation. Consequently, the time complexity is proportional to the number of evaluations needed. However,

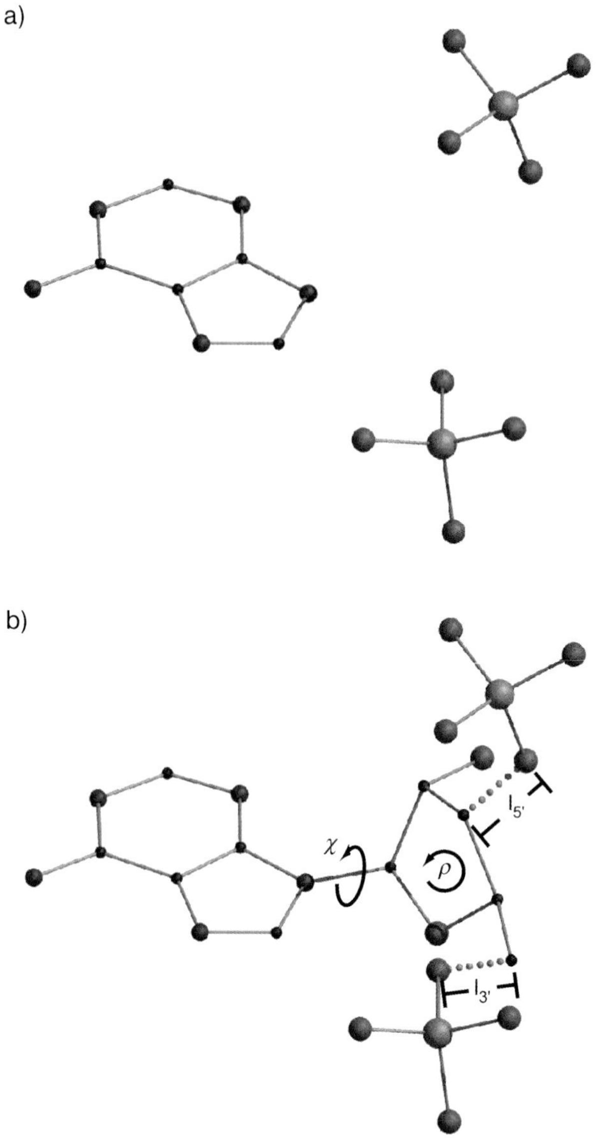

Figure 23 Ribose construction. (a) A base and two phosphate groups. (b) The ribose structure is appended to the base, and is parameterized by ρ and χ. The lengths of the implicit interconnections (shown in dashed lines), $l_{5'}$ and $l_{3'}$, respectively, represent the covalent C5′–O5′ and C3′–O3′ bonds, which quantify the precision of the construction.

we recently derived an optimal parameter estimation of Eq. (7), which solves the ribose optimization in constant time.

Theoretically, the backbone construction can be applied once all variables of the CSP are assigned values – when all bases and phosphate groups are in place. After the backbone construction, however, there is no guarantee that the final model is free from steric conflicts. Therefore, in practice, as soon as the two phosphate groups adjacent to a base are in place, the backbone is built for this nucleotide and the steric clashes verified. This, as in the former backtracking, allows us to prune the search tree.

4.2.2 Probabilistic Backtracking

Exhaustive searches for all valid RNA 3-D structures are useful to analyze possible alternative folding of an RNA. However, sometimes only one or few valid models are desired. We recently developed a probabilistic search algorithm that generates valid structures faster and with an increased diversity rate than that of the deterministic backtracking.

We select a random path from the root node of the search tree to any leaf. If at any variable assignment along the random path a constraint is not satisfied, a fixed-size regular backtracking is run until a consistent node is found, which resumes the probabilistic search. If the fixed-size regular backtracking cannot find a consistent node, then a new random path from the root node is selected.

4.2.3 "Divide and Conquer"

As for many problems, the "divide and conquer" paradigm has proven useful in RNA tertiary structure prediction as well. A "divide and conquer" algorithm splits a complex problem in many smaller and simpler to resolve subproblems. The solutions of the smaller problems are then combined in complete solutions of the larger problem.

RNA structures can be split into smaller fragments (as we saw in Section 3). Each fragment can be built independently and then merged in complete tertiary structures. In the CSP context, solving a fragment means locally enforcing the constraints in its scope: solving subset $W \subset V$ involves the verification of the constraints defined on W only. The solutions of W become the values of the domain of a new variable, say w, which is added to the CSP. The new CSP, CSP′, is defined over the variables $V' = (V \backslash W) \cup \{w\}$, the new domain for w and the new constraints $C' = C \backslash C_w$.

The search space sizes of CSP and CSP′ can be compared. Let $D = \{d_1, \ldots, d_n\}$ be the domains for the n variables in V and as defined in the CSP. $E = \{e_1, \ldots, e_k\}$ are the domains for the k variables in W, $E \subset D$. The domain for the new variable, w, is d_w, the solutions of W. If S and S' are, respectively, the search space sizes of CSP and CSP′, then from Eq. (4) we obtain:

$$S' < S \Leftrightarrow \frac{\prod_{i=1}^{n} d_i}{\prod_{i=1}^{k} e_i} \times d_w < \prod_{i=1}^{n} d_i \Leftrightarrow d_w < \prod_{i=1}^{k} e_i$$

showing the trivial result that CSP′ has a smaller search space than CSP if the search space size of CSP′ is larger than $|d_w|$, which is generally the case in the presence of actual structural constraints. To save even more time, when fragments of a tertiary structure correspond to RNA motifs, as described in Section 3, their solutions can directly be taken from the X-ray crystal structures [53]. We sometimes refer to this practice as RNA homology modeling.

4.3 MC-Sym at Work

Here we show how MC-Sym can be used to generate RNA tertiary structure models. The input description file has sections to describe the sequence, the nucleotide conformations, the nucleotide interactions, the constraints, and execution arguments.

Figure 24 shows an example for the modeling of a tandem W/W base pairs: C=G stacked with A–U. Figure 24(a) shows the secondary structure of this simple fragment, made of two strands: "a" 5′-AC-3′ and "b" 5′-GU-3′. Figure 24(b) shows the spanning tree of the RNA graph, as chosen by the user, to build the fragment.

In the input (Figure 24c), the "sequence" section defines the two strands and introduces a global numbering system for the nucleotides. The "residue" section defines the nucleotide conformations, D_v, and sampling sizes; here, 10 different C3′-*endo anti* conformations. The nucleotide interactions are defined in the "connect" and "pair" statements, specified using the *LW*+ nomenclature. The "connect" statement is used for adjacent nucleotides in the sequence. In the example, one of the two base stack interactions is included in the spanning tree and five different stacking transformations will be assigned. The "pair" statement is used for the two Watson–Crick base pairs. Here, seven different Watson–Crick transformations will be assigned.

The "backtrack" statement defines the spanning tree, instructing MC-Sym about the order in which the nucleotides will be inserted in the models. Here, a1 is selected as the global referential, then b2 is Watson–Crick to a1, a2 is stacked with a1 and, finally, b1 is Watson–Crick to a2.

The domain specifications are translated by MC-Sym into queries to the appropriate nucleotide conformation or interaction database. The results of the logical queries define the domains. For instance, the query for the conformation of the cytosine a2 could match entries #5 and #8 in the conformation database shown in Table 2, resulting in the conformation domain $\{S_5, S_8\}$. Similarly, the Watson–Crick query for the a2–b1 interaction could match database entries #1, #3 and #8 in Table 3, resulting in the transformation domain $\{T_1, T_3, T_8\}$.

As indicated in Tables 2 and 3, the conformation and transformation domains come from X-ray crystal structures. The atomic coordinates for the

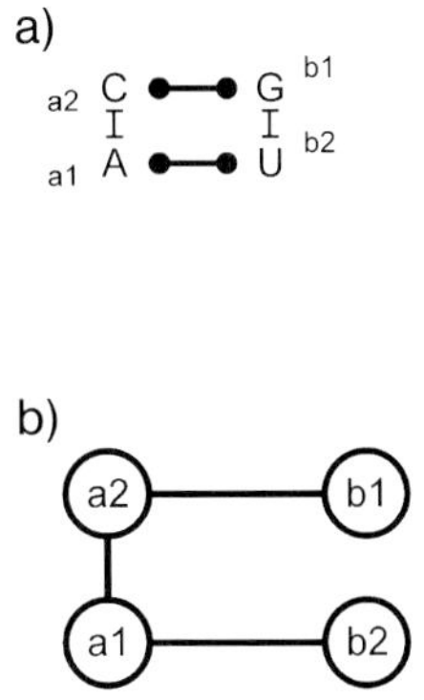

c)
```
sequence ( r a1 AC )
sequence ( r b1 GU )

residue (
  a1 { C3'_endo && anti } 10
  a2 { C3'_endo && anti } 10
  b1 { C3'_endo && anti } 10
  b2 { C3'_endo && anti } 10
)

connect (
  a1 a2 { stack } 5
)

pair (
  a1 b2 { W / W && cis } 7
  a2 b1 { W / W && cis } 7
)

myRNA = backtrack (
  ( a1 b2 )
  ( a1 a2 b1)
)

[...]
```

Figure 24 Tandem $|W/W|$ base pairs. (a) RNA graph. (b) Spanning tree. (c) MC-Sym input. The "stack" keyword is used as a wildcard matching any of the four stacking types: upward, downward, inward or outward.

Table 2 *MC-Sym* nucleotide conformation database snippet

#	3-D coordinates set [a]	Symbols list	Origin [b]
1	S_1	A, C3′-*endo*, *anti*	1EVV 'A'23
2	S_2	U, C3′-*endo*, *anti*	1FFK '0'55
3	S_3	G, C2′-*endo*, *anti*	1EHZ 'A'18
4	S_4	A, C2′-*exo*, *anti*	1EVV 'A'35
5	S_5	C, C3′-*endo*, *anti*	1JJ2 '0'361
6	S_6	U, C3′-*endo*, *syn*	1FFK '0'10
7	S_7	A, C4′-*exo*, *anti*	1JJ2 '0'407
8	S_8	C, C3′-*endo*, *anti*	1EHZ 'A'13
[...]			

[a] Each S_i contains the 3-D coordinates of the nucleotide.
[b] PDB ID and numbering of the nucleotide.

conformations are directly extracted. The transformation between two nucleotides, A and B, is extracted by computing the relative transformation that places B's frame in A's local frame (Figure 25). If ${}^{O}M_A$ and ${}^{O}M_B$ are respectively the relative transformations that place the frame of A and B in

Table 3 *MC-Sym* nucleotide interaction database snippet

#	3-D coordinate set [a]	Symbol list	Origin [b]	
1	T_1	C-G, *Ws / Ww*, *cis*, antiparallel	1JJ2	'0'284, '0'367
2	T_2	U-C, adjacent, upward	1EHZ	'A'59, 'A'60
3	T_3	C-G, *Ww / Ww*, *cis*, antiparallel	1EVV	'A'27, 'A'43
4	T_4	A-C, adjacent	1FFK	'0'337, '0'338
5	T_5	A-C, outward	1EVV	'A'6, 'A'7
6	T_6	G-U, *Ww / Ws*, *cis*, parallel	1EHZ	'A'4, 'A'69
7	T_7	G-A, *Ss / Hh*, *cis*, antiparallel	1FFK	'0'2865, '0'2891
8	T_8	C-G, *Ww / Ww*, *cis*, antiparallel	1JJ2	'0'154, '0'182
[...]				

[a] Each T_i corresponds to the relative transformation matrix of the interaction.
[b] PDB ID and numbering of the two interacting nucleotides.

the global frame, O, then we extract the relative transformation ${}^{A}M_B$ so that ${}^{O}M_B = {}^{O}M_A \times {}^{A}M_B$. Isolating for ${}^{A}M_B$, we obtain ${}^{A}M_B = {}^{O}M_A^{-1} \times {}^{O}M_B$. We save the ${}^{A}M_B$ matrix in the database so that it can be reproduced for any other pair of bases in any local frame.

The MC-Sym database contains nearly 3000 nucleotide conformations and nearly 20 000 base interactions; hence, the domain size argument next to each conformation and interaction. It is the task of the modeler to assign domain sizes so that the conformational space of a given tertiary structure is correctly addressed – not too small to miss valid models and not too large to avoid prohibitive search space sizes.

4.3.1 **Modeling a Yeast tRNA-Phe Stem–Loop**

In Figure 26, we present a modeling example for the yeast tRNA-Phe T-stem–loop tertiary structure. The secondary structure of the stem–loop is shown in Figure 26(a). The stem and hairpin loop are modeled independently, and the results of each modeling merged. Figure 26(b–d) shows the three inputs. The first describes the structure of the first four base pairs of the stem. The second describes the hairpin loop, closed by the last base pair of the stem. The last merges the resulting fragments and, thus, models the entire stem–loop. Figure 26(e–g) illustrates the spanning trees defined by the three inputs.

The "res_clash" and "adjacency" statements parameterize the steric clashes and adjacency constraints, respectively. The "explore" statement launches the CSP solver. The RNA graph of the loop is divided into two sections by the W/H U54–A58 base pair, leaving the sequence adjacency implicit to the construction between G57 and A58, and between C60 and C61. Figure 26(h) shows one solution.

4.3.2 Modeling a Pseudoknot

To model the tertiary structure of the FIV pseudoknot (see Section 3.1.1), a novel methodological protocol based on mass spectroscopy and computer modeling was designed by Fabris and his coworkers. The experimental data were generated using multiplexing solvent-accessibility probes and chemical bifunctional crosslinkers with a characterization by an electrospray ionization Fourier transform mass spectrometry method (ESI-FTMS) [68]. The chemical and enzymatic probes cleave at specific sites or attack specific chemical groups that are exposed to the solvent. The secondary structure, detailed protection maps and inter-nucleotide distance information were then input to MC-Sym, which generated a set of consistent tertiary structures. Finally, the modeled structures were refined by energy minimization using the Crystallography and NMR System (CNS) [69].

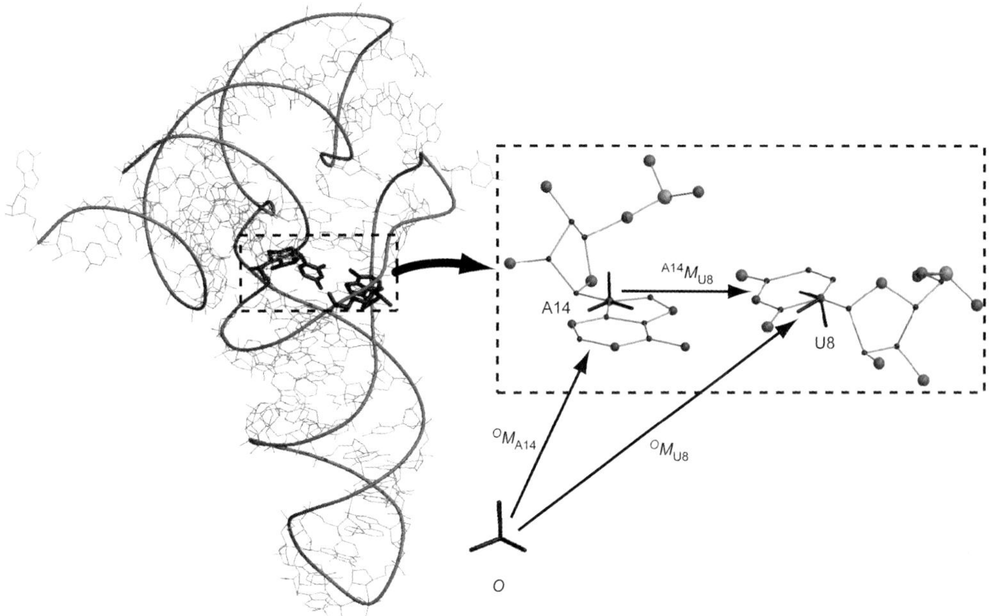

Figure 25 Extraction of the A14–U8 H/W base pair transformation in the yeast tRNA-Phe X-ray crystal structure (PDB ID 1EVV). A thread follows the strands by the phosphorus atoms. Hydrogen atoms are not shown. The base pair is zoomed and shown with the frames ($^{O}M_{A14}$ and $^{O}M_{U8}$) defined in the global frame, O. The transformation $^{A14}M_{U8} = {}^{O}M_{A14}^{-1} \times {}^{O}M_{U8}$ is extracted.

a)

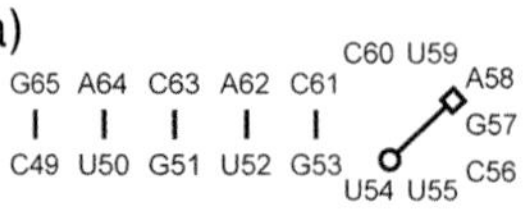

b)

```
sequence (r 49 CUGU)
sequence (r 62 ACAG)
residue (
  49 52 { C3'_endo && anti } 10
  62 65 { C3'_endo && anti } 10
)
connect (
  50 51 { stack } 6
  62 63 { stack } 6
  64 65 { stack } 6
)
pair (
  49 65 { Ww / Ww && cis } 5
  50 64 { Ww / Ww && cis } 5
  51 63 { Ww / Ww && cis } 5
  52 62 { Ww / Ww && cis } 5
)
stem = backtrack (
  (49 65 64 50 51 63 62 52)
)
res_clash (
  stem
  fixed_distance 1.0
  all no_hydrogen
)
adjacency (stem 1.0 4.0)
explore (
  stem
  rmsd (1.0 base_only no_hydrogen)
  file_pdb ("stem-%04d.pdb" zipped)
)
```

c)

```
sequence (r 53 GUUCGAUCC)
residue (
  53 61 { C3'_endo || C2'_endo } 10
)
connect (
  53 57 { } 10
  58 60 { } 10
)
pair (
  53 61 { Ww / Ww && cis } 5
  54 58 { W / H } 8
)
loop = backtrack (
  (53 61)
  (53 54 58 59 60)
  (54 55 56 57)
)
res_clash (
  loop
  fixed_distance 1.0
  all no_hydrogen
)
adjacency (loop 1.0 4.0)
explore (
  loop
  rmsd (1.0 base_only no_hydrogen)
  file_pdb ("loop-%04d.pdb" zipped)
)
```

d)

```
sequence (r 49 CUGUGUUCGAUCCACAG)
stem = library (
  file_pdb ("stem-%04d.pdb.gz")
)
loop = library (
  file_pdb ("loop-%04d.pdb.gz")
)
connect (
  52 53 { stack } 6
)
stem_loop = backtrack (
  stem
  place (52 53 loop)
)
res_clash (
  stem_loop
  fixed_distance 1.0
  all no_hydrogen
)
adjacency (stem_loop 1.0 4.0)
explore (
  stem_loop
  rmsd (1.0 base_only no_hydrogen)
  file_pdb ("stem_loop-%04d.pdb" zipped)
)
```

e)

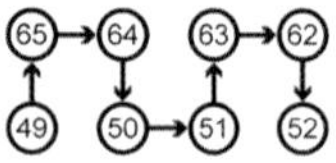

f)

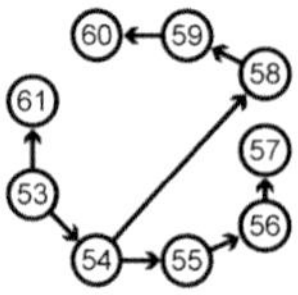

g)

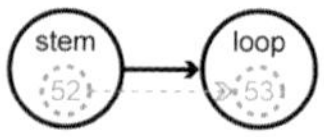

h)

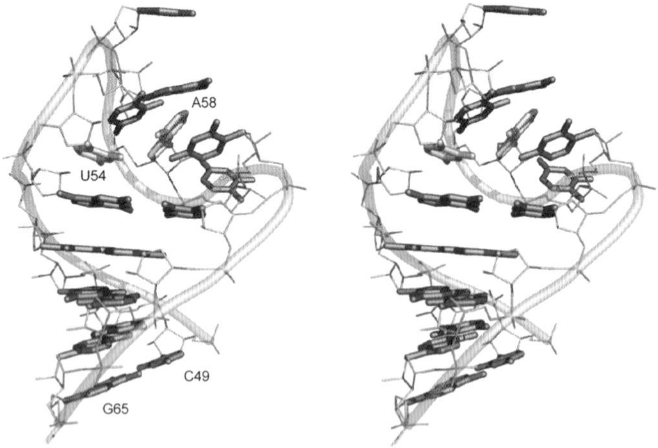

4.3.3 Cycles of Interactions

Traversal of a spanning tree to build tertiary structures implies a conceptual problem, as was pointed out in Section 4.2. A spanning tree does not cover all the arcs of the RNA graph. User constraints must be added to the input to make sure the dropped interactions are satisfied. However, such *ad hoc* constraints are difficult to define and compute. Lemieux and Major have designed a novel building approach to decompose an RNA graph in a series of minimum cycles of interactions [70], whose solutions can be combined by superimposing common arcs. The RNA graph in Figure 24, for instance, is a minimum cycle of four nucleotides. The product of the four transformation matrices is the identity matrix, representing an additional constraint to ensure the consistency of the cycle and the satisfaction of the four interactions.

These minimum interaction cycles could well be used as first-class objects in stochastic graph grammars [71, 72] to represent the tertiary structure of related RNAs and of their sequence alignment. This is similar, but yet more expressive, than stochastic context-free grammars employed to represent RNA secondary structures [73,74].

5 Perspectives

Accurate prediction of RNA tertiary structures from sequence alone is still an unresolved problem. In the meantime, formalizing RNA attributes, searching for higher-order levels of structural organization and modeling their tertiary structure represent current efforts towards better understanding of the RNA architectural principles. In addition, formalizing RNA structural knowledge in computer programs offers the possibility to apply it in a systematic and objective manner, allowing us to generate new and experimentally testable data.

The recent resolution of several RNA structures by X-ray crystallography, NMR, as well as by computer modeling, has allowed us to observe repeated RNA fragments and to infer their function. We are starting to understand the sequence constraints imposed by the tertiary structure of these fragments, and to discover local and global folding rules. In the coming years, as we can

Figure 26 Yeast tRNA T-stem–loop. (a) Secondary structure. (b) MC-Sym input for the stem fragment. (c) MC-Sym input for the loop fragment. (d) MC-Sym input for merging both fragments. (e) Spanning tree of (b). (f) Spanning tree of (c). (g) Spanning tree of (d). (h) Stereoview of a final model generated by MC-Sym. The bases are shown in blue; the U54–A58 W/H base pair in lighter blue. The backbone is shown in yellow. The thread follows the phosphodiester chain. Hydrogen atoms are not shown.

expect agreement on an RNA ontology (nomenclatures and formalisms), we might assist in the deployment and implementation of these folding rules into accurate RNA tertiary structure prediction algorithms. An RNA ontology will enable the interoperability of research results. Consequently, as we identify the RNAs of key cellular processes, determine their structure and characterize their role, we will be in a better position to manipulate them. As a result, we should observe an increase in the number and an improvement in the accuracy of RNA-based molecular medicine techniques.

References

1 Berman, H. M., J. Westbrook, Z. Feng, G. Gilliland, T. N. Bhat, H. Weissig, I. N. Shindyalov and P. E. Bourne. 2000. The Protein Data Bank. Nucleic Acids Res. **28**: 235–42.

2 Gendron, P., S. Lemieux and F. Major. 2001. Quantitative analysis of nucleic acid three-dimensional structures. J. Mol. Biol. **308**: 919–36.

3 Yang, H., F. Jossinet, N. B. Leontis, L. Chen, J. Westbrook, H. M. Berman and E. Westhof. 2003. Tools for the automatic identification and classification of RNA base pairs. Nucleic Acids Res. **31**: 3450–60.

4 Altona, C. and M. Sundaralingam. 1972. Conformational analysis of the sugar ring in nucleosides and nucleotides. A new description using the concept of pseudorotation. J. Am. Chem. Soc. **94**: 8205–12.

5 Kraulis, P. J. 1991. MOLSCRIPT: a program to produce both detailed and schematic plots of protein structures. J. Appl. Crystallogr. **24**: 946–50.

6 Merritt, E. A. and M. E. Murphy. 1994. Raster3D version 2.0. A program for photorealistic molecular graphics. Acta Crystallogr. D **50**: 869–73.

7 Saenger, W. 1984. *Principles of Nucleic Acid Structure*. Springer, New York, NY.

8 Olson, W. K. 1980. Configurational statistics of polynucleotide chains. An updated virtual bond model to treat effects of base stacking. Macromolecules **13**: 721–8.

9 Major, F., M. Turcotte, D. Gautheret, G. Lapalme, E. Fillion and R. Cedergren. 1991. The combination of symbolic and numerical computation for three-dimensional modeling of RNA. Science **253**: 1255–60.

10 Gautheret, D., F. Major and R. Cedergren. 1993. Modeling the three-dimensional structure of RNA using discrete nucleotide conformational sets. J. Mol. Biol. **229**: 1049–64.

11 Paul, R. P. 1981. *Robot Manipulators: Mathematics, Programming and Control*. MIT Press, Cambridge, MA.

12 Duarte, C. M., L. M. Wadley and A. M. Pyle. 2003. RNA structure comparison, motif search and discovery using a reduced representation of RNA conformational space. Nucleic Acids Res. **31**: 4755–61.

13 Hershkovitz, E., E. Tannenbaum, S. B. Howerton, A. Sheth, A. Tannenbaum and L. D. Williams. 2003. Automated identification of RNA conformational motifs: theory and application to the HM LSU 23S rRNA. Nucleic Acids Res. **31**: 6249–57.

14 Murray, L. J. W., W. B. Arendall III, D. C. Richardson and J. S. Richardson. 2003. RNA backbone is rotameric. Proc. Natl Acad. Sci. USA **100**: 13904–9.

15 Schneider, B., Z. Morávek and H. M. Berman. 2004. RNA conformational classes. Nucleic Acids Res. **32**: 1666–7.

16 Serganov, A., Y. R. Yuan, O. Pikovskaya, et al. 2004. Structural basis

for discriminative regulation of gene expression by adenine- and guanine-sensing mRNAs. Chem. Biol. **11**: 1729–41.

17 GABB, H. A., S. R. SANGHANI, C. H. ROBERT AND C. PRÉVOST. 1996. Finding and visualizing nucleic acid base stacking. J. Mol. Graph. **14**: 6–11.

18 LEONTIS, N. B. AND E. WESTHOF. 2001. Geometric nomenclature and classification of RNA base pairs. RNA **7**: 499–512.

19 LEMIEUX, S. AND F. MAJOR. 2002. RNA canonical and non-canonical base pairing types: a recognition method and complete repertoire. Nucleic Acids Res. **30**: 4250–63.

20 LEE, J. C. AND R. R. GUTELL. 2004. Diversity of base-pair conformations and their occurrence in rRNA structure and RNA structural motifs. J. Mol. Biol. **344**: 1225–49.

21 LEONTIS, N. B., J. STOMBAUG AND E. WESTHOF. 2002. The non-Watson–Crick base pairs and their associated isostericity matrices. Nucleic Acids Res. **30**: 3479–531.

22 WALBERER, B. J., A. C. CHENG AND A. D. FRANKEL. 2003. Structural diversity and isomorphism of hydrogen-bonded base interactions in nucleic acids. J. Mol. Biol. **327**: 767–80.

23 LESCOUTE, A., N. B. LEONTIS, C. MASSIRE AND E. WESTHOF. 2005. Recurrent structural RNA motifs, isostericity matrices and sequence alignments. Nucleic Acids Res. **33**: 2395–409.

24 BAN, N., P. NISSEN, J. HANSEN, P. B. MOORE AND T. A. STEITZ. 2000. The complete atomic structure of the large ribosomal subunit at 2.4 Å resolution. Science **289**: 905–20.

25 WIMBERLY, B. T., D. E. BRODERSEN, W. M. CLEMONS JR., R. J. MORGAN-WARREN, A. P. CARTER, C. VONRHEIN, T. HARTSCH AND V. RAMAKRISHNAN. 2000. Structure of the 30S ribosomal subunit. Nature **407**: 327–39.

26 HARMS, J., F. SCHLUENZEN, R. ZARIVACH, et al. 2001. High resolution structure of the large ribosomal subunit from a mesophilic eubacterium. Cell **107**: 679–88.

27 LEONTIS, N. B. AND E. WESTHOF. 2003. Analysis of RNA motifs. Curr. Opin. Struct. Biol. **13**: 300–8.

28 WAUGH, A., P. GENDRON, R. ALTMAN, et al. 2002. RNAML: a standard syntax for exchanging RNA information. RNA **8**: 707–17.

29 JOSSINET, F. AND E. WESTHOF. 2005. Sequence to structure (S2S): display, manipulate and interconnect RNA data from sequence to structure. Bioinformatics **1**: 3320–1.

30 ULLMAN, J. R. 1976. An algorithm for subgraph isomorphism. J. ACM **23**: 31–42.

31 LEONTIS, N. B., J. STOMBAUGH AND E. WESTHOF. 2002. Motif prediction in ribosomal RNAs. Lessons and prospects for automated motif prediction in homologous RNA molecules. Biochimie **84**: 961–73.

32 NAGASWAMY, U. AND G. E. FOX. 2002. Frequent occurrence of the T-loop RNA folding motif in ribosomal RNAs. RNA **8**: 1112–9.

33 NISSEN, P., J. A. IPPOLITO, N. BAN, P. B. MOORE AND T. A. STEITZ. 2001. RNA tertiary interactions in the large ribosomal subunit: the A-minor motif. Proc. Natl Acad. Sci. USA **98**: 4899–903.

34 OGLE, J. M., F. V. MURPHY IV, M. J. TARRY AND V. RAMAKRISHNAN. 2002. Selection of tRNA by the ribosome requires a transition from an open to a closed form. Cell **111**: 721–32.

35 KLEIN, D. J., T. M. SCHMEING, P. B. MOORE AND T. A. STEITZ. 2001. The kink-turn: a new RNA secondary structure motif. EMBO J. **20**: 4214–21.

36 PLEY, H. W., K. M. FLAHERTY AND D. B. MCKAY. 1994. Model for an RNA tertiary interaction from the structure of an intermolecular complex between a GAAA tetraloop and an RNA helix. Nature **372**: 111–3.

37 JAEGER, L., F. MICHEL AND E. WESTHOF. 1994. Involvement of a GNRA tetraloop in long-range RNA tertiary interactions. J. Mol. Biol. **236**: 1271–6.

38 BÉLANGER, F., M. G. GAGNON, S. V. STEINBERG, P. R. CUNNINGHAM AND L. BRAKIER-GINGRAS. 2004. Study of the functional interaction of the 900 tetraloop of 16S ribosomal RNA with helix 24 within the bacterial ribosome. J. Mol. Biol. **338**: 683–93.

39 TEN DAM, E., C. W. A. PLEIJ AND D. E. DRAPER. 1992. Structural and functional aspects of RNA pseudoknots. Biochemistry **47**: 11665–76.

40 GIEDROC, D. P., C. A. THEIMER AND P. L. NIXON. 2000. Structure, stability and function of RNA pseudoknots involved in stimulating ribosomal frameshifting. J. Mol. Biol. **298**: 167–85.

41 MORIKAWA, S. AND D. H. L. BISHOP. 1992. Identification and analysis of the *gap–pol* ribosomal frameshift site of feline immunodeficiency virus. Virology **186**: 389–97.

42 YU, E. T., Q. ZHANG AND D. FABRIS. 2005. Untying the FIV frameshifting pseudoknot structure by MS3D. J. Mol. Biol. **345**: 69–80.

43 RUFFNER, D. E., G. D. STORMO AND O. C. UHLENBECK. 1990. Sequence requirements of the hammerhead RNA self-cleavage reaction. Biochemistry **29**: 10695–702.

44 PLEY, H. W., K. M. FLAHERTY AND D. B. MCKAY. 1994. Three-dimensional structure of a hammerhead ribozyme. Nature **372**: 68–74.

45 SCOTT, W. G., J. T. FINCH AND A. KLUG. 1995. The crystal structure of an all-RNA hammerhead ribozyme: a proposed mechanism for RNA catalytic cleavage. Cell **81**: 991–1002.

46 LEGAULT, P., C. G. HOOGSTRATEN, E. METLITZKY AND A. PARDI. 1998. Order, dynamics and metal-binding in the lead-dependent ribozyme. J. Mol. Biol. **284**: 325–35.

47 LEMIEUX, S, P. CHARTRAND, R. CEDERGREN AND F. MAJOR. 1998. Modeling active RNA structures using the intersection of conformational space: application to the lead-activated ribozyme. RNA **4**: 739–49.

48 WEDEKIND, J. E. AND D. B. MCKAY. 1999. Crystal structure of a lead-dependent ribozyme revealing metal binding sites relevant to catalysis. Nat. Struct. Biol. **6**: 261–8.

49 DAVID, L., D. LAMBERT, P. GENDRON AND F. MAJOR. 2001. Leadzyme. Methods Enzymol. **341**: 518–40.

50 PINARD, R., D. LAMBERT, J. E. HECKMAN, et al. 2001. The hairpin ribozyme substrate binding domain: a highly constrained D-shaped conformation. J. Mol. Biol. **307**: 51–65.

51 PINARD, R., D. LAMBERT, G. POTHIAWALA, F. MAJOR AND J. M. BURKE. 2004. Modifications and deletions of helices within the hairpin ribozyme–substrate complex: an active ribozyme lacking helix 1. RNA **10**: 395–402.

52 LESCOUTE, A. AND E. WESTHOF. 2006. Topology of the three-way junctions in folded RNAs. RNA **12**: 83–93.

53 KHVOROVA, A., A. LESCOUTE, E. WESTHOF AND S. D. JAYASENA. 2003. Sequence elements outside the hammerhead ribozyme catalytic core enable intracellular activity. Nat. Struct. Biol. **10**: 708–12.

54 MASSIRE, C. AND E. WESTHOF. 1998. MANIP: an interactive tool for modeling RNA. J Mol. Graph. Model. **16**: 197–205.

55 HILEY, S. L. AND R. A. COLLINS. 2001. Rapid formation of a solvent-inaccessible core in the *Neurospora* Varkud satellite ribozyme. EMBO J. **20**: 5461–9.

56 HOFFMANN, B., G. T. MITCHELL, P. GENDRON, F. MAJOR, A. A. ANDERSEN, R. A. COLLINS AND P. LEGAULT. 2003. NMR structure of the active conformation of the Varkud satellite ribozyme cleavage site. Proc. Natl Acad. Sci. USA **100**: 7003–8.

57 TAMURA, M. AND S. R. HOLBROOK. 2002. Sequence and structural conservation in RNA riboses. J. Mol. Biol. **320**: 455–74.

58 OLIVIER, C., G. POIRIER, P. GENDRON, A. BOISGONTIER, F. MAJOR AND P. CHARTRAND. 2005. Identification of a conserved RNA motif essential for She2p recognition and mRNA localization to the yeast bud. Mol. Cell. Biol. **25**: 4752–766.

59 CHARTRAND, P., X.-H. MENG, R. H. SINGER AND R. M. LONG. 1999. Structural elements required for the localization of *ASH1* mRNA and of a green fluorescent protein reporter particle *in vivo*. Curr. Biol. **9**: 333–6.

60 LONG, R. M., W. GU, E. LORIMER, R. H. SINGER AND P. CHARTRAND. 2000. She2p is a novel RNA-binding protein that recruits the Myo4p–She3p complex to *ASH1* mRNA. EMBO J **19**: 6592–601.

61 MALHOTRA, A., R. K. TAN AND S. C. HARVEY. 1990. Prediction of the three-dimensional structure of *Escherichia coli* 30S ribosomal subunit: a molecular mechanics approach. Proc. Natl Acad. Sci. USA **87**: 1950–4.

62 MAJOR, F. 2003. Building three-dimensional ribonucleic acid structures. IEEE Comput. Sci. Eng. **Sep/Oct**: 44–53.

63 HENTENRYCK, P. V. 1989. *Constraint Satisfaction in Logic Programming*. MIT Press, Cambridge, MA.

64 DECHTER, R. AND D. FROST. 2002. Backjump-based backtracking for constraint satisfaction problems. Artificial Intell. **136**: 147–88.

65 MAJOR, F., D. GAUTHERET AND R. CEDERGREN. 1993. Reproducing the three-dimensional structure of a tRNA molecule from structural constraints. Proc. Natl Acad. Sci. USA **90**: 9408–12.

66 LEMIEUX, S., S. OLDZIEJ AND F. MAJOR. 1998. Nucleic acids: qualitative modeling. In SCHLEYER, P. V. R., N. L. ALLINGER, T. CLARK, et al. (eds.), *Encyclopedia of Computational Chemistry*. Wiley, Chichester: 000–00.

67 BAZARAA, M. S. AND C. M. SHETTY. 1979. *Nonlinear Programming Theory and Algorithms*. Wiley, New York, NY.

68 YU, E. T. AND D. FABRIS. 2003. Direct probing of RNA structures and RNA-protein interactions in the HIV-1 packaging signal by chemical modification and electrospray ionization Fourier transform mass spectrometry. J. Mol. Biol. **330**: 211–23.

69 BRÜNGER, A.T., P. D. ADAMS, G. M. CLORE, et al. 1998. Crystallography & NMR system: a new software suite for macromolecular structure determination. Acta Crystallogr. D **54**: 905–21.

70 LEMIEUX, S. AND F. MAJOR. 2006. Automated extraction and classification of RNA tertiary structure cyclic motifs. Nucl. Acids Res. **34**: 2340–6.

71 NAGL, M. 1987. Set theoretic approaches to graph grammars. In EHRIG, H., M. NAGL, G. ROZENBERG AND A. ROSENFELD (eds.), *Graph-grammars and their Application to Computer Science*. Springer, Berlin: 41–54.

72 JONES, C. V. 1993. An integrated modeling environment based on attributed graphs and graph-grammars. Decision Support Syst. **10**: 255–75.

73 SAKAKIBARA, Y., M. BROEN, R. HUGHEY, I. S. MIAN, K. SJÖLANDER, R. C. UNDERWOOD AND D. HAUSSLER. 1994. Stochastic context-free grammars for tRNA modeling. Nucleic Acids Res. **22**: 5112–20.

74 DOWELL, R. D. AND S. R. EDDY. 2004. Evaluation of several lightweight stochastic context-free grammars for RNA secondary structure prediction. BMC Bioinformatics **5**: 71.